T0227622

A Multidisciplinary Introduction
to Desalination

RIVER PUBLISHERS SERIES IN EARTH
AND ENVIRONMENTAL SCIENCES

Series Editors

Medani P. Bhandari
Atlantic State Legal Foundation
New York, USA

Hanna Shvindina
Sumy State University
Ukraine

Krishna Prasad Oli
Government of Nepal, Nepal
and
Sichuan University, China
and
Akamai University, Hawaii, USA

Indexing: All books published in this series are submitted to the Web of Science Book Citation Index (BkCI), to CrossRef and to Google Scholar.

The "River Publishers Series in Earth and Environmental Sciences" is a series of comprehensive academic and professional books which focus on Earth and Environmental and Geo-Sciences. The series focuses on topics ranging from theory to policy and technology to applications.

Books published in the series include research monographs, edited volumes, handbooks and textbooks. The books provide professionals, researchers, educators, and advanced students in the field with an invaluable insight into the latest research and developments.

Topics covered in the series include, but are by no means restricted to the following:

- Sustainable Development
- Climate Change Mitigation
- Protected Area Management
- Institutional Architectures of Biodiversity Conservation
- Environmental System Monitoring and Analysis
- Migration / Immigration
- Flood Management
- Conflict Management
- Sustainability: Greening the World Economy
- Desalination and water treatment

For a list of other books in this series, visit www.riverpublishers.com

A Multidisciplinary Introduction to Desalination

Editor

Alireza Bazargan

Research and Development Manager
Noor Vijeh Company (NVCO)

and

Assistant Professor, Department of Civil Engineering
K.N. Toosi University of Technology
Tehran, Iran

LONDON AND NEW YORK

Published 2018 by River Publishers
River Publishers
Alsbjergvej 10, 9260 Gistrup, Denmark
www.riverpublishers.com

Distributed exclusively by Routledge
4 Park Square, Milton Park, Abingdon, Oxon OX14 4RN
605 Third Avenue, New York, NY 10017, USA

First issued in paperback 2023

A Multidisciplinary Introduction to Desalination / by Bazargan, Alireza .

Routledge is an imprint of the Taylor & Francis Group, an informa business

Publisher's Note
The publisher has gone to great lengths to ensure the quality of this reprint but points out that some imperfections in the original copies may be apparent.

While every effort is made to provide dependable information, the publisher, authors, and editors cannot be held responsible for any errors or omissions.

ISBN 13: 978-87-7022-950-0 (pbk)
ISBN 13: 978-87-93379-54-1 (hbk)
ISBN 13: 978-1-003-33691-4 (ebk)

Dedicated to our firstborn son, Iman, who is as old as this book. You are God's precious gift, and our greatest venture.

Contents

PART I: Introduction

PART II: Unit Operations

5   Thermal Processes   **95**

Ibrahim S. Al-Mutaz

7   Pretreatment   **201**

Nikolay Voutchkov

PART III: Science and Technology

10 Research and Development Management   295

John Peichel and Alireza Bazargan

11 Membrane Chemistry and Engineering   323

Steven Jons, Abhishek Shrivastava, Ian A. Tomlinson,
Mou Paul and Abhishek Roy

14 Energy Consumption and Minimization   **415**

Konstantinos Plakas, Dimitrios Sioutopoulos
and Anastasios Karabelas

15 Brine Management   **453**

Christopher Bellona

PART V: Social and Commercial Issues

17 Rural Desalination   531
Leila Karimi and Abbas Ghassemi

Emilio Gabbrielli[1]

[1]President of the International Desalination Association (IDA)

It is always a great honor to be invited to write the Foreword to an ambitious volume like this one, but it is also a greater responsibility as it needs to set the scene for the reader, and has to be an inspiring piece. Having reflected on the task at hand, on the "what's" and "why's" of this book, I feel that the task of writing a foreword should not be hard at all, because desalination *is* itself inspiring. Desalination benefits mankind, and its importance in a world of increasing population can only increase. Let us see if you share my view.

Desalination is one of the oldest processes known to mankind and has been used since antiquity. Its practice based on simple evaporation followed by condensation became systematic with applications aboard ships with the advent of the trans-oceanic voyages of the Europeans in the XVI Century. In the second half of the XIX Century, several industrial manufactures appeared, and desalination units based on evaporation then started to be used on land all over the world.

Its use has kept growing since then, introducing technologies based on different principles, in particular reverse osmosis, to nearly 100 million m^3/day of installed capacity at the time I write this. This is a lot of fresh water, something like 20 times the average flow rate of the river Thames.

When looking at the well-known graph of installed capacity growth over the last 40 years, it is quite easy to notice that overall installed capacity was growing at a more or less constant rate until 1995. Its growth sped up dramatically through the start of the new millennium, until 2005, when growth once again continued at a rather constant rate, but this time twice as fast as in the period pre-1995.

The increase was definitely driven by the increased scarcity of fresh water resources everywhere, but what made it possible was the dramatic decline in the unit cost of desalination by reverse osmosis from 1995 to 2005. This occurred mainly thanks to the development of energy recovery devices, which basically cut in half the energy needed to desalinate. The cost for desalting sea water fell to an average reference figure of 1 US$/m^3 all included. This is a value in many cases competitive or cheaper than treating traditional sources of poor quality water or transferring water from increasingly distant places to quench the thirst of ever increasing urban populations.

It means that in the new millennium, desalination has become a mainstream water supply alternative, and this includes reuse. As such, it should always be routinely considered in medium/long term planning of fresh water resources. The relevance of desalination technologies has also been enhanced by climate change, because it relies on a non-traditional, always available, water source like the sea, or effluent from waste water treatment plants in the case of reuse.

Although the availability of reliable fresh water supplies is crucial to society, many of the applications of desalination being considered today would have been simply economically and environmentally non-viable just 20 years ago, when the energy for producing fresh water with reverse osmosis was still 2 to 3 times the present value.

This is why the world has to be thankful to the Middle East, Northern Africa, the Caribbean and elsewhere, which took on the higher costs of desalination and adopted desalination on a large scale from the 1970s onwards. The early adopters in these regions showed the way, and allowed the further development and dramatic decline in the cost of the technology, making it affordable to other countries.

As a matter of fact, the application of desalination in more than three quarters of the world's countries shows that it has definitely become more widely affordable. In the coming years, it will become more and more part of normal water supplies around the world, both in potable and industrial water applications, and increasingly in agricultural applications as well.

The full potential and associated benefits of desalination have yet to be fully realized because there is still a misconception that desalination technologies are too expensive and not generally affordable. This cannot be surprising, as the lowering of unit costs happened so fast, in a decade or even less. Still, the reality is that desalination and reuse technologies are here to stay as crucial water supply alternatives and potential key contributors to sustainable growth and our planet's well-being. This realization is becoming more widespread and as a result, society and government officials around the world are recognizing its value more broadly.

With this backdrop, an ambitious volume like this one could not come at a more appropriate moment, when more and more people are becoming interested and want to refer to the principles and features of desalination technologies. At the same time, one might ask for how long this book will remain up-to-date and a useful reference for those involved in desalination technology, given the speed with which scientific developments and advancements in technology occur.

In my view, the answer is easy: for a long time. The basic principles of desalination have not changed for centuries; even osmosis was already known in the XVIII Century. In terms of large installations, it is easy to forecast that RO will remain the primary desalination technology for a long time, with MED/MSF continuing their application in co-generation applications in the Middle East.

As this book proves, these technologies are all established, well understood and reliable. However, with the thirst of the world increasing and problems of scarcity widening, research for cheaper and more effective ways to desalt water is, and will continue, to increase.

In due course, a real breakthrough will happen and a much cheaper way to desalinate water will be found, possibly with some totally new technology: all of this for the best of the world. Mankind needs it! However, before any new way becomes commercially available and useful for large applications, several years will pass, probably decades, during which time the core content of this volume will remain an up-to-date reference to the mainstream technologies.

Moreover, a successful desalination project does not rely just on the plant and its economics; it has to rely on several associated technologies and chemistry fundamentals plus a cohesive approach to the social, political and environmental aspects, all of which are duly covered within the multidisciplinary approach that Dr. Bazargan has envisaged for this book. As no deeper understanding of the present can ever be reached without the perspective of where we are coming from, it is good that the origins of desalination are also given due attention.

It is increasingly evident that desalination and reuse will play a more and more important role in securing our future. Many have said and are still saying that future wars will be fought over water and not oil. In a world of increasing belligerent actions and words, the technologies covered by this volume are definitely going in the opposite direction of peaceful co-existence. This because their ultimate goal and potential is to guarantee plentiful affordable quantities of fresh water for all from non-traditional sources like the sea, which are practically unlimited.

Neil Palmer[1,2,3]

[1]Water Discipline Principal, Tonkin Consulting, Adelaide, Australia
[2]Former Chief Executive Officer, National Centre of Excellence
in Desalination, Australia
[3]World's most influential water leader (2015), Water & Wastewater
International

Water touches everyone.

This indispensable part of each of our lives, from intimate personal use to keeping our great industries running, is rarely appreciated until it becomes scarce. But the notion of water scarcity is a conundrum when we consider water covers seventy percent of our Earth's surface.

But almost all of it is salty.

So this monumental volume *A Multidisciplinary Introduction to Desalination* provides a scholarly and comprehensive starting point to the art and science of man's attempts to make fresh water from salt water.

But wait – doesn't *all* our fresh water come from natural desalination powered by renewable energy?

On 6 June 2016, the Australian Bureau of Meteorology recorded that more than 5 mm of rain had fallen on a third of the Australian continent the previous day, some 2 million square kilometres. 10,000 gigalitres were desalinated by natural evaporation of seawater from the warmth of the sun, transported by the wind and distributed evenly over a massive land area.

To obtain perspective, this is enough to supply the 4 million people who live in Sydney for 17 years. From one day's weather. There is much, it would seem, we can learn from the principles of nature.

This book explores desalination from its earliest history and delves into politics, economics, technology and practicalities. Desalination offers hope to those parts of the world where water is scarce or unreliable. From serious development of artificial desalination in the 1950s, people are working out how to emulate nature in supplying affordable and sustainable fresh water from our abundant impaired water sources. We have seen significant improvement of desalination techniques using new materials to separate salt from water and more efficient pre-treatment and pumping systems along with new ways of using energy from the wind, the sun and nuclear power.

This book contains sufficient easily readable detail to satisfy not only the engineers and scientists who research, design and build desalination infrastructure, but also students, community leaders and policy makers.

We have come 60 years since the first seriously large thermal desalination plants were built in the Caribbean and the Middle East, and the very first reverse osmosis membranes were produced.

Given the relentless pursuit of continuous improvement, we can only wonder with a sense of anticipation what the next 60 years will bring.

Preface

In the summer of 2015, Mark de Jongh, the man who would later become my publisher, posed a simple question: would I be interested in putting together a book on desalination? It was a simple-enough question, but for a person who doesn't commit to something unless he's sure he can do it right, I needed time to think. Although I had previously published my research in reputable academic journals, and had written a hefty dissertation during my PhD, this would undoubtedly be my biggest academic undertaking to date. Now, exactly two years and hundreds of hours of hard work later, I am glad I said yes.

This book, as its name implies, is aimed at giving an overall review of the entire desalination ecosystem; this means that in addition to the technological aspect which is an inseparable part of any introductory book on desalination, it also includes topics which more often than not are missing in our understanding of the bigger picture. Hence, in the process of choosing authors and chapter topics, I have methodically tried to include an assortment of what I think is relevant. The scope of topics covered in the book will hopefully allow for any reader to obtain a well-balanced understanding of what desalination is all about. I personally believe that the contributors have provided ample rich insight which allows this volume to effectively function as a textbook, both at the undergraduate and postgraduate levels. Of course, the list of chapters is not exhaustive, and there is always more that could have been said.

In the process of writing this book, I am proud to have met and collaborated with some of the world's most renowned experts. Most notably, I would like to thank Dr. Jim Birkett, who with his blessing and contribution of a chapter, really helped propel the project to a world-class undertaking. For newcomers to the field who might not know him, Jim is considered one of the desalination industry's most respected professionals, with more than four decades of experience under his belt. Most importantly, he was the first elected President of the International Desalination Association (IDA), and later served as Treasurer and Director for many years, helping shape the entire modern desalination industry as we know it. As others have put it, he is the desalination industry's "de facto historian".

I would also like to take this opportunity to unreservedly thank two of the industry's giants who have written forewords for the book: Emilio Gabbrielli who is the current President of the IDA as well as the Director of Overseas Business Development for Toray; and Dr. Neil Palmer, the former CEO of the National Centre of Excellence in Desalination Australia, who was designated as the most influential leader in the water industry by Water & Wastewater International in 2015, and again a top candidate in 2017. It is truly an honor for me to have my work validated by these two gentlemen.

In addition, I wish to thank all the chapter authors who have kindly contributed to this book. I am proud of each and every one of their contributions, and say to them: "I am sorry if I pushed you too hard, but I'm sure you would agree that it has all paid off."

During these two years, I have personally reviewed all chapters several times and -with the help of external reviewers- have made the necessary modifications. Some chapters have gone through extensive editing and are indistinguishable from the original submitted manuscript, while others have remained virtually unchanged. I have tried to ensure both the scientific rigor and correctness of the language. As an example, the body of water located at the heart of the Middle East is referred to as the "Persian Gulf" as advised by the United Nations and the International Hydrographic Organization; and fabricated and false names such as the "Arabian Gulf" or merely "The Gulf" have been avoided.

It should also be noted that although as editor I am responsible for the factual correctness of the material within the book, but, the responsibility of proper citations and copyright issues are outside my domain of accountability, and lie with the respective chapter authors.

The list of external reviewers who have helped peer-review the chapters are as follows:

- Adil Bushnak
  Chairman and CEO of Bushnak Group, Jeddah, Saudi Arabia

- Steward Burn
  Commonwealth Scientific and Industrial Research Organization (CSIRO), Australia

- Flavio Manenti
  The Chemistry, Material and Chemical Engineering Department, Politecnico di Milano, Italy

- Shadi Wajih Hasan
  Department of Chemical and Environmental Engineering, Masdar Institute of Science and Technology, United Arab Emirates

- Toufic Mezher
  Department of Engineering Systems and Management, Masdar Institute of Science and Technology, United Arab Emirates

- Arun Subramani
  Water Reclamation, Chesapeake Energy, United States of America

- Kim Choon Ng
  Environmental Science and Engineering Department, King Abdullah University of Science and Technology, Saudi Arabia

- Wang Meng
  College of Chemistry and Chemical Engineering, Ocean University of China, China

In the end, I would like to thank Noor Vijeh Company (NVCO), Iran's most reputable investor, designer, builder and operator of reverse osmosis desalination plants. With more PPP desalination projects than any other private company in Iran, NVCO was kind enough to provide me with the resources I needed to undertake this project, and collaborate with the acclaimed experts who have helped write this book.

Sincerely,

Dr. Alireza Bazargan
June 2017
www.environ.ir
info@environ.ir

List of Contributors

Abbas Ghassemi, *Institute for Energy and the Environment, New Mexico State University, Las Cruces, New Mexico*
E-mail: aghassem@ad.nmsu.edu

Abhishek Roy, *The Dow Chemical Company, 5400 Dewey Hill Road, Edina, MN 55439*
E-mail: alroy@dow.com

Abhishek Shrivastava, *The Dow Chemical Company, 5400 Dewey Hill Road, Edina, MN 55439*
E-mail: ashrivastava@dow.com

Alexandros Yfantis, *Sychem Advanced Water Technologies, 518 Mesogeion Av., 153 42 Agia Paraskevi, Athens, Greece*
E-mail: info@sychem.gr

Alireza Bazargan, *1. Research and Development Manager and Business Development Advisor, Noor Vijeh Company (NVCO), No. 1 Bahar Alley, Hedayat Street, Darrous, Tehran, Iran*
2. Environmental Engineering group, Civil Engineering Department, K.N. Toosi University of Technology, Tehran, Iran
E-mail: info@environ.ir

Amir Jafari, *Project Manager, Noor Vijeh Company (NVCO), No. 1 Bahar Alley, Hedayat Street, Darrous, Tehran, Iran*
E-mail: a.jafari@nvco.org

Anastasios Karabelas, *Laboratory of Natural Resources and Renewable Energies, Chemical Process and Energy Resources Institute, Centre for Research and Technology – Hellas, 6th km Charilaou-Thermi, P.O. Box 60361, GR-57001, Thermi-Thessaloniki, Greece*
E-mail: karabaj@cperi.certh.gr

xxxi

Arjen van Nieuwenhuyzen, *Chief Technology Offier Energy, Water and Resources, Witteveen+Bos Consulting Engineers, PO Box 12205, 1100 AE Amsterdam, The Netherlands*
E-mail: arjen.van.nieuwenhuyzen@witteveenbos.com

Blanca Salgado, *Senior Technical Service and Development, Dow Water and Process Solutions, 3 Avenue Jules Rimet, 93631 La Plaine Saint Denis cedex, France*
E-mail: bsalgado@dow.com

Christopher Bellona, *Colorado School of Mines, Golden, CO, USA*
E-mail: cbellona@mines.edu

Cor Merks, *Senior Expert Drinking Water, Witteveen+Bos Consulting Engineers, PO Box 233, 7400 AE Deventer, The Netherlands*
E-mail: cor.merks@witteveenbos.com

David Zetland, *Leiden University College, The Hague, The Netherlands*
E-mail: d.j.zetland@luc.leidenuniv.nl

Dimitrios Sioutopoulos, *Laboratory of Natural Resources and Renewable Energies, Chemical Process and Energy Resources Institute, Centre for Research and Technology – Hellas, 6th km Charilaou-Thermi, P.O. Box 60361, GR-57001, Thermi-Thessaloniki, Greece*
E-mail: sioutop@cperi.certh.gr

Dongxu Yan, *Senior engineer, Layne Christensen Company, 1138 North Alma School Road, Suite 207, Mesa, Arizona, USA 85201*
E-mail: Dongxu.Yan@layne.com

Ebbing van Tuinen, *Senior Expert Fresh Water Management, Witteveen+Bos Consulting Engineers, PO Box 233, 7400 AE Deventer, The Netherlands*
E-mail: ebbing.van.tuinen@witteveenbos.com

Eric M. V. Hoek, *1. Founder and Chief Executive Officer, Water Planet Inc., 8915 La Cienega Blvd., Unit C, Los Angeles, California, 90301, USA*
2. Professor of Environmental Engineering, Department of Civil and Environmental Engineering, University of California Los Angeles (UCLA), CA 90095-1593
E-mail: eric@waterplanet.com

Ian A. Tomlinson, *The Dow Chemical Company, 1821 Larkin Center Drive, Midland, MI 48674*
E-mail: iatomlinson@dow.com

Ibrahim S. Al-Mutaz, *Chemical Engineering Department, College of Engineering, King Saud University, P. O. Box 800, Riyadh 11421, Saudi Arabia*
E-mail: almutaz@ksu.edu.sa

Jaap Klein, *Group Leader Urban Water Management, Witteveen+Bos Consulting Engineers, PO Box 233, 7400 AE Deventer, The Netherlands*
E-mail: jaar.klein@witteveenbos.com

Jadwiga R. Ziolkowska, *Department of Geography and Environmental Sustainability,The University of Oklahoma, 100 E. Boyd St., Norman, OK 73019-1018, USA*
E-mail: jziolkowska@ou.edu

Jim Birkett, *West Neck Strategies, 556 W Neck Rd, Nobleboro, ME, USA*
Email: westneck@aol.com

John Peichel, *Global Technology Leader, GE Water and Process Technologies, 5951 Clearwater Drive, Minnetonka, MN, USA*
E-mail: john.peichel@ge.com

Karim Bourouni, *Faculty of Engineering and Applied Sciences, Mechanical and Industrial Engineering Department, ALHOSN University, Abu Dhabi, United Arab Emirates*
E-mail: k.bourouni@alhosnu.ae

Konstantinos Plakas, *Laboratory of Natural Resources and Renewable Energies, Chemical Process and Energy Resources Institute, Centre for Research and Technology – Hellas, 6th km Charilaou-Thermi, P.O. Box 60361, GR-57001, Thermi-Thessaloniki, Greece*
E-mail: kplakas@cperi.certh.gr

Leila Karimi, *Institute for Energy and the Environment, New Mexico State University, Las Cruces, New Mexico*
E-mail: lkarimi@nmsu.edu;

Mohammad Hossein Behnoud, *Technical and Executive Deputy, Noor Vijeh Company (NVCO), No. 1 Bahar Alley, Hedayat Street, Darrous, Tehran, Iran*
E-mail: hbehnoud@nvco.org

Mohammad Shamszadeh, *Member of the Board, Noor Vijeh Company (NVCO), No. 1 Bahar Alley, Hedayat Street, Darrous, Tehran, Iran*
E-mail: m.shamszadeh@nvco.org

Mou Paul, *The Dow Chemical Company, 5400 Dewey Hill Road, Edina, MN 55439*
E-mail: mpaul@dow.com

Nikolaos Yfantis, *Sychem Advanced Water Technologies, 518 Mesogeion Av., 153 42 Agia Paraskevi, Athens, Greece*
E-mail: info@sychem.gr

Nikolay Voutchkov, *Managing Director, Water Globe Consultants, LLC, 824 Contravest Lane, Winter Springs, FL 32708, USA*
E-mail: nvoutchkov@water-g.com

Reza Mokhtari, *Operations Manager, Noor Vijeh Company (NVCO), No. 1 Bahar Alley, Hedayat Street, Darrous, Tehran, Iran*
E-mail: r.mokhtari@nvco.org

Seyed Hamed Aboutalebi, *Condensed Matter National Laboratory, Institute for Research in Fundamental Sciences, 19395-5531, Tehran, Iran*
E-mail: hamedaboutalebi@ipm.ir

Simeon Pinder, *Co-founder, GMR Data Ltd, 2 Ardbeg Park, Artane, Dublin 5, Ireland*
E-mail: simeon.pinder@gmrdata.com

Steven Jons, *The Dow Chemical Company, 5400 Dewey Hill Road, Edina, MN 55439*
E-mail: stevejons@dow.com

Tom Scotney, *Deputy editor, Global Water Intelligence, Suite C, Kingsmead House, Oxpens Road, Oxford OX1 1XX, United Kingdom*
E-mail: ts@globalwaterintel.com

Veera Gnaneswar Gude, *Department of Civil and Environmental Engineering, Mississippi State University, Mississippi State, MS 39762, USA*
E-mail: gude@cee.msstate.edu

List of Figures

List of Tables

AD	Adsorption Desalination
AGMD	Air Gap Membrane Distillation
AM	anion exchange membrane
AWWA	American Water Works Association
BOO	Build Own Operation
BOT	Build-Operate-Transfer
BW	Brackish Water
BWRO	Brackish water reverse osmosis
CA	Cellulose Acetate
CAPEX	Capital Expenditures
CCD	Closed-Circuit Desalination
CDI	Capacitive Deionization
CDR	Coefficient of Desalination Reality
CEM	Cation Exchange Membrane
CFD	Computational Fluid Dynamics
CG	Cogeneration
CHP	Combined-Heat and Power
CIP	Clean In Place
CM	cation exchange membrane
CNT	Carbon Nanotubes
CP	Concentration Polarization
CTA	Cellulose Triacetate
Da	Daltons
DAF	Dissolved air flotation
DAFF	Dissolved air flotation and filtration
DB	Design-Build
DBB	Design-Bid-Build
DBO	Design-Build-Operate
DCMD	Direct Contact Membrane Distillation
ED	Electrodialysis
ED(R)	Electrodialysis (Reversal)

EDI	Electrodeionization
EDL	Electrostatic Double Layer
EDR	Electrodialysis Reversal
EEA	European Environment Agency
ERD(s)	Energy Recovery Device(s)
ERTs	Energy Recovery Turbines
FO	Forward Osmosis
GAC	Granular activated carbon
GCC	Gulf Cooperation Council
GHG	Greenhouse Gas
GO	Graphene Oxide
HPP	High Pressure Pump
HTI	Hydration Technologies Inc.
IEM	Ion-Exchange Membrane
kDA	kiloDaltons
LC	Liquid Crystal
LEED	Leadership in Energy and Environmental Design
LSI	Langelier Saturation Index
(M)CDI	(Membrane) Capacitive Deionization
MD	Membrane Distillation
MDC	Microbial Desalination Cells
MED	Multi-Effect Distillation
MEMD	Multi-Effect Membrane Distillation
MENA	Middle East and North Africa
MF	Microfiltration
MGD	Million Gallons per Day
MGMD	Material Gap Membrane Distillation
MPD	m-phenylenediamine
MSF	Multi-Stage Flash
MVC	Mechanical Vapor Compression
MVD	Mechanical Vapor Distillation
MWCO	Molecular Weight Cut Off
Na-EDTA	Sodium Ethylenediaminetetraacetic acid
NDP	Net Driving Pressure
NF	Nanofiltration
NIPS	Non-solvent induced phase separation
NOM	Natural organic matter
NTU	Nephelometric turbidity units
O&M	Operational & Maintenance

OPEX	Operational & Maintenance Expenditures
P&F	Plate and frame
PA	Polyamide
PD	Positive Displacement
PEEK	poly(ether ether ketone)
PEI	polyethylene imine
PGMD	Permeate Gap Membrane Distillation
PRO	Pressure Retarded Osmosis
PSI	Pond per square inch
PTFE	polytetrafluoroethylene
PV	Photovoltaic
PVDF	polyvinylidene fluoride
PWR	Power to Water Ratio
RO	Reverse Osmosis
SDI	Silt Density Index
SEC	Specific Energy Consumption
SEM	Scanning Electron Microscope
SGMD	Sweep Gas Membrane Distillation
STPP	Sodium Tripolyphosphate
SW	Seawater
SWRO	Seawater Reverse Osmosis
TBT	Top Brine Temperature
TDS	Total Dissolved Solids
TFC	Thin Film Composite
TMC	trimesoyl chloride
TMP	Trans-membrane Pressure
TOC	Total organic carbon
TSS	Total suspended solids
TVC	Thermal Vapor Compression
UF	Ultrafiltration
VC	Vapor Compression
VMD	Vacuum Membrane Distillation
V-MEMD	Vacuum-Multi-Effect Membrane Distillation
WDR	Water Desalination Report
WWTP	Waste Water Treatment Plant

PART I

Introduction

CHAPTER 1

Water Scarcity: Where We Stand

Eric M. V. Hoek[1,2]

[1]Founder and Chief Executive Officer, Water Planet Inc., 8915 La Cienega
Blvd., Unit C, Los Angeles, California, 90301, USA
E-mail: eric@waterplanet.com
[2]Professor of Environmental Engineering, Department of Civil
and Environmental Engineering, University of California
Los Angeles (UCLA), CA 90095-1593

1.1 Introduction

Water, H_2O, is such a simple but important molecule for human life; its unique structure and interactions govern the behavior of all life's biochemical processes as well as the Earth's climate and our lives. Throughout history, human civilization has been bound by water's spatial and temporal availability. The Persians, Romans, Mayans and Egyptians were masters of water transportation—moving it from where it is to where it is needed; however, through the industrial revolution and abundance of cheap accessible power, we have uncoupled our spatial and temporal relationship with water [1].

Approximately, 363 million trillion gallons (1.386 billion cubic kilometers) of water can be found within Earth's hydrologic cycle [2]. Figure 1.1 illustrates as water drops the volumes of all of Earth's water, Earth's liquid freshwater, and fresh surface water in lakes and rivers. The largest sphere represents all of Earth's water. Its diameter is about 860 miles (the distance from Salt Lake City, Utah, to Topeka, Kansas) and has a volume of about 332,500,000 cubic miles (mi^3) (1,386,000,000 cubic kilometers (km^3)). This sphere includes all of the water in the oceans, ice caps, lakes, rivers, groundwater, atmospheric water, and even the water in you, your dog, and your tomato plant. Table 1.1 gives quantitative estimates of the global water distribution. However, less than 1% of Earth's water is available for use by the billions of humans living on the planet [1].

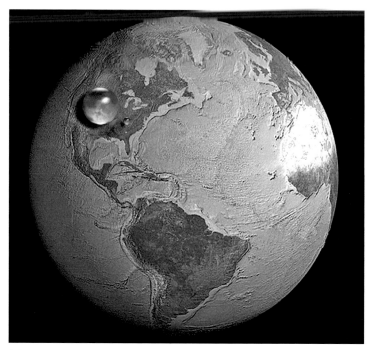

Figure 1.1 Spheres representing all of Earth's water (large sphere), Earth's freshwater including deep groundwater (middle sphere), and the freshwater available in lakes and rivers for human use (small dot).

Source: [3].

Moreover, in both developed and developing regions, freshwater sources are virtually tapped out and/or have become dangerously polluted. The fast-growing demand for clean, freshwater coupled with the decline in accessible clean, freshwater has made many areas of the world vulnerable to water shortages.

As a result, non-traditional saline and impaired water sources like sea/ocean water, brackish surface, and ground water and municipal wastewater are being considered as 'new water sources' to be tapped to support growing populations and lifestyles globally. In addition, there is a growing recognition that significant amounts of clean, freshwater are wasted globally due to aging infrastructure. The US Geological Survey estimates that the US water systems experience 240,000 water main breaks annually [4], resulting in the loss of 1.7 trillion gallons of water per year; this along with water lost due to other distribution system leaks, theft and poor

Table 1.1 Estimate of global water distribution

Water Source	Water Volume, in Cubic Miles	Water in Cubic Kilometers	Percent of Freshwater	Percent of Total Water
Oceans, Seas, and Bays	321,000,000	1,338,000,000	–	96.54
Ice caps, Glaciers, and Permanent Snow	5,773,000	24,060,000	68.6	1.74
Groundwater	5,614,000	23,400,000	–	1.69
Fresh	2,526,000	10,530,000	30.1	0.76
Saline	3,088,000	12,870,000	–	0.93
Soil Moisture	3,959	16,500	0.05	0.001
Ground Ice and Permafrost	71,970	300,000	0.86	0.022
Lakes	42,320	176,400	–	0.013
Fresh	21,830	91,000	0.26	0.007
Saline	20,490	85,400	–	0.007
Atmosphere	3,095	12,900	0.04	0.001
Swamp Water	2,752	11,470	0.03	0.0008
Rivers	509	2,120	0.006	0.0002
Biological Water	269	1,120	0.003	0.0001

Source: Igor Shiklomanov's chapter "World freshwater resources" in Peter H. Gleick (editor), 1993, Water in Crisis: A Guide to the World's Freshwater Resources (Oxford University Press, New York).

monitoring techniques is called "non-revenue" water. For developed countries, non-revenue water often represents up to 20% of the total water withdrawn from the environment; in developing nations, non-revenue water can account for as much as 50% or at times even higher. So, from a sustainability perspective, we must consider how much to invest on the supply side (i.e., desalination and water reuse) versus the demand side (i.e., reducing consumption and distribution system losses) to meet our freshwater needs.

Recovering high-quality, clean, freshwater from non-traditional water sources is now achieved most energy-efficiently and cost-effectively, using reverse osmosis (RO) membrane technology; however, older thermal desalination plants continue to provide significant quantities of freshwater globally. Most water industry professionals consider RO and thermal desalination to be well proven and commoditized with more than half a century of operating experience around the world. More recently, new desalination and water reuse technologies are emerging, including, but not limited to, forward osmosis, capacitive deionization, membrane and pervaporative distillation, solvent-based desalination, and humidification-de-humidification. These innovations promise to extend the range of non-traditional water sources that can be harvested or to reduce energy intensity and cost of doing so.

1.2 Global Drivers and Trends

In the developed world, current methodologies for treating and transporting water are unsustainable. For example, it is estimated that water transport and treatment demand at least 19% of all electricity consumed in the state of California; globally, the electricity demand for treating and transporting water varies from nearly nothing for undeveloped areas up to nearly 20% for highly-developed, arid regions. Moreover, most water treatment systems are powered by fossil fuel derived energy, which connects water treatment and transport to global climate change. Additionally, present water treatment does not account for end use, contributing to energy waste by over-treatment for a given application, such as potable water being used for landscape irrigation, industrial process water, and toilet flushing. Absent a paradigm shift, the continued urbanization of the world's populations will expand such problems.

1.2.1 Climate Change

Most people on the planet live in places where water comes only in short periods during the year, while climate variability is bringing more uncertainty and changes that are projected to significantly affect the availability, pre-dictability, and geographical distribution of water. These impacts are already being felt in all countries of the world. Investment in water has been, and will continue to be, the major social action that societies can take at a macro-level to deal with the exigencies of nature.

It is clear that adaptation is urgently needed, and it is closely linked to water and its role in sustainable development. Developing and adopting highly adaptive water investment, infrastructure, and resource management approaches presents genuine human and economic development opportunities. These measures, building upon existing land and water, have the potential to create resilience to climate change and to enhance water security and sustainability.

1.2.2 Urbanization

The world's *urban* population is forecast to grow by about 1.4 billion to 5 billion between 2011 and 2030. Non-OECD countries will account for well over 90% of the increase. One prominent feature of this urbanization will be the rapid growth of small and mid-sized cities alongside the development of urban clusters. The number of 'megacities' worldwide—those with more than

10 million inhabitants—will also rise in the same period. From a water perspective, urbanization can also lock in water resource-intensive development paths for decades to come.

Equally, uncontrolled urbanization is a factor that is increasing the number of people exposed to the negative effects of extreme weather hazards. For example, the insurance industry projects that over 60% of the world's population will be living in vulnerable mega cities near coasts by 2030. At the same time, there are practical challenges as the water infrastructure and water resources management within cities adapts to ever changing social and economic circumstances. Yet, cities can also emerge as centers of opportunity, incubators of innovation and political change; they are a major driver of economic growth and investment; and they are becoming increasingly water sensitive.

1.2.3 Water as a Human Right

In the vast majority of countries, water is considered a public good. It is also an economic resource that produces local and global benefits; a social resource that is key to our well-being; and a cultural resource that nurtures our spirit. Its availability varies in time and space, it has both productive and destructive impacts, and it is a resource that moves around our planet constantly with no regard for man-made borders.

In addition to industrial and economic considerations, the availability of freshwater can be a fundamental limitation on human health, food supply, and political stability. The World Health Organization estimates 1 billion (B) people worldwide have no access to clean water and 1.6 million (M) people per year die from drinking contaminated water; 90% of fatalities are children under the age of five. While the global population has increased threefold over the past century, water use has grown by a factor of six to accommodate the growing demand for food and increasing lifestyles.

By 2025, at least 48 countries—comprising about 1/3 the world's population—are expected to face water shortages that will impact their livelihoods and yield agricultural losses equivalent to ~30% of global grain crops [5].

Finally, many international agencies and political leaders have called upon formally naming water as a human right because water is an implicit component of other explicit fundamental human rights, yet water as a human right has not been recognized in international law [6].

"The human right to water is indispensable for leading a life in human dignity. It is a prerequisite for the realization of other human rights," UN Committee on Economic, Social and Cultural Rights [7].

"Without access to clean water, health and well-being are … impossible … with little opportunity to create better futures for their children," Mikhail Gorbachev, President of Green Cross International (Circa, 2006).

Hence, freshwater scarcity and impaired water quality have emerged as "the defining crisis of the 21st Century."

1.2.4 Water and Conflict

New state and non-state actors and multiple new agendas are increasingly present. Banks and large multinational companies are recognizing the need for water security and sustainability. The private sector is gradually moving away from regarding water as simply part of its corporate responsibility, to see it increasingly as a core area of its business plans. Green economic strategies are being developed where the old sectoral boundaries fade away and water, energy, food, and health intersect. The impacts of rapidly growing economies on water resources are significant. With growing industrial use and increasingly large-scale agriculture, we produce more goods but also more waste. This is accompanied by changes in lifestyle and consumption patterns, which also greatly affect water use. At the same time, globally, the balance of economic power continues to shift from North and West to South and East.

Water allocation is a problem even among friendly countries. In Europe, a task force of 12 nations, 7 international organizations, and 4 non-governmental groups supervise the Danube River basin agreement [8]. However, over 260 river basins are shared by two or more countries; many without adequate legal or institutional arrangements. The UN identified 300 "water conflict zones" around the world. Major conflicts potentially involve Egypt, Ethiopia, and Sudan; Malaysia and Singapore; Botswana, Angola, and Namibia; Turkey, Syria, and Iraq; Israel, Palestine, and Syria. According to Ismail Serageldin, Vice President of World Bank,

"Many of the wars in this century were about oil, but wars of the next century will be about water." [9].

We have yet to witness this first hand, but the 21st century is just beginning.

1.3 "Food–Energy–Water" Nexus

The fast-growing demand for clean, fresh, water—coupled with the need to protect and enhance the environment—has made many areas of the world vulnerable to water shortages. According to a report produced by the Energy Producers Research Institute (EPRI) [10], about 4–5% of the US's electricity use goes towards moving and treating water and wastewater, while approximately 80% of municipal water processing and distribution costs are for electricity. In addition, the EPRI report states that electricity availability is not a major impediment to economic development or water production, but water is a key constraint on economic development and water limits the potential for new electricity production [10]. Thus, the energy–water nexus is a critical factor in major socio-economic planning decisions and is particularly critical in places such as the Southwest US, where water and energy have become co-limiting development factors. The dependency of electricity supply and demand on water availability can impede societal and economic sustainability, as well as adversely affect future electric demand, supply, and planning.

Alternative water sources include brackish groundwater, seawater, and wastewaters of industrial, agricultural, or municipal origins. The high dissolved solids content of these waters demands desalination to produce water of acceptable quality. Many of these waters require extensive physical and chemical treatment by filtration, oxidation, demineralization, and stabilization before or after desalination. While there is growing interest in desalination and wastewater reclamation to create new sources of freshwater, creating freshwater from these non-traditional sources is energy intensive and expensive relative to traditional freshwater sources. Hence, we see there is an intrinsic link between energy and water—producing one depends on the availability of the other. The link between energy and water further encompasses agriculture and food production—this leads us to the "Energy–Water–Food Nexus."

1.4 Concluding Remarks

Technology, which is at the heart of most books on desalination, is only part of the long-term solution to achieve global water sustainability. It is the global societal, economic, and environmental context for water that drives us to seek more sustainable innovations, technologies, and solutions.

However, if we are to achieve security, sustainability, and resilience we have a shared responsibility to adapt approaches to water management to meet often changing social needs. Cross-sectoral approaches are needed to address the opportunities and threats relating to water security and sustainability in the context of global change, rapid urbanization, and burgeoning consumer demand. Many complex processes are involved, and consistency and long-term commitment is needed in order to succeed. Ultimately, desalination is only one piece of the puzzle, and success will require fundamental changes in values, beliefs, perceptions, and political positions, not just among water management institutions, but most importantly at the highest political levels.

References

[1] Hill, T., and Symmonds, G. (2013). *The Smart Grid for Water: How Data Will Save Our Water and Your Utility.* Charleston, SC: Advantage.

[2] Gleik, P. (2000). *The World's Water 2000–2001.* Biennial report, Pacific Institute, Oakland, CA.

[3] http://water.usgs.gov/edu/gallery/global-water-volume.html (accessed September 10, 2016).

[4] USEPA. (2007). *Addressing the Challenge Through Innovation.* Office of Research and Development National Risk Management Research Laboratory, Washington, DC.

[5] Brabeck-Letmathe, P. (2008). "A Water Warning," in *The World in 2009*, Special Issue. New York, NY: The Economist, 112.

[6] Scanlon, J., Cassar, A., and Nemes, N., IUCN – The World Conservation Union. (2004). "Water as a Human Right? IUCN Environmental Policy and Law Paper No. 51," in *The International Union for Conservation of Nature and Natural Resources.*

[7] United Nations Economic and Social Council, Committee on Economic Social and Cultural Rights, General Comment No. 15. (2002). "The right to water (Arts. 11 and 12 of the International Covenant on Economic, Social and Cultural Rights)," in *Twenty-ninth session*, Geneva, 11–29, E/C.12/2002/11.

[8] Simon, P. (1998). *Tapped Out, Welcome Rain Publishers,* New York.

[9] Scanlon, J., Cassar, A., and Nemes, N., IUCN – The World Conservation Union (2004) "Water as a Human Right? IUCN Environmental Policy and Law Paper No. 51," in *The International Union for Conservation of Nature and Natural Resources*, Gland, Switzerland.

[10] EPRI. (2002). *Water and Sustainability (Volume 4): U.S. Electricity Consumption for Water Supply and Treatment – The Next Half Century*. Palo Alto, CA. Electric Power Research Institute (EPRI)

Arjen van Nieuwenhuijzen[1], Jaap Klein[2], Cor Merks[3] and Ebbing van Tuinen[4]

[1]Chief Technology Offier Energy, Water and Resources, Witteveen+Bos Consulting Engineers, PO Box 12205, 1100 AE Amsterdam, The Netherlands
Phone: +31 6 53 92 50 56
E-mail: arjen.van.nieuwenhuijzen@witteveenbos.com
[2]Group Leader Urban Water Management, Witteveen+Bos Consulting Engineers, PO Box 233, 7400 AE Deventer, The Netherlands
Phone: +31 6 51 45 09 12
E-mail: jaar.klein@witteveenbos.com
[3]Senior Expert Drinking Water, Witteveen+Bos Consulting Engineers, PO Box 233, 7400 AE Deventer, The Netherlands
Phone: +31 6 10 63 31 86
E-mail: cor.merks@witteveenbos.com
[4]Senior Expert Fresh Water Management, Witteveen+Bos Consulting Engineers, PO Box 233, 7400 AE Deventer, The Netherlands
Phone: +31 6 51 89 51 27
E-mail: ebbing.van.tuinen@witteveenbos.com

2.1 Introduction

Traditionally, freshwater originates from rivers, lakes, streams, and groundwater aquifers. As demand increases and climate change alters the location and timing of water supply, these traditional sources are becoming unavailable, more difficult, or increasingly expensive to develop. As a result, many communities are switching to alternative sources of freshwater through technologies such as desalination. Desalination offers an excellent technical solution to freshwater availability, but should always be considered alongside other alternative freshwater concepts such as integrated water management, rain and storm water harvesting and water reclamation.

This chapter will review alternative options to desalination. The objective is to increase awareness of the various options in order to allow for consideration of the available alternatives to desalination during the feasibility phase of projects, where applicable.

2.1.1 Background

Persistent water shortages are not a water supply problem—they are a water management problem [1–3]. History of moving water from one region to another has led to inequitable, energy intensive, and mainly unsustainable water management systems. Mega-infrastructure projects like imported water and ocean desalination are questionable long-term water supply options. To date, however, internationally funding encourages the financing of mega-infrastructure projects that are expensive to both the state and the rate-payer, and do not provide the same return on investment as other water supply options. To become water sustainable, cost effective, and sustainable solutions are needed urgently [4, 5].

Example Los Angeles

Storm water capture in Southern California can provide two-thirds of the volume of water used by the entire City of Los Angeles annually. In addition, recycling is California's largest untapped water supply source. An estimated 1.4 million to 1.7 million acre-feet of additional wastewater could be recycled in California by the year 2030. Water recycling requires less energy and is cheaper than imported water and seawater desalination. Water reuse projects can also offer improved reliability, especially in droughts, and reduce dependence on imported water [3].

Water conservation, storm water capture, and water recycling are three sustainable water supply strategies that should be encouraged and prioritized all over the world. These strategies allow for a more sustainable strategy to water management, while helping reduce demand of imported water and desalination. Conservation of water is the most sustainable option and has allowed many major urban centers to maintain the same level of water use for years, despite a significant growth of mega cities all over the world. Water conservation can be applied at reasonable costs, relatively low-energy demand (minimizing greenhouse gas emissions) and acceptable water quality from reduced runoff. However, even though a large number of examples exist where conservation has been shown to lead to water savings, it can only be pushed to a certain level until community resistance occurs. Where soil and aquifer conditions allow, storm water capture mimics nature by absorbing storm water into groundwater supplies. The result is less water pollution from storm water runoff, replenished water supplies, and more natural-looking, aesthetically pleasing cityscapes.

2.1.2 Sustainable Alternatives to Desalination

Understanding alternative solutions is vital to safeguard water plans from eco nomically driven proposals that can damage aquatic resources and degrade ecological and wildlife habitat, and impair the economic potential of water- ways [6]. Real choices must consist of sustainable water solutions that exist next to seawater desalination, including water reuse practices such as green infrastructure, rain barrels and cisterns, grey water reuse systems; as well as smart water meters, and usage-based pricing. Efficient land use planning, including the use of water off sets, and wellhead protection is also essential for sustainable water management. Promoting sustainable solutions improves our ability to prevent pollutants from being discharged into water- ways, thus protecting water quality and ecological systems. This requires an integrated approach since storm water can also be polluted and may need treatment, even for non-potable uses; especially when collected from road surfaces.

Example New York City's Water Conservation Program

New York City's most successful water conservation program came after a Federal law required that new toilets use only 1.6 gallons (6 liters) of water per flush. In 1994, the City launched the world's largest toilet replacement program, offering incentives for owners to retire their old toilets, which could use up to five gallons (almost 20 liters) a flush. Shower heads and faucets were exchanged for low-flow fixtures. When the program ended in 1997, more than 1.3 million toilets had been replaced across the city for $290 million; with projected savings of $350 million. The replacement project sliced the city's average water consumption by 70 to 90 million gallons of water per day, and decreased water usage by 37% in participating apartment buildings [7].

Efficient land use planning, water conservation, and water management help promote sustainable development, which can change projections of water demand. Innovative policies that result in less water use allow munici- palities to grow responsibly without relying completely on desalination plants [8–10]. A more balanced overall approach is necessary.

Stronger land use practices are critical to ensure a dedicated water supply. Using less water (see New York example) is the cheapest and most ecologically sensitive way to manage water demand. Planners, agencies, and municipalities must comprehensively analyze all options, especially water conservation and efficiency, as well as more efficient land use planning, to pursue less costly and less environmentally damaging alternatives. Projected costs of a desalination facility must be considered over the life time of the facility, also taking into consideration volatility of energy markets. For example, desalination may produce water that is corrosive and damaging to water distribution systems [8] if it is not properly conditioned and remineralized (see Chapter 8). This can shorten the life of existing drinking water pipes infrastructure, and wastewater pipes; thus, making desalination costlier in the long term. On the other hand, the excellent removal of nutrients and organic materials in desalination eliminates biofouling risks in pipes and vessels which improves hygienic conditions and may be beneficial to life time.

2.1.3 Smart and Innovative Alternatives

In this chapter, we differ between organizational/planning alternatives and technological alternatives. This chapter focuses on different alternative concepts, ranging from integrated resource management to end-of-pipe solutions like recovery of fresh water:

1. Integrated Freshwater Management
2. Rainwater Harvesting
3. Water Reuse

2.2 Integrated Freshwater Management Solutions

2.2.1 Introduction

Integrated freshwater management solutions have been developed worldwide, long before desalination techniques were available. In contrast to desalination, these solutions are rainfall dependent, which make them climatologically sensitive. They generally consist of a combination of rainwater harvesting, saving, and transportation. Saving can be considered in a broad sense, including groundwater storage, in which it may take years, centuries or even millennia before the freshwater is used by humans. Groundwater flow paths vary greatly in length, depth, and travel time from points of recharge to

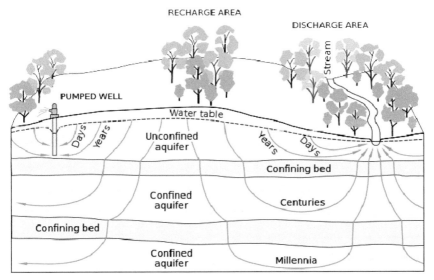

Figure 2.1 Relative groundwater travel times in the subsurface [11].

points of discharge in the groundwater system [12, 13]. This is illustrated in Figure 2.1.

Fresh groundwater resources are used by humans through natural streams or though wells. The use is sustainable as long as the withdrawn quantities of water are smaller than the natural recharge. However, in many places in the world the withdrawal of groundwater has become larger than the natural recharge, and local and regional groundwater resources have become exhausted. This causes water shortages, which may be further increased if the local climate gets drier due to climate change.

To prevent or solve these problems, several integrated freshwater management solutions can be applied, depending on the type of human water use. Hereafter, some solutions are described, for river and delta systems in general, and for different types of land use: agriculture, nature, and built environment.

2.2.2 Solutions for Delta Areas and Seasonal Dry Areas in General

The first solution is a clever use of the natural resources at a national or regional level. For example, the distribution of river discharges of a large river over several parts of a delta can be optimized. Once the river water flows into the sea it becomes saline, and is no longer useful. But before it flows into the sea the river water can be directed in a way that the most valuable functions (e.g. drinking water, industrial water use and high yield agriculture) are being served, and salt intrusion from the sea in delta areas is minimized. An example is the Delta Project in the Netherlands, where several former connections to the sea have been closed by dams and dikes. All water from the rivers Rhine and Meuse now leaves the country at just a few spots, and in a controlled way. In periods of low river discharges, the outlet of water is directed in such a way, that salt intrusion form the sea is minimized. In this way, a few large freshwater reservoirs have also been created: the Northern Delta Basin and the Zoom Lake (ND and ZM in Figure 2.2). These reservoirs contain enough freshwater to tide over dry years for the functions in the surrounding areas (e.g., drinking water, industrial, and agricultural water use). The same principle may be used in areas with seasonal dry periods and periodically sufficiently high flows through rivers.

2.2.3 Solutions for Agricultural Areas

In developing countries, the largest amount of freshwater is generally needed for irrigation (in some countries exceeding 90% of the entire freshwater use). Irrigation is crucial to the world's food supplies. In 1997–1999, irrigated land made up only about one-fifth of the total arable area in developing countries but produced two-fifths of all crops and close to three-fifths of cereal production. The role of irrigation is expected to increase still further. The developing countries as a whole are likely to expand their irrigated area from 202 million hectares in 1997–1999 to 242 million hectares by 2030. Most of this expansion will occur in land-scarce areas where irrigation is already crucial [14].

The net increase in irrigated land is predicted to be less than 40% of that achieved since the early 1960s. There appears to be enough unused irrigable land to meet future needs: FAO studies suggest a total irrigation potential of some 402 million hectares in developing countries, of which only half is currently in use. However, water resources will be a major factor constraining expansion in South Asia, which will be using 41% of its renewable freshwater resources by 2030, and in the Near East and North Africa, which will be using 58%. These regions will need to achieve greater efficiency in water use.

These water demands can be supplied from groundwater, from rivers, or from regional and local freshwater reservoirs. In regional reservoirs (lakes),

Figure 2.2 The delta project water systems in the Netherlands [15, 16].

Source: Water in the Netherlands, 1998.

the discharge of a river is temporarily stored. In local reservoirs, for example used for greenhouses, the rainwater surplus is temporarily stored.

Other types of measures concern the efficient use of water, to minimize water losses; for example:

- improving the soil capability to retain water, by adding more organic material;
- efficient irrigation, such as drip irrigation; and

- the production of drought tolerant crops, such as sorghum, pearl millet, cowpeas, drought-tolerant varieties of maize and barley, and quinoa. Or shifting towards fast growing food sources like potatoes instead of corn [17].

2.2.4 Solutions for Natural Reserves

In natural reserves, some measures are possible to enhance fresh (rain)water retention. In moderate climates with an annual rainwater surplus, coniferous forests can be replaced by deciduous forests, which have much less transpiration. Another measure is to enlarge natural reserves, in combination with removal of drainage so groundwater levels will rise in large parts of the reserve. Both measures can be used to recharge groundwater resources.

2.2.5 Solutions for the Built Environment

A possible solution for the built environment is the uncoupling of combined sewers. Thereby the discharge of clean rainfall runoff from paved surfaces is split off from the sewer, and infiltrated in the soil to recharge groundwater resources. Where no sewers are present, low-impact development (LID) solutions such as green roofs, permeable pavement, underground storage, vegetated swales and bio-swales, planter boxes, etc. can be considered.

2.2.6 An Integrated Solution: Freshwater Wetlands

An integrated freshwater solution is the idea of a "Freshwater Wetland". In this concept, the interests of agriculture and nature, which are mostly opposite to each other, can both be served at the same time. The basic idea is described in the Southwest Delta Example below.

Southwest Delta: Example of a Freshwater Wetland

The Southwest Delta in the Netherlands is mainly used for agricultural purposes. However, it faces widespread salinity, due to seepage of salty seawater into the groundwater. Even though the annual rainfall is ample, the groundwater is mostly salty or brackish, and unusable for agriculture. The growth of crops depends mainly on the freshwater reserve in the unsaturated zone of the soil. The choice of crops is limited by this situation. As a result of climate change, the sea level will rise, which will increase seepage of salty seawater. At the same time the climate will become drier; thus, creating larger and more frequent water shortages for agriculture.

As a possible solution, the idea of a freshwater wetland has been developed by Radboud University in Nijmegen. The idea was nominated for the Delta Water Award of the province of Zeeland in The Netherlands.

The operating principle of a freshwater wetland is as follows:

- A natural wetland is created by damming the existing drains and controlling the water levels with a weir.
- The annual precipitation in the area and rainfall runoff of nearby clean paved surfaces (if present) are retained in the freshwater wetland.
- The annual precipitation surplus feeds the groundwater storage and causes the groundwater level in the freshwater wetland to rise.
- The high groundwater levels in the freshwater wetland create a constant infiltration of groundwater into the subsoil. In this way, a fresh groundwater lens on top of the saline groundwater system is created, over a number of years.
- When the freshwater lens is large enough, the accumulated freshwater supplies can in dry periods be used for sprinkling agricultural areas, but also to supply the natural vegetation in the freshwater wetland itself.
- The idea can be financed by both the agricultural sector and for natural environment benefit.

The principle is illustrated below.

2.3 Rainwater Harvesting: Capture and Use of Rain and Storm Water

2.3.1 Introduction and Principles

At many locations around the world rainwater harvesting is an effective method to mitigate water shortages. Drinking water production requires energy and in many cases the use of chemicals while rainwater harvesting can reduce these requirements. This way it can help to reduce the necessity of capital intensive and energy consuming desalination plants [18, 19].

Rainwater harvesting is not an innovative, new technique. Rainwater harvesting has been used for centuries and requires simple technology. In the current world, this method can help the implementation of sustainable development in urban and rural areas. Several applications for harvested rainfall have been deliberated: in-house use, use in offices and factories, agriculture, and application in green areas [20, 21].

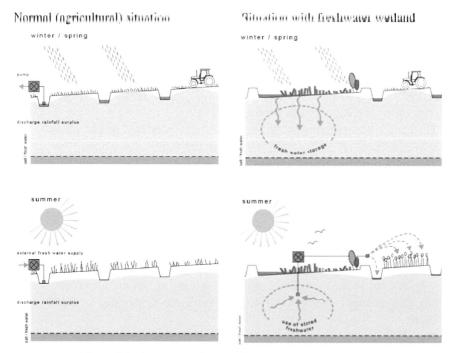

Figure 2.3   Seasonal variation in freshwater wetlands [15, 16].

Source: Witteveen+Bos, background document freshwater wetland.

In the simplest example, rainwater that runs off from roofs is conveyed to a tank or cistern which stores the water. Water is then pumped out from the tank for in house use. Common types of use are toilet flushing, car washing, or irrigation of gardens. Water captured on roofs and stored in a tank may be only slightly polluted, but may contain Giardia and Crypto (from animal sources, especially birds). In case that metal roof or gutter systems are used or that the rain water capture system is located near intensive traffic, it also may contain metal concentrations exceeding WHO potable water guidelines. This means the water is not suitable for applications that demand drinking water quality, unless treatment by physical-chemical methods, filtering, disinfection and monitoring of water quality are applied.

To increase the availability of water, other types of paved areas (roads, etc.) can also be connected. Hence, clustered systems instead of individual systems (one house) are introduced. To reduce the installation costs, storage tanks can be replaced by natural storage in ponds, reservoirs or soil, for

Figure 2.4 Example of a multipurpose (rain) water storage in densely populated areas, Rotterdam [22].

example sandy aquifers. Space limitations (in urban areas) and soil conditions have to always be considered; nonetheless, many examples have shown that even in large urban conglomerations alternative storage facilities can be incorporated in densely populated areas (see Figure 2.4 as an example).

The applicability of rainwater harvesting depends on the local climate and the storage volume. Generally, in areas with longer dry seasons or only incidental rainfall, the possibilities are limited.

2.3.2 Benefits and Challenges of Rainwater Harvesting

Before planning and installing a rainwater harvesting system, the benefits, needed efforts, and risk should be considered. These analyses will show if installing a system is useful in a specific case.

The benefits related to rainwater harvesting are [20, 21, 23]:

- Rainwater harvesting reduces the demand of drinking water from the network. To assess the impact on drinking water consumption and the economic benefits in a specific case a water balance and/or business case are useful instruments.

- Rainwater harvesting reduces storm water runoff. Installing these systems might contribute to lowering flooding risks since rainwater is stored instead of discharged. However, since local (or household) storage facilities are generally small (1–10 m³) the overall effect seems to be limited and is only very local.
- Rainwater harvesting is a sustainable and ecologically friendly choice. Its energy consumption is very low and hardly any chemicals are needed. There are variations as per the location and application of the water, and therefore, analyses (such as life cycle analysis) are required before decisions are made. A recent review has concluded that the median energy intensity of theoretical studies pertaining to rainwater harvesting (0.20 kWh/m³) was considerably lower than that found in actual empirical studies (1.40 kWh/m³). The authors concluded that perhaps the theoretical assessments of energy intensity did not consider all energy uses such as for pump start up and standby, or that the difference may be overcome by better system design and operation [5].
- Rainwater harvesting systems are generally easy to install. This makes it possible to use this technique not only in rich counties, but also in developing areas as long as the institutional and organizational basis is properly organized.
- These systems are easy to understand and operate.
- Rainwater harvesting contributes to a positive, sustainable image of the owner. For companies this can be an important additional consideration for installing such a system.
- Rainwater harvesting systems are in most cases individual installations. In other cases, community-based collection systems are installed as they have been shown to have greater benefits compared to individual household tanks. In either case, the owner(s) will be less dependent (if at all) on public systems. Both solutions make the system suitable for remote areas.

On the other hand, there are restrictions and challenges in planning and implementation of rainwater harvesting systems:

- Rainwater harvesting systems have limited capacity. This capacity depends on the type and size of the catchment area, the volume of the tank and the local climate (weather patterns). As mentioned before: in a water balance, the potential reduction of drinking water requirements from the network use can be assessed.

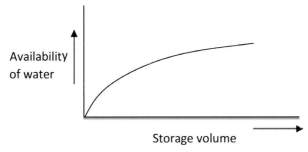

Figure 2.5   Relation between storage volume and availability of water (or drinking water savings).

- Larger tanks require larger investment. Additionally, there is a limit to the efficiency of larger tanks. Figure 2.5 illustrates the relation between storage capacity and the availability of water from a rainwater system.
- Debris, dust, bird droppings, leaves etc. will influence water quality. According to standards in developed countries, the water quality will render it unsuitable as a reliable source of potable water. A more reliable quality can be achieved by regular cleaning of roofs, and filtering the water. For developing countries, such a treatment may result in a more reliable water quality than other types of sources (public water network, rivers, and groundwater) [24].
- Time-related precipitation is not constant. In a proper design you should be able to adjust the storage volume to the requirements for dry periods or a dry year. Sizing should be considered in relation to the goal and reliability of supply. The actual design should incorporate different levels of reliability and use. But not all situations can be included in design calculations. Therefore: consider how to deal with extreme drought and for which period and frequency of drought? Are alternative sources of water available? Is the system a temporary option?
- At locations where the access to freshwater is easy and drinking water production requires little effort: is installation of a rainwater harvesting system a wise choice from an economical point of view?
- Maintenance: this aspect is highly important if you have a choice between supply from a public network or installing an individual rainwater harvesting system. Be cautious that even the most low tech systems need proper operation and maintenance.

- Capturing rainwater will reduce the runoff from the area. For larger projects, and in dry climates this might have a negative impact on water systems (ponds and streams) or green areas that originally received water from the terrain where rainwater harvesting is applied.

2.3.3 Components of a Rainwater Harvesting System

Figure 2.6 shows the components of the rainwater harvesting system. The first component is the area that captures the water (catchment area). In this case, the catchment area is the roof. For better water quality, it is important for the roof to be clean. The roof material and the roof slope determine the percentage of water that is discharged to the conveyance system. The run off coefficient expresses this by relating the amount of runoff to the amount of precipitation received. For flat roofs the run off coefficient will be lower than for buildings with a steep roof.

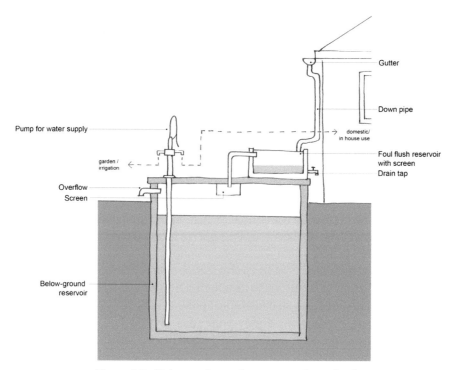

Figure 2.6   Rainwater harvesting system at house level.

The next component is the conveyance system. The conveyance system is constructed with drain pipes and other elements can be introduced to prevent contamination:

- A filter, in some cases in two steps: in a first step a screen to filter leaves and debris and a second one to filter smaller particles.
- A first flush outlet. This outlet prevents the first part of the rainwater from entering the tank. The first part is in most cases the most contaminated. A first flush outlet or treatment (sieves, sedimentation, etc.) should be considered if water quality problems are expected. Disadvantages of flush outlets are the smaller volume that enters the tank and the introduction of a new element (for example an automatic device). If additional equipment is required you should consider subsequent need for maintenance.

The tank or cistern is the largest element in the system. A tank can be part of a building or it can be built as a separate element. All kinds of materials can be used for construction as long as the materials are inert (do not influence water quality). If the water is used for gardens or irrigation, an excavated pond in the ground can be an economic alterative. For other applications, the water should stay clean and a closed tank is needed.

The tank should have an overflow for extreme water events. This overflow should have sufficient capacity to convey water to the storm water system.

The delivery system conveys water from the tank to tap points and other locations where the water is applied. Usually a small pump is part of the delivery system. If the water is used as a source for drinking water, advanced treatment of the water is required in this step (such as sand filtration and UV disinfection) which can result in relatively expensive systems requiring significant levels of maintenance to ensure water quality.

2.3.3.1 Rainwater harvesting using natural components

Generally, materials such as plastics, concrete, and steel are used to construct rainwater harvesting systems in urban areas. However, it is good to note that parts of the system can be replaced by natural materials. Examples can be found in both rural and urban areas:

- The construction of ponds or small earth dams to capture run off from a small valley. The water is stored behind the dam and used for irrigation which increases agricultural production. This is an example of a land-based harvesting system.

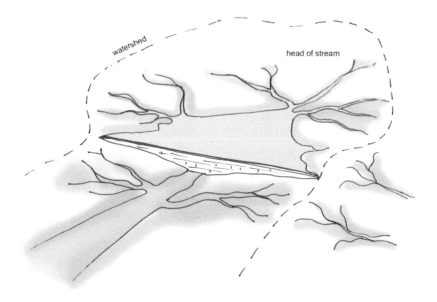

Figure 2.7   Schematic of a watershed.

- If the soil is sufficiently permeable, the captured water will infiltrate. The infiltrated water can then be used for different applications. However, in densely populated urban areas the underground is already used for multiple purposes (cables, pipes, heat storage, transportation) so space may be scarce for water storage.
- Artificial aquifer storage and recovery (ASR). In this concept, wells are used to infiltrate and store water in deeper, sandy aquifers. In dry periods the water is pumped out from the aquifer and used. Such systems have their own benefits and challenges.

2.3.4 Calculating the Potential of Rainwater Harvesting

Water balances can be used to calculate and assess the potential applicability and dimensions of rainwater harvesting systems. The calculations show the relation between the catchment area, climate characteristics, water demand and the needed storage volume.

As an example, Dutch rainfall data will be discussed. Generally, several years of daily rainfall data are needed to assess storage volume and the potential drinking water savings. In a hypothetical situation, the roof is 120 m^2.

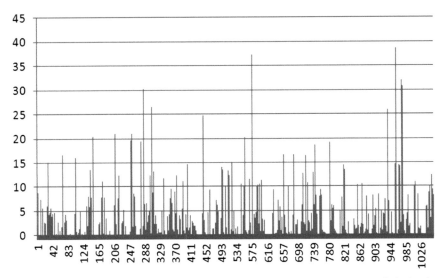

Figure 2.8 Daily precipitation during 2013, 2014, and 2015. The horizontal axis is the day number, and the vertical axis is the daily precipitation in mm.

The water demand for toilet flushing is 100 l/day and a 1000-l tank is installed.

In the calculations, the percentages of rain that runs off from the roof is important. In the literature, coefficients between 0.7 and 0.95 are used. It should be mentioned that in reality this coefficient is not a constant factor between 0.7 and 0.95. If the rainfall is small, there will be little or no run off. On the other hand, during a heavy rainfall event, more than 95% will flow from the roof. To account for this effect, losses can be introduced in the calculations, just as in storm water calculations. The loss will be approximately 0.5 mm for sloped roofs constructed with smooth materials. For flat roofs the loss can be 2 mm or more. The example shows that the run off factor or losses are important for the simulation results

According the calculations, rainwater harvesting will cover 89% of the water demand for toilet flushing. More detailed calculations with a loss of 0.5 mm result in a coverage of 86%. If the building had a flat roof (with a loss of 2 mm) the coverage would decrease to 76%. This shows that the type of roof significantly influences the functioning of the system.

A smaller tank will result in lower water savings. If a 500 liter tank is installed instead of one with a capacity of 1000 l, the percentage of toilet-flush water covered with rainwater will decrease from 86 to 72%.

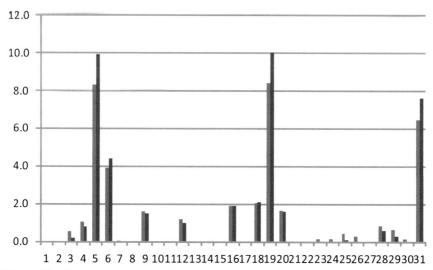

Figure 2.9   Calculated run off during May 2014. (red: calculated with 0.5 mm loss; blue calculated with a run off coefficient of 0.8). The horizontal axis is the day number, and the vertical axis is the flow, liters/day.

Points to pay attention to when dimensioning the system:

1. Be careful when you introduce rainwater harvesting in the design of storm water systems. The tanks may be full at the moment the rain event starts. In that case, rainwater harvesting will not result in additional retention capacity or allow for smaller pipes in the storm water network.
2. Climate change might influence the capacity of the system. For example, if longer dry periods are expected a larger tank may be needed to store water during the wet days, and supply the required water volumes during the dry periods.

2.3.5 Additional Information

More information about rainwater harvesting can easily be found. Authorities (national, municipalities, etc.), consumer organizations and groups promoting sustainable development provide information about the considerations, planning, implementation and maintenance of rainwater harvesting systems. Their internet sites give access to free calculation tools for dimensioning systems and cost calculation.

Other information is available from producers of rainwater harvesting system. This information includes installation guidelines [21, 23, 25, 26].

For in-house use, there may be (national) regulations related to the use of captured rainwater. In many cases, these regulations distinguish individual and clustered or public systems. Before installing a rainwater harvest system these regulations should be considered. For example, in some parts of America, harvesting rainwater for drinking (even if it is treated on site) is prohibited, although other applications such as toilet flushing and using the rainwater to water the lawn are promoted.

2.4 Reclamation of Fresh Used Water (Greywater Reuse)

2.4.1 Introduction

As pressures on freshwater resources grow around the world, and as new sources of supply become increasingly scarce, expensive or politically controversial, efforts are underway to identify new ways of meeting water needs. Of special note are efforts to reduce water demand by increasing the efficiency of water use and to expand the usefulness of alternative sources of water previously considered unusable. As an alternative to desalination, the reuse of used water streams can also be considered. Preferably low polluted used water streams, like grey water, are used for this alternative, but even full strength wastewater could be turned into clean freshwater as long as suitable technologies are applied to guarantee health-related water quality concerns.

2.4.2 Greywater Reuse

Greywater, defined slightly differently in different parts of the world, generally refers to the wastewater generated from household uses like bathing and washing clothes. This waste water is distinguished from more heavily contaminated "black water" from toilets [27]. In many utility systems around the world, grey water is combined with black water in a single domestic wastewater stream. Yet greywater can be of far higher quality than black water because of its low level of contamination and higher potential for reuse. When grey water is reused either on-site or nearby, it has the potential to reduce the demand for new water supply, reduce the energy and carbon footprint of water services, and meet a wide range of social and economic needs.

In particular, the reuse of greywater can help reduce demand for costlier high-quality potable water coming from the network.

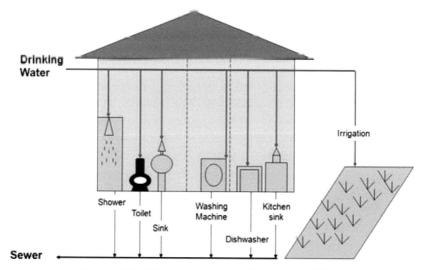

Figure 2.10   Water use in a single-family home [28].

By appropriately matching water quality to water need, the reuse of used water can replace the use of potable water in non-potable applications like toilet flushing and landscaping. For instance, many homes have one set of pipes that bring drinking water in for multiple uses and another that takes water away. In this system, all devices that use water and all applications of water use a single quality of water: highly treated potable drinking water. This water is used once and then it enters a sewer system to be transported and treated again, in places where wastewater treatment occurs. In most modern wastewater systems, treated wastewater is then disposed of into the ocean or other water bodies, voiding its reuse potential. This system wastes water, energy, and money by not matching the quality of water to its use.

A water reuse system, on the other hand, captures water that has been used for some purpose, but has not come into contact with high levels of contamination, e.g., sewage or food waste. This water can be reused in a variety of ways. For instance, water that has been used once in a shower, washing machine, or bathroom sink can be diverted outdoors for irrigation. In this case, the demand for potable water for outdoor irrigation is reduced and the streams of wastewater produced by the shower, washing machine, and sink are reduced. When the systems are designed and implemented properly, probable public health concerns regarding using different water qualities can be confidently appeased.

Attention to the public health impacts of water reuse is of major importance in scaling up grey water solutions in areas where regulations around water reuse are not well enforced. Water quality issues related to bacteria, salts, and detergents have to be addressed when applying these kinds of solutions.

Greywater technologies, uses, and policies vary widely around the world. Here we provide a broad overview of the state of greywater implementation and policy, as well as examine the potential of grey water to reduce the water and energy intensity. Key issues that must be addressed for greywater to be accepted and implemented at larger scales are discussed.

2.4.3 Used Water Treatment Technologies

Used water systems range from simple low-cost devices that divert used water to direct reuse, such as in toilets or outdoor landscaping, to complex treatment processes incorporating sedimentation tanks, bioreactors, filters, pumps, and disinfection [27]. Some used water systems are home-built, do-it-yourself piping and storage systems, but there are also a variety of commercial used water systems available that filter water to remove hair, lint, and debris, and remove pollutants, bacteria, salts, pharmaceuticals, and even viruses from used water. However, these decentralized in-house treatment systems have many bottlenecks such as relatively high energy demand, less controlled water quality, operational and maintenance issues; not to mention that at the end of the day they may not be safe and sustainable.

Used water treatment can be categorized into three main categories:

1. Diversion systems, which do not store used water (but may filter and disinfect it) before immediate reuse;
2. Physiochemical used water treatment systems, which allow used water to be stored, and treated with filtration and disinfection processes;
3. Biological used water treatment systems, which use biological water processing technologies and approaches.

The following sections discuss the technologies, and land requirements for used water systems.

2.4.3.1 Physical–chemical used treatment systems

Used water treatment systems that involve storing used water must treat the water to reduce the bacteria and other microorganisms that can multiply in stagnant water. Physical and chemical treatment systems primarily utilize disinfection and filtration to remove contaminants while biological treatment uses biological activity (usually enhanced with aeration) and membrane bioreactors.

Table 2.1 provides a list of selected common used water treatment and reuse technologies and some of their respective advantages and disadvantages. Within aerobic biological treatment, many different process concepts may be used, such as activated sludge process, granular sludge, sequencing batch reactors, trickling filters, biofilm systems, or moving bed reactors.

Table 2.1   Technologies applied in water reuse

Treatment Technique	Description	Pros	Cons
Disinfection	Chlorine, ozone, or ultraviolet light can all be used to disinfect greywater.	Highly effective in killing bacteria if properly designed and operated, low operator skill requirement.	Chlorine and ozone can create toxic byproducts, ozone and ultraviolet can be adversely affected by variations in organic content of greywater.
Activated carbon filter	Activated carbon has been treated to open up millions of tiny pores between the carbon atoms. This results in highly porous surfaces with areas of 300–2,000 square meters per gram, These filters thus are widely used to adsorb odorous or colored substances from gases or liquids.	Simple operation, activated carbon is particularly good at trapping organic chemicals, as well as inorganic compounds like chlorine.	High capital cost, many other chemicals are not attracted to carbon such as sodium, alcohols, etc. This means that an activated carbon filter will only remove certain impurities. It also means that, once all of the bonding sites are filled, an activated carbon filter stops working.
Sand filter	Beds of sand or in some cases coarse bark or mulch which trap contaminants as greywater flows through.	Simple operation, low maintenance, low operation costs.	High capital cost, reduces pathogens but does not eliminate them, subject to clogging and flooding if overloaded.

(*Continued*)

Table 2.1 Continued

Treatment Technique	Description	Pros	Cons
Aerobic biological treatment	Air is bubbled to transfer oxygen from the air into the greywater. Bacteria present consume the dissolved oxygen and digest the organic contaminants, reducing the concentration of contaminants.	High degree of operations flexibility to accommodate greywater of varying qualities and quantities, allows treated water to be stored indefinitely.	High capital cost, high operating cost, complex operational requirements, does not remove all pathogens.
Membrane bioreactor	Uses aerobic biological treatment and filtration together to encourage consumption of organic contaminants and filtration of all pathogens.	Highly effective if designed and operated properly, high degree of operations flexibility to accommodate greywater of varying qualities and quantities, allows treated water to be stored indefinitely.	High capital cost, high operating cost, complex operational requirements.

2.4.3.2 Biological used water treatment systems

Some used water treatment systems use aerobic biological treatment. These systems can often be scaled up or down, depending on the quantity of greywater produced. The treatment technologies include membrane filters to remove contaminants, bacteria, and viruses along with aerobic biological treatment. Aerobic biological treatment involves aeration to increase dissolved oxygen and activate bacteria to consume the oxygen and digest the organic contaminants. Some aerobic treatment systems include corrugated plastic sheets or other media for bacteria to attach to and grow on. One common method of aerobic biological treatment uses a rotating biological contactor that cycles discs in and out of greywater tanks.

Biological used water treatment also includes membrane bioreactors (MBR, see Table 2.1), which have become common in wastewater treatment since the early 2000s. The breakthrough for the MBR came in the early 1990s when the separation membrane was directly immersed into the bioreactor. Until then, MBRs required a great deal of pressure (and therefore energy) to maintain filtration. The submerged membrane relies on bubble aeration to mix the effluent and limit clogging of the membrane pores. The energy demand of the submerged system can be up to two orders of magnitude lower than previous bioreactors [29]. Aeration is considered as one of the major parameters on process performances, both hydraulic and biological.

The lower operating cost obtained with the submerged membrane along with the steady decrease in the membrane cost encouraged an exponential increase in MBR in wastewater plants. There are now a range of MBR systems commercially available, most of which use submerged membranes, although some external modules are available. Membranes typically consist of hollow fibers and flat sheets.

2.5 Concluding Remarks

Traditionally, freshwater originates from rivers, lakes, streams, and ground-water aquifers. In many places around the world, increasing demand on freshwater and the impacts of climate change on water availability are reducing the security of water access. Globally, many regions have reached a point at which existing water resources are already being over-used, as evidenced by the depletion of groundwater aquifers and rivers which no longer reach the sea. As a result, many communities are switching to alternative sources of freshwater like brackish and seawater desalination. Desalination offers an excellent technical solution to freshwater availability, but should always be challenged by the possibilities of alternative freshwater concepts like integrated water management, rain and storm water harvesting, and water reclamation.

This chapter showed the alternatives to desalination which can be considered in the feasibility phases where, when and for whom applicable.

Integrated water management solutions are locally applicable where soil and landscape conditions allow it. Integrated solutions should be considered as a basis for freshwater supply and conservation in early stages of feasibility and initiation phases.

Rain and storm water harvesting systems could be applied on the scale of individual houses or more centralized as community systems in urban areas. Specific site situations always determine the possibilities for applicability.

Used water reclamation is a third possible alternative strategy in terms of the local water savings. Small demonstration projects and new, more flexible, water policies have demonstrated the successful use of used-water at multiple scales. However, there are also a variety of challenges when incorporating reused water in homes, farms, and businesses. Currently, these challenges have hampered broad implementation of water reuse and immense opportunities exist in this regard. In effect, the more such alternative methods for supplying freshwater are explored, the less the responsibility will rest on the shoulders of the desalination industry.

References

[1] Joint Research Centre (JRC). (2012). *The water and energy nexus, in Science for Water*. Thematic report No. JRC71148, Publication Office of the European Union, Luxembourg.

[2] UN Water. (2015). *Statistics*. Available at: http://www.unwater.org/ statistics/statistics-detail/pt/c/211827/ (last accessed November 2015).

[3] Klein G., Krebs, M., Hall, V., O'Brien, T., Blevins B. (2005). *California's Water–Energy Relationship*. Final Staff Report, California Energy Commission, Sacramento, CA.

[4] Goyder Institute. (2016). Available at: http://goyderinstitute.org/

[5] Urban Water Security Research Alliance. (2012). *5 Years of Urban Water Research in South East Queensland 2007–2012*. Queensland: Urban Water Security Research Alliance

[6] Wallach, M., et al. (2012). *Promoting Sustainable Water Solutions That Save Water, Energy, and Money; Alternatives to Hudson River Desalination*. Newburgh, NY: Hudson Valley Program.

[7] Vieira, A. S., Beal, C. D., Ghisi, E., Stewart, R. A. (2014). Energy intensity of rainwater harvesting systems: a review. *Renew. Sustain. Energy Rev.* 34, 225–242, ISSN 1364-0321. doi: 10.1016/j.rser.2014.03.012

[8] Plany, C. (2010). *A Greater, Greener New York*. Available at: www.nyc.gov/html/planyc2030/html/publications/publications.shtml

[9] Cooley, H., Gleick, P., and Wolff, G. H. (2006). *Desalination with a Grain of Salt: A California Perspective*. Oakland, CA: Pacific Institute.

[10] United Nations World Water Assessment Programme (UN WWAP). (2014). *The United Nations World Water Development Report 2014: Water and Energy*. Paris: UNESCO.

[11] Winter, T. C., Harvey, J. W., Franke, O. L., and Alley, W. M. (1998). Ground Water And Surface Water A Single Resource, Circular 1139. Sioux Falls, SD: U.S. Geological Survey.

[12] Strootman architects. (2013). *Design of fresh water alternatives.* by order of Deltaplan higher sandy grounds in The Netherlands.

[13] Netherlands Hydrological Society. (1998). *Water in the Netherlands.* Netherland: Netherlands Hydrological Society.

[14] FAO. (2002). *World agriculture: towards 2015–2030.* Rome: Food And Agriculture Organization Of The United Nations,

[15] Witteveen+Bos and Radboud University Nijmegen. (2014). *Background document Fresh water wetland, Delta Water Award*, 3rd edn. Nijmegen: Radboud University.

[16] Witteveen+Bos and ORG-ID. (2014). *Implementation plan for fresh water measures.* By order of Deltaplan higher sandy grounds in The Netherlands.

[17] Janssens, S. R. M., Wiersema, S. G., Goos, H. J. Th., Wiersma, W. (2013). *The value chain for seed and ware potatoes in Kenya: Opportunities for development.* Wageningen: Wageningen University Research Report.

[18] Kloss. C. (2008). *Managing Wet Weather with Green Infrastructure.* Municipal Handbook. Rainwater Harvesting Policies.

[19] UKK Environmental Department. (2010). *Harvesting Rainwater for Domestic Uses: An Information Guide.* Bristol: Environment Agency (UK).

[20] Khoury-Nolde, N. (2010). *Rainwater Harvesting*, Rainwater conference, Germany.

[21] EPA. (2004). *Code of practise for aquifer storage and recovery.* Salisbury: EPA.

[22] Musch, J. (2014). Waterplein, Rotterdam Stadspark, skateplek, amfitheater, schoolplein en nieuw Deltawerk ineen. de Architect.

[23] The Global Development Research Center. An Introduction to Rainwater Harvesting. Available at: http://www.gdrc.org/uem/water/rainwater/introduction.html (accessed 7 June 2017).

[24] Gould, J. and Nissen-Petersen, E. (1999). Rainwater catchment systems for domestic supply: design, construction and implementation. Intermediate Technology Publications: Southampton Row, London.

[25] United Nations Environment Programme (1997). Source Book of Alternative Technologies for Freshwater Augmentation in Latin America and the Caribbean. Available at: http://www.oas.org/dsd/publications/unit/oea59e/ch10.htm (accessed 7 June 2017).

[26] Texas A&M University, Rainwater Harvesting. Available at: https://wateruniversity.tamu.edu/rainwater-harvesting/ (accessed 7 June 2017).

[27] Allen, L., Christian-Smith, J., Palaniappan, M. (2010). *Overview of Used water Reuse: The Potential of Used water Systems to Aid Sustainable Water Management*. Oakland, CA: Pacific Insitute.

[28] Cohen, Y. (2009). *Southern California Environmental Report Card*. Available at: www.environment.ucla.edu/reportcard/article4870.html

[29] Judd, S. (2008). *The MBR Book Principles and applications of membrane bioreactors in water and wastewater treatment*. Oxford: Elsevier.

CHAPTER 3

Fundamentals of Desalination Technology

Alireza Bazargan[1,2] and Blanca Salgado[3]

[1]Research and Development Manager and Business Development Advisor, Noor Vijeh Company (NVCO), No. 1 Bahar Alley, Hedayat Street, Darrous, Tehran, Iran
[2]Environmental Engineering group, Civil Engineering Department,
K.N. Toosi University of Technology, Tehran, Iran
Phone: +98(21)22760822
E-mail: info@environ.ir
[3]Senior Technical Service and Development, Dow Water and Process Solutions, 3 Avenue Jules Rimet, 93631 La Plaine Saint Denis cedex, France
E-mail: bsalgado@dow.com

3.1 Introduction

As seen in the opening chapter of the book, although the amount of water on our planet is vast, only a small portion is considered to be naturally available for direct human use. The majority of water contains excessive inorganic salts, i.e., is saline. Desalination is defined by the Merriam–Webster dictionary as "to remove salt from something, such as water." It should be noted that technically speaking, desalination by removing salts from the water is different from desalination by removing the water from the salts, but for practical purposes, the end product is freshwater either way. So, in order to produce freshwater from saltwater, it must be desalinated. If wastewaters contain a high salt content, they could also be treated with desalination technology for reuse.

Before we discuss desalination, a question which may arise is: why do we need to go through all the trouble of desalinating saltwater into freshwater? Can't we just add flavoring to counter the unfavorable taste of seawater and drink it?

The answer to these questions lies in the way our bodies are built. Although consuming moderate amounts of salt is essential for our physiological well-being, having too much or too little entering our system is damaging.

Our cell membranes prevent the free passage of salt, but allow for easy transfer of water, and through diffusion, they attempt to equalize the concentration of salts on both sides of their walls. For the human body, an osmolarity of 300 mOsm/L (roughly equivalent to less than 9 g of sodium chloride per 1000 g of bodily fluid) is referred to as *isotonic* [1, 2]. If our bodily fluids are isotonic, all is good, and our cells will neither shrink nor swell. Seawater is *hypertonic*, and its salinity is more than four times that of our bodily fluids such as blood. Hence, consuming seawater will lead to our cells transferring water out from their insides in order to dilute their surroundings and correct the imbalance. The net transfer of water out from the cells will cause shrinkage and damage.

Meanwhile, our bodies excrete excess salts and waste in the form of urine to return our bodily fluids to their desirable state. However, the human kidney cannot produce urine which is as salty as seawater. This means that if we drink seawater, the body will urinate more water than it has actually drank, leading to dehydration, and an accumulation of salts. Essentially, by consuming seawater, the body incurs a net loss of H_2O through urination, leading to a depletion of bodily fluids, muscle cramps, dry mouth, thirst, nausea, and even delirium. In other words, no matter how thirsty you are, drinking seawater will only make you thirstier. In the case of extreme intake of seawater, if adequate amounts of freshwater are not drunk to reverse the effects of excess salts, coma, organ failure and death ensue [3]. Desert animals which must conserve water, such as kangaroos, have efficient kidneys that can excrete urine more than twice as concentrated as seawater. Unlike sea mammals such as seals and whales, we do not have kidney's that can produce highly concentrated urine. In effect, the reason that seawater is toxic to humans is that our kidneys cannot produce urine more saline than seawater [4, 5]. In addition, unlike albatrosses, penguins, marine iguanas and a variety of other animals, we do not have salt glands. Salt glands contain many tubules which radiate outward from a canal at the center. Active transport of salts via the sodium–potassium pump, removes salt from the blood into the gland, and excretes it as a concentrated solution. Hence, salt glands maintain salt balance in the animals' bodies and allow marine vertebrates to drink seawater [4, 5].

A prime example, shown in Figure 3.1, are sea turtles who like some species of crocodiles excrete salts through their tear ducts, giving an impression of "crying". With this in mind, the expression "crocodile tears" becomes more meaningful. In essence, although the crocodile is shedding what seem to be tears, it is reducing its body's salt content, not weeping for an injustice or due to pain!

Figure 3.1 Excretions from the salt glands of a sea turtle giving the impression of "crying" [6].

It is interesting to add that high concentrations of sugars are also *hypertonic*. This is one of the reasons (if not the main reason) why dates and honey resist bacterial decay [7]. In effect, bacteria in a highly sugary medium will become dehydrated and die.

The salt content of water is usually expressed in milligrams per liter (mg/L), parts per million (ppm), parts per thousand (permille, ‰), or simply percent. Since pure water does not conduct electricity and it is the ions (dissolved salts) that make the water a conductor, a measurement of conductivity can be used as an indicator of salinity. Conductivity is measured with an electrical conductivity meter (EC meter) and the dissolved salts can subsequently be estimated using conversion factors. The salinity can also be calculated by measuring the water's specific gravity which is also affected by the water's salt content.

Alternative salinity units used by oceanographers (rather than desalination experts) are the Practical Salinity Unit (PSU), Knudsen salinity and the TEOS-10 absolute salinity. If extremely high precision measurements are required, researchers can use standard seawater samples from the "International Association for the Physical Sciences of the Oceans (IAPSO)" for calibration and standardization. Figure 3.2 depicts the annual mean sea surface salinity for the world's oceans. It is good to note that the salt content of even the saltiest sea is eclipsed by the amount of H_2O which has dissolved it.

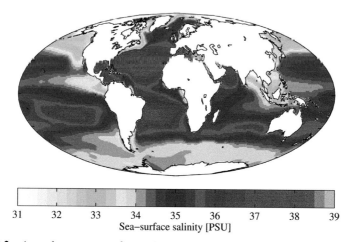

Sea–surface salinity [PSU]

Figure 3.2   Annual mean sea surface salinity from the World Ocean Atlas 2009 [8], plotted using the MATLAB software [9]. The Practical Salinity Unit (PSU) defines salinity in terms of a conductivity ratio, and is dimensionless. By definition, a water sample with a Practical Salinity of 35 has the same conductivity as a potassium chloride (KCl) solution containing 32.4356 g of KCl per kilogram of solution, at $15°C$ and 1 atmosphere pressure.

Figure 3.3 shows the proportion of the salts to the water, and their typical composition.

Figure 3.3 may help somewhat understand the challenge facing the desalination industry. Take the case for the Persian Gulf, which can more or less be representative of the industry's challenge. A cubic meter of seawater (1000 kg) taken from the northern Persian Gulf coast will contain approximately 40–44 kg dissolved salts. Based on recent contracts for the purchase of desalinated water, these salts and other unwanted constituents must be removed for less than 1 Euro per cubic meter (meaning less than 0.1 Euro cents per liter). This presents a financial challenge that—with the help of innovations and energy minimization devices—the desalination industry has been able to meet. It should be noted that desalination costs are very much dependent on local conditions, such as legislative or environmental issues as well as the cost of fuel/electricity. Hence, two seemingly identical plants situated in different locations could have dramatically different capital and operating costs.

Traditionally, the financing, construction oversight, plant operation and facility maintenance were responsibilities of governments. But recent developments and trends in the desalination industry have resulted in increased involvement of the private sector. Today, well more than one-third of new

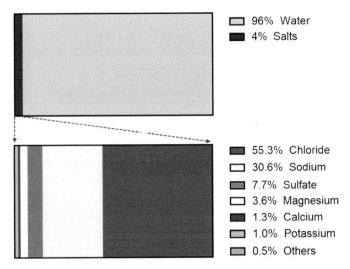

Figure 3.3 The proportion of salt to water (*top*) and the salt constituents (*bottom*) for a sample with a salinity of approximately 4% (the sample was taken off the coast of the Persian Gulf near NVCO's plant in Bandar Abbas, Iran). Note that the values presented here are 'typical' and will vary with time and location.

desalination plants around the world are privately financed. The recently popular model allows for governments to focus on maintaining and policing regulatory frameworks regarding quality standards, services and sustainability [10].

The magnitude of water consumption should also be touched upon. Considering average values for water consumption around the world, a cubic meter of desalinated water (1000 L) would be enough for about half-a-dozen people in one day. Though in countries such as the USA and Australia a cubic meter would only be enough for two people, while in Cambodia and Uganda, it would be used by more than 20 people.

3.2 Definitions

In order to better understand desalination technology and the associated issues covered in this book, it may be useful to provide a definition of some terminology often encountered.

The **total dissolved solids** (TDSs) is defined as the residual solids remaining after a solution has first been filtered to remove suspended material, and then evaporated. In simple terms, since the soluble salts do not evaporate during boiling, the material remaining at the bottom of the container after the water has evaporated, are the total dissolved solids. The TDS is a measure of the water's salinity. Saline waters can be classified into the following categories, as a function of the concentration of their TDS:

- Brackish water: water which has more salinity than freshwater, but not as much as seawater. The most common natural sources of brackish water are wells, rivers and dams with an approximate TDS concentration ranging from about 1000–10,000 mg/L. A TDS lower than 1000 mg/L would categorize the water as "fresh", and waters with TDS values exceeding 10,000 mg/L to as high as 30,000 mg/L are considered to be highly salty brackish.
- Seawater: although varied, these waters have TDS concentrations above 30,000 mg/L. Salinities can widely differ as a function of the considered sea or ocean. Closed seas and restricted gulfs are typically more saline (such as the Persian Gulf with a TDS of 40,000–45,000 mg/L), while open oceans have lower salinity values (such as approximately 35,000 mg/L for the Atlantic Ocean). The world's saltiest body of water is the Don Juan Pond in Antarctica with a TDS exceeding 400,000 mg/L.

The **recovery** of a desalination process is the portion of the input water which is converted into freshwater. High recovery minimizes the feed water requirements and hence pumping and pretreatment costs. However, it is important to note that due to problems such as scale formation and precipitation, the recovery never reaches 100%. In fact, modern seawater reverse osmosis (RO) desalination plants usually have recoveries less than 50%.

Rejection in a desalination process is defined as the percentage of salts within the feed water which is removed from the product (freshwater). For example, if a desalination plant takes in seawater at 40,000 ppm and produces a freshwater product of 400 ppm, then it is said to have a rejection of (40,000 – 400)/40,000 = 0.99 or 99%. This means that 99% of the TDS in the incoming water has been rejected and sent out with the brine. The **brine** is the portion of the input water which is not desalinated. The salt concentration of the brine is always higher than the input water, because it contains all the salts, including those which have been removed from the freshwater product.

In order to prepare the water for entering the desalination process, **pretreatment** is critical. The main purpose of physicochemical pretreatment processes (clarification, coagulation, and/or flocculation, filtration, ultrafiltration, etc.) are to limit the amount of suspended solids, turbidity, colloids, and organic matter in the water which enter the desalination unit.

Fouling is the accumulation of unwanted solids—either in the form of living organisms (**biofouling**) or non-living substances (inorganic or organic)—on surfaces. Colloids may result in a form of fouling known as particulate fouling. **Scaling** (also known as crystallization fouling or precipitation fouling) is caused by the crystallization and deposition of inorganic salts (such as calcium carbonate), oxides, and hydroxides. **Composite fouling** is when several types of fouling occur simultaneously [11].

The concentration of **total suspended solids (TSS)** is the amount of solids undissolved in a liquid. It is made up of both organic and inorganic particles. The amount of TSS in a water stream generates **turbidity**, which is a suspension of fine particles that scatters or absorbs light rays. Water with high turbidity looks murky and is perceived by the public to be unclean.

Finally, after the desalination process has produced freshwater, since the product is always slightly acidic, has a very low buffering capacity, and contains very little calcium and magnesium hardness, **post-treatment** processes are required for its conditioning. As part of the post-treatment, disinfection is employed in order to maintain the water quality throughout the distribution network and inhibit the growth of unwanted microorganisms.

3.3 Unit Operations

Modern desalination plants are composed of a series of operations, collectively resulting in the production of freshwater from saline sources. As depicted in Figure 3.4, the plant can be generally divided into three main process steps, all of which are vital for the successful and reliable production of freshwater: pretreatment (discussed in Chapter 7), desalination (discussed in Chapters 5 and 6), and post-treatment (discussed in Chapter 8).

The quality of the source water and the required final product specifications are very important in determining the details of the process. For example, if the produced water is to be used by a poultry farm, the TDS concentration should ideally be kept under 1000 mg/L (although up to 3000 mg/L is acceptable), while if it is to be used by a horse stable, TDS values could be

Figure 3.4   The unit operations of a desalination plant.

as high as 7000 mg/L [12]. According to the World Health Organization, water to be consumed by humans is classified as "excellent" if its TDS value is below 300 mg/L while a TDS value higher than 1200 mg/L is unacceptable.

Moreover, in addition to local conditions, the desalination technology chosen will affect the pre- and post-treatment steps to be employed. For example, desalination units that use evaporative/thermal technology often produce water which is purer than that of RO. Hence, freshwater produced via evaporative/thermal technology will require more extensive post-treatment to render it noncorrosive. On the other hand, since RO technology is more susceptible to problems such as the formation of unwanted biological films on membranes, they often require a more extensive pretreatment.

The most basic desalination process that comes to mind is the simple still. This classic distillation process is composed of boiling followed by condensation. Boiling is the rapid vaporization of liquid occurring when it is heated to its boiling point. When water is boiled, the vapor does not include any salts which are left behind. The vapor can then be liquefied (condensed) to form freshwater. Although the simple still can be used for water desalination and disinfection (because most bacteria do not survive under boiling conditions), it is only employed in small capacities as an emergency water desalination and/or treatment method because of the high energy it consumes.

There are currently more than 16,000 desalination plants in the world [13] with a combined desalination capacity of nearly 90 billion liters/day (which is rapidly increasing every year) [14]. Over the past five decades, desalination has experienced dramatic growth due to reduced capital and operating costs as well as improved reliability and performance. Modern desalination equipment that are currently commercially available have capacities ranging from as low as several liters per day to about 100 million liters per day per single operating unit. In many cases, several operating units are placed alongside each other for a higher total desalination capacity. For example, Saudi Arabia's Ras Al-Khair plant on the Persian Gulf, the world's largest, will produce more than 1 billion liters of desalinated water per day after it is completed. Hence, over such a broad range of values and equipment sizes, it is dangerous to make too many generalizations.

Although pretreatment and post-treatment are vital components of any desalination plant, the actual heart of the plant is the desalination unit itself (after pretreatment and before the post-treatment). Desalination technologies can be classified in a variety of ways including their energy source (Figure 3.5(a)), separation method (Figure 3.5(b)), maturity of technology (Figure 3.5(c)), what is removed from the feed stream (Figure 3.5(d)), etc.

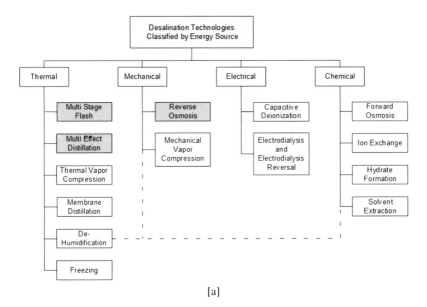

[a]

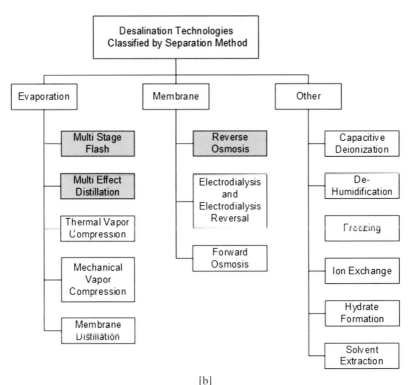

[b]

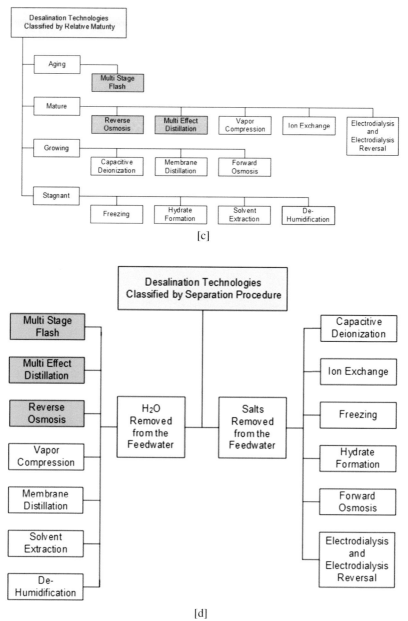

Figure 3.5   Classification of desalination technologies. The green boxes denote the major technologies that are currently used on a wide commercial scale. Reverse Osmosis (RO), Multi-Stage Flash (MSF), and Multi-Effect Distillation (MED) make up 94–95% of the world's current desalination capacity, with RO leading the way.

Although RO is currently the world's go-to technology with the lion's share of new plants under construction, the majority of the installed capacity in the middle east–the world's largest desalination market–still uses thermal technologies. Overall, RO, MSF (Multi-Stage Flash), and MED (Multi-Effect Distillation, also known as Multi-Effect Evaporation, MEE) lead the market, with roughly 65, 21, and 7% of the global installed capacity respectively. Between them, these technologies comprise approximately 94–95% of the world's desalination capacity. The share of RO has continued to grow and currently comprises over 70% of the total global capital expenditures (under construction). Electrodialysis (ED)/Electrodialysis Reversal (EDR) come after the big three, with an approximate market share of 3% [13]. It should be noted that ED/EDR is only suitable for brackish water and is not used for seawater desalination.

For the remainder of the chapter, a short description of each process in Figure 3.5 will be provided. Note that this list is not exhaustive, and that there are numerous other experimental and novel desalination processes (such as microbial desalination cells [15]) which are not described herein. Although new technologies may be promising in the early development stages, they frequently fail in the commercial arena. For a review of state-of-the art desalination research, refer to Chapter 12.

3.4 Desalination Technologies

3.4.1 Thermal (and Evaporative) Technologies

For years, seawater reverse osmosis (SWRO) technology used more electrical energy than thermal methods. In addition, thermal methods showed a high degree of reliability and guaranteed performance under varied conditions. Hence, traditionally, the majority of desalination plants around the world used thermal technologies. But with the boom of RO technology during the past decade, and the dramatic fall of its electrical energy usage, the thermal desalination market share has diminished.

3.4.1.1 Multi-Stage Flash Distillation (MSF)

Multi-stage flash (MSF) was the favored process for the desalination industry with a market share close to 60% of the total world production capacity at the end of the 20th century. This process will be extensively explained in Chapter 5, and here, a quick summary is provided [16–18].

The MSF system is similar to simple stills, while it takes advantage of the fact that under lower pressures, the energy required for boiling water is reduced. First is the preheating step, in which incoming saline water passes through consecutive heat exchangers where its temperature increases. Next, the water is heated to the maximum process temperature followed by its release into the first vacuum chamber where sudden evaporation or "flashing" takes place. Since evaporation occurs from the bulk of the water rather than the surface of a hot heat exchanger, scale formation is reduced. The water vapor formed in the flashing chamber condenses into freshwater on the cooling coils (these are the same coils that the feed water passes through during the preheating step). The portion of the feed water which is not evaporated in the first chamber makes its way to the second chamber. The operation is repeated in subsequent flashing stages, each of which has a lower operating pressure than the one before. More complex schemes include partial recirculation of the concentrate.

Multi-stage flash systems are almost exclusively constructed in conjunction with power plants. In such power-and-water (co-generation) configurations, high pressure steam generated in boilers is fed to turbines to generate electric power. The lower pressure steam drawn off from the turbine is then fed into the MSF plant to provide the thermal energy needed for desalination. Technologically speaking, MSF still remains as a highly reliable option for large installations.

3.4.1.2 Multi-Effect Distillation (MED)

Multi-effect distillation is the second-most popular thermal desalination method in terms of installed capacity world-wide. For years it lagged behind MSF in terms of unit size and customer acceptance, but the technology is currently enjoying something of a rebirth [10].

The design concept of MED can be explained as follows: saline water is sprayed onto tube bundles which contain hot steam and act as heat exchangers. The steam condenses inside the tubes, and as a result of heat transfer via the tube walls, part of the saline water film on the outside of the tube bundle evaporates. The vapor produced on the outside of the bundles is devoid of salts and goes on to the next stage. In the next stage, this vapor heats the insides of tube bundles as it condenses into desalinated water. Again, saline water (this time at a lower pressure) is sprayed on the outside of the bundles where it evaporates and the process is repeated. This process continues for several stages known as "effects"; from 5 to 20 effects in a typical large plant [19]. The pressure difference between stages allows the fact that only the first effect is heated, with no need for reheating between steps. An increase in the number of effects can increase the internal energy recovery [20]. MED can be designed to operate efficiently at a lower temperature than MSF. This means that it can use lower grade steam (reduced cost) as heat input.

A parameter known as the **Gained Output Ratio (GOR)** is defined as the kilograms (or pound) of desalinated water produced per kilogram (or pound) of steam used for heating. If a simple still were used, the GOR could never exceed unity. However, due to the design of the MED system where only the first effect is heated, the GOR can exceed unity. Considering process inefficiencies, the GOR of an MED plant is approximately 0.87 times the number of its effects [10]. In MSF desalination, the GOR is primarily a function of the temperature range (between the first and last stage) and not the number of stages as in the case for MED.

A similar parameter called the **Performance Ratio (PR)** is defined as the kilograms of water produced per 2326 kJ of heat consumed (or the pounds of water produced per 1000 BTU of heat consumed). 1000 BTU and 2326 kJ are representative of the average enthalpy of a unit of steam [21]. In other words, these parameters are a measure of the amount of desalinated water produced for the amount of heat consumed. It is worth mentioning that at or about the normal atmospheric pressure boiling point of water the GOR and PR are more or less the same (typical of older systems). These values diverge as the temperature (i.e., the quality and hence the cost) of the input steam varies.

3.4.1.3 Thermal Vapor Compression (TVC)

In vapor compression (VC), also referred to thermal vapor compression (TVC), the heat necessary to evaporate water is obtained in a thermo-compression chamber. VC units are built in a variety of configurations, but in general, gas under high pressure (motive steam) is forced through a nozzle, creating a vacuum which draws feedwater from a reservoir. The mixture of the motive steam and the drawn water is then compressed at the diverging outlet of the thermo-compressor to yield desalinated water [16–19]. When properly designed and operated, the quantity of water produced is several times the quantity of the steam introduced at the nozzle, and hence the GOR of TVC units is considerably greater than 1.

As standalone units, VC has high energy consumption but occupies a lower space when compared to MSF and MED. However, the combination of MED and TVC into hybrid desalination systems has become rather popular [22, 23]. This is due to the fact that whenever thermal energy is available

In the form of medium-pressure steam, MED–TVC is the most economically advantageous steam-heated process. When coupled, the steam at the outlet of the TVC is supplied to the tube bundles of the first MED stage. Hence, the latent heat of the sucked water in the TVC is recycled into the MED system, leading to energy savings. Although less popular, other alternatives are also possible. For example, the exhausted vapor from the last MED stage could be extracted and fed into the TVC.

3.4.1.4 Mechanical Vapor Compression (MVC)

In MVC, saline water is sprayed onto the front of a heat exchanger surface where it evaporates and cools the exchanger surface. The un-evaporated portion of the feed water (containing all the salts) leaves as water from the brine outlet. Meanwhile, a partial vacuum is created by a compressor which pulls off the vapor. The vapor is compressed and subsequently applied to the back side of the original heat exchanger surface. This high-pressure causes the vapor to condense on the back side of the heat exchanger, re-warming it so that it can evaporate more vapor on the front side. In this way, the heat of vaporization can be recycled within the system. This concept is in many ways similar to the technology used in heat pumps and refrigeration. Original MVC processes were developed more than 100 years ago, and have ever since been used for small applications [10, 24]. MVC is suitable for situations where very high purity water is required with very high percentage recovery.

3.4.1.5 Membrane Distillation (MD)

Membrane distillation is classified under thermal technologies rather than membrane technologies, because the evaporation of the water via heating is the main component of the process which separates the water from the salts. The MD process can be classified into various configurations, the four classic configurations being [25–27]:

- The Direct Contact MD: where a thin membrane is located between hot saline water and cold fresh water. On the hot side, water is evaporated and passes through the membrane and finally condenses on the cold side of the membrane.
- The Air Gap MD: where vapor penetrates through the membrane and an air gap; and is finally cooled and condenses on a cold plate where desalinated water is produced.
- The Sweeping Gas MD: where the vapor emerges from the hot liquid, moves through the membrane, and a stripping gas such as air or nitrogen is used for carrying the produced vapor to a separate section to be condensed.

- The Vacuum MD: where the vapor emerges from the hot liquid, moves through the membrane, and is evacuated and moved to a separate device/section, where it is condensed.

3.4.1.6 Freezing

When water freezes, dissolved salts do not stay in the formed ice crystals. Hence, under controlled conditions, saline water can be desalinated via freezing. For this, before the entire mass of water is frozen, the mixture is washed and rinsed to remove the salts in the remaining water or adhering to the ice. The ice is then melted to produce fresh water. There are various process configurations for freezing desalination such as vacuum freezing desalination (VFD) and secondary refrigerant fluid (SRF) among others. In VFD, cooled saline water is sprayed into a vacuum chamber; some of the water flashes off as vapor, removing heat from the water and causing ice to form. In SRF, a liquid hydrocarbon refrigerant such as propane or butane is vaporized in direct or indirect contact with the saline water; with this vaporization, a slurry of ice is produced [28, 29]. Over 60 years of intermittent research on freeze-desalination has not lead to any commercially successful process due to high costs and excessive energy consumption. Freezing desalination is characterized under thermal technologies, because technically speaking, heating and freezing are two sides of the same coin.

3.4.2 Membrane Technologies

Unlike thermal technologies which employ heat to separate the saline water and its salts, membrane technologies use "**semipermeable membranes**" to get the job done. By definition, a semipermeable membrane is one which allows the passage of certain molecules and/or ions, but restricts others. For desalination purposes, this means the passage of water molecules while the dissolved salts are obstructed (or vice versa). The rate with which species pass through a semipermeable membrane depends on various factors such as the pressure, concentration, temperature, solubility, and size of the solute, etc.

3.4.2.1 Reverse Osmosis (RO)

By definition, **osmosis** is the flow of liquid between two aqueous solutions confronted by a semipermeable membrane, from the less concentrated solution to a more concentrated solution, until equilibrium is achieved.

Assume that we have two aqueous solutions: one with a higher concentration of salts (solution A) and one with a lower concentration (solution B). If these solutions are placed side-by-side, with a semipermeable membrane separating the two, a concentration gradient will exist. This concentration difference is a driving force for water from solution B to seep through and enter solution A. This will continue until the forces acting on the system (pressure difference, concentration gradient, etc.) balance each other out. However, if an external pressure is applied to solution A, water will flow in the opposite or "reverse" direction of the natural osmosis process. This means that the solution with a higher salt concentration (solution A) will become even more concentrated as water (without the salts) flows across the membrane to the dilute side (solution B). This is the basis of RO desalination technology [30].

Among all technologies, the majority of new desalination plants around the world currently use RO, which is comprised of the following three process streams:

1. Feed water, defined as the water entering the membrane process. Membrane feed water is obtained after pretreatment of the raw water (coming from wells, surface sources, the sea, or public water supplies). Since RO membranes are designed to separate ionic species (rather than other impurities such as colloids or suspended solids), it is extremely important to clean the water upstream of the membrane process, in order to avoid excessive fouling. Pretreatment processes such as clarification, coagulation and/or flocculation, media filtration, microfiltration, ultrafiltration, chlorination and dechlorination, antiscalant dosing, and biocide dosing, are put in place ahead of the membrane elements [31].

2. Concentrate, brine or reject, is defined as the heavily loaded stream (with dissolved solids or particulates, or both) resulting from the membrane process. As mentioned earlier, the concentrate stream is the portion of the feed stream which does not pass through the membrane. Chapter 15 discusses the brine stream in more detail.

3. The permeate, or filtrate, is the portion of the feed stream which has passed through the membrane—the desalinated product. Since the semipermeable membrane is very effective at removing larger species, the permeate is devoid of large ions and molecules. However, a minute amount of smaller species can dissolve across the membrane along with the water. Hence, the permeate is never *completely* devoid of salts, although for practical purposes this is seldom a problem.

It is good to note that although it may seem that RO is just another filtration process involving very fine pores, on a molecular level the separation is not a physical process, but rather the chemical interaction and diffusion of water across the membrane material itself [10].

3.4.2.2 Forward Osmosis (FO)

Forward osmosis is a membrane based separation process, like RO, which relies on the semi-permeable character of a membrane in salt removal. However, unlike RO, the driving force for separation is osmotic pressure, not hydraulic pressure. By using a solution which is more concentrated than the saline feed on one side of the membrane (draw solution), the saline water from the other side will be induced to flow across, leaving the salts behind. In other words, the osmotic pressure of the draw solution should be greater than the saline feed water. The flow across the membrane occurs because of the natural tendency of the system to send water from the less concentrated side to the more concentrated side (osmosis). The draw solution is selected in a way so that it can be easily separated from the desalinated water, usually through heating. The draw solution is then re-concentrated and recycled back to extract more water. Although FO requires more energy (in the form of heat and electricity) than RO, it theoretically has a higher desalination flux. There is currently very limited experience for this technology on a commercial scale [32, 33].

3.4.2.3 Electrodialysis (ED) and Electrodialysis Reversal (EDR)

Electrodialysis is a membrane process which uses electrical energy (rather than pressure) in order to separate the salts from the water. In an ED system, synthetic membranes that are selectively permeable to positively (cations) or negatively (anions) charged ions, but not to water, are used. A set of anionic and cationic membranes forms a cell. The saline water is pumped through spacers in the middle of the cell while an electrical current creates a potential across electrodes. The positively charged cations in the solution, diffuse through the respective membrane, and migrate towards the cathode. Likewise, the negatively charged anions migrate toward the anode. Hence, the cell spacer channel will become devoid of salts and desalinated water is produced.

When numerous cells are placed next to each other, a "stack" will be made up of alternating anion- and cation-permeable membranes with flow channels between them. Cations pass through the negatively charged cationic exchange membrane but are retained by the positively charged anion exchange membrane. Similarly, anions pass through the anionic exchange membrane but are retained by the cationic exchange membrane. The outcome of this process is that one cell becomes depleted of ions while the adjacent cell becomes enriched in ions. There are three main parameters in the design of an ED system: electrical current, membrane type and the number of cells. While the electrical current and membrane type affect the product purity, the number of cells affects the production rate [34].

Electrodialysis Reversal (EDR) is a variation of the ED process, which works in the same way, except that the polarity of the DC power is reversed two to four times per hour [35]. When the polarity is reversed, the freshwater and concentrate cells are reversed, and so are the chemical reactions at the electrodes. This polarity reversal helps clean the surface of the membranes from scale accumulated since the last reversal. Although ED/EDR technology is not applied for seawater desalination (due to inefficiencies at high salt concentrations), this process has a market share in specialty applications and brackish water desalination where the TDS concentration is lower than seawater.

3.4.3 Other Technologies

There are numerous other desalination technologies which do not fit under the thermal or membrane desalination categories. However, most of these alternatives have not gained traction for large scale implementation. Here, a number of these technologies will be presented, although there are various other experimental options which will not be touched upon.

3.4.3.1 Hydrate Formation

In this process, the saline water is mixed with a hydrocarbon which forms hydrates or clathrates. In a hydrate, a hydrocarbon molecule is enclosed in a molecular cage of water, forming a solid ice-like phase. On a molecular scale, all the salts are excluded from the cage/hydrate structure. Subsequently, the hydrocarbon is removed from the hydrate, leaving behind freshwater devoid of salts. This technology is still under experimental development [36, 37].

3.4.3.2 Ion Exchange (IX)

Ion exchange technologies are often used for water softening among other applications and can be described as the interchange of ions between a solid phase (such as a chemical resin) and a liquid phase (such as saline water), leading to the purification of the liquid. Resins can be made of naturally-occurring inorganic materials or synthetic materials. When ion exchange technology is applied for desalination, the process removes Na^+ and Cl^- ions from the feed water, replacing them with other less-problematic ions, thus producing fresh water. No heat is required for this process, as it operates at ambient temperature. Ion exchange technology can be very advantageous in removing boron from saline water. When resin beads are saturated, they are washed/regenerated so they can regain their efficiency [38, 39]. Rapid, complete, and cheap regeneration of the resin is a challenge. Hence, rather than using IX directly for desalination, it is usually used for polishing water which has already been desalinated for increased quality.

Electro De-Ionization is currently the most advanced variation of ion exchange technology. It is in fact a process which incorporates both IX and ED into a single technology to produce ultrapure water. The EDI process uses very little amounts of chemical (less than 95% of conventional IX) processes. However, EDI cannot be used for treating high TDS water and requires a desalination process upstream. In effect, EDI makes freshwater even more pure. Each EDI chamber is composed of ion exchange resins, packed between a cationic exchange membrane and an anionic exchange membrane. When water flows through an EDI cell, some ions are scavenged by the resin, which are then pulled off the resin and drawn through the membranes towards the respective electrodes. In this way, these charged ions are continuously removed and transferred across the membrane, leaving behind purified water [40, 41]. Electrodeionization can be considered a continuously regenerated IX system.

3.4.3.3 Capacitive Deionizaiton (CDI)

Capacitive DeIonization also referred to as electrochemical demineralization or electrosorption of salt ions is an emerging technology capable of desalinating brackish water. A CDI cell consists of a separator (such as a channel or porous dielectric material) placed between a pair of porous electrodes typically composed of carbon. As the water flows through the separator, the electrodes are charged (typically 1–1.4 V) causing salt ions to migrate from the water towards the electrodes where they are caught in the pore surfaces and electrical double layers. Hence, the water is desalinated. Salt ions are electrostatically held in the double layer until a discharging step is applied, in which the external power supply is cut-off or its polarity reversed. During the discharge step, the ions are released in the form of a brine stream, and the electrodes are readied once again for desalination [42, 43].

3.4.3.4 Solvent Extraction

In this method, a saline solution is brought into contact with a solvent, thereby producing two distinct phases; a first phase containing the solvent and freshwater extracted from the saline solution, and a second phase containing a highly concentrated residue of the saline solution. The phases are heated before (or during) contact to enhance the directional dissolution of water into the solvent. The two phases are then separated. After separation, the first phase is cooled to precipitate the water from the solvent. This desalination technology has the advantage of being able to use low-value heat, generated from terrestrial heat sources, the ocean, the sun, or as waste heat from other processes [44–46]. Complete, rapid, and energy efficient regeneration of the solvent is required. The carry-over of solvent into the product stream must be avoided for both health and aesthetic reasons.

3.4.3.5 De-Humidification

In the Air De-Humidification (Dehydration) process, humid air from the surroundings enters into the system and its moisture is extracted as freshwater. For this, equipment can be used which either cool a surface below the dew point of the ambient air, concentrate water vapor through use of solid or liquid desiccants, induce and control convection in a tower structure, etc. Patented devices vary in scale from small units suitable for one person's daily needs to structures as big as multi-story office buildings in coastal and humid region [47]. This process is not commercialized on a large scale yet and is only used in small capacity units with high costs.

Also, instead of acquiring humid air for the surrounding environment, a humidification step could first be incorporated. Under such a process, a carrier gas first extracts moisture as it come into direct contact with saline water. After humidification of the carrier gas, the humid gas is sent to another section to extract the moisture as fresh water [48]. The challenge with such processes is on the dehumidification side which should be rapid, cheap, and recycle the heat of condensation back to the humidification side.

3.4.4 Hybrid Systems

Two or more desalination processes may be combined together in "hybrid systems" in order to improve the process efficiency or desalinated water quality [49]. A common mistake is to confuse hybrid systems with power-and-water (co-generation) plants which combine electricity production and desalination in order to improve the overall efficiency of the system. Power and water plants are discussed in more detail in Chapter 14.

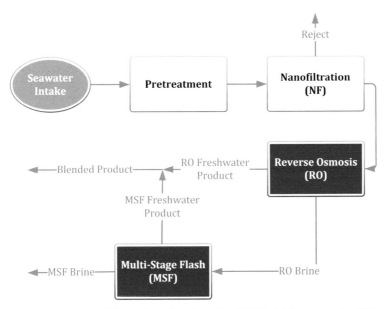

Figure 3.6 A possible configuration for a RO–MSF hybrid system, with NF.

The most popular hybrid processes which have gained commercial acceptance are the MED–TVC (as discussed) and the RO–IX. The RO–IX is used for the production of very pure water. In this system, a high proportion of the hardness is removed by RO, followed by the ion exchange system which removes the remaining ions, reaching a TDS of less than 1 [50].

Many other hybridization combinations are being investigated by researchers at the lab or pilot scale. For example, hybrid MSF–RO plants have potential advantages of a low power demand, improved water quality and lower running and maintenance costs compared to stand-alone RO or MSF plants. Figure 3.6 shows a MSF-RO hybrid scheme (NF stands for nanofiltration which removes the larger species and divalent ions from the water). The system could have instead been designed so that the concentrate (blowdown) from the MSF plant was fed into the RO system as the feed water. Many other hybrid alternatives are possible [51].

3.5 Conclusion

In this chapter in addition to the fundamentals of desalination, a summary of the various technologies has been reviewed. Most of the topics that have been mentioned in this chapter are thoroughly explained in subsequent sections of the book. Although various technologies have been touched upon, it is good to once again note that the overwhelming majority of the desalination plants around the world use RO, MSF, or MED technology, with RO dominating the market in recent years.

References

[1] Sherwood, L., Klandorf, H., and Yance, P. (2013). *Animal Physiology: From Genes to Organisms*, 2nd ed. Boston, MA: Brooks/Cole, Cengage Learning.

[2] Poon, C. Y. (2006). "Tonicity, Osmoticity, Osmolality, and Osmolarity," in *Remington: The Science and Practice of Pharmacy*, D. B. Troy, ed, 21st edn Philadelphia: Lippincott Williams & Wilkins, 250–265.

[3] Giuggio, V. (2012). *What if you drink saltwater?* HowStuffWorks.com. 2012. Available from: http://science.howstuffworks.com/science-vs-myth/what-if/what-if-you-drink-saltwater.htm [cited 23 December, 2015].

[4] Schmidt-Nielsen, K. (1960). The Salt-Secreting Gland of Marine Birds. *Circulation* 21, 955–967. Available at: http://circ.ahajournals.org/content/21/5/955.full.pdf

[5] Williams, M. F. (2011). "Marine Adaptations in Human Kidneys," in *Was Man More Aquat. Past?* eds M. Vaneechoutte, A. Kuliukas, M. Verhaegen. Sharjah: Bentham Science Publishers 148–155.

[6] Scholz, M. (1998). Meeresschildkröte bei Eiablage [Internet]. Pazifik, Puerto Vallarta, Jalisco, Mexiko. Available from: https://en.wikipedia.org/wiki/File:Sea_turtle_head.jpg

[7] Ghabanchi, J., Bazargani, A., Daghigh Afkar, M., et al. (2010). In vitro assessment of anti-*Streptococcus mutans* potential of honey. *Iran. Red Crescent Med. J.* 12, 61–64. Available at: http://www.sid.ir/en/VEWSSID/J_pdf/88120100111.pdf

[8] National Centers for Environmental Information. (2009). *World Ocean Atlas 2009*. Available at: http://www.nodc.noaa.gov/OC5/WOA09/pr_woa09.html [cited 22 December, 2015].

[9] Plumbago. (2012). *Annual mean sea surface salinity from the World Ocean Atlas 2009*. Available at: https://en.wikipedia.org/wiki/File:WOA09_sea-surf_SAL_AYool.png [cited Dec. 22, 2015].

[10] Birkett, J. (2011). *Desalination at a Glance*.

[11] Sheikholeslami, R. (1999). Composite fouling—inorganic and bio-logical: A review. *Environ. Prog.* 18, 113–122. Available at: http://dx.doi.org/10.1002/ep.670180216

[12] Soltanpour, P. N. and Raley, W. L. () Livestock series manage-ment: livestock drinking water quality no. 4.908. Available at: http://veterinaryextension.colostate.edu/menu2/Cattle/04908Livestock Drinkingwaterquality.pdf

[13] Voutchkov, N. (2012). Desalination: Technology Trends & Market Insights. *Desalin. Expert. Exch.*

[14] Koschikowski, J. Solar Seawater Desalination. Fraunhofer - Institute for Solar Energy Systems ISE. Available at: http://www.fraunhofer.cl/ content/dam/chile/en/documents/8-Koschikowski-Fraunhofer-Chile-150527.pdf

[15] Brastad, K. S., and He, Z. (2013). Water softening using microbial desalination cell technology. *Desalination* 309, 32–37. Available at: http://www.sciencedirect.com/science/article/pii/S001191641200519X

[16] Al-Sahali, M., and Ettouney, H. (2007). Developments in thermal desalination processes: Design, energy, and costing aspects. *Desalina-tion*. 214, 227–240. Available at: http://www.sciencedirect.com/science/ article/pii/S0011916407003657

[17] El-Dessouky, H. T., and Ettouney, H. (2002). Fundamentals of salt water desalination, 1st edn. Amsterdam: Elsevier.

[18] Maria Antony Raj, M., Kalidasa Murugavel, K., Rajaseenivasan, T., et al. (2016). A review on flash evaporation desalination. *Desalin. Water Treat* 57, 13462–13471.

[19] Khawaji, A. D., Kutubkhanah, I. K., Wie, J.-M. (2008). Advances in sea-water desalination technologies. *Desalination* 221, 47–69. Available at: http://www.sciencedirect.com/science/article/pii/S0011916407006789

[20] Shen, S. (2015). Thermodynamic Losses in Multi-effect Distillation Process. *IOP Conf. Ser. Mater. Sci. Eng.* 88, 12003.

[21] Tonner, J. (2008). *Barriers to Thermal Desalination in the United States.* Denver.

[22] Al-Mutaz, I. S., and Wazeer, I. (2015). Current status and future direc-tions of MED–TVC desalination technology. *Desalin. Water Treat.* 55, 1–9.

[23] Alamolhoda, F., KouhiKamali, R., Asgari, M. (2016). Parametric sim-ulation of MED–TVC units in operation. *Desalin. Water Treat.* 57, 10232–10245.

[24] Gebel, J. (2014). *Thermal Desalination Processes. Desalination*. Hoboken, NJ: John Wiley & Sons, Inc., 39–154.

[25] Alkhudhiri, A., Darwish, N., and Hilal, N. (2012). Membrane distillation: A comprehensive review. *Desalination* 287, 2–18. Available at: http://www.sciencedirect.com/science/article/pii/S0011916411007284

[26] Khayet, M. (2011). Membranes and theoretical modeling of membrane distillation: a review. *Adv. Colloid Interface Sci.* 164, 56–88. Available at: http://www.sciencedirect.com/science/article/pii/S000186861 0001727

[27] Drioli, E., and Ali, A., and Macedonio, F. (2015). Membrane distillation: Recent developments and perspectives. *Desalination* 356, 56–84.

[28] Williams, P. M., Ahmad, M., Connolly, B. S. (2013). Freeze desalination: An assessment of an ice maker machine for desalting brines. *Desalination* 308, 219–224. Available at: http://www.sciencedirect.com /science/article/pii/S0011916412004201

[29] Williams, P. M., Ahmad, M., Connolly, B. S., et al. Technology for freeze concentration in the desalination industry. *Desalination* 356, 314–327. Available at: http://www.sciencedirect.com/science/article/pii/ S0011916414005463

[30] Malaeb, L., and Ayoub, G. M. (2011). Reverse osmosis technology for water treatment: State of the art review. *Desalination* 267, 1–8. Available at: http://www.sciencedirect.com/science/article/pii/S001191641 0006351

[31] The Dow Chemical Company. (2009). Dow Water & Process Solutions, FILMTEC^{TM} Reverse Osmosis Membranes Technical Manual. Available at: http://www.dowwaterandprocess.com

[32] Lutchmiah, K., Verliefde, A. R. D., Roest, K., et al. (2014). Forward osmosis for application in wastewater treatment: a review. *Water Res.* 58, 179–197. Available at: http://www.sciencedirect.com/science/article/pii/ S0043135414002358

[33] Akther, N., Sodiq, A., Giwa, A., et al. (2015). Recent advancements in forward osmosis desalination: a review. *Chem. Eng. J.* 281, 502–522. Available at: http://www.sciencedirect.com/science/article/pii/S1385894 715007688

[34] Moura Bernardes, A. (2014). "General Aspects of Membrane Separation Processes BT" in *Electrodialysis and Water Reuse: Novel Approaches*, eds. A. Moura Bernardes, A. M. Siqueira Rodrigues, and J. Zoppas Ferreira. Berlin, Heidelberg: Springer, 3–9. Available at: http://dx.doi.org/10.1007/978-3-642-40249-4_2

[35] Lee, H.-J., Song, J.-H., and Moon, S.-H. (2013). Comparison of electrodialysis reversal (EDR) and electrodeionization reversal (EDIR) for water softening. *Desalination*. 314, 43–49. Available at: http://www.sciencedirect.com/science/article/pii/S0011916413000064

[36] Kang, K. C., Linga, P., Park, K., et al. (2014). Seawater desalination by gas hydrate process and removal characteristics of dissolved ions (Na+, K+, Mg2+, Ca2+, B3+, Cl–, SO42–). *Desalination* 353, 84–90. Available at: http://www.sciencedirect.com/science/article/pii/S001191641 4004731.

[37] Bao, Z., Xie, Y., Yang, L., et al. (2015). Advances in hydrate-based seawater desalination (水合物海水淡化技术研究进展). *Mod. Chem. Ind*. 10, 40–44.

[38] Schoeman, J. J., and Steyn, A. (2000). *Defluoridation, Denitrification and Desalination of Water Using Ion-exchange and Reverse Osmosis Technology*. Report to the Water Research Commission.

[39] Guyer P. (2013). An Introduction to Ion Exchange Techniques for Water Desalination. Createspace Independent Publishing Platform.

[40] Arar, Ö., Yüksel, Ü., Kabay, N., et al. (2014). Various applications of electrodeionization (EDI) method for water treatment—A short review. *Desalination* 342, 16–22. Availableat: http://www.sciencedirect.com/ science/article/pii/S0011916414000745

[41] Alvarado, L., and Chen, A. (2014). Electrodeionization: principles, strategies and applications. electrochim. *Acta* 132, 583–597. Available at: http://www.sciencedirect.com/science/article/pii/S0013468614 007087

[42] Suss, M. E., Porada, S., Sun, X., et al. (2015). Water desalination via capacitive deionization: what is it and what can we expect from it? *Energy Environ. Sci*. 8, 2296–2319. Available at: http://dx.doi.org/10.1039/C5EE00519A

[43] Porada, S., Zhao, R., van der Wal, A., et al. (2013). Review on the science and technology of water desalination by capacitive deionization. *Prog. Mater. Sci*. 58, 1388–1442. Available at: http://www.sciencedirect.com/science/article/pii/S0079642513000340

[44] Fowler M. J. Construction of prototype system for directional solvent extraction desalination. Massachusetts Institute of Technology; 2012.

[45] Sanap, D. B., Kadam, K. D., Narayan, M., et al. Analysis of saline water desalination by directed solvent extraction using octanoic acid. Desalination [Internet]. 2015; 357: 150–162. Available from: http://www.sciencedirect.com/science/article/pii/S0011916414006110

[46] Bajpayee, A., Luo, T., and Muto, A., et al. (2011). Very low temperature membrane-free desalination by directional solvent extraction. *Energy Environ. Sci.* 4, 1672–1675. Available at: http://dx.doi.org/10.1039/C1EE01027A

[47] Wahlgren, R. V. (2001). Atmospheric water vapour processor designs for potable water production: a review. *Water Res.* 35, 1–22. Available from: http://www.sciencedirect.com/science/article/pii/S0043135400002475

[48] Ghalavand, Y., Hatamipour, M. S., and Rahimi, A. (2014). Humidification compression desalination. *Desalination* [Internet]. 341, 120–125. Available at: http://www.sciencedirect.com/science/article/pii/S0011916414001131

[49] Marcovecchio, M. G., Mussati, S. F., Aguirre, P. A., et al. Optimization of hybrid desalination processes including multi stage flash and reverse osmosis systems. *Desalination* [Internet]. 182, 111–122. Available at: http://www.sciencedirect.com/science/article/pii/S0011916405004248

[50] Al Abdulgader, H., Kochkodan, V., and Hilal, N. (2013). Hybrid ion exchange—Pressure driven membrane processes in water treatment: a review. *Sep. Purif. Technol.* 116, 253–264. Available at: http://www.sciencedirect.com/science/article/pii/S1383586613003456

[51] Hamed, O. A. (2005). Overview of hybrid desalination systems—current status and future prospects. *Desalination* 186, 207–214. Available at: http://www.sciencedirect.com/science/article/pii/S0011916405006934

CHAPTER 4

Water Chemistry and Desalinated Water Quality

Alireza Bazargan[1,2], Amir Jafari[3] and Mohammad Hossein Behnoud[4]

[1]Research and Development Manager and Business Development Advisor, Noor Vijeh Company (NVCO), No. 1 Bahar Alley, Hedayat Street, Darrous, Tehran, Iran

[2]Environmental Engineering group, Civil Engineering Department, K.N. Toosi University of Technology, Tehran, Iran

Phone: +98(21)22760822

E-mail: info@environ.ir

[3]Project Manager, Noor Vijeh Company (NVCO), No. 1 Bahar Alley, Hedayat Street, Darrous, Tehran, Iran

E-mail: a.jafari@nvco.org

[4]Technical and Executive Deputy, Noor Vijeh Company (NVCO), No. 1 Bahar Alley, Hedayat Street, Darrous, Tehran, Iran

E-mail: hbehnoud@nvco.org

4.1 Introduction

The water entering a desalination plant can come from various sources. The source water will contain various unwanted substances, and by definition, the desalination process will aim to reduce the concentration of these constituents—in particular the salts. Of course, the first logical step for reducing the impurities (both chemical and microbial) of the product water is preventing or reducing their presence in the source water. In some cases, this may be possible, while in other cases, not as much. For example, if the source water for a thermal desalination plant comes from a power plant (combined power and water), then there are not many options in terms of controlling the entering impurities to the desalination plant. A counterexample would be desalination plants that take their water by subsurface intakes or deep open ocean intakes, which by default will prevent certain contaminants from entering the desalination plant.

For a given water source, the quality of desalinated water varies considerably from plant to plant. This variation depends on the type of desalination technology used (Multi-Stage Flash (MSF), and Multi-Effect Distillation (MED), Reverse Osmosis (RO), etc.) as well as the specific design and operation of the desalination process. For example, if reverse osmosis technology is used, using a single-pass system will yield freshwater with higher total content of dissolved solids (TDS) compared to a two-pass system [1]. Nevertheless, one thing is for certain: in all cases, desalinated water is slightly acidic, has a very low buffering capacity, and contains very low amounts of calcium and magnesium hardness [2]. Post-treatment processes are almost always employed to condition the water and adjust its quality to the desired level. This chapter will focus on the fundamental issues regarding the quality of desalinated water while post-treatment processes will be discussed in Chapter 8.

In this section, some of the various chemicals of concern in the desalinated water will be briefly introduced. The importance of these constituents in desalinated water has been recognized by the World Health Organization [3]. It should also be noted that desalination processes are very effective at the removal of pathogens (viruses, bacteria, etc.), and the discussion on disinfection will be presented in Chapter 8 along with post-treatment technologies.

4.1.1 Boron

One of the most discussed elements of importance for health- and environment-related issues in desalinated water is boron. Boron is an essential element for plant growth and is required in small amounts. However, if concentrations exceed the required values, the element becomes herbicidal. Hence, in areas where the desalinated water is to be used for irrigation purposes of boron-sensitive crops such as citrus fruit, strawberries, and avocado, controlling the residual boron increases in importance. In areas where there is not a lot of rainfall to wash away the boron accumulated on the surface soil layer over time, the detrimental effects of the residual boron are more pronounced. The most tolerant plants can survive a boron concentration as high as 4 mg/L in the irrigation water; however many plants experience detrimental effects at concentrations above 0.5 mg/L, with very sensitive plants tolerating no more than 0.3 mg/L [4]. In the most recent edition of the WHO Guidelines for Drinking Water Quality the maximum value for boron in drinking water has increased to 2.4 mg/L [5]. The newest revision has come as a result of reviewing toxicological data and studies in areas with high background exposure, concluding that boron is not as dangerous to human health as previously thought.

A point to consider is that RO systems are less effective at removing boron than other inorganic constituents. For example, the concentration of calcium in the permeate of RO units after the desalination of seawater is comparable to that of boron, even though the concentration of calcium is two orders of magnitude greater than boron in the source water. Boron is generally not a problem in thermal plants because of its removal to levels below 0.3 mg/L during the distillation process [6].

In the case of reverse osmosis, if the boron concentration in the desalinated water is too high, special high-boron-rejection membranes should be used. Alternatively, the pH of the water should be increased in the pretreatment step. This is because pH increase enables conversion of boric acid ($B(OH)_3$) into borate, which has a larger molecule and stronger charge, and is therefore, easier for RO membranes to reject. The boric acid–borate system is a buffer with a peculiar chemistry. It is actually so complex that there are at least ten different equilibrium reactions in solution. Nonetheless, the main equilibrium concerning us is as follows [7]:

$$4B(OH)_3 \leftrightarrows B_4O_7{}^{2-} + 5H_2O + 2H^+$$

4.1.2 Bromide

Seawater contains large amounts of bromine, in the form of bromide. Even with relatively effective removal of the element (e.g., >95%), the desalinated water may contain up to several milligrams of bromide per liter. Due to the similarity between removal mechanisms, the concentration of bromide and chloride in the desalinated water will be proportional to their concentration in the source water. Inorganic bromide may also be present in brackish and freshwater sources, particularly those affected by seawater intrusion

The acceptable daily intake of bromide is 1 mg/kg of body weight. Assuming that 20% of a person's daily bromide intake comes from drinking water, then the acceptable range for bromide concentration in 2 L of water drunk by a 60-kg person each day, would approximately be 6 mg/L [5].

If desalinated water is oxidized (for example via ozonation under appropriate conditions), the bromide can form bromate (BrO_3^-), which is carcinogenic and has a much lower acceptable threshold value of 10 µg/L [5]. This is particularly important because desalinated water produced for subsequent packaging may incorporate an ozonation step prior to bottling, which could increase the bromate levels beyond the allowable guideline value (this is more probable if the desalination process only has a single pass RO system). The production of chlorine via seawater electrolysis also generates considerable amounts of bromate. However, there are indications that the potential risk of bromate is currently overestimated, and that future revisions of the WHO guidelines may become less strict [3].

4.1.3 Calcium and Magnesium

The concentration of calcium in seawater is approximately one third of that of magnesium. These essential nutrients are very efficiently removed from the water via desalination, and thus will need to be reintroduced by a remineralization process. After remineralization, a minimum Ca^{2+} concentration of 20 mg/L has been proposed in some countries (equivalent to 50 mg/L as $CaCO_3$) [2]. Most countries in the world do not have a minimum requirement for magnesium.

It is important to note that the principal source of nutrients and minerals (such as calcium and magnesium) is the diet of an individual, and the minerals in drinking water play only a secondary supplemental role. Nonetheless, due to the deficiency of calcium and magnesium in many developing countries, and more specifically in some sectors of the population such as women and the elderly, supplementary sources are welcomed. Supplementary calcium intake is known to reduce the risk of osteoporosis, while magnesium deficiency has been associated with ischemic heart disease and metabolic syndrome, indicating a prediabetic condition [3]. If desalinated water is to be used for irrigation, the addition of minerals may be needed for plant growth, and damage to crops after irrigation with extremely pure desalinated water has revealed a need for post-treatment [8].

4.1.4 Fluoride and Other Supplements

The addition of dietary supplements in foods and drinks is common such as vitamin D in milk, vitamin C in drinks, and iron, B vitamins and folic acid in bread. The supplements added to drinking-water are limited in scope and include: fluoride for strengthening dental enamel and reducing the incidence of tooth decay (dental caries), ferric iron–ethylenediaminetetraacetic acid complex in some dietary iron-deficient areas, and iodine in some areas with high incidence of goitre in the Russian Federation [3]. As a matter of principal, drinking water should not be relied upon as the main contributor of nutrients needed by the body—such function is typically provided by food.

Low fluoride intake has been associated with poor health of bones and teeth. On the other hand, drinking excess amounts of fluoride results in dental fluorosis. If fluoride is to be added as a supplement to water, an optimal value of 1 mg/L has been suggested [9], with the WHO health-based guidelines suggesting a maximum value of no more than 1.5 mg/L [5]. The local conditions, such as the climate, the diet of the population (sugar consumption levels), and dental care (such as the use of fluoride-containing toothpaste), influence the decision on whether or not to add fluoride to drinking water [3]. Drinking water produced by thermal desalination of seawater has low fluoride content (approximately 0.5 mg/L), while RO technology produces water with a typical fluoride content of 0.9 mg/L. Note that the addition of fluoride and other supplements are often the responsibility of the distributing utility, not the desalination plant.

Fluoridation of water is a matter of policy and differs from area to area, ranging from less than 1% of the population in Denmark receiving optimally fluoridated drinking water, to 100% of the population in the city of Washington, DC. For instance, Figure 4.1 depicts the percentage of the population receiving optimally fluoridated water in the various provinces of Australia (one of the most extensively fluoridated countries in the world). The data is extracted from a report published by the British Fluoridation Society in 2012 [9].

Figure 4.1 Percentage of the Australian population receiving optimally fluoridated water (2012). Green areas: 91–100% of the population; yellow areas: 80–90% of the population; red areas: less than 80% of the population.

4.1.5 Organics

Organic chemicals constitute a broad range of substances ranging from harmless chemicals that at most may cause foul taste and odor, to hazardous toxins, which are dangerous for human health. Most notably, the cyanotoxin microcystin-LR, which arises from freshwater cyanobacterial blooms, has a provisional WHO guideline value of 1 μg/L [3]. Seawater can also contain toxins such as domoic acid and saxitoxin, which if concentrated in tissue of shellfish and other marine species could result in food poisoning. However, it should be noted that most marine algae do not produce large amounts of cyanotoxins and therefore are usually not a practical concern.

Organic chemicals can be in the form of natural organic materials (NOMs) such as humic and fulvic acids and the byproducts of algal and seaweed growth. Due to the relatively large size of organic molecules, which results in low volatility and high molecular weight, they are usually very efficiently removed by both thermal and membrane desalination processes. Volatile organics such as geosmin with odor thresholds measured in the range of nanograms per liter, may pose problems for thermal processes. On the other hand, solvent-type low molecular weight neutral organics can pass through RO membranes. A group of nonliving, planktonic particles known as transparent exopolymer particles (TEPs) are ubiquitous in marine and freshwater environments and have been recognized as an active agent in biofilm formation and membrane fouling. Since biofouling is one of the main hurdles for efficient operation of membrane-based technologies, TEP have been under scrutiny for the past two to three decades. Due to their assorted species, size, concentrations, and chemical composition, they are only partially removed by current pretreatment technologies [10].

If the intake of the desalination plant is placed near sewage treatment plants or industrial factories, further complications may ensue. Shipping activities near the plant may also be detrimental. In addition, contamination of the source water by petroleum hydrocarbons may occur in areas near oil extraction and petrochemical plants. Since volatile substances such as benzene, toluene, ethylbenzene, and xylene and solvents such as chloroform, carbon tetrachloride, trichloroethene, and tetrachloroethene may carry over into the product water in thermal desalination processes, these substances need to be vented and removed. Meanwhile, if such substances are present in large quantities, they may damage and destroy the RO membranes. Oil spills can affect not only product water quality but also the production rate.

Overall, the prevention of source water contamination by unwanted organic chemicals is the best method to inhibit them from appearing in the final product.

Potassium is an essential nutrient whose deficiency will result in health problems. However, drinking water does not constitute a considerable portion of a person's daily potassium intake, which is usually obtained through a variety of other sources—mainly food. Potassium concentration in seawater is usually in the range of 400–500 mg/L. Desalination is very effective at removing potassium cations (>98% removal). Hence, the residual amount of potassium in desalinated water is several orders of magnitude below the daily dietary requirement which exceeds 3000 mg/day. Potassium does not have a upper health-based limit as per the WHO guideline for drinking water [3].

Sodium is abundant in seawater, with a concentration in the range of 10,000–14,000 mg/L. Sodium is an essential nutrient required for proper function of human physiology; and although there is no upper health-based limit as per the WHO guideline for drinking water, the taste threshold is in the region of 200–250 mg/L [5]. As for sodium chloride (common table salt), it is widely accepted that the sodium ion is primarily responsible for the salty taste. The chloride ion plays a modulatory role; as a case in point, as the anion increases in size (e.g., from chloride to acetate or gluconate) the relative salty taste declines [11]. If a sodium cation is accompanied with a bitter anion, the bitterness may predominate and overshadow the saltiness. Major gaps exist in the understanding of the salt taste reception. The most popular hypothesis at the moment is that epithelial sodium channels (ENaC) allow sodium to move from outside the taste receptor cell, where it has been dissolved in saliva or water, into the taste cell. The increase of sodium ions inside the taste cell subsequently causes the release of neurotransmitters that signal the salty taste to the brain [11]. Other cations such as potassium also taste salty, although their taste is not as distinct as sodium.

The atmosphere contains various gases, including nitrogen, oxygen, argon, and carbon dioxide. When air and water come into contact, there is an exchange of molecules until equilibrium is reached. As carbon dioxide from the air dissolves in water, it reacts to form carbonic acid with the following equilibria [12]:

$$CO_{2(g)} \leftrightarrows CO_{2(aq)}$$
$$CO_{2(aq)} + H_2O \leftrightarrows H_2CO_3$$

By releasing a proton, the carbonic acid forms bicarbonate.

$$H_2CO_3 \leftrightarrows H^+ + HCO_3^-$$

Further dissociation of the bicarbonate forms carbonate ions:

$$HCO_3^- \leftrightarrows H^+ + CO_3^{2-}$$

The above reactions are reversible, and form a chemical equilibrium known as the carbonate system. In essence, the carbonate system is a weak acid–base system consisting of carbon dioxide, carbonic acid, bicarbonate, carbonate ions, and their complexes [13].

The following general equation describes the carbonate system:

$$CO_2 + H_2O \leftrightarrows \underbrace{H_2CO_3}_{\text{carbonic acid}} \leftrightarrows H^+ + \underbrace{HCO_3^-}_{\text{bicarbonate}} \leftrightarrows 2H^+ + \underbrace{CO_3^{2-}}_{\text{carbonate}}$$

The gas phase forms an integral part of the carbonate system. Any alteration in the partial pressure of CO_2 in the surrounding atmosphere creates a state of non-equilibrium between the gas and the aqueous phase which in turn causes an exchange of CO_2, a change in the pH, and adjustment of the species' concentrations until a new equilibrium is established. Under normal conditions of 1 atm and 25°C, the equilibrium concentration of $CO_{2(aq)}$ is 0.54 mg/L [2]. The concentration of CO_2 in unconditioned desalinated water is often less than 0.54 mg/L. Hence, CO_2 is absorbed form the atmosphere, causing a decrease in pH. In addition, since some carbonate minerals are insoluble, their precipitation affects the system. Thus, in order to predict the response of the system to external influences, a simultaneous consideration of all three phases (solid, liquid, and gas) is required.

The inorganic species of carbon within water provide buffering capacity with respect to pH. In addition, the pH is an important factor which controls various geochemical reactions such as the solubility of carbonates. Many key biochemical reactions such as photosynthesis and respiration also interrelate with the pH and carbonate system. Hence, the carbonate system is of importance for a variety of reasons [13].

Overall, six values can be used to describe the carbonate system [12]:

1. The concentration of dissolved carbon dioxide, $[CO_2]$
2. The concentration of bicarbonate ions, $[HCO_3^-]$
3. The concentration of carbonate ions, $[CO_3^{2-}]$

4. The pH value, i.e. the concentration of protons, [H$^+$], or hydroxyl ions, [OH$^-$]

5. The total inorganic carbon content, C_T, which is the sum of the concentrations of the species in the carbonate system.

$$C_T = [CO_2] + [H_2CO_3] + \left[HCO_3^-\right] + \left[CO_3^{2-}\right]$$

An imaginary species, $H_2CO_3{}^*$, is often defined as the sum of [CO$_2$] and [H$_2$CO$_3$]:

$$C_T = [H_2CO_3{}^*] + \left[HCO_3^-\right] + \left[CO_3^{2-}\right]$$

Since the concentration of carbonic acid is extremely small (three orders of magnitude less than the concentration of CO$_2$) some have neglected it from the above equations.

6. The total alkalinity, Alk, which is equivalent to all the bases that can accept a proton when the water is titrated:

$$\text{Alk} = \left[OH^-\right] + \left[HCO_3^-\right] + 2\left[CO_3^{2-}\right] - \left[H^+\right]$$

Alkalinity can also be defined as the ability of a solution to neutralize acids to the equivalence point of carbonate and bicarbonate. This is the amount of [H$^+$] required for reducing the pH to where all the bicarbonate and carbonate have been converted to CO$_2$ (approximately a pH of 4.3) [14].

Under equilibrium conditions, the carbonate system can be characterized by two of the six quantities above. By measuring the total alkalinity and the pH value, the remaining four values can be calculated using the following equations:

$$\left[H^+\right] = 10^{-pH}$$

$$[CO_2] = \frac{\left[HCO_3^-\right]\left[H^+\right]}{k_1}$$

$$\left[CO_3^{2-}\right] = \frac{\left[HCO_3^-\right]k_2}{\left[H^+\right]}$$

$$\left[OH^-\right] = \frac{k_w}{\left[H^+\right]}$$

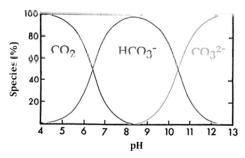

Figure 4.2　The relation between the pH and the concentration of species within the carbonate system (desalinated water, room temperature).

in which k_1, k_2, and k_w are the first dissociation constant of carbonic acid, second dissociation constant of carbonic acid, and dissociation constant of water, respectively. These constants are experimentally determined, and are available in the literature under various conditions (temperature, ionic strength of the water, etc.). By rearranging the equation for total alkalinity, and by inserting the equations above, the concentration of bicarbonate can be calculated as follows:

$$\left[HCO_3^-\right] = \frac{Alk\,[H^+] + [H^+]^2 - k_w}{[H^+] + 2k_2}$$

As mentioned earlier, the species within the carbonate system are unfixed, and as shown in Figure 4.2, it is the pH of the water that determines their relative abundance.

The addition of an acid or base to the system alters the pH and in turn changes the concentrations of the species that constitute the system. The temperature and ionic strength of the liquid are also influential on the species' concentrations. Higher temperatures and higher ionic strengths shift all the curves in Figure 4.2 to the left. For example, 65°C seawater (which has a much higher ionic strength than desalinated water) exhibits a peak for the bicarbonate ion at a pH of approximately 7 [13].

4.3 Water Hardness

Magnesium and calcium are the main components of "hard water". Although other elements, namely strontium, aluminum, barium, iron, manganese, and zinc also cause hardness in water, their concentrations are relatively small and their contribution to "total hardness" is insignificant [15]. The original definition of hardness relied on the capacity of water to precipitate soap. The graying of laundry, as well as the curd which causes "bathtub ring" are examples of calcium and magnesium soap precipitates. The scales formed from hard water (usually calcium carbonate) cause a variety of problems such as leaving behind water spots on glassware, silverware, and faucets. Scale in pipes is problematic and in appliances, pumps, valves, and water meters it causes accelerated wear of moving parts [15]. With the increase of temperature, scale formation is enhanced. Hence in boilers and water-heaters, scales create insulation layers over the heat exchangers, which in turn hinder heat-transfer and increase heating costs.

There are two types of water hardness: temporary and permanent hardness. Temporary hardness is due to the presence of calcium and magnesium bicarbonates. When water containing these compounds are heated, they decompose, leaving behind insoluble carbonates which precipitate. Hence temporary hardness can be removed from the water by heating:

$$Ca(HCO_3)_{2(aq)} \xrightarrow{heat} CO_{2(g)} + H_2O_{(l)} + CaCO_{3(s)}$$

$$Mg(HCO_3)_{2(aq)} \xrightarrow{heat} CO_{2(g)} + H_2O_{(l)} + MgCO_{3(s)}$$

On the other hand, hardness caused by calcium and magnesium salts other than bicarbonates (such as sulfates, chlorides, etc.) is not significantly affected by heating and is therefore called permanent hardness.

Hardness can be reported in various units. However, since calcium carbonate is the de facto source of the majority of hardness in water, levels of water hardness are often referred to in terms of *hardness as CaCO$_3$* and reported in "mg/L as CaCO$_3$".

In order to calculate hardness as CaCO$_3$ we need to use "equivalent" values. By definition, an equivalent is the amount of one substance that reacts with one mole of another substance in a given chemical reaction. An equivalent can also be defined as the amount of substance needed to supply (or reach with) one mole of hydrogen ions (H$^+$) in an acid–base reaction, or, the amount of substance needed to supply (or react with) one mole of electrons in a redox reaction. For practical purposes, an equivalent can be calculated from the number of moles of an ion in a solution, multiplied by the valence of that ion.

So, assuming that we want to calculate the hardness caused by substance A in terms of *hardness as CaCO$_3$* then:

$$equivalent\ CaCO_3 = equivalent\ substance\ A$$

Since $CaCO_3$ displays a valence of 2:

$$\text{moles } CaCO_3 \times 2 = \text{moles A} \times \text{valence A}$$

From the definition of moles:

$$\frac{\text{mass } CaCO_3}{\text{molecular weight } CaCO_3} \times 2 = \frac{\text{mass A}}{\text{molecular weight A}} \times \text{valence of A}$$

Dividing both sides by unit volume, the numerators of the fractions become concentrations. Rearranging for the concentration of $CaCO_3$ on one side:

concentration $CaCO_3$

$$= (\text{molecular weight } CaCO_3) \left(\frac{\text{concentration A}}{\text{molecular weight A}} \right) \left(\frac{\text{valence A}}{2} \right)$$

By substituting the molecular weight of $CaCO_3$ (100.09 mg/mmol $CaCO_3$), using the absolute value for the valence (because the sign is not of importance here), and repositioning the brackets, the formula for converting the concentration of substance A into its *hardness as $CaCO_3$* (mg/L as $CaCO_3$) then becomes:

$$\text{hardness as } CaCO_3 = \frac{mg}{L} \text{ as } CaCO_3$$

$$= (\text{concentration A}) \left(\frac{100.09}{\text{molecular weight A}} \right) \left| \frac{\text{valence A}}{2} \right|$$

As an example, if a water sample is found to have a Ca^{2+} concentration of 20 mg/L, then its calcium hardness in terms of $CaCO_3$ can be calculated as follows:

$$\left(20 \frac{mg}{L} Ca^{2+} \right) \left(\frac{100.09 \frac{mg}{mmol} CaCO_3}{40.08 \frac{mg}{mmol} Ca^{2+}} \right) \left| \frac{2}{2} \right| = 50 \frac{mg}{L} CaCO_3$$

So "20 mg/L calcium" and "50 mg/L of calcium as $CaCO_3$" are equivalent – they are simply two different ways of expressing the calcium content of a given solution. The conversion factor between the two is always 100.09/40.08, or 2.5. Meaning that the calcium concentration multiplied by 2.5 will yield the calcium concentration as $CaCO_3$. Similarly, the conversion factor between magnesium ion concentration and magnesium hardness as $CaCO_3$ is the ratio of the molecular weight of calcium carbonate to the molecular weight of magnesium, 100.09/24.31, or 4.12.

Now if the water sample mentioned above also had an Mg^{2+} concentration of 6 mg/L then its magnesium hardness in terms of $CaCO_3$ would be:

$$\left(6\ \frac{mg}{L} Mg^{2+}\right)\left(\frac{100.09\frac{mg}{mmol}\ CaCO_3}{24.31\frac{mg}{mmol}\ Mg^{2+}}\right)\left|\frac{2}{2}\right| = 6 \times 4.12 = 24.7\ \frac{mg}{L}\ CaCO_3$$

The *total hardness* of a sample is the sum of *all* the hardness as $CaCO_3$, including that from the calcium ions *and* the magnesium ions. So, by assuming that no other sources of hardness are present to an appreciable degree, the total hardness in terms of $CaCO_3$ for the above water sample would be: $50 + 24.7 \approx 75$ mg/L as $CaCO_3$.

Interestingly, there is no universal standard for designating which water is "soft" and which is "hard", and various organizations have provided their own guidelines. As an example of dissimilarity, Table 4.1 displays the differences between the US Geological Service and by the UK's Drinking Water Inspectorate. Evidently, a water sample with a total hardness of 75 mg/L as $CaCO_3$ would be classified as "moderately soft" in the UK, but "moderately hard" by the US Geological Service.

4.4 Sodium Adsorption Ratio

The sodium, calcium, and magnesium concentrations of irrigation water are important because they may alter the physical properties of the soil, most importantly its permeability. As the sodium content of the water increases (relative to calcium and magnesium), the infiltration rate in the soil decreases, meaning that the water does not enter the soil rapidly enough. This is due to the lack of sufficient calcium to counter the dispersive effects of sodium, which separates the soil into fine particles. These particles in turn fill the small pore spaces, and seal the conduits through which the water flows [18]. Hence, in order to assess water's suitability for irrigation purposes, an index called the sodium adsorption ratio (SAR) is defined as:

Table 4.1   Various guidelines for defining "hard" water

Classification	US Geological Service [16] (mg/L as $CaCO_3$)	UK Drinking Water Inspectorate [17] (mg/L as $CaCO_3$)
Soft	0–60	0–50
Moderately soft	–	51 100
Slightly hard	–	101–150
Moderately hard	61–120	151–200
Hard	121–180	201–300
Very hard	Over 180	Over 300

$$SAR = \frac{Na^+}{\sqrt{\frac{Ca^{2+}+Mg^{2+}}{2}}}$$

in which Na^+, Ca^{2+}, and Mg^{2+} are the concentrations expressed in milliequivalents per liter for each ion. By using the definition of milliequivalents, the SAR can be calculated by using mass concentrations as follows, where Na^+, Ca^{2+} and Mg^{2+} are the concentrations expressed in mg/L.

$$meq = \frac{mg \times valency}{MW} \rightarrow SAR = \frac{Na^+}{2.1\sqrt{3\times Ca^{2+}+5\times Mg^{2+}}}$$

The effects of the water on the soil depending on its SAR value are reported in Table 4.2. The higher the SAR, the less suitable the water is for irrigation. Evidently, there is no universal standard and different guideline SAR values for irrigation water have been proposed. It should also be noted that different soils and different crops are affected to different extents by irrigation water. For example, nuts and citrus are extremely sensitive, while beets and barley are very tolerant. This may explain the difference in interpretation of the SAR values to some extent.

When SAR values are relatively high, physical and/or chemical methods can be employed to mitigate risks [18]. Chemical methods are those that change the water or soil chemistry in order to improve infiltration. One example is the addition of amendments, such as gypsum, $(CaSO_4 \cdot 2H_2O)$ which increase the calcium concentration (thus reducing the SAR). Other possible chemical amendments include calcium nitrate, calcium chloride, and acid-forming substances such as ferric sulfate and sulfuric acid (the acids will dissolve the calcium carbonate in the soil and increase calcium concentration). Alternatively, the desalinated water

Table 4.2   The interpretation of the SAR values from two different references. Since there are many other indicators for a water's suitability for agricultural purposes such as the Adjusted SAR, Residual Sodium Carbonate, Permeability Index, Kelly's Ratio, etc. it is wiser to interpret the SAR value alongside other indicators [19]

SAR [20]	SAR [21]	Implication
<5	<3	No negative effects (low risk).
5–13	3–6	Minor negative effects (medium risk).
>13	6–9	Significant negative effects (high risk).
–	>9	Not suitable for irrigation.

can be blended with other water sources in order to increase calcium and magnesium content and improve the SAR.

Physical methods include activities which help improve and maintain suitable infiltration rates, such as physically reducing the soil compaction by deep tillage. Deep tillage physically tears, shatters, and rips the soil, and should be done when soils are dry enough to crack. Another method for improving water infiltration is leaving behind crop residues and organic matter in the field. The most suitable residues are those that have a high lignocellulose content and do not decompose or break down rapidly. For maximum effectiveness, large quantities of residues (10–30% by soil volume in the upper 15 cm of the soil) may be needed [18].

4.5 Acidity and Buffering Capacity

As stated earlier, due to the dissolution of CO_2 form the surrounding atmosphere (formation of carbonic acid), desalinated water is slightly acidic. The acidity of water is indicated by its pH, which is the negative logarithm of its active H^+ concentration. It is noteworthy that popular drinks such as lemonade and soda have H^+ concentrations several orders of magnitude more than desalinated water, and so the acidity of the desalinated water is not a hazard for drinking purposes. Nonetheless, acidity does negatively impact corrosion within the distribution network.

Although desalinated water is acidic, due to its low buffering capacity, significant pH changes may occur due to even small chemical reactions downstream. Hence in order to truly define the acid-base characteristics of the water, information regarding the alkalinity, total inorganic carbon, and the pH is required.

By definition, the buffering capacity of a solution is the number of moles of strong acid or strong base needed to change the pH of 1 L of the solution by 1 pH unit. The higher the buffering capacity, the more resistant the water pH is to change. Due to the various processes that occur within the distribution network, maintaining an appropriate buffering capacity for desalinated water is essential, in order to avoid excessive corrosion or sedimentation of minerals downstream.

The increase of the total inorganic carbon (C_T), and consequently the increase of alkalinity, enhances buffering capacity. This is because when a strong acid comes into contact with the water and releases protons (H^+) in the system, the carbonate and bicarbonate ions react with the protons dampening the effect of the acid. Similarly, when a strong base comes into contact with

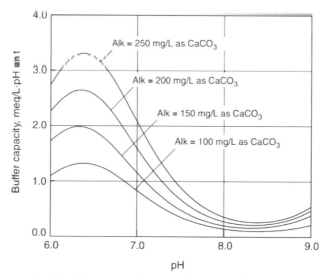

Figure 4.3   The influence of pH and alkalinity on buffering capacity [22].

the water, the released hydroxide ions (OH^-) react with the bicarbonate and $H_2CO_3{}^*$ [2].

The influence of pH on buffer capacity is depicted in Figure 4.3 for alkalinity concentrations ranging from 100 to 250 mg/L as $CaCO_3$. A minimum buffer capacity is observed at around a pH of 8.3. At lower pH values, the alkalinity has a pronounced effect on the buffering capacity.

4.6 Corrosivity

Corrosion has been defined as a galvanic process by which metals deteriorate through oxidation [23]. Various types of corrosion exist, a list of which is presented in Table 4.3 [22]. When iron is oxidized (corrodes), it forms a red-brown hydrated metal oxide commonly known as rust. The rust is flaky and can detach from the surface, exposing fresh metal vulnerable to further reaction (Figure 4.4). In contrast, the oxides of metals such as aluminum, chromium, and nickel form a thin protective layer which acts as an impenetrable barrier preventing further oxidation. Hence "stainless steels" employ significant percentages of these metals, making them resistant to corrosion.

Desalinated water is corrosive, and for decades, the need to stabilize desalinated water to inhibit corrosion has been known. Corrosion can lead to various problems including the release of metal ions into the water, decrease

Table 4.3 The various types of corrosion. Adapted and modified from [22]

Corrosion Type	Description
Crevice corrosion	This type of corrosion is localized and occurs at narrow openings and/or spaces between two material surfaces. By causing an exhaustion of inhibitors, depletion of oxygen, creation of acidic conditions, and the accumulation of aggressive ions, the local chemistry within a crevice may change leading to this type of corrosion.
Microbiologically induced corrosion	Certain bacteria and other microorganisms can enhance corrosion kinetics by producing acids and/or accelerating the rate of redox reactions.
Pitting corrosion	In this type of corrosion, a small portion of the metal becomes a permanent anode and the surrounding areas act as the cathode. This type of corrosion is localized and leads to cavities. Factors leading to pitting corrosion are: manufacturing defects, condition of the metal surface, high concentration of aggressive ions, acidity, and low oxygen concentration.
Tuberculation	When the surface of steel or cast iron is oxidized, a tubercle will form on the surface under which corrosion proceeds. The outside of the tubercle becomes the cathode while the inside becomes anodic. Microorganisms can also enhance this process.
Uniform corrosion	This occurs when metals react with the constituents in water and corrode uniformly. The presence of aggressive ions, principally chloride, sulfide, and sulfate are important factors.
Electrolytic corrosion	When a stray direct electrical current from an outside source comes into contact with a metal structure and is then removed, the metal surface acts as an anode and corrodes. The strength of the electrical current, the pH, the conductivity and velocity of the flow, the amount of oxygen reaching the surface, and other internal and external conditions are factors that affect this type of corrosion.
Erosion	Erosion is caused by mechanical action of flowing corrosive liquid (or gas), abrasion by suspended particles in a slurry, bubbles or droplets possibly formed due to a pressure drop, cavitation and impingement.
Galvanic corrosion	When two dissimilar metals are placed in contact with each other, the difference in the electrical potential of the two metals, their relative areas, and the conductivity of water can result in galvanic corrosion. Under such conditions, one metal acts as the anode and the other as the cathode. The potential difference between the dissimilar metals is the driving force.
Stress corrosion cracking	The combination of tensile stress and a corrosive environment can lead to this type of corrosion. Residual stresses caused by welding and improper alignment are examples of possible causes of tensile stress. Stress corrosion cracks are propagated by the synergistic effects of corrosion and stress.

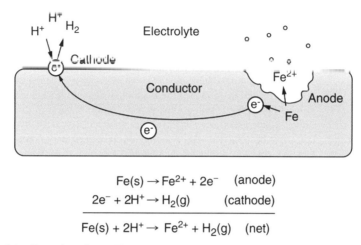

$$Fe(s) \rightarrow Fe^{2+} + 2e^- \quad \text{(anode)}$$
$$2e^- + 2H^+ \rightarrow H_2(g) \quad \text{(cathode)}$$
$$\overline{Fe(s) + 2H^+ \rightarrow Fe^{2+} + H_2(g) \quad \text{(net)}}$$

Figure 4.4 Corrosion of metallic iron in contact with water [22]. Note that in the presence of oxygen, an alternative cathodic reaction is possible: $\frac{1}{2}O_2(g) + 2H^+ + 2e^- \longrightarrow H_2O(l)$.

of distribution system integrity, and increased maintenance costs. A "red water" event, when corroded materials suddenly detach from the sidewall and are released into the water, is the most common symptom of corrosion. At the dawn of the 21st century, the AWWA estimated that more than $300 billion would be required in the following 20 years to upgrade damages caused by corrosion in water distribution systems in the USA [24]. Other estimates put the damages at $19 billion per year [22].

Corrosion-inhibiting chemicals such as silicates, orthophosphate, or polyphosphate could be added to desalinated water. If of sufficient quality, these chemicals are safe and do not have adverse impacts on human health [3]. Although these chemicals have been known to reduce the occurrence of "red water" events (visible appearance of ferric hydroxide particles created by corrosion) there are still doubts regarding their suitability. For example, studies have shown that phosphorous may increase the total microbial count in the water. In addition, the phosphorous may be reduced to toxic PH_3 [2]. Hence, the addition of corrosion-inhibiting chemicals has received some skepticism in recent years, and some plants have decided to inhibit corrosion through other means such as remineralization.

The oxygen content of the water is also of importance. Saturation with dissolved oxygen can accelerate corrosion of metallic components such as iron in the distribution network. On the other hand, the depletion of oxygen can also promote corrosion, in addition to making the water taste flat and undesirable [4, 22].

4.7 Indexes

4.7 Indexes

There are numerous indexes in the literature which aim to quantify a water's "aggressive" and "corrosive" tendencies. Among them are the *Stiff and Davis Stability Index, Riddick Corrosion Index, Driving Force Index*, and the *Singley Index*, each of which have their own advantages and disadvantages.

As an example, the *Larson Ratio*, describes the corrosivity of water towards mild steel and is developed from the relative corrosive behavior of chlorides and sulfates to the protective properties of bicarbonate [6]:

$$\text{Larson Ratio (LR)} = \frac{[Cl^-] + 2\left[SO_4^{2-}\right]}{[HCO_3^-]}$$

in which $[Cl^-]$, and $[HCO_3^-]$ are the concentrations of chloride, sulfate, and bicarbonate, respectively, all expressed in mol/L. When the Larson ratio is greater than 1, the water is considered to have "strong" corrosion potential [6]. Variations of the Larson Ratio exist, with some references including alkalinity (in the form of bicarbonate plus carbonate) in the denominator of the *Larson–Skold* index, and indicating that values higher than 1.2 denote high corrosion [1]. Care should be taken as the LR is far from accurate when applied to waters with low TDS, i.e. desalinated water [25]. Technically speaking, the Larson Ratio is not a "saturation index" as it refers to the faster rates of corrosion in metals due to conductivity effects, rather than the solubility of $CaCO_3$. Nonetheless, the Larson Ratio indicates that with the increase of bicarbonate ions (or the decrease of chloride and sulfate ions, possibly due to a second RO pass), the water's corrosivity decreases. It is noteworthy that the protective oxide and hydroxide films formed on surfaces during corrosion can chemically dissolve in a solution containing chloride and sulfate. Hence the presence of these ions has been associated with increased corrosion rates [26]. Ions such as chloride can also enhance corrosion by acting as intermediaries in rust formation [27].

The Aggressiveness Index (AI) is an empirical relationship initially developed for assessing the potential of water to cause corrosion in asbestos-concrete pipes, and is defined as:

$$\text{Aggressiveness Index (AI)} = pH + \log(\text{Alk} \cdot \text{Cah})$$

where Alk is the total alkalinity (mg/L as $CaCO_3$), and Cah is the calcium harness (mg/L as $CaCO_3$). Water is classified as non-aggressive (noncorrosive) when the AI value is 12 or greater [22]. Values between 10 and 12 are considered moderately aggressive, and values less than 10 are associated with highly aggressive water. Unfortunately, since AI does not incorporate the effects of temperature and ionic strength, its application is limited.

The most popular method for measuring desalinated water's aggressiveness, the *Langelier Saturation Index (LSI)*, is calculated as the difference between the saturation pH (pH_s) and the measured (actual) pH of the water. The pH_s is defined as the pH at which the calcium and bicarbonate of a given water sample would be in equilibrium with calcium carbonate. Hence if the following equation is considered:

$$CaCO_3 + H^+ \leftrightarrows Ca^{2+} + HCO_3{}^-$$

then, the pH_s is the pH at which the above equation is at equilibrium. If the pH is lower than the pH_s, this would mean that the concentration of H^+ is higher than equilibrium. Thus the above chemical equation will shift to the right (and dissolve calcium carbonate). Hence a negative LSI means that the water is aggressive, while a positive LSI denotes deposition of calcium carbonate. In other words, the LSI can be interpreted as the change in pH required for bringing the water to equilibrium:

$$\text{Langelier Saturation Index (LSI)} = pH - pH_s$$

The values of pH_s can be calculated from reaction kinetics, or by using a variety of different empirical relations with different levels of accuracy [28]. Since various empirical relations for the LSI (as well as numerous LSI calculators online) use different equations, care should be taken as using them can result in significant differences. An example of an empirical relation for when calcium hardness and total alkalinity values are in the range of 10–1000 mg/L as $CaCO_3$ is given below [29]:

$$LSI = pH - \left[10.0754 + 2.432636e^{\left(-\frac{T}{86.89927}\right)} - 0.2006e^{(-0.004624 \times TDS)} - \log(Cah) - \log(Alk)\right]$$

in which T is the temperature (°C), TDS is the total dissolved solids (mg/L), Cah is calcium hardness (mg/L as $CaCO_3$) and Alk is the total alkalinity (mg/L as $CaCO_3$). Note that as the temperature increases, calcium carbonate solubility in water decreases. Calcium carbonate is one of the few examples that have an inverse solubility relationship with temperature.

Although the Langelier Saturation Index provides an estimate of the thermodynamic driving force regarding the dissolution/precipitation of calcium carbonate, it cannot predict *how much* calcium carbonate (i.e., what concentration) will actually precipitate. From a corrosion-control perspective, the optimum LSI should be slightly positive, to promote coating of the pipe wall by calcium carbonate without leading to excessive deposition. It should be noted that the LSI is not a definite measure of precipitation potential, for instance, precipitation of calcium carbonate has been observed even at neutral or slightly negative LSI values due to localized pH variations (for example at the pipe wall generated by cathodic reduction of oxygen) [22]. In addition, it is noteworthy that the presence of corrosion inhibiting chemicals is neglected in the calculation of the LSI. Although the American Water Works Association (AWWA) and various experts have explicitly called for the abandonment of the LSI due to its inaccuracy to predict water corrosivity [2], it still remains the most popular saturation index. Figure 4.5 displays a scale of LSI values.

The calcium carbonate precipitation potential (CCPP) is a measure of the amount of calcium carbonate that will precipitate (mg/L) from a solution. Hence, the more positive the CCPP is, the more calcium carbonate will precipitate. If the CCPP is negative, then the water will not produce precipitate, is corrosive/aggressive, and will solubilize any calcium carbonate which may be present in the system. Typically, a CCPP value of 4–10 mg/L as $CaCO_3$ is optimum. The calculation of the CCPP by hand is not a straightforward and simple process, and requires inconvenient iterations. Examples of the calculations can be found in the literature [22]. Alternatively, software such as PHREEQ can be used to calculate CCPP values easily [30].

Figure 4.5 The interpretation of LSI values. Negative values indicate corrosiveness, while positive values indicate a tendency for calcium carbonate precipitation. Waters with LSI values between −0.5 and +0.5 are generally considered balanced (more stringently between −0.2 and +0.2). From a corrosion-control perspective, the optimum LSI value should be slightly positive.

The CCPP is among the most accurate measures for assessing desalinated water aggressiveness, and has been gaining popularity in recent years. However, the CCPP is based on the assumption that only one form of calcium carbonate will deposit. This index also neglects the possibility of kinetic barriers to deposition [22]. In addition, the CCPP is a pure calcite-related index, meaning that it has a limited capability in predicting corrosion due to other factors, such as high concentrations of chloride or sulfate ions. Nevertheless, reviews have attested to the CCPP's usefulness and have gone as far as stating that it is the best suited index to describe the water's saturation state with respect to calcite [2]. Figure 4.6, adopted from Metcalf & Eddy, Inc. [22], displays calcium carbonate precipitation and dissolution.

The *Ryznar Stability Index* is a variation of the LSI, which puts more weight on the saturation pH as follows:

$$\text{Ryznar Stability Index (RSI)} = 2\text{pH}_s - \text{pH}$$

where pH_s is the saturation pH (as defined before), and pH is the actual pH of the water. The interpretation of RSI values is dependent on the pH_s and should be adjusted accordingly [22]. Generally, higher values (above 7) mean that the water will dissolve existing $CaCO_3$ deposits on the inner walls of the pipe

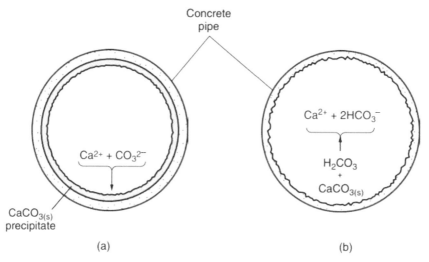

Figure 4.6   (a) Calcium carbonate precipitation, CCPP and Langelier Saturation Index (LSI) meaningfully greater than zero; (b) Calcium carbonate dissolution and damage to the pipes, CCPP and LSI meaningfully less than zero. Adapted and modified from [22].

and cause corrosion, whereas lower values (below 6) denote oversaturation and precipitation of $CaCO_3$.

Over a 10-year period, Puckorius and Brooke [31] developed and verified their own index, similar to that of Ryznar, but with a modification to improve it. By replacing pH with an equivalent pH (pH_e) the Puckorius or Practical Saturation Index (PSI) attempts to incorporate an estimate of buffering capacity of the water into the index, and is defined as follows [32]:

$$\text{Puckorius Scaling Index (PSI)} = 2pH_s - pH_e$$

in which pH_s is the saturation pH (as defined before), and pH_e is found using the following empirical relation derived from 12–15 years' worth of concentrated cooling water data:

$$pH_e = 4.54 + 1.485 \times \log(\text{Alk})$$

where Alk is the total alkalinity (mg/L as $CaCO_3$). According to the index, $CaCO_3$ begins to precipitate at PSI values less than 6.0.

Finally, the Stiff and Davis Stability Index (S&DSI) is a modification of the LSI that is used for higher salinity water (TDS > 10,000 ppm). A positive S&DSI denotes precipitation whereas a negative value shows corrosion. The index is expressed as:

$$S\&DSI = pH + Log[Ca] + \log[Alk] - K$$

In which pH is the actual pH of the water sample, the calcium and alkalinity are in moles per liter, and K is a constant which can be found from the relevant curves. Alternatively, instead of the curves, the following curve-fitting equations could be used [33]:

If $\mu < 1.2$ then $K = 2.022e^{\frac{(\text{Ln}(\mu) \ | \ 7.544)^2}{102.6}} - 0.0002T^2 + 0.00097T + 0.262$

If $\mu > 1.2$ then $K = -0.1\mu - 0.0002T^2 + 0.00097T + 3.887$

Where T is the temperature (degrees Celsius) and μ is the molar ionic strength of the sample calculated as $\mu = \frac{1}{2}\sum(C_iZ_i)$ in which C_i is the molar concentration of an ion and Z_i is the ion's valence.

4.8 Conclusion

This chapter has discussed the chemicals of concern in the desalinated water. Based on the guidelines published by the WHO, the most important constituents of desalinated water include: boron, bromine, calcium, magnesium, fluorine and other supplements, organic chemicals, potassium, and sodium. In addition, by drawing upon fundamentals of water chemistry, the carbonate system has been introduced and it has been explained that the system is a function of the concentration of dissolved CO_2, the concentration of bicarbonate ions, the concentration of carbonate ions, the pH value, the total inorganic carbon content, and the total alkalinity.

Further, the chapter indicated that although elements such as strontium, aluminum, barium, iron, manganese, and zinc also cause hardness, calcium and magnesium are the main components defining the hardness of the water. These elements are also important in the calculation of the sodium adsorption ratio (SAR) which is one of the several indicators used to measure the water's suitability for irrigation purposes.

Desalinated water is always slightly acidic with a low buffering capacity. These parameters have been presented followed by discussion on desalinated water corrosiveness. Indexes used for defining the water's corrosiveness have been reviewed, and it has been explained that although the LSI is the most widely used parameter to measure water corrosivity, at present the CCPP is considered by experts as the best suited indicator for water's saturation state with respect to calcite.

Overall, although unconditioned desalinated water is free of unwanted chemicals and pathogens and is suitable for drinking purposes, if it is to be distributed via a network, conditioning and post-treatment processes (as will be discussed in Chapter 8) are undoubtedly needed.

References

[1] Withers, A. (2005). Options for recarbonation, remineralisation and disinfection for desalination plants. *Desalination* 179, 11–24.

[2] Birnhack, L., Voutchkov, N., and Lahav, O. (2011). Fundamental chemistry and engineering aspects of post-treatment processes for desalinated water—A review. *Desalination* 273, 6–22.

[3] World Health Organization (2011). *Safe Drinking-Water from Desalination. Guidance on Risk Assessment and Risk Management Procedures to Ensure the Safety of Desalinated Drinking-Water*. Geneva: WHO.

[4] Duranceau, S., Wilder, R., and Douglas, S. (2012). Guidance and recommendations for post-treatment of desalinated water. *J. Am. Water Works Assoc*. 104, E510–E520.

[5] World Health Organization (2011). *Guidelines for Drinking-Water Quality*, 4th Edn. Geneva: WHO.

[6] Delion, N., Mauguin, G., and Corsin, P. (2004). Importance and impact of post treatments on design and operation of SWRO plants. *Desalination* 165, 323–334.

[7] Thorsten, C. (2013) *Chemistry of the Borate-Boic Acid Buffer System. CR Sci. LLC.* Available at: http://www.crscientific.com/experiment4.html [accessed August 9, 2016]

[8] Yermiyahu, U., Tal, A., Ben-Gal, A., Bar-Tal, A., Tarchitzky, J., and Lahav, O. (2007). Rethinking desalinated water quality and agriculture. *Science*, 318, 920–921.

[9] The British Fluoridation Society (2012). *One in a Million: The Facts about Water Fluoridation. Manchester.* Available at: http://bfsweb.org/onemillion/onemillion.htm

[10] Bar-Zeev, E., Passow, U., Romero-Vargas Castrillón, S., and Elimelech, M. (2015). Transparent exopolymer particles: from aquatic environments and engineered systems to membrane biofouling. *Environ. Sci. Technol.* 49, 691–707.

[11] Institute of Medicine (US) Committee on Strategies to Reduce Sodium Intake (2010). "Taste and flavor roles of sodium in foods: a unique challenge to reducing sodium," in *Strategies to Reduce Sodium Intake United States*, eds J. Henney, C. Taylor, and C. Boon (Washington, DC: National Academies Press).

[12] Al-Rawajfeh, A. E. (2004). *Modelling and Simulation of CO2 Release in Multiple-Effect Distillers for Seawater Desalination.* Halle: Martin Luther University of Halle-Wittenberg.

[13] Kalff, J. (2002). *Limnology: Inland Water Ecosystems.* Upper Saddle River, NJ: Prentice Hall.

[14] Bang, D. P. (2012). *Upflow Limestone Contactor for Soft and Desalinated Water: Based on Calcite Dissolution Kinetics and PHREEQC Built-in Excel Models.* Delft: Delft University of Technology.

[15] Minnesota Rural Water Association (2009). *Minnesota Water Works Operations Manual.* 4th Edn. Elbow Lake, MN: Minnesota Rural Water Association.

[16] The U.S. Geological Survey Water Science School (2015). *Water Hardness.* Available at: https://water.usgs.gov/edu/hardness.html

[17] Drinking Water Inspectorate (2011). *Water Hardness. UK Department for Environment Food and Rural Affairs.* Available at: http://dwi.defra.gov.uk/consumers/advice-leaflets/hardness_map.pdf

[18] FAO (1994). "Infiltration problems," in *Water Quality for Agriculture*, eds R. S. Ayers, and D. W. Westcot (Rome: Food and Agriculture Organization of the United Nations).

[19] Ramesh, K., and Elango, L. (2012). Groundwater quality and its suit-ability for domestic and agricultural use in Tondiar river basin, Tamil Nadu, India. *Environ. Monit. Assess.* 184, 3887–3899.

[20] Horneck, D. A., Ellsworth, J. W., Hopkins, B. G., et al. (2007). Managing Salt-Affected Soils for Crop Production. Available at: https://catalog.extension.oregonstate.edu/pnw601

[21] Agriculture NSW Water Unit (2014). *Primefact 1344: Interpreting Water Quality Test Results.*

[22] Metcalf & Eddy, Inc., Asano, T., Burton, F., et al. (2007). Industrial Uses of Reclaimed Water. *Water Reuse.* New York, NY: McGraw-Hill.

[23] Averill, B., and Eldredge, P. (2006). *Electrochemistry-General Chemistry: Principles, Patterns.* Upper Saddle River, NJ: Prentice Hall.

[24] McNeill, L. S., and Edwards, M. (2001). Iron pipe corrosion in distribution systems. *J. Am. Water Works Assoc.* 93, 88–100.

[25] DeBerry, D. W., Kidwell, J. R., and Malish, D. A. (1982). *Corrosion in Potable Water System: Final Report.*

[26] Shetty, S., Nayak, J., and Shetty, A. N. (2015). Influence of sulfate ion concentration and pH on the corrosion of Mg-Al-Zn-Mn (GA9) magnesium alloy. *J. Magnes. Alloy.* 3, 258–270.

[27] Neville, A. (1995). Chloride attack of reinforced concrete: an overview. *Mater. Struct.* 28, 63–70.

[28] Wojtowicz, J. A. (2001). A revised and updated saturation index equation. *J. Swim. Pool Spa Ind.* 3, 28–34.

[29] Hach Company (2015). *Determination of Langelier Index in Water: DOC316.52.93108.*

[30] United States Geological Survey (2015). *PHREEQC (Version 3)—A Computer Program for Speciation, Batch-Reaction, One-Dimensional Transport, and Inverse Geochemical Calculations.* Available at: http://wwwbrr.cr.usgs.gov/projects/GWC_coupled/phreeqc/

[31] Puckorius, P. R., and Brooke, J. M. (1991). A new practical index for calcium carbonate scale prediction in cooling tower systems. *Corrosion* 47, 280–284.

[32] Revie, R. W. (2011). *Uhlig's Corrosion Handbook*, 3rd Edn. New York, NY: Wiley.

[33] Leitz, F., and Guerra, K. (2013). Water Chemistry Analysis for Water Conveyance, Storage, and Desalination Projects. U.S. Department of the Interior Bureau of Reclamation Technical Service Center. Available at: https://www.usbr.gov/tsc/techreferences/mands/mands-pdfs/WQeval _documentation.pdf

PART II

Unit Operations

Ibrahim S. Al-Mutaz[1]

[1]Chemical Engineering Department, College of Engineering,
King Saud University, P. O. Box 800, Riyadh 11421, Saudi Arabia
E-mail: almutaz@ksu.edu.sa

5.1 Introduction

Thermal desalination processes involve the evaporation and condensation of water. Thermal desalination processes simulate the natural water cycle (hydrological cycle), where the salty water is heated by the sun, producing water vapor that is in turn condensed to form fresh water (rain). This principle is applied in thermal desalination technologies. So in thermal desalination, thermal energy or heat is applied to the water present in a boiler or evaporator to drive water evaporation, this water vapor is condensed later in a condenser by exchanging heat.

Thermal desalination processes are often called distillation.[1] It is one of the most ancient ways of converting salty water into potable water. Today, the main thermal desalination technologies are divided into three groups: multi-effect distillation (MED), multi-stage flash (MSF), and vapor compression (VC).

The MED process is considered the oldest desalination process with references and patents existed since the 1840s [1]. However recent developments in the MED process have once again brought the process to the forefront and in competition with other technologics [1]. The MED process is more thermodynamically efficient than the MSF with low-temperature multi-effect distillation, LT-MED being the most efficient of all thermal

---

[1]The term "distillation" is generally accepted in the field of desalination. It does not have the same meaning that is normally reserved by chemical engineers for processes involving the separation of two or more volatile liquids which are mutually soluble. The term "evaporation" is the correct term that should be used to indicate the process of separating pure water from aqueous solutions of nonvolatile materials.

distillation processes. Hence, MED has a lower power consumption and a higher performance ratio compared to MSF [2].

Multi-stage flash (MSF) is based on the principle of flash evaporation; where the heated seawater boils when exposed to reduced pressure. Generally, seawater is heated by low pressure steam up to 90–120°C [3], and flows into the flashing chamber, where it boils and evaporates, i.e., flashes. The unflashed brine is then passed to the next flash chamber also known as the next stage, exposed to further flashing at lower pressure. In this way the water can be flashed in a sequence of stages without supplying additional heat. In each stage, the pressure should be maintained below the pressure corresponding to the saturation temperature of the feedwater. The MSF process is an inventive process where vapor formation takes place within the bulk of the liquid instead of the surface of hot tubes. In other thermal processes, submerged tubes of heating steam are usually used to perform evaporation. This always results in scale formation on the tube.

Vapor compression distillation uses the heat generated by the compression of vapor to evaporate saline water. Compression of the vapor increases both the pressure and temperature, and hence, water vapor quality. Then, the latent heat of condensation generates additional vapor. The effect of compressing water vapor can be done by two methods. Water vapor can be compressed by means of a mechanical device, so called mechanical vapor compression (MVC). Water vapor may also be compressed by an ejector system motivated by high-pressure steam. This form is called thermal vapor compression (TVC).

Other thermal desalination processes including freezing, solar stills and humidification–dehumidification are only found on pilot or experimental scale. They can also be utilized alongside other desalination processes.

This chapter focused on the detailed description of the MED and MSF thermal desalination processes, including design features. A general review will also be presented on the vapor compression distillation (VCD) process. The principle of other thermal desalination processes (freezing, solar stills, and humidification-dehumidification) will be discussed briefly.

5.2 Multi-Effect Distillation (MED)

The term "multi" implies the multiple reuse of the thermal energy of steam in the successive processes of evaporation and condensation. MED was formerly called multiple effect boiling (MEB).

In the MED process, the seawater enters the first evaporator (called effect) and is heated up to the boiling temperature by external steam, after being preheated in tubes. Seawater is normally sprayed onto the surface of the evaporator to promote rapid evaporation. The un-evaporated brine from the first effect is then fed to the second effect where further evaporation takes place. This time, the brine is heated by the vapor formed in the first effect. The process is then repeated in subsequent effects where the temperature and pressure are lowered from each effect to the next [3].

Based on the configuration of the heat transfer surface, the MED plant can be classified as vertical climbing film tube, rising film vertical tube, or horizontal tube falling film. Another form of classification is based on brine flow direction regarding the vapor direction from one effect to the other. These include the forward configuration, backward configuration, and parallel feed configuration.

5.2.1 Types of MED Tube Arrangements

Based primarily on the arrangement of the heat exchanger tubing there are three arrangements. They are as follows [4].

5.2.1.1 Horizontal tube arrangement

In this arrangement, the tube bundles are arranged horizontally in the vessel as shown in Figure 5.1. The feed water is sprayed over the outside surfaces of the tubing, and the inside of the tubing contains the heat to vaporize the feed water. The vapor generated in each effect is directed to the next lower pressure effect.

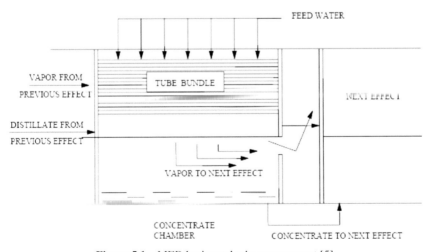

Figure 5.1   MED horizontal tube arrangement [5].

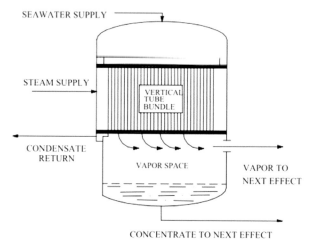

Figure 5.2   MED vertical tube bundle arrangement [5].

5.2.1.2 Vertical tube arrangement

The vertical tube bundle arrangement is depicted in Figure 5.2. The feed water enters at the top of the effect and flows on the inside surface of the tube. The heat for vaporization is on the outside surface of the tubing. The advantage of this design over the horizontal tube arrangement is higher heat transfer rates. Higher heat transfer rates result from having a thin film on both the inside and outside surfaces of the heat exchanger tubing. One drawback to this design, however, is the difficulty of ensuring that good flow distribution is achieved for each tube.

5.2.1.3 Vertically stacked tube bundles

The tubing arrangement in the vertically stacked unit is depicted in Figure 5.3. For this design, the concentrate flows down between effects; thus, eliminating the need for pumping. As with the vertical unit described above, the feed water is fed to the inside surface of the tubing, and the heating for vaporization is on the outside surface of the tube bundle. This drawing depicts two sets of bundles, but the unit can consist of many sets.

5.2.2 Conventional MED Process

An MED system mainly consists of a sequence of single effect evaporators and a final condenser. Other additional equipment includes a venting system, distillate flashing boxes and seawater feeding and rejecting facilities.

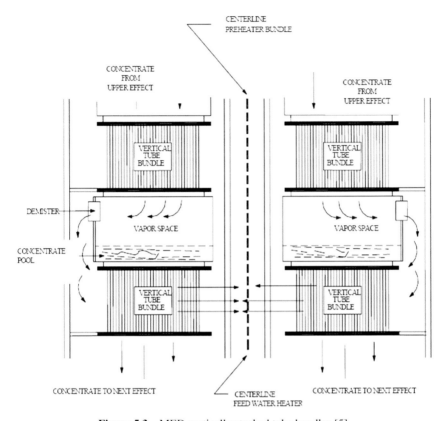

Figure 5.3   MED vertically stacked tube bundles [5].

Freshwater is obtained as a result of the evaporation of water in each evaporator. Each evaporator mainly consists of a horizontal heat transfer tube bundle, demister, and vapor space. Schematics of an MED system with the forward, backward, and parallel/cross feed configurations are shown in Figures 5.4, 5.5, and 5.6, respectively. These configurations differ in the flow direction of the evaporating brine and heating steam. In each scheme, saturated steam from an external source enters into the tube side in the first effect as heating medium to heat the feed seawater. The condensate from the first effect is returned to the external source such as a boiler.

In the forward feed arrangement as shown in Figure 5.4, feedwater is supplied to the first effect. Part of the feed is evaporated in the first effect while the remaining part known as the brine, enters into the second effect. In the second effect, some part of the brine evaporates while the remnant enters

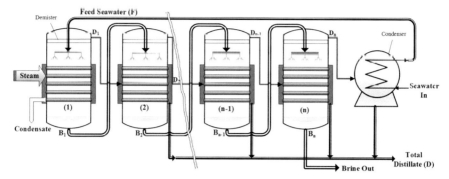

Figure 5.4   Schematic of the forward feed MED configuration [6].

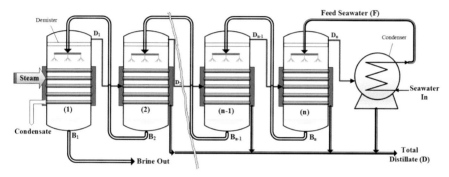

Figure 5.5   Schematic of the backward feed MED configuration [6].

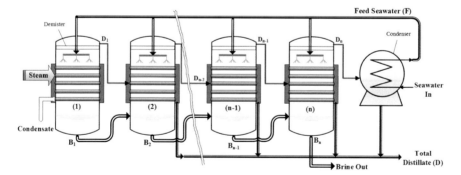

Figure 5.6   Schematic of the parallel/cross feed MED configuration [6].

as a feed source to the next effect and this process continues to the last effect. The cooling seawater enters into the condenser, where it exchanges heat with the vapor coming from the last effect. So, in the forward feed configuration, feed and vapor enter the effects and flow in the same direction. The main advantage of the forward-feed configuration is its ability to operate at high top brine temperatures.

In the backward feed configuration, shown in Figure 5.5, feed seawater after passing through the end condenser, enters into the last effect which has the lowest temperature and pressure within the system. The brine leaving the last effect flows through successive effects toward the first effect. The brine leaving the first effect is sent back to the sea. Temperature and pressure increase as we move from the last effect to the first effect. In this layout, feed and vapor that enter the effects have opposite flow directions.

In the backward feed arrangement, brine with the highest concentration is subjected to the effect which has the highest temperature in the system, which is one of the major drawbacks of this configuration. It can cause severe scaling problems in the evaporator. Another disadvantage of this arrangement is the increase in maintenance cost and pumping power.

Figure 5.6 shows the scheme of parallel/cross feed MED systems. In this configuration, the feed is distributed almost equally to all effects. Brine from each effect enters into the next one. Feed and vapor enter the effects and flow in the same direction in this configuration. The flow pattern is the same as in the forward feed configuration. This configuration has a simple design compared to the other two layouts. Another feature of this arrangement is that, no pumps are required for moving the un-evaporated brine between the effects because brine flows from higher to lower pressure effects. The parallel/cross feed MED system is the most commonly used on the industrial scale.

5.2.3 Multi-Effect Distillation with Thermal Vapor Compression (MED–TVC)

The combination of TVC with conventional MED processes results in a competitive and energy-efficient MED-TVC system. The MED-TVC process has attracted research interest recently because of the development or large systems reaching over 15 MIGD (million imperial gallons per day).

Among different types of MED systems, the most attractive MED configuration is the horizontal tubes falling-film evaporation (HT–FF–MED) with coupled TVC [7]. The HT–FF–MED with TVC has become an attractive choice for the Middle East region because of such factors as ease of operation with any heat source available, low scale formation, and high performance ratios (PR) [8].

5.2.3.1 MED-TVC process description

A schematic diagram of a MED–TVC system with n effects is shown in Figure 5.7. The main components of the system are the condenser, steam jet ejector which acts as a thermal compressor and falling film evaporators also known as effects [9].

As shown in Figure 5.7, the feed seawater leaves the condenser at T_f where it exchanges heat with the vapors formed in the last effect. The greater portion of this feed seawater is rejected back to the sea as cooling water while the remaining feed seawater (F) is divided equally between the effects. The cooling water is used to remove the excess heat added to the system by the hot motive steam. The generated vapor flows in the direction of falling pressure, i.e., from left to right while feed seawater is sprayed in the perpendicular direction.

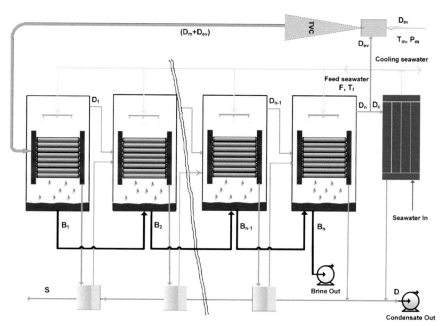

Figure 5.7   Schematic of a MED–TVC process with n number of evaporators [10].

Motive steam from the external source goes through the thermal vapor compressor which entrains and compresses a portion of the vapor formed in the last effect. Compressed steam from the thermal vapor compressor is supplied into the tube side in the first effect as heating medium to heat the feed seawater (F) which is sprayed onto the outside surface of the tube. The heating steam form the thermal vapor compressor is condensed inside the tubes which warms up the feed seawater to the boiling temperature of the first effect (T_1) which is also known as the top brine temperature. Part of the feed evaporates and generates an amount of vapor (D_1), which is directed to the second effect as the heat source at a lower temperature and pressure than the previous effect. The temperature of the vapor formed in the first effect (T_{v_1}) is less than the boiling temperature (T_1) by the boiling point elevation (BPE).

A wire mist separator known as wire mesh demister is also used in each effect for the removal of entrained brine droplets. Part of the condensate from the first effect goes back to its source while the remnant enters the first flashing box. In order to the utilize the energy of the brine leaving the first effect (B_1), it is directed to the next effect and so on, until it reaches the last effect at a lower pressure than the preceding effect.

The vapor is formed inside effects $2 - n$ by two different mechanisms: first, by boiling over the surface of the tubes and second by flashing inside the liquid bulk. Flashing also takes place in the flashing box due to the condensation of the distillate in the second effect and so on. The purpose of the flashing boxes is to recover the heat from the condensed freshwater. In the last effect, the produced vapor (D_n) is split into two streams. One stream (D_{ev}) is entrained and compressed by the thermal vapor compressor and the other stream (D_c) is directed to the condenser where it increases the temperature of the feed seawater from T_{cw} to T_f.

5.3 Multi-Stage Flash (MSF)

The concept of the MSF desalination process was introduced in the 1950s. The design was based on the old multi-effect system known as the once-through system [11]. The first installation was constructed by Westinghouse company in Kuwait in late 1956. This particular system was not a true flash desalination configuration. The first patent on MSF desalination was published in 1957 by Professor Robert Silver [12]. Several developments and improvements took place in the following years and plant capacity increased from 500 m³/day in the 1960s to 75,000 m³/day in the 1990s. It is expected that the size of MSF units will continue to increase beyond 80,000 m³/day (17.5 MIGD) up to 100,000 m³/day per unit, with the possibility of using several units in parallel. The large production capacity of MSF is among its main advantages.

The flashing chamber is the heart of the MSF desalination process. Flashing takes place in each stage, caused by reduction in pressure. The generation and condensation of vapor take place in the same stage. MSF desalination plants consist of several (multi-stage) flashing chambers. Other components include, brine pools, brine transfer devices, demisters, distillation trays, tube bundles of the condenser or pre-heater, and water boxes to transfer recycled brine from one stage to the next.

Multi-stage flash plants can be operated as single-purpose desalination processes where only water is produced or as dual-purpose processes (cogeneration plants) where desalination and power generation (electricity) occur side by side. Dual-purpose plants are widely used for co-production of water and power. In dual purpose plants, high pressure and temperature steam expands in steam turbines (thus producing work) before being supplied to the MSF desalination plant.

The main components of MSF desalination plants include: deaerator, decarbonator, vacuum system, and evaporator. MSF technology is now considered a mature and reliable technology.

5.3.1 MSF Configurations

Multi-stage flash commercial desalination systems have two major process configurations: once-through MSF (MSF-OT) and brine recirculation MSF (MSF-BR) as shown in Figures 5.8 and 5.9. In MSF-OT, feed seawater passes throughout the process once through at a time while in the MSF-BR process a small amount of seawater feed is mixed with a major recycling flow of rejected brine from the last stage.

Once-through MSF plants consist of an evaporation section (heat recovery section), a brine heater and the condenser tubes arrangement. Brine recirculation MSF consists of a brine heater, heat recovery section, and heat rejection section. The role of the rejection section is to remove the surplus thermal energy from the plant, thus cooling the distillate product and the concentrated brine to the lowest possible temperature. For large scale desalination processes, the brine circulation system is most widely used. MSF plants normally

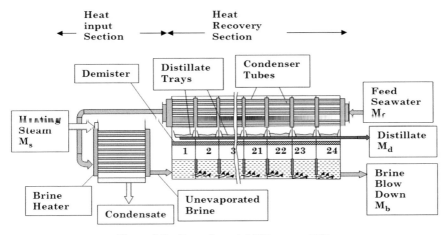

Figure 5.8 Once-through MSF process [13].

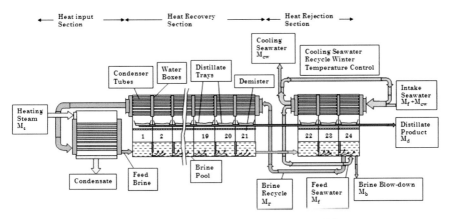

Figure 5.9 Brine recirculation MSF process [13].

consist of 15–25 stages. The once through configuration is generally limited to small scale plants.

In the once through MSF system the whole amount of treated seawater flows through the condenser tube in heat rejection, heat recovery to the brine heater, and into the flash chamber. All of the brine which exists in the last stage is rejected as blowdown without any recirculation. While in the brine recirculation system, as its name indicates, part of the rejected brine is circulated back again into the system.

The main disadvantage of the once trough MSF system is the requirements to pretreat the entire feed seawater. This consumes large quantities of chemicals and hence more operational cost in addition to the large seawater feed entering the evaporator. There is more chance for the generation of non-condensable gases (NCG) which may be associated with corrosion potential and an increase in size and consumption of the vacuum system.

However, once through arrangements have simple process control, low-scaling tendency due to low salinity of brine, lower pumping energy requirements, and the elimination of the heat rejection section and the recirculation pump. On the other hand, the brine recirculation system has some advantages such as less make-up to be treated and variable seawater flow to suit different operation needs. The main disadvantages of the brine recirculation configuration are the scale formation, corrosion problems and high brine concentration due to the increase of seawater temperature and concentration.

The general features of the MSF plants include the following [13]:

- Stable and reliable operation through ensuring an adequate heat transfer area, suitable materials, and proper corrosion allowance.
- Multi-stage flash desalination units can operate with dual-purpose power generation plants. Design of the co-generation plants allows for flexible operation during peak loads for power or water. In the Middle East, the peak loads for electricity and water occur during the summer time due to the high ambient temperature, which is associated with massive use of indoor air-conditioning units and increase in domestic and industrial water consumption. The opposite is true for the winter season with mild temperatures and limited use of indoor heating units.
- The majority of MSF plants use the brine circulation design, which is superior to the once through design. Brine circulation results in a higher conversion ratio, uses a smaller amount of chemical additives, and gives good control on the temperature of the feed seawater.
- Cross-tube design has simpler manufacturing and installation properties compared to the long tube arrangement.
- All auxiliaries are motor-driven and have better operating characteristics than turbine driven units; even for large brine circulation pumps.
- Additive treatment is superior to acid treatment, because acidic solutions may enhance corrosion rates of tubing, shells, and various metallic parts.
- Proper system design should allow for load variations between 70 and 110% of the rated capacity.

There are three tube arrangements in MSF plants, these are:

Vertical tube arrangements: There are a few MSF plants which use the vertical tube configuration, but it is not commonly used. Figure 5.10 shows the MSF vertical tube design.

Cross-tube arrangements: The majority of commercial plants use cross-tube arrangements, where heat transfer tubes are laid at right angles to the flow direction of the flashing brine. In the cross-flow arrangement as shown in Figure 5.11, the tube bundle is generally located in the middle of the flash chamber and each stage tube bundle is connected by water boxes external to the vessel. This arrangement minimizes the cost of shells but requires a large number of water boxes (two for each stage) and tube sheets (one per stage).

Long-tube arrangements: In the long-tube configuration, the tubes are parallel to the flow direction of flashing brine, and they pass through several flash stages as shown in Figure 5.12. Flashing brine is indicated by green arrows. In the long-tube configuration, the flow is parallel to the tube bundle which crosses each stage partition wall but in the opposite direction to the

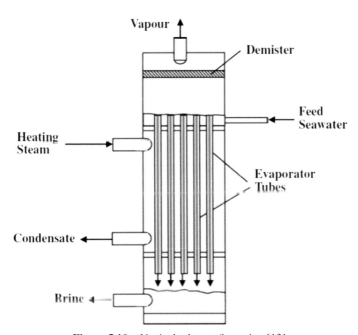

Figure 5.10   Vertical tube configuration [13].

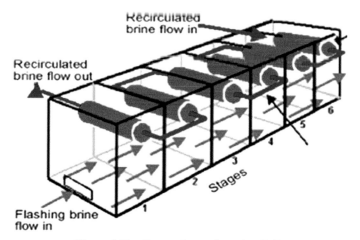

Figure 5.11   Cross-tube configuration [14].

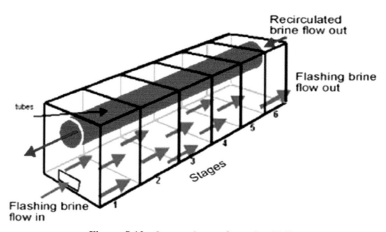

Figure 5.12   Long-tube configuration [14].

recirculating brine. Usually, this arrangement requires more stages than cross-tube but fewer tube sheet and water boxes. The main disadvantage of the long-tube arrangement is the difficulty of removing the tubes which will affect many stages involved.

The tube arrangements have an effect on the number of MSF stages. A typical MSF plant may have 16–24 stages. In case of long-tube, the number of stages could reach 54 stages [15, 16]. The increased number of stages increases the overall efficiency of heat recovery in the plant and decreases its operating costs, while increasing the capital cost of the plant. The long-tube arrangement is a favorable option when a higher performance ratio is wanted (12–15). The choice between any of these configurations for a plant depends on the designer's view point and the manufacturer and owner's experience. Table 5.1 shows a comparison between cross tube and long tube MSF plants.

5.3.3 MSF Process Description

In MSF–BR, raw seawater is screened and chlorinated before passing through the heat exchanger tubes in the heat rejection section where it is heated by the flashing brine. Part of the heated seawater (makeup water) is chemically pretreated with either acid (sulfuric acid) or antiscalant additives are used to suppress the formation of alkaline scale in the heat transfer tubes.

Table 5.1 Comparison between cross-tube and long-tube MSF plant parameters, for a unit producing 1000 cubic meters of freshwater per hour

Tube Arrangement	Cross-Tube	Long-Tube
Total number of tubes	4	4
Plant availability (%)	95	97
Feed treatment	Anti-scale/Ball-cleaning system	Acid
Number of stages	20	48
Top Brine temperature (°C)	112	113
MP steam pressure, bar	6	6
MP steam temperature (°C)	261	261
MP steam Enthalpy (KJ/kg)	2,980	2,980
MP Steam Consumption/ unit (kg/h)	1,900	1,600
LP steam pressure (bar)	2	2
LP steam temperature (°C)	161	161
LP steam Enthalpy (kJ/kg)	2,792	2,792
LP Steam Consumption/ unit (kg/h)	122,000	68,000
Gain Output ratio (GOR)	8.1	14.4
Electric absorption of BRP/ unit (kW)	1,620	1,620
Other electric absorption/ unit, (kW)	1,246	848
Total electric absorption/ unit (kW)	2,866	2,468
Antiscale dosing rate on makeup seawater (ppm)	3	0.5

The dissolved oxygen content of the makeup water is stripped to a level below 20 ppb in a deaerator to minimize corrosion. For acid-treated MSF desalination plants, a decarbonator unit in the form of a packed tower is employed to remove the carbon dioxide. Carbon dioxide may affect the heat transfer performance and corrodes heat transfer tubes. Further reduction of oxygen content to a level below 10 ppb is carried out by a chemical scavenger treatment such as sodium bisulfite. The makeup water is then mixed with the recirculation brine from the heat rejection section to form the feed.

The feed is then introduced to a series of heat exchanger tubes in a heat recovery section. The feed temperature rises gradually due to the transfer of heat with the vapor generated by flashing of brine. Vapor condenses outside the heat transfer tube and is collected on the distillate tray. The feed water then goes to the brine heater where its temperature rises to the maximum plant operating temperature (top-brine temperature, TBT). Heated feed enters the flash chamber, and is maintained slightly below the saturation vapor pressure of water. A fraction of its water content flashing into steam goes through the demister and condenses on the outside of the tubes as distillate water which is collected in trays and passes through all the stages via special inter-stage transfer orifices. The unflashed brine enters the second stage, which is at a lower temperature, and pressure.

The same steps are repeated, until the last stage in the heat rejection section, where part of the brine is rejected as blow down. The NCG formed by flashing are vented to the atmosphere. The distillate water is subjected to posttreatment which includes a recarbonation and or alkalization for increasing carbonate and calcium content and pH correction before it is sent to the consumers.

5.4 Vapor Compression Distillation (VCD)

The vapor compression distillation (VCD) process is generally used for small- and medium-scale desalination units. The heat for evaporating the water comes from the compression of vapor rather than the direct exchange of heat from steam produced in a boiler or other heat sources. In VCD, the compression may be achieved by applying thermal energy (TVC) or mechanical energy (MVC). In both thermal and mechanical systems, after vapor is generated from the saline solution, it is compressed and then condensed to obtain potable water. In MED and MSF processes an external source of heat is used to warm up the incoming saline water, whereas the energy needed to heat the saline water in the VCD process comes from thermal or mechanical source which compress the generated vapor. VCD systems can operate at very high salt concentrations. VCD with one effect is as beneficial as a 15–20 effect MED method [17]. Vapor compression distillers use about 31.70 kWh/m³ (0.12 kWh/gallon) of distilled water produced [18].

According to Howe [19], the VCD process can be represented on the temperature-entropy diagram as shown in Figure 5.13. Line 1'–1 is a line of constant temperature, and results in superheated vapor at state point 1 This vapor is compressed to state point 2 and is then passed to the heating jacket of the evaporator where it is condensed along the constant pressure line 2–3 to produce liquid at point 3. The liquid distillate then passes into the liquid-to-liquid heat exchanger where it is cooled to state point 4, and is finally discharged at this temperature. Meanwhile, the portion of the brine left after the vapor formation also passes into a liquid-to-liquid heat exchanger and is cooled from point 1' to point 4 before being discharged. The incoming saline water, entering the heat exchanger at a temperature less than at point 4, is heated to a temperature slightly less than point 1. The added heat required

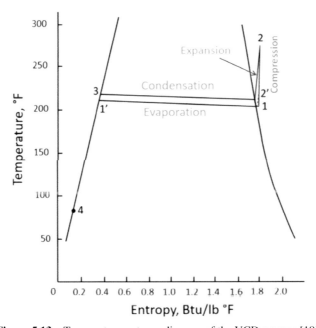

Figure 5.13 Temperature-entropy diagram of the VCD process [19].

to change the temperature of the incoming water to the boiling point, 1, is derived from the condensation of a small portion of the vapor in the heating jacket of the evaporator. It is often assumed that the incoming saline water enters the evaporator at temperature 1. The compression of the vapor is also assumed to be isentropic rather than isothermal, as in the ideal reversible case, because isothermal compression has not as yet been achieved in any practical device. It should be noted that compression from point 1–2 is needed to overcome the boiling point elevation due to the salinity of the evaporating liquid, and that the further compression from 2' to 2 provides the temperature difference $t_3 - t_1$, for heat transfer in the evaporator–condenser.

5.4.1 Mechanical Vapor Compression (MVC)

The MVC system is suitable for providing desalinated water to small communities. The capacity of the MVC systems is in the range of 500–5000 m^3/day. The operation of MVC systems faces two main drawbacks, mainly the limitation on the compressor capacity and the mechanical wear of the moving parts of the compressor [20].

The MVC uses heat exchangers and the vapor compressor to recover some of the latent heat that is generated during condensation [21]. The MVC system uses a vapor compressor to drive the process and heating elements to start and maintain it. Heated seawater is sprayed onto a heat exchanger in an evaporator chamber. The water changes to steam when it hits the surface of the evaporator heat exchanger tubes. The steam is then taken up in the vapor compressor where it is pressurized and sent through the evaporator heat exchanger tubes. When the steam is compressed it raises the temperature of the steam which is higher than the seawater temperature in the evaporator chamber. The steam undergoes another phase change as it condenses into water. The condensed water passes through another heat exchanger, where it heats the incoming seawater that is sprayed in the evaporator. The condensed water is now discharged out of the unit as distilled water. The heating elements are used to heat the water to allow the evaporation to start and to maintain the correct temperatures. Most of the energy consumed in the MVC process is through the compressor work, as the compressor not only increases the pressure but it reduces the vapor pressure in the evaporator. Figure 5.14 shows a schematic diagram of a MVC distillation unit.

The mathematical model for the MVC can be found at Ettouney and El-Dessouky [13].

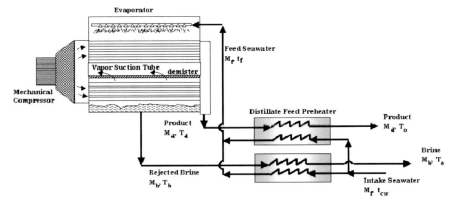

Figure 5.14 Mechanical vapor compression (MVC) [21].

5.4.2 Thermal Vapor Compression (TVC)

During TVC, vapor from a boiling chamber is recompressed to a higher pressure and hence energy is added to the vapor. For this purpose, steam jet vapor compressors (thermo-compressors) are used. The thermo-compressor uses a certain quantity of live steam, so-called motive steam, to increase the pressure of the vapor emerging from the heating chamber, and enabling the vapor to be reused for heating the "mother" liquid (seawater). In other words, since the pressure increase of the vapor results in a condensation temperature increase, the vapor can be used to heat the incoming seawater. Without the thermo-compressor, the vapor would be at the same temperature as the seawater, and no heat transfer could take place.

Thermal vapor compression operates according to the jet pump principle. It has no moving parts, ensuring a simple and effective design that provides the highest possible operational reliability.

The use of a thermal vapor compressor has the same steam/energy saving effect as an additional evaporation effect. The TVC process is used on an industrial scale in the MED system. The combined system is particularly attractive due to its high performance ratio, low number of effects, good flexibility to load variation, simple geometry, and absence of moving parts. Figure 5.15 shows a flow diagram of the TVC distillation process.

The mathematical model of TVC process can be found at Ettouney and El-Dessouky [11].

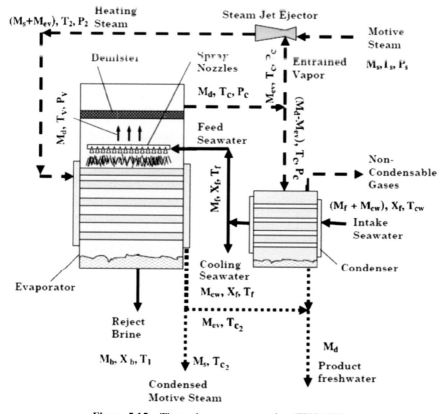

Figure 5.15   Thermal vapor compression (TVC) [23].

5.5 Other Thermal Processes

Besides the conventional thermal distillation process (MED, MSF, and VCD), other thermal desalination processes are possible. These processes include freezing, solar desalination, humidification dehumidification, and membrane distillation. They can also be utilized with other desalination processes. The following is a brief description.

5.5.1 Freezing

Desalination by freezing is similar to thermal desalination in the fundamental concept of relying on a phase change to achieve separation. However, desalination by freezing has not gone beyond the pilot stage although it was among the oldest methods of separation. It takes advantage of the relatively low enthalpy of phase change of freezing. Freezing of water at atmospheric conditions requires 334 kJ/kg whereas evaporation would require 2,326 kJ/kg.

The basis of freeze desalination technologies is to change the phase of water from liquid to solid (ice crystals). The ice crystals do not include salts and are essentially pure water. Desalination by freezing consists of three discrete steps: ice formation, ice cleaning, and ice melting. Seawater cooling is supplied by means of refrigeration, either mechanical or thermal (absorption cooling). Once crystallization of the water has occurred, the ice crystals need to be separated from the saline solution and washed to produce pure water. The difficulties of freeze desalination include implementing the proper washing and separation of the crystals without premature melting and/or recontamination with salt.

Different configurations of freeze desalination systems that have been developed include direct freezing and indirect freezing. In the indirect freezing process, saline water does not come into contact with the refrigerant directly. Ice is formed on a surface by mechanical refrigeration or other methods. The indirect freezing process is seldom utilized since its energy requirements are relatively high due to the resistance of the surface between the saline water and refrigerant. Also, the equipment is complex, expensive, and difficult to operate and maintain. So the direct exchange of heat between the feed seawater and refrigerant is more favorable. Direct freezing includes vacuum freezing processes and secondary refrigerant freezing.

In general, freeze desalination has five basic operations: (i) precooling the feed stream; (ii) partial freezing of the feed stream; (iii) separation of the ice–brine mixture; (iv) melting the associated ice; and (v) heat rejection.

Figure 5.16 is a schematic of a direct refrigeration method which uses water as a refrigerant and mechanically compresses the resulting water vapor. This method is called the vacuum-freezing vapor-compression method [24]. In this process, the incoming saline water is cooled in a heat exchanger and then sprayed into a freezing chamber. The slurry of ice and brine is fed to the wash column, where the ice and brine are separated. The ice is then transferred to a melting unit. The water vapor originating in the freezing chamber is compressed (and thus heated) and discharged to the melting unit. The compressed water vapor transfers heat to the ice crystals within the melting unit where the ice is melted and the vapor is condensed to form the product water. An auxiliary refrigeration coil is necessary in the system to remove the equivalent of the mechanical energy supplied and heat influx due

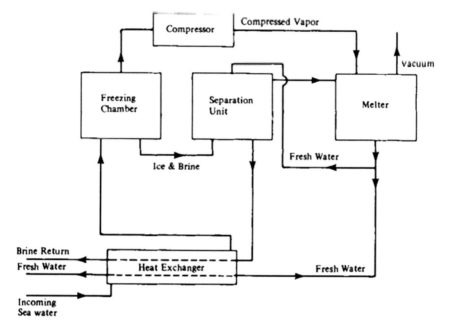

Figure 5.16   The vacuum-freeze vapor-compression method [24].

to heat leakage. Again, some of the product water from the melter is by-passed for washing. The main energy supplied is that required to drive the compressor while the heat discarded is extracted by means of the auxiliary refrigeration system (not shown). The system design is simple and requires minimum ancillary equipment. On the other hand, the compressor design is difficult, owing to the large specific volume of the water vapor at this low temperature.

Another interesting variation of the direct refrigeration method is the direct evaporation and condensation of an immiscible refrigerant in contact with the saline water [24]. This process is called the secondary refrigerant process. The refrigerant used must be practically insoluble in water. This method uses isobutane as the refrigerant and is schematically illustrated in Figure 5.17. In the freezing chamber, liquid isobutane is dispersed in the saline water by internal jet sprays. The liquid isobutane flashes into vapor, due to lower pressure, thus removing heat from the saline water and causing some of the water to freeze forming an ice-brine slurry. The slurry is led to the wash column, where the ice is separated and washed and then transferred to the melting unit. The isobutane vapor leaving the freezing chamber goes

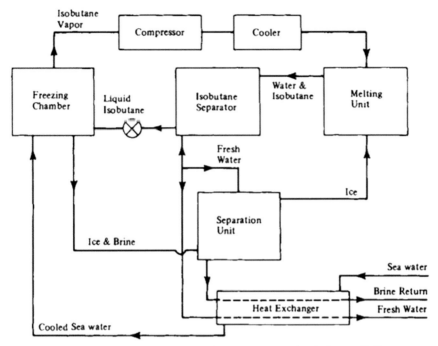

Figure 5.17 The secondary refrigerant method using isobutene [24].

to a compressor where it is compressed (and thus heated) and then to the melting unit. Here, the compressed isobutane vapor transfers heat to the ice, and a mixture of water and condensed isobutane results. The water and the immiscible refrigerant are then separated and the product water is discharged through the heat exchanger. The water necessary for counter washing is again by-passed from the main line. The main energy supplied in this system is that required to drive the isobutane compressor, and heat is discharged by the isobutane cooling coil to the surroundings.

A schematic of the indirect freezing process is shown in Figure 5.18 [25]. The feed seawater is first pumped through a heat exchanger to reduce its temperature, then sent into the freezing chamber, where it is cooled further to the temperature at which the ice crystals are formed. The ice and brine slurry is pumped to a wash column where the ice and brine are separated. The ice is transported to the melter, where ice is melted by the heat released from condensation of the compressed refrigerant. A small part of the product fresh-water is bypassed to the wash column and is used to wash the ice crystals and the major part is passed through the heat exchanger to cool the feed seawater.

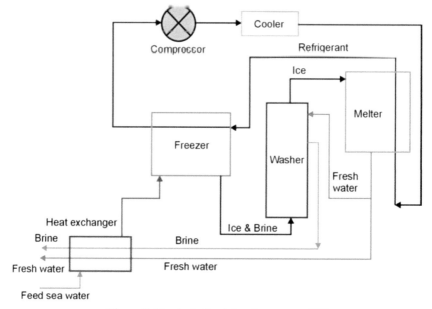

Figure 5.18   An indirect freezing process [25].

It is then discharged for storage. The brine from the wash column is returned to the heat exchanger to cool the feed seawater and is discarded.

The main advantages of freezing include [24]:

1. Lower heat exchange per unit of product because the latent heat of fusion is approximately one seventh the latent heat of evaporation;
2. The low freezing temperature increases the solubility of scale forming compounds commonly present in saline water;
3. The corrosion rate of metallic surfaces is greatly reduced due to the low temperatures; and
4. The low temperature allows the use of inexpensive plastics for many components.

The main disadvantages include:

1. The low freezing temperature requires operation under relatively high vacuum for some processes;
2. An external refrigeration system is required in some of the processes to extract the heat leakage and other energy inputs; and
3. The difficulty of separating the ice crystals from the brine and the related loss of product water during washing as compared to the separation of water vapor from a brine solution.

5.5.2 Solar Desalination

Solar energy is free of cost, and with clever ingenuity and design, it can be highly beneficial [26]. The utilization of solar energy in desalination is an old issue. For example, Della Porta, mentioned the use of solar energy for desalination in the early 17th century. The earliest solar powered desalination plant was established in 1872 in Chile. It produced about 14.8 m^3/day (3900 gpd) of desalinated water. One of the largest solar stills is installed in Gwadar, Pakistan, on an area of 9100 m^2 producing about 27 m^3/day (7120 gpd).

The amount of solar energy received varies from each location to the next depending on the distance and incidence angle that radiation had to travel until it reaches the receiver. The Middle Eastern countries are among parts of the world that receive high solar radiation for long durations in the year. The average daily radiation received in the region is about 5200 kcal/m^2. This is equal to an annual rate of 30 × 10^{15} kWh, more than six times the world's estimated oil reserves (1 MWh = 0.086 tons of oil = 0.112 barrels of oil).

Solar energy can be converted to heat at low temperatures by direct sorption or to heat at high temperatures by sorption after focusing. This heat can be used directly in desalination processes, by so called passive methods. Solar energy could also be converted into electrical energy by the use of heat engines. Electricity can be generated by thermoelectric or photovoltaic devices to be used in desalination, by so called active solar energy utilization.

Solar stills are the simplest form of solar desalination. They use solar energy directly to heat saline water up to its evaporation. The vapors formed are condensed and collected to obtain the product. Water in these stills evaporates at a temperature below its boiling points. Normally evaporation occurs at 50–60°C (122–140°F). The average still production rate is 3.3–4.1 l/m^2 (0.08–0.10 gal/ft^2)

Figure 5.19 shows the schematics of a double-slope symmetrical basin still (also known as a roof type or greenhouse type) [27]. First, water is directly heated by solar radiation in a closed enclosure covered with glazing/glass. The produced vapor, condenses on the cool glazing which is slightly inclined, and is collected in the gutters. In this figure the distribution of the solar energy falling on a basin still system is shown. T_a, T_b, T_g, and T_w in the figure are ambient temperature, basin temperature, glass temperature, and water temperature, and α'_b, α'_g, and α'_g are; the solar fluxes absorbed by the basin liner, glass cover, and the water mass, respectively. h_{wg}, h_{ga}, and h_{wb} are the heat transfer coefficients from the water surface to the glass, from the glass to the environment, and from the water to the basin liner, respectively. h_{ew} is the coefficient of heat loss by evaporation from the water surface,

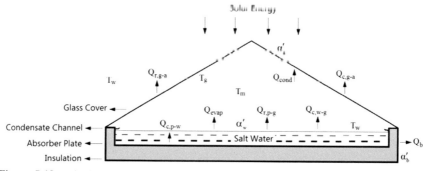

Figure 5.19 Distribution of the solar energy falling on a double slope symmetrical basin still [27].

P_g is the glass saturated partial pressure, and P_w is the water saturated partial pressure.

The following is a simplified mathematical model for the solar still [28]. The energy balance on the glass can be given as:

$$q_r + q_e + q_c = q_{c,ca} + q_{r,c-sky}$$

or

$$h_e \left(t_w - t_g\right) + h_r \left(t_w - t_g\right) + h_c \left(t_w - t_g\right) = h_{c,ca} \left(t_g - t_a\right)$$
$$+ h_{r,c-sky} \left(t_g - t_a\right)$$

Which can be shortened as

$$h_t \left(t_w - t_g\right) = h_o \left(t_g - t_a\right) \tag{5.1}$$

Where:

$$h_c = \frac{q_e}{\left(t_w - t_g\right)}$$

$$h_c = \frac{q_c}{\left(t_w - t_g\right)}$$

$$h_r = \frac{q_r}{\left(t_w - t_g\right)}$$

$$h_{c,ca} = \frac{q_{c,ca}}{\left(t_g - t_a\right)}$$

$$h_{r,c-sky} = \frac{q_{r,c-sky}}{\left(t_g - t_a\right)}$$

$$h_t = h_e + h_r + h_c$$
$$h_o = h_{c,ca} + h_{r,c-sky}$$

The energy balance on the water mass results in:

$$h_t (t_w - t_g) + k_b (t_w - t_a) = -MC_p \frac{dt_w}{dr}_\lambda \qquad (5.2)$$

Eliminating t_s out of Equations (5.1) and (5.2) one gets:

$$(t_w - t_a) \left(\frac{h_t h_o}{h_o + h_t} + K_b \right) = -MC_p \frac{dt_w}{dr} \qquad (5.3)$$

Writing an equation for the water mass balance, then:

$$\frac{h_e}{h_{fgw}} (t_w - t_g) = -dM/dr$$

The glass temperature may be eliminated by using the following relation derived from Equation (5.2):

$$(t_w - t_g) = \frac{h_o}{h_o + h_t} (t_w - t_a)$$

Hence,

$$\frac{h_e}{h_{fgw}} \left(\frac{h_o}{h_o + h_t} \right) (t_w - t_a) = -dM/dr \qquad (5.4)$$

Divide Equation (5.3) by Equation (5.4), to get:

$$\frac{dt_w}{dm/M} = \frac{h_t h_o + k_b(h_t + h_o)}{c_p h_e h_o} h_{fgw} = \lambda \qquad (5.5)$$

where λ = constant in the steady state.

The solution of Equation (5.5), subject to the boundary conditions:

$$t_w = t_{wo}; \; M = M_o$$

results in:

$$M = M_o - m_w = M_o e^{-\frac{1}{\lambda}(t_{wo} - t_w)}$$

which can be rearranged to:

$$\frac{m_w}{M_o} = 1 - e^{-\frac{1}{\lambda}(t_{wo} - t_w)} \qquad (5.6)$$

From Equation (5.6), it is clear that water distillate depends primarily on the initial water mass M_o (and hence the water depth, δ), initial water temperature (t_{wo}), and drop in water temperature $(t_{wo} - t_w)$.

Different solar stills have been designed to improve operating efficiencies. These designs have tried to achieve low heat capacity, low air content, vapor tight cover, water tight basin, and good insulation around the basin. Increasing the incident radiation and using multi-effect thin film diffusion stills will also improve the still's performance.

Direct solar thermal desalination has so far been limited to small-capacity units, which are appropriate in serving small communities in remote areas having scarce water. Solar-still design can generally be grouped into four categories: (i) basin still, (ii) tilted-wick solar still, (iii) multiple-tray tilted still, and (iv) concentrating mirror still. Figure 5.20 shows a variety of solar stills.

Indirect solar thermal desalination methods involve two separate systems: the collection of solar energy by a solar collecting system, coupled to a conventional desalination unit [29]. Processes include solar pond-assisted

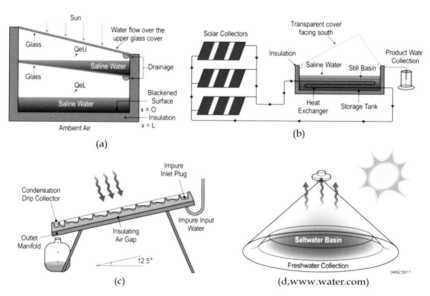

Figure 5.20   Diagrams of various solar stills: (a) double-basin solar still, (b) single-basin solar still coupled with flat plate collector, (c) multi-steps tilted solar still, and (d) micro-solar still [29].

desalination, and solar thermal systems such as solar collectors, evacuated-tube collectors, and concentrating collectors (CSP) driving conventional desalination processes such as MSF and MED.

Concentrating solar thermal power technologies are based on the concept of concentrating solar radiation to provide high-temperature heat for electricity generation within conventional power cycles using steam turbines, gas turbines, or Stirling and other types of engines [27]. For concentration, most systems use glass mirrors that continuously track the position of the sun. The four major concentrating solar power (CSP) technologies are parabolic trough, Fresnel mirror reflector, power tower, and dish/engine systems. Figure 5.21 shows diagrams of these systems.

Multistage evaporation and MSF can be operated indirectly by solar energy. Low- or high-temperature steam is generated by solar collectors of different forms. The steam is then used to provide heat for the distillation unit.

Flat plate collectors usually operate at a temperature below 100°F. While focused collectors can produce steam in the temperature range of 150–455°C (300–800°F).

Alternatively, the produced steam can be used to drive a turbine. Vapor compression, VC, or RO desalination plants can be run by this turbine. Flat plate collectors, solar ponds, or focused collectors can process heated fluid to drive a heat engine for an RO plant.

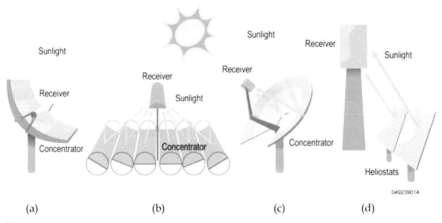

Figure 5.21 Solar concentrating systems, (a) parabolic trough, (b) Fresnel lenses, (c) dish engine, and (d) power tower [29].

Photovoltaic cells have been used since 1955 to convert light from radiation into electricity. They can also be used for desalination applications, in order to produce electricity to operate pumps for RO plants, or direct use in ED stacks.

However, the application of solar energy in desalination faces two important problems. These are the low efficiency of energy conversion and the inefficient storage of energy. So most desalination processes are limited in capacity and work during certain hours in the day time. Photovoltaic cells often have an energy conversion efficiency of 9–15% [30].

Solar-powered desalination systems do not differ much from the conventional desalination systems. They often consist of feed water pretreatment, solar collectors, electric power generation, a distillation or membrane unit with an energy storage subsystem, brine disposal subsystem, and product water storage and delivery subsystem. The basic criteria for choosing among the available subsystems is to obtain freshwater with the minimum possible cost.

The following solar desalination processes have been intensively studied:

1. Solar stills.
2. Reverse osmosis (RO) plants operated by electricity produced by photovoltaic cells.
3. Electrodialysis operated by electricity produced by photovoltaic cells.
4. Multiple effect evaporation plants operated by steam of high temperature produced by focused collectors.
5. Multistage flash plants operated by steam of high temperature produced by focused collectors.

Vapor compression and freezing are not often run with solar energy. They need auxiliary sources of energy. In fact, sometimes, their non-solar energy consumption is greater than the solar energy requirements [31].

5.5.3 Humidification–Dehumidification

The humidification–dehumidification (HDH) process is a type of thermal desalination system in which a heat recovery system operates with moist air. This process is also called humid air distillation. The HDH process is based on the fact that air can be mixed with large quantities of water vapor and the air carrying capability increases with increasing temperature. 1 kg of dry air saturated with vapor can carry an additional 0.26 kg of water vapor (about

208 kJ/kg) when its temperature increases from 30 to 80°C. The amount of vapor that air can hold depends on its temperature [29]. So the basic idea in the HDH process is to mix air with water vapor and then extract water from the humidified air with a condenser. For example, seawater is added into an air stream to increase its humidity. Upon cooling the humid air water vapor contained in the air is condensed and collected as freshwater. So moist air is used as a means of transporting water vapor from the humidifier (evaporator) to the dehumidifier (condenser). However, HDH processes have had very limited commercial and industrial success.

Some advantages of the HDH processes are: low-temperature operations, ability of combining with renewable energy sources such as solar energy, modest level of technology, and high productivity rates. Two different cycles are available for the HDH processes: HDH units based on the open-water closed-air cycle, and HDH units based on open-air closed-water cycle [27].

Figure 5.22(a) shows an open-water closed-air cycle. In the process, seawater enters the system, is heated in the solar collector, and is then sprayed into the air in the evaporator. Humidified air is circulated in the system and when it reaches the condenser, a certain amount of water vapor starts

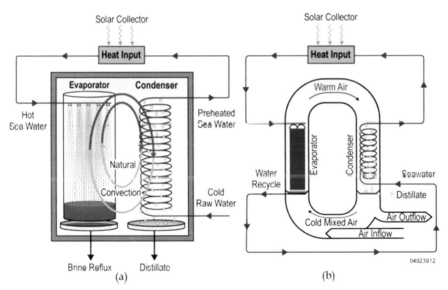

Figure 5.22 Humidification–dehumidification systems: (a) open-water closed-air cycle, and (b) open-air closed-water cycle [29].

to condense. Distilled water is collected in a container. Some of the brine can also be recycled in the system to improve the efficiency, and the rest is removed [32].

Figure 5.22(b) shows an open-air closed-water cycle, which is used to emphasize recycling the brine through the system to ensure a high utilization of the salt water for freshwater production. As air passes through the evaporator, it is humidified and by passing through the condenser, water vapor is extracted [33]. The simplest form of the HDH cycle is illustrated in Figure 5.23 [34]. Humidification–dehumidification processes have been modelled, for example by Huaigang and Shichang [35].

5.5.4 Membrane Distillation

Membrane distillation (MD) is a relatively newer membrane-thermal technology which was introduced in the late 1960s [36]. In the membrane distillation process, a porous membrane separates the hot saline water from the cold de-mineralized water, and a vapor pressure difference is formed across the membrane through which the water vapor passes from the saline side to the de-mineralized side.

Membrane distillation is characterized as a thermal-membrane separation process since it has vapor pressure—resulting from heating—as the driving

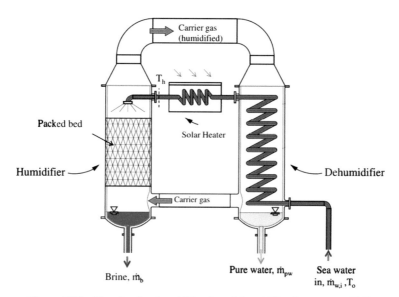

Figure 5.23   The simplest humidification–dehumidification process [34].

force for water vapor to pass through the pores of a hydrophobic membrane. The main difference between the membrane distillation process and other membrane separation processes is the driving force of mass transfer. The hydrophobic nature of the membrane prevents liquid solutions from entering its pores due to surface tension forces. As a result, liquid/vapor interfaces are formed at the entrance of the membrane pores.

Membranes used in the MD process require certain characteristics. They should be porous, and should not be wetted by the process liquid [37]. In addition to that, no capillary condensation should take place at the membrane and the vapor phase should be the only phase transported through. The membrane must not alter the vapor equilibrium of the different components in the process liquid. At least one side of the membrane will be in direct contact with the process liquid.

Membrane distillation systems can be classified into four configurations, according to the nature of the cold side of the membrane as follows [38]:

5.5.4.1 Direct Contact Membrane Distillation (DCMD)

An aqueous solution colder than the feed solution is maintained in direct contact with the permeate side of the membrane. Both the feed and permeate aqueous solutions are circulated tangentially to the membrane surfaces by means of circulating pumps or are stirred inside the membrane cell. In this case the trans-membrane temperature difference induces a vapor pressure difference. Consequently, volatile molecules evaporate at the hot liquid/vapor interface, cross the membrane pores in the vapor phase and condense in the cold liquid/vapor interface inside the membrane module. A configuration with a liquid gap (LGDCMD) is a variant, in which a stagnant cold liquid, frequently distilled water, is kept in direct contact with the permeate side of the membrane.

5.5.4.2 Air Gap Membrane Distillation (AGMD)

A stagnant air gap is interposed between the membrane and a condensation surface. In this case, the evaporated volatile molecules cross both the membrane pores and the air gap to finally condense over a cold surface inside the membrane module.

5.5.4.3 Vacuum Membrane Distillation (VMD)

A vacuum is applied in the permeate side of the membrane by means of a vacuum pump. The applied vacuum pressure is lower than the saturation pressure of volatile molecules to be separated from the feed solution. In this case, condensation takes place outside of the membrane module.

5.5.4.4 Sweep Gas Membrane Distillation (SGMD)

A cold inert gas sweeps the permeate side of the membrane carrying the vapor molecules and condensation takes place outside the membrane module. In this configuration, due to the heat transfer from the feed side through the membrane, the sweeping gas temperature in the permeate side increases considerably along the membrane length. A variant termed Thermostatic Sweeping Gas Membrane Distillation (TSGMD) has been proposed recently. In this mode, the increase in the gas temperature is minimized by using a cold wall in the permeate side.

Figure 5.24 shows various membrane distillation process configurations [3].

Because AGMD and DCMD do not need an external condenser, they are best suited for applications where water is the permeating flux. SGMD and VMD are typically used to remove volatile organic or dissolved gas from an aqueous solution. Of the four configurations, DCMD is the most popular for membrane distillation laboratory research, with more than half of the published references for membrane distillation based on DCMD [39].

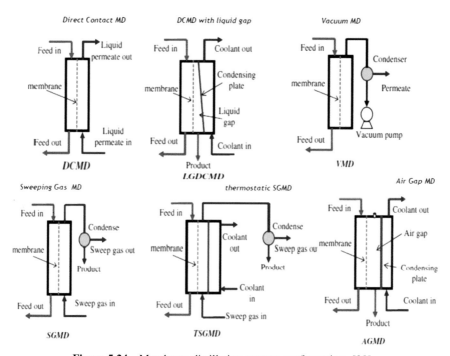

Figure 5.24   Membrane distillation process configurations [38].

However, AGMD is more popular in commercial applications, because of its high energy efficiency and capability for latent heat recovery.

Membrane distillation has two configurations that are the tubular module and plate and frame module. The tubular module consists of a hollow fiber membrane mainly prepared from Polypropylene (PP), Polyvinylidenefluoride (PVDF), and PVDF–PTFE (Polytetrafluoroethylene) composite material. The plate and frame model consists of a flat sheet membrane mainly prepared from PP, PTFE, and PVDF.

Since the original idea of membrane distillation was with the DCMD configuration, the following is a brief description of DCMD process [40]. Mass and heat transfer occur in the same direction from the hot side to the cold side. Figure 5.25 shows a simplified diagram of the DCMD process. Both the feed and permeate solutions are circulated tangentially to the membrane surfaces by means of circulating pumps and stirred inside the membrane cell using a magnetic stirrer. The temperature of the feed decreases along the feed side until it reaches the boundary layer and becomes the boundary temperature. Heat is conducted through the membrane to the cold side. In the permeate side, cold flow temperature increases across the boundary layer to the boundary temperature at the membrane surface. The driving force is the vapor pressure difference between the feed temperature at the membrane surface and the permeate temperature at the other membrane surface. This difference is less than the vapor pressure difference between the feed and permeate temperature. This phenomenon is called temperature polarization.

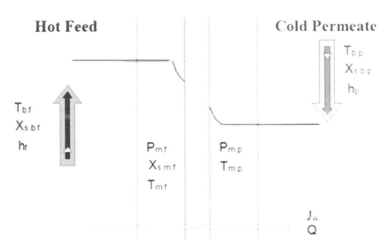

Figure 5.25 A simplified diagram of the DCMD process.

So, mass transfer takes place when the hot feed vaporizes from the liquid/gas interface, and the vapor is moved across the membrane pores from the hot side to the cold side by the vapor pressure difference. Eventually, the condensation of vapor takes place on the cold side (permeate). The controlling factors of mass transfer in the membrane distillation process are vapor pressure difference and permeability.

5.6 Operational Experience of Thermal Desalination Processes

The following is a brief overview of the operational experience of the thermal desalination processes. It focuses on MSF and MED–TVC processes since they have had commercial success. This section will review the pretreatment and scale control, efficiency of thermal desalination processes, design experience of large MSF and MED-TVC plants, impact of NCG and materials selection.

5.6.1 Pretreatment and Scale Control

Scale formation represents a major operational problem encountered in thermal desalination plants. There are two types of scales: alkaline scale and non-alkaline scale. Alkaline scale consists of calcium carbonate and magnesium hydroxide separately or in mixtures. Non-alkaline scale is mainly due to calcium sulphate.

The scale formed in evaporators running below 80°C, is primarily calcium carbonate. As heat is applied to the brine, bicarbonate alkalinity decomposes to form carbonate ions:

$$2HCO_3^- + heat \leftrightarrows H_2O + CO_2 + CO_3^{2-}$$

The carbonate ions then react with calcium in the seawater to form calcium carbonate

$$Ca^{2+} + CO_3^{2-} \leftrightarrows CaCO_3$$

The solubility of calcium carbonate $CaCO_3$ decreases as temperature rises, pressure is reduced, and CO_2 is released. At higher brine temperatures, the carbonate further decomposes to yield hydroxyl ions.

$$CO_3^{2-} + H_2O \leftrightarrows CO_2 + 2OH^-$$

The magnesium in the water can then react with the hydroxyl ions to form magnesium hydroxide precipitate.

$$Mg^{2+} + 2OH^- \leftrightarrows Ma(OH)_2$$

Formation of magnesium hydroxide usually proceeds slowly, but accelerates in the presence of nucleation sites. Figure 5.26 shows the percentage distribution of alkaline scale as a function of temperature [41].

For prediction of alkaline scaling tendency of seawater, various models and indices have been developed. As explained in Chapter 4, the most common index is the Langelier Saturation Index (LSI). Langelier [42] proposed a qualitative index indicating if the solution is under-saturated or supersaturated with $CaCO_3$. The Langelier Saturation Index is defined as:

$$LSI = pH - pH_S$$

with pH as the actual pH value and pH_S as the pH value of calcium carbonate saturation given by various estimation methods. For a more detailed discussion, refer to Chapter 4.

Another type of scale that may form in seawater distillers is calcium sulfate (non-alkaline scale). These types of deposits are rarely found if/when the desalination unit is operated well below the solubility limit of calcium sulfate as shown in Figure 5.27.

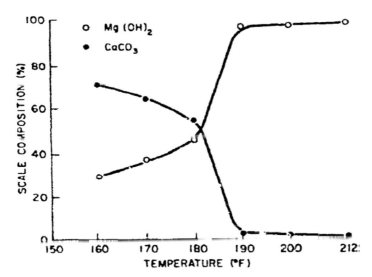

Figure 5.26 Percentage distribution of alkaline scale forms as a function of temperature [41].

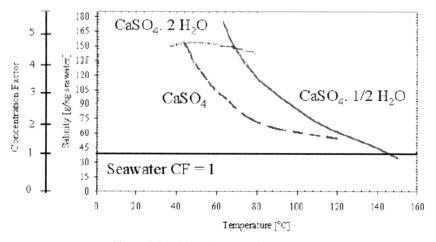

Figure 5.27   Phase diagram of CaSO₄ [42].

Operating lines are shown in terms of solubility limits of anhydrate calcium sulfate, di-hydrate calcium sulfate (gypsum), and hemi-hydrate calcium
sulfate. For seawater salinity of 36,000 ppm and inlet temperature of 25°C,
the maximum brine salinity for all effects should be kept at 52,000 ppm;
while the top brine temperature is equal to 65°C and the brine blowdown
temperature is equal to 38°C.

A rough estimate of calcium sulfate scaling can be found using the
Skillman sulfate solubility index [43]. The Skillman index is a ratio between
the actual concentration, S_{actual} (meq/L), of either calcium or sulfate and its
theoretical or equilibrium concentration whichever is the limiting species.
The limiting species is the one with the lower meq/L concentration:

$$\text{Skillman Index} = \frac{S_{actual}}{\left(\sqrt{x^2 + 4K_{sp}} - x\right) \times 10^3} \tag{5.7}$$

where x is the absolute value of the excess common-ion concentration of
calcium and sulfate ions (mg/L) which can be found from:

$$x = \left|2.5\left[Ca^{2+}\right] - 1.04\left[SO_4^{2-}\right]\right| \times 10^{-5} \tag{5.8}$$

The solubility product constant (K_{sp}) for calcium sulfate for any temperature
can be found from various referenced [44].

An important strategy to reduce the formation of inorganic scale relies on
the control of the operating conditions and on the addition of small quantities

of scale prevention additives (inhibitors). The top brine temperature is limited to 70°C, the concentration of the brine is restricted to a salinity of 70 g/kg, and the feed water mass flow rate per unit tube length (wetting rate) is set between about 0.032 and 0.14 kg/s m [45].

Decarbonation can be accomplished in a vacuum deaerator or an atmospheric decarbonator followed by a vacuum deaerator. Separate decarbonation with gas release to the atmosphere is generally preferable for large plants to reduce the size of the vacuum system and decrease corrosion in the vent system.

Feedwater is typically deaerated (the deaerator is often integrated in the evaporator) and treated with polyelectrolytes for scale and foam control, plus sodium bisulphate for scavenging oxygen (after the deaerator), and residual chlorine.

Scale in heat exchanger tubes is additionally controlled by on-line sponge ball cleaning systems. Antiscale additive dosing and ball cleaning systems ensure an operating period between acid cleaning of up to several years.

Addition of a threshold chemical such as polyphosphate retards alkaline scale formation by preferentially combining calcium and magnesium into soluble complexes. These compounds are usually added to the extent of 2–4 ppm to the seawater ahead of the deaerator and allow operation up to 190°F without severe scaling if properly controlled. Unfortunately, these chemicals are not wholly stable in water solutions and revert to sludge with time. Conventional acid cleaning techniques are used to remove the sludge fouling that forms.

Calcium sulfate scale can form in seawater evaporators when its inverse solubility limit is exceeded. This limits evaporator temperature and restricts the brine concentration factor in distillation plants. Three forms of calcium sulfate prevail. Precipitation temperature depends upon concentration and generally commences near 235°F for the anhydrite form at 1.0 concentration. The anhydrite is the most difficult of the three types to remove because it has no water hydration. It is usually removed by mechanical means, reaming, hydrojets, or special complexing chemicals.

The only proven method of control has been to limit the concentration and temperature before $CaSO_4$ becomes insoluble. The calcium may be removed from saline waters by chemical methods which leave the non-troublesome magnesium content relatively undisturbed. An ion exchange removal of the calcium can use the evaporator blowdown brine as a regenerant if there is sufficient difference in selectivity between the resin and the solution at

Table 5.2   Chemical cleaning agents

Category	Chemicals Commonly Used	Typical Target Contaminant(s)
Acid	Citric acid Hydrochloric acid (HCl)	Inorganic scale
Base	Caustic (NaOH)	Organics
Oxidants/disinfectants	Sodium hypocholorite (NaOCl) Hydrogen peroxide Chlorine gas (Cl_2)	Organics; biofilms
Surfactants	Various	Organics; inert particles

the concentrations of the incoming feed water and out-going brine solution. The various types of chemical cleaning agents used are summarized in Table 5.2.

5.6.2 Efficiency of Thermal Desalination Processes

In practice, the efficiency of thermal desalination plants is usually measured by either the gain output ratio (GOR) or performance ratio (PR).

The gain output ratio (GOR) is defined as the mass ratio of the produced water to that of steam consumed (both have the same units). The GOR does not account for the steam enthalpy difference across the desalting unit, the supply steam quality (temperature and pressure), or the pumping work. It is a measure of how much thermal energy is consumed in a desalination process:

$$ GOR = \frac{kg_{product}}{kg_{steam}} = \frac{lb_{product}}{lb_{steam}} $$

The GOR is the most commonly used criterion for performance evaluation of thermal desalination systems. It is applicable to thermal performance comparison of the desalination systems driven by the same heat source conditions. The value of the GOR generally ranges from 1 to 10. Lower values are typical of applications where there is a high availability of low-value thermal energy. Higher values, as high as 18, have been associated with situations where local energy prices are high, when the local value or need for water is high, or a combination of both.

In desalination, performance ratio is defined as the number of pounds of distillate produced per 1000 Btu of heat input. This translates into the number of kg of distillate produced by approximately 2330 kJ heat input. As 1000 Btu is approximately equal to the latent heat of condensation of one pound of water, this ratio is then very nearly equal to the ratio of the mass of water produced in the multi-effect plant per unit mass of saturated steam supplied to the first effect. In distillation plants driven by steam from a fossil fuel boiler, the amount of fuel consumed in the boiler is inversely proportional to the performance ratio. The higher the performance ratio the lower the amount of fuel consumed.

PR is sometimes preferred for the comparison of different evaporators because it is independent of the external steam conditions. The value of the GOR varies with the pressure of the steam.

The relation between PR and GOR can be found as

$$\text{PR} = \frac{\text{Distillate}}{\left(\frac{\text{Heat Input}}{2326}\right)} = \frac{2326}{\left(\frac{\text{Heat Input}}{\text{Distillate}}\right)} = \frac{2326\ \text{Distillate}}{\Delta h_\text{s}\ \text{Steam}} = \frac{2326\ \text{GOR}}{\Delta h_\text{s}}$$

(5.9)

where Δh_s is the enthalpy difference of the steam across the unit. When steam is supplied as saturated vapor and leaves as saturated liquid, then

$$\text{Heat Input} - \text{Steam}\ \lambda_\text{s}$$

(5.10)

Where λ_s is the steam latent heat, and the PR becomes

$$\text{PR} = \frac{2326}{\lambda_\text{s}} \frac{\text{Distillate}}{\text{Steam}} = \frac{2326}{\lambda_\text{s}} \text{GOR}$$

(5.11)

5.6.3 Design Experience of Large MSF and MED–TVC Plants

Most of the large MSF plants operate within the context of dual-purpose facilities for the simultaneous production of power and water. Such co-generation arrangement uses either backpressure or extraction condensing turbines. Co-generation cycles which were used till 1982 were employing extraction condensing turbines with a power-to-water ratio ranging between 10.2 and 17.5 MW/migd [15]. Typical power-to-water ratios for dual MSF desalination plants are shown in Table 3.3. The power-to-water ratio is defined as megawatt of power generated per million gallons/day of fresh water produced. The values in the table were estimated with knowledge of the achievable GOR in MSF desalination plants and the turbine efficiency in power plants. Values of GOR are often between 8 and 10 with a practical maximum of 12, which corresponds to about 55 kWh/m³ of thermal energy. As clearly shown in Table 5.3, a backpressure steam coupling system requires half the power plant capacity compared to an extraction steam coupling system.

Table 5.3   Typical power-to-water ratios for dual MSF desalination plants

Technology	Power-to-Water Ratio[a]
Steam turbine BTG	5
Steam turbine EST	10
GT-HRSG	8
Combined cycle BTG	16
Combined cycle EST	19

[a]Power-to-water ratio = MW power generated per mgd of water produced.

BTG, backpressure turbine generator; EST, extraction steam turbine; GT, gas turbine; HRSG = heat recovery steam generator.

From 1983 onwards backpressure turbines were used in all new co-generation plants. Backpressure turbines give lower power-to-water ratios (high water demand) and they are also characterized by high thermal efficiencies. They make the best use of low-grade heat that would otherwise be rejected by the power plant cycle.

Multi-stage flash desalination plants could be installed in single deck or double deck configurations. The tendency to install desalination plants in double deck has been abandoned as the maintenance of these units proved to be more difficult. On the other hand there are still several double deck distillers installed worldwide. Some facts and features of MSF include the following [46]:

- It is a major thermal desalination process—90% of all thermal production. Thus, it is among the most commonly used desalination technologies.
- It is the most robust of all desalination technologies.
- It can process water at a very high rate with relatively low maintenance.
- It is capable of very large yields. Plants with design capacities of 600,000–880,000 m^3/day are in operation in Saudi Arabia and the UAE.
- It operates using a cascade of chambers, or stages, each with successively lower temperature and pressure, to rapidly vaporize water, which is condensed afterward to form freshwater. The number of stages may be as high as 40.
- Multi-stage flash operates at top brine temperatures (TBT) of 90°C–120°C—the highest temperature to which the seawater is heated in the brine heater by the low-pressure steam in a cogeneration system. Higher temperatures than this lead to scaling, the precipitation, and formation of hard mineral deposits such as manganese oxides, aluminum hydroxide, and calcium carbonate.
- Its capital and energy costs are quite high.
- 25–50% recovery takes place in high-temperature recyclable MSF plant.

- It gives high-quality product water. The TDS of the product is less than 50 mg/L.
- Minimal pretreatment of feedwater required.
- Plant process and cost are independent of salinity level.
- MSF is an energy-intensive process.
- Multi-stage flash has a large footprint in terms of land and materials.
- Corrosion problems arise if materials of lesser quality are used.
- It has slow startup rates.
- Its maintenance requires shutdown of the entire plant.
- High level of technical knowledge required.

Multi-stage flash distillers are characterized by a wide range of design features and performance characteristics [47]. Distiller production capacity ranges from 2.5–10.0 migd. Performance ratios vary between 5.6 and 10.6 kg/2326 kJ (2.39–4.57 kg/1000kJ). All MSF distillers are operated with brine recirculation modes and the majority is with cross flow configuration. The number of stages varies from 15 to 40.

Multiple stage flash has a current modular capacity up to 90,000 m^3/d which can treat very salty water up to 70,000 mg/l. For example, the Ras Alkhair MSF plant has eight identical cross tube MSF evaporator with brine recirculation. At 100% of design rating, each desalination unit produces 92,582 m^3 of distillate per day. The evaporator in the Ras Al Khair plant measures 123 m long, 33.7 m wide, and weighs 4,150 tons. The product has less than 25 ppm TDS. MSF has high energy use: from 3 to 5 kWh/m^3 electricity and 233 to 258 MJ/m^3 heat required.

As an example of the design details of MSF evaporator tubes in different plant sections, Table 5.4 shows tube information in the Al-Jobail II MSF Plant. Table 5.5 summarizes the major design parameters of the two modes of MSF operation; namely, low temperature operation (LTO) at 90.6 °C and high temperature operation (HTO) at 112.8°C.

The MED–TVC design is capable of operation at higher temperatures to reduce the specific heat transfer area (process capital) and reduce the required

Table 5.4 Evaporator tube information in Al-Jobail II MSF plant

Item	Brine Heater	Recovery Sect.	Rejection Sect.
Tube material	66 Cu/30 Ni/2 Fe/2 Mn	90–10 Cu/Ni	Titanium
Tube wall thickness (mm)	1.245	1.245	0.711
Tube OD (mm)	39.0	39.0	29.0
Tube length (m)	14.3	19.9	19.9

Table 5.5   Operation design parameters of the Al-Jobail II MSF plant

Item	LTO	HTO
Max. Brine Temp. (°C)	90.6	112.8
Max. Brine Conc., (ppm)	64,900	61,800
Distillate Capacity, (m^3/hr)	985.0	1163.3
Performance Ratio, (kg/mJ)	3.44	4.09
Average Energy Consumption (kW)	3873.3	3729.2
Brine Velocity in Tube (m/s)	1.98	1.58
Recovery Fouling, m^2 (K/W)	0.000176	0.000146
Brine Heater Fouling, m^2 (K/W)	0.000176	0.000176
Scale Cont. Add. Dosing Rate (ppm)	5	7
Scale Control Additive Type	Polyphosphate	Polymers

compression ratio (specific power consumption). Increasing the compressor capacity is necessary to increase the unit capacity. It is possible to design compressor units with two stages (higher efficiency, more expensive) or units operating in parallel.

Depending on the arrangement of the effects, MED plants can be horizontal or vertical (multi-effect stack, MES). Higher capacity MED plants are generally horizontal because of their stability and their operational and maintenance simplicity. Vertical MED plants have lower capacities. They can be in a simple-stack arrangement, in which the evaporators are piled one on top of the other, or a double-stack arrangement, in which the effects are piled in two groups; for example, the effects 1, 3, 5, etc., are piled on top of each other in one group, while effects 2, 4, 6, etc., are piled on top of each other in another group, parallel to the first. The main difference between horizontal and vertical arrangements is that, in the latter, the brine flows under gravity from the effects at higher temperature towards the bottom with no additional pumping between stages. Morsy et al. [48] compared a horizontal and vertical MED plant and found that the heat transfer area in the horizontal configuration was roughly double that required by the vertical configuration.

An example of a recent MED-TVC plant is the Yanbu MED-TVC unit which was designed to produce 2,841 t/h, 45°C distillate with 293 t/h HP steam of 65 bar absolute and 525°C condition [49]. Received HP steam is reduced in the steam reducing station down to 17 bar absolute, 209.3°C, and it flows into the steam transformer. Entering LP steam is then condensed and returned back to the power station at the temperature of 85°C. In the meantime, the steam transformer receives latent heat from the LP steam and it

Table 5.6 Yanbu Project important design parameters [49]

Parameters	Flow (t/h)	Pressure (barA)	Temp. (°C)	TDS (ppm)	Enthalpy (kJ/kg)
HP steam	293.0	65.0	525.0	–	3,476.9
LP steam	371.5	17.0	209.3	–	2,809.4
Condensate Return	293.0	9.0	85.0	–	355.9
Seawater Intake	20,000.0	4.0	33.0	45,000	130.6
Cooling Water Discharge	10,055.0	3.0	43.0	45,000	170.2
Distillate Production	2,841.0	6.0	45.0	–	188.4
Brine Discharge	7,104.0	2.0	46.1	63,000	178.7

produces saturated process steam at the pressure of 15 bar absolute. Table 5.6 shows the Yanbu project's important design parameters.

This plant is designed to intake 33°C seawater with a top brine temperature of 63°C and the last effect vapor temperature of 45°C, returning the concentrate back to the sea at 43°C. When the seawater temperature increases to 35°C, 2°C of temperature shift in the evaporator occurs, so basically seawater consumption will remain the same. Figure 5.28 shows the overall mass and energy balances of the whole plant [49].

5.6.4 Impact of Non-Condensable Gases (NCG)

Non-condensable gases are essentially oxygen (O_2), nitrogen (N_2), argon (Ar), and carbon dioxide (CO_2). They form from the evaporating brine in

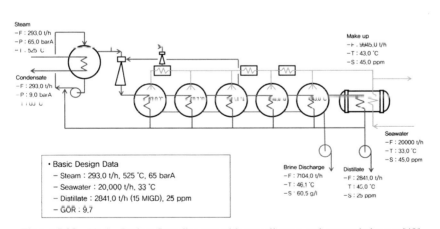

Figure 5.28 Yanbu Project flow diagram with overall mass and energy balances [49].

Table 5.7   Concentration of main gases dissolved in seawater [50]

Gas	Concentration mole/m^3	ppm
CO$_2$	0.005	0.22
O$_2$	0.240	7.70
N$_2$	0.450	12.6
Ar	0.010	0.40
HCO$_3^-$	3.062	187.1

desalination distillers and affect condensation, energy consumption, operation, and material lifetime of the distillers. The nature of MED operation causes the NCG concentration to increase upstream.

There are two sources of NCG in desalination systems. The first one is a simple penetration of gases through small holes and pores in the containers' flanges, connections, pipes, etc. This is the main source of air penetration. The second is based on gases dissolved in the feed water. Gases such as O$_2$, N$_2$, and CO$_2$ are dissolved in surface water. The first two have a low concentration in water, but CO$_2$ can have more significant concentrations when accounting for its role in the carbonate cycle. The dissolved gas content of sea water is given in Table 5.7 [50]. As water evaporates, the concentration of the dissolved gases increases above the saturation level, and they escape from the liquid phase. It is also assumed here that the moles of gases released in time ($n_{i,\text{gas}}$) are directly proportional to the concentration of the gas in the seawater and the amount of vapor being generated, $\dot{M}_v$.

$$n_{i.\text{gas}} = \frac{(\text{concentration of gas } i)}{\rho_{\text{sw}}} \dot{M}_v \, \Delta t \qquad (5.12)$$

Prevention of NCG in the vapor stream is associated with proper manufacturing, assembling and monitoring of all the outer surfaces to minimize leaks. Acidulation of the feed brine, followed by vacuum to remove most gases from the brine, may considerably reduce the problem.

The presence of the NCG not only impedes the heat transfer process but also reduces the temperature at which steam condenses at the given pressure. This occurs partially because of the reduced partial pressure of vapor in a film of poorly conducting gas at the interface. To help conserve steam economy, venting is usually cascaded from the steam chest of one preheater to the steam chest of the adjacent one. The effects that operate above atmospheric pressure are usually vented to the atmosphere. The NCG are always saturated with vapor. The vent for the last condenser must be connected to vacuum-producing equipment to compress the NCG to atmosphere. This is usually a steam jet ejector if high-pressure steam is available. Steam jet ejectors are relatively inexpensive but also quite inefficient. Since the vacuum is maintained on the last effect, the un-evaporated brine flows by itself from effect to effect and only a blow down pump is required on the last effect.

It was found that the CO_2 release rates decrease from the first effect to the last effect mainly due to the reduction in the evaporation temperature and pressure [53]. About 0.006% of CO_2 is released in MED distillers relative to the distillate production compared to about 0.020–0.026% in the MSF distillers [54]. The release rates in MSF distillers are much higher than MED because the top brine temperatures are much higher and the surface area of release is a lot more due to flashing and bubbling.

Semiat and Galperin [55] studied the influence of NCG on heat transfer in an MED seawater desalination plant. They showed that 1 w% NCG can decrease the heat transfer coefficient by 10%. Conversely, Wu and Vierow [56] concluded in their study that an air fraction up to 20 w% in air/steam mixtures impeded the overall heat transfer rate of a condenser tube only in a minor way.

Based on figures published by Al-Rawajfeh et al. [57] an estimate of the fraction of CO_2 in the flash vapor is 50 wppm. Depending on the evacuation scheme, NCGs can accumulate downstream of the evaporator. This may lead to NCG fractions higher than 1 w% in the second or third pass of downstream effects in large systems. Jernqvista et al also estimated the rate of non-condensable venting as a fixed amount equal to 0.5% of the vapor flow entering each effect [58]. Generally, in any thermal desalination plant, the volumetric concentration of NCG is about 4% [59].

Effects of the NCG include the following:

- Reducing the heat transfer coefficient because of their low thermal conductivity. Designers view this reduction as an additional thermal resistance to heat transfer. Therefore, additional heat transfer area is incorporated to handle the presence of NCG.
- Enhancing corrosion reactions due to the presence of O_2 and CO_2.
- Reducing the partial pressure of the condensing vapor and hence the condensation temperature.
- In the MSF process, gas accumulation beyond the design specifications would reduce the brine recycle temperature entering the brine heater. This will require use of a larger amount of heating steam, which will reduce the process's thermal performance ratio.

- In the MVC process, the presence of NCG would affect the performance of the compressor. Their presence would reduce the amount of compressed vapor and consequently the thermal load would decrease as well as the amount of product water.

5.6.5 Material Selection

The selection of equipment for a given desalination project is based on the type of the treatment process for which the equipment is intended, its efficiency in terms of energy use, its cost, its ease of operation and maintenance, the size and capacity of the individual equipment units available on the market, and its useful life and track record for similar applications. Another important factor for equipment selection is the quality of the materials from which the equipment is built and their suitability for the ambient environment to which the equipment is exposed.

The first generation of MSF plants largely employed carbon steel as a material for construction. The life of the equipment was planned for a maximum of 15 years. The operational experience with these plants has revealed that the industrial life can be substantially longer than originally anticipated and several rehabilitation and refurbishment projects have been launched to extend the original life for 20 years and beyond.

Developments in material technology and a more in-depth knowledge of the corrosion aspects in seawater have resulted in the economical adoption of nobler materials. It is expected that the second generation of large MSF desalination plants installed in the last 10 years will last for more than 40 years with minimum maintenance.

Improved-performance construction materials are used in place of conventional materials. For example, duplex stainless steel has replaced 316L stainless steel. Duplex stainless steel provides higher corrosion resistance and has a longer service life. Similarly, titanium tubing provides superior performance compared to copper-nickel tubing, in low temperature sections of MSF and MED.

The following is a list of typical materials used in MSF plants [13]:

- Titanium is used for evaporator tubes at temperatures below 80°C with small wall thickness of 0.5 mm.
- Cu/Ni 70/30 with a wall thickness of 1–1.2 mm is used to construct tubes at operating temperatures above 80°C. Its main disadvantage results from contamination of the rejected brine with the copper element, which has effects on the environment.

- Cu/Ni 90/10 with 1 mm wall thickness is used to construct tubes in lower temperature evaporators. Also, it is used for cladding carbon steel in water boxes as well as partition walls and floors in evaporators.
- Aluminum brass tubing for evaporators operating below 80°C.
- Stainless steel, SS 316L, is used to manufacture partition walls, distillate trays, and evaporator internals (limited to 80°C, pitting and crevice corrosion).
- Carbon steel is used for construction of steam and condensate piping. Cladding materials such as stainless steel, CuNi 90/10, or polyurethane are used together with carbon steel to construct evaporator partition walls and floors, dearator shells, water boxes, and seawater, distillate and brine piping.

A new generation of stainless steel[2], austenitic-ferritic or duplex[3] grades, is entering the desalination sector due to a number of reasons. The duplex stainless steels have good resistance to corrosion, especially to stress corrosion cracking (SCC). They have twice the strength of austenitic grades, they are less costly due to lower contents of mainly nickel and molybdenum, and they are excellent engineering materials. The reason for not being so common in the past is that the modern versions, with good engineering properties, did not even exist when stainless steel was introduced to the desalination industry, i.e. during the 1970s.

5.6.6 Maintenance Procedures

To achieve a desired high level of plant availability and life, it is imperative that a maintenance program be included in the initial planning stage and adhered to during operation. Periodic maintenance will ensure long and

---

[2]Stainless steel is an alloy of iron with a minimum of 10.5% chromium. Chromium produces a thin layer of oxide on the surface of the steel known as the 'passive layer'. This prevents any further corrosion of the surface. Increasing the amount of chromium gives an increased resistance to corrosion. Stainless steel also contains varying amounts of carbon, silicon and manganese. Other elements such as nickel and molybdenum may be added to impart other useful properties such as enhanced formability and increased corrosion resistance.

[3]Duplex steels have a microstructure which is approximately 50% ferritic and 50% austenitic. This gives them a higher strength than either ferritic or austenitic steels. They are resistant to stress corrosion cracking. So called "lean duplex" steels (LDX) are formulated to have comparable corrosion resistance to standard austenitic steels but with enhanced strength and resistance to stress corrosion cracking. "Superduplex" steels have enhanced strength and resistance to all forms of corrosion compared to standard austenitic steels. They are weldable and have moderate formability. They are magnetic but not as much as the ferritic, martensitic and precipitation hardening (PH) grades due to the 50% austenitic phase.

reliable system operation. Maintenance also helps in keeping steady operation of the process, and parameters at the desired values. For example, keeping desired operating temperature levels with a minimum of make-up heat, requires primarily the minimization of venting losses, the maintenance of high heat transfer coefficients in the feed preheaters, and the proper balancing of seawater flow through the product and blowdown coolers.

During maintenance, the flashing chambers in MSF or the evaporation effects in MED are opened to ambient air. Therefore, air removal is one of the main activities in the startup procedure. The startup ejectors have higher capacity and are capable of processing large air volumes over a short period of time.

Usually maintenance involves both scheduled and preventive maintenance programs. Preventative maintenance programs with proper documentation can improve a thermal desalination plant's availability significantly, as well as reduce the cost of replacing defective equipment on a sudden shutdown. The following significant maintenance activities are normally covered in thermal desalination plants:

- Inspection and cleaning of brine heater water boxes
- Pressure test of the brine heater
- Detection and plugging/replacement of faulty tubes
- Examining of the side walls of the water boxes and tubes for hard scale precipitation
- Removal of any debris and/or foreign materials from the water boxes

The periodic maintenance inspections and checks are classified into the following periods (time intervals):

1. Every Day
2. Every 6 Months
3. Every Year
4. Every 2 Years

The listing is obtained from manufacturers, as well as from the recorded experience with similar past applications. The following is an example of possible listings.

1. Every Day

Brine Heater

- Check for leakage through the flange joints and any other connections.
- Check the condensate level.
- Inspect the tube bundles

Condensate Extraction Pump

- Check for noise and vibration.
- Check that the pump/motor is operating at the correct load by checking the current absorbed by the motor, the pump suction/discharge pressure and pump speed.
- Check the oil level.
- Check the bearing temperature.

Condensate Extraction Pump, De-Superheater

- Check the oil level.
- Check the bearing temperature.

2. Every 6 months

- Check the alignment of the pump and motor and adjust with shim plates, if necessary.
- If misalignment occurs frequently, inspect the discharge flange connection, piping supports and confirm effective support of load.
- Check mechanical seal for excessive leakage.
- Check oil level.
- Dismantle and check for damaged/worn gear teeth and O-ring. Renew if necessary.
- Check oil for deterioration. Exchange or supplement, if necessary.

3. Every Year

Brine heater

- Inspect water box for corrosion. Repair if necessary.
- Inspect tube bundles and plug or replace any faulty tube.
- Check gaskets for cracking or tears. Renew if thickness is less than 80% of original gasket.

Pump

- Check shaft casing, wear ring, impeller and shaft for general function, tightness, damage or wear.
- Check impeller, suction part and bearing spider for blockage and clean.

De-Superheater

- Check nozzle and spring.
- Replace actuator valve stem packing.
- Inspect main valve diaphragm and replace, if necessary.
- Check valve fittings and clean.

4. Every Two Years

Pump

- Remove the rotating element and inspect it thoroughly for wear. Replace the parts as required.
- Check clearance of the impeller.
- Remove any deposit of scaling.
- Clean up the scaling water line.
- Check the bearing and replace, if necessary.
- Check the wear ring and replace, if necessary.
- Replace the O-ring.
- Replace the gasket.
- Check the oil seal and replace if necessary.

According to Khan [60], a typical maintenance outline for an MSF plant is shown in Table 5.8.

Table 5.8 Typical maintenance outline for an MSF plant [60]

Unit/Internals	Time Intervals
Evaporators	
Inter surface painting	Yearly
Condenser tubes	Yearly
Demisters	Yearly as required
Brine level indicator	Monthly/Constantly
Distillate level indicator	
Brine Heater	
Condensate level indicator	Monthly/Constantly
Condenser tubes	Yearly as required
Air Ejectors and Condensers	
Steam strainer	Monthly
Condenser tubes	Yearly
Throat nozzles	Yearly
Condensate level indicator	Monthly/Constantly
Vent Condenser	
Condenser tubes	Yearly
Condensate level indicator	Monthly
Seawater Supply Pump	
Strainer	Weekly
Coupling	Monthly
Lube oil	Quarterly checking/yearly changing
Gland packing	Constantly
Overhaul	Yearly

Table 5.8   Continued

Pumps (brine circulation, distillate, seawater recirculation, brine heater, condensate and ejector condensate)	
Coupling	Monthly
Lube oil	Quarterly checking/yearly changing
Gland packing	Constantly
Overhaul	Yearly
Turbine for Brine Recirculation Pump	
Lube oil/grease	Monthly checking/yearly changing
Overhaul	Yearly
Tripping system	Yearly
Lube oil/ pump	Yearly
Steam Reducing Station	
Strainer	Monthly/yearly
Gland packing	Constantly/yearly
Chemical Dosing (polyphosphate)	
Strainer	Daily
Agitator	Yearly
Pumps	Yearly
Pumps (Lube oil/grease)	Regularly (change biannually)
Caustic Soda Dosing	
Strainer	Weekly
Agitator	Yearly
Pumps	Yearly
Pumps (Lub oil/grease)	Regularly (change biannually)
Acid Cleaning	
Pumps	Yearly
Strainer	Yearly
Tube Cleaning	
Strainer	Monthly
Recirculating pump	Monthly
Balls	Weekly
High and Low Voltage Motors	
Insulation resistance	Yearly
Stator	Yearly
Rotor	Yearly
Air gap measurement	Yearly
Bearings	Yearly
Vibration	Yearly
Control Valves	
Calibration and adjustment	Yearly
Air leakage	Yearly
Lubrication	Yearly
Vibration	Yearly

Also Khan [60] recommended that some important physical (flow, pressure, level, vibration, noise, wall thickness) or chemical characteristics (TDS, Fe, CO_2, O_2, H_2S content) can be monitored and the data analyzed on a regular basis. Monitoring and analysis costs can be high, but the costs can be offset by the improved reliability.

5.6.7 Evaporator Start-Up

The startup of a desalination unit is a complex operation [61]. It is the hardest part in running the plant far beyond keeping constant operation or even adjusting the operation mode. Evaporator start-up procedure is based on predefined preparation steps prior to beginning the start-up procedures. The startup procedure will be conducted by a series of automatic or manual steps and actions depending on the control and instrumentation system of the plant. Figure 5.29 shows the units involved in the general start up sequence [61].

A list of minimum requirements that shall be met prior to start-up is as follows [61]:

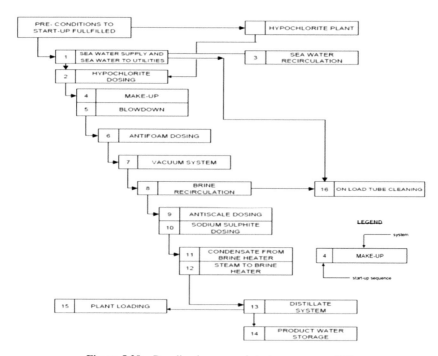

Figure 5.29   Desalination general start up sequence [61].

- All instruments must be put in service (drained, vented, etc.) and visually inspected.
- All compressed air supply valves must be opened and visually inspected.
- All water pipes must be filled up and vented.
- Evaporator and brine heater tube bundles must be filled and vented.
- Pump suction lines up to the first delivery shut off valve must be filled and vented.
- Brine heater shell must be filled with water up to a nominal level.
- All manual valves must be open or closed according to their duty.
- Steam pipework must be heated-up.
- Vacuum must be raised in the evaporator and deaerator chamber by the hogging ejector and phase ejectors must be put in operation.

General guidelines to startup of a desalination plant based on MED technology is given by Manent et al. [62]. During the start-up of the MED unit, non-condensable gases are evacuated from the evaporator effects, final condenser, steam transformer and deaerator by the startup ejector. When starting from atmospheric temperature and pressure, the start-up ejector is utilized to evacuate the system piping, the evaporator, the final condenser and the steam transformer to achieve vacuum conditions quickly.

References

[1] Al-Shammiri, M., and Safar, M. (1999). Multi-effect distillation plants: state of the art. *Desalination* 126, 45.
[2] Ophir, A., and Lokiec, F. (2005). Advanced MED process for most economical seawater desalination. *Desalination* 182, 187.
[3] Khawaji, A. D., Kutubkhanah, I. K., and Wie, J. (2008). Advances in seawater desalination technologies. *Desalination*, 221, 47.
[4] Al-Mutaz, I. S. (2015). Features of multi-effect evaporation desalination plants, *Desalination Water Treat.* 54, 3227–3235.
[5] Watson, I. C. Jr., Morin, O. J., and Henthorne, L. (2003). *Desalting Handbook for Planners*, 3rd Edn. Washington, DC: United States Department of the Interior, Bureau of Reclamation, Technical Service Center.
[6] Al-Mutaz, I. S., and Wazeer, I. (2014). Comparative performance evaluation of conventional multi-effect evaporation desalination processes. *Appl. Ther. Eng.* 73, 1194–1203.

[7] Ameri, M., Mohammadi, S. S., Hosseini, M., and Seifi, M. (2009). Effect of design parameters on multi-effect desalination system specifications. *Desalination* 245, 266–283.

[8] El-Dessouky, H. T., and Ettouney, H. M. (1999). Multiple-effect evaporation desalination systems. thermal analysis. *Desalination* 125, 259–276.

[9] Al-Mutaz, I. S., and Wazeer, I. (2015). Current status and future directions of MED-TVC desalination technology. *Desalination Water Treat.* 55, 1–9.

[10] Al-Mutaz, I. S., and Wazeer, I. (2014). Development of a steady-state mathematical model for MEE-TVC desalination plants. *Desalination* 351, 9–18.

[11] Al-Wazzan, Y., and Al-Modab, F. (2001). Seawater desalination in Kuwait using multistage flash evaporation technology—historical overview. *Desalination* 134, 257–267.

[12] Silver, R. S. (1970). "Multi-stage flash distillation-The first 10 years", in *Proceedings of the 3rd International Symposium On Fresh Water from the Sea,* 191–206. Athens.

[13] Ettouney, H. M., and El-Dessouky, H. (2002). *Fundamentals of Salt Water Desalination*. Amsterdam: Elsevier.

[14] Sommariva, C. (2010). *Desalination and Advance Water Treatment Economics and Financing*. Rehovot: Balaban Publishers.

[15] Al-Mutaz, I. S., and Al-Namlah, A. M. (2004). Characteristics of dual purpose MSF desalination plants. *Desalination* 166, 287–294.

[16] Al-Mutaz, I. S., and Al-Namlah, A. M. (2005). "Optimization of operating parameters of msf desalination plants," in *Proceedings of the IDA World Congress on Desalination and Water*, Singapore.

[17] Spiegler, K. S. (1977). *Salt Water Purification*, 2nd Edn. New York, NY: John Wiley and Sons.

[18] Kucera, B. (2004). The use of distillation technology in the bottled water industry. *Water Condition. Purific.* 43–44.

[19] Howe, E. D. (1974). *Fundamentals of Water Desalination*. New York: Marcel Dekker, Inc. 344.

[20] Aly, N. H., and El-Fiqi, A. K. (2003). Mechanical vapor compression desalination systems-a case study. *Desalination* 158, 143–150.

[21] Saidura, R., Elcevvadia, E. T., Mekhilefb, S., Safarib, A., and Mohammedc, H. A. (2011). An overview of different distillation methods for small scale applications. *Renew. Sustain. Energy Rev.* 15, 4756–4764.

[22] Ettouney, H. (2006). Design of single-effect mechanical vapor compression, *Desalination* 190, 1–15.

[23] El-Dessouky, H., and Ettouney, H. (1999). Single-effect thermal vapor-compression desalination process: thermal analysis. *Heat Trans. Eng.* 20, 52–68.

[24] Tleimat, B. W. (1980). "Freezing methods," in *Principles of Desalination*, Chap. 7, 2nd Edn, eds K. S. Spiegler, and A. D. K. Laird (New York: Academic Press), 359–400.

[25] Lu, Z., and Xu, L. (2002). Freezing desalination process. *Desalination Water Resour. Ther. Desalination Process.* 2, 275–290.

[26] Al-Mutaz, I. S., and Al-Ahmed, M. I. (1989). Evaluation of solar powered desalination processes. *Desalination* 73, 181–190.

[27] Deniz, E. (2015). "Solar-powered desalination", in *Desalination Updates* ed. R. Y. Ning ().

[28] Al-Mutaz, I. S. (1989). "Techno-economic study of solar desalination," in *Proceedings of the Fifth Scientific Conference of the Scientific Research Council*, Baghdad.

[29] Al-Karaghouli, A. A., and Kazmerski, L. L. (2011). "Renewable Energy Opportunities in Water Desalination," in *Desalination, Trends and Technologies*, ed. M. Schorr (Rijeka: InTech Open).

[30] Lior, N. (1981). "Principles of desalination," in *Proceeding of the Second SOLERAS Workshop*, Denver, CO.

[31] Khoshain, B. H. (1985). 200 m^3/day solar sea water desalination pilot plant. *Solar Wind Technol.* 2: 173.

[32] Mathioulakis, E., Belessiotis, V., and Delyannis, E. (2007). Desalination by using alternative energy: review and state-of-the-art. *Desalination* 203, 346–365.

[33] Bohner, A. (1989). Solar desalination with a higher efficiency multi effect process offers new facilities. *Desalination* 73, 197–203.

[34] Narayan, G. P. (2011). "Status of Humidification Dehumidification Desalination," *Proceedings of the IDA World Congress on Desalination and Water Reuse*, Perth, Australia.

[35] Huaigang, C., and Shichang, W. (2007). Modelling and experimental investigation of humidification-dehumidification desalination using a carbon-filled-plastic shell-tube column. *Chin. J. Chem. Eng.* 15, 478–485.

[36] Alklaibi, M., and Lior, N. (2005). Membrane-distillation desalination: status and potential. *Desalination* 171, 111–131.

[37] Camacho, L. M. (2013). Advances in membrane distillation for water desalination and purification applications. *Water* 5, 94–196.

[38] Khayet, M., and Matsuura, T. (2011). *Membrane Distillation: Principles and Applications.* Amsterdam: Elsevier.

[39] Curcio, E., and Drioli, E. (2005). Membrane distillation and related operations: a review. *Sep. Purif. Rev.* 34, 35–86.

[40] Al-Mutaz, I. S., Al-Motek, A. S., and Wazeer, I. (2016). Variation of distillate flux in direct contact membrane distillation for water desalination. *Desalination Water Treat. J.* 62, 86–93.

[41] Dqoly, R., and Glater, J. (1972). Alkaline scale formation in boiling sea water brines. *Desalination* 11, 1–16.

[42] Langelier, W. F. (1936). The analytical control of anti-corrosion water treatment. *J. Amer. Water Works Assoc*, 28, 1500–1521.

[43] Skillman, H. L., McDonald, J. P. Jr., and Stiff, H. A. Jr. (1969). A simple, accurate, fast method for calculating calcium sulfate solubility in oil field brine, *Paper No. 906-14-I, Spring Meeting of the Southwestern District*, Division of Production, American Petroleum Institute, Lubbock, TX.

[44] Linke, W. F. A. (1965). *Solubilities of Inorganic and Metal- Organic Compounds*, 4th Edn. New York, NY: Van Nostrand-Reinhold.

[45] Desportes, C. (2005). "MED desalination and scale control," in *Proceeding of the DME Seminar, Fouling und Scaling in der Meerwasserentsalzung*, Ludwigshafen.

[46] Thye, J. F. (2010). Desalination: can it be greenhouse gas free and cost competitive? Report of MEM Masters Project. Haven, CT: Yale School of Forestry and Environmental Studies.

[47] Hamed, O. A. (2001). "Overview of design features and performance characteristics of major saline water conversion corporation (SWCC) MSF plants," in *Proceedings of the WSTA 5th Gulf Water Conference*, Doha.

[48] Morsy, H., Larger, D., and Genthner, K. (1994). A new multiple-effect distiller system with compact heat exchangers. *Desalination* 96, 59–70.

[49] Dossan, Yanbu Phase 2 MED Expansion, "Mass Balance Calculation Report", QC20-J-031.

[50] Seifert, A., and Genthner, K. (1991). A model for stagewise calculation of non-condensable gases in multi-stage evaporators. *Desalination*, 81, 333–347.

[51] http://en.wikipedia.org/wiki/Henry's_law

[52] Glade, H., and Genthner, K. (). "Analysis, approaches, and models for the release of carbon dioxide, nitrogen, oxygen, and argon in evaporators," in *Common Fundamentals and Unit Operations in Thermal Desalination Systems*, Vol. 1. (Bremen: University of Bremen).

[53] Al-Rawajfeh, A. E. (2010). $CaCO_3$-CO_2-H_2O system in falling film on a bank of horizontal tubes: model verification. *J. Ind. Eng. Chem.* 16, 1050–1058.

[54] Schausberger, P., Nowak, J., and Medek, O. (2009). "Heat transfer in horizontal falling film evaporators," in *Proceedings of the IDA World Congress, Atlantis*, The Palm.

[55] Semiat, R., and Galperin, Y. (2001). Effect of non-condensable gases on heat transfer in the tower MED seawater desalination plant. *Desalination* 140, 27–46.

[56] Wu, T., and Vierow, K. (2006). Local heat transfer measurements of steam/air mixtures in horizontal condenser tubes. *Int. J. Heat Mass Trans.* 49, 2491–2501.

[57] Al-Rawajfeh, A. E., Glade, H., Qiblawey, H. M., and Ulrich, J. (2004). Simulation of CO_2 release in multiple effect distillers. *Desalination* 166, 41–52.

[58] Jernqvista, A., Jernqvista, M., and Alyb, G. (2001). Simulation of thermal desalination processes. *Desalination* 134, 187–193.

[59] Office of Saline Water (OSW) (1970). *Distillation Digest*. Research and Development Progress Report No. 538. Washington, DC: United States Department of the Interior.

[60] Khan, A. H. (1986). *Desalination Processes and Multistage Flash Distillation Practice*, Amsterdam: Elsevier Science Publisher B.V.

[61] Ghiazza, E., and Chiola, G. (1999). "MSF evaporator start up; from manual to fully automatic operation," in *Proceedings of the IDA World Congress Desalination and Water Reuse*, San Diego, CA.

[62] Manenti, F., Masi, M., and Santucci, G. (2013). Start-up operations of MED desalination plants. *Desalination* 329, 57–61.

CHAPTER 6

Membrane Desalination Technologies

Dongxu Yan[1]

[1]Senior engineer, Layne Christensen Company,
1138 North Alma School Road, Suite 207, Mesa, Arizona, USA 85201
Phone: 602-345-8550
Fax: 602-345-8632
E-mail: Dongxu.Yan@layne.com

6.1 Introduction to Membrane Desalination Technologies

Membrane separation technology is one of the two major types of technologies that are widely used globally for desalination. The other major type is thermal-based technologies, which has been discussed in the previous chapter. As discussed in Chapter 3, for years membrane technologies required more energy than thermal methods. In addition, thermal methods showed a high degree of reliability and guaranteed performance under varied conditions. Hence, traditionally, the majority of desalination plants around the world used thermal technologies. But with the boom of membrane technologies [in particular reverse osmosis (RO)] during the past decade, including their increased reliability and the dramatic fall of their energy requirement, they have overtaken the market. Today, nearly 70% of the world's 90 million m^3/day desalination capacity uses membrane processes.

Membrane desalination technologies involve using semipermeable membranes and a driving force to separate salts from water. The driving force can be differential osmotic pressure, differential voltage, or differential temperature.

6.2 Reverse Osmosis

6.2.1 Introduction to RO Membrane Technology

Reverse osmosis membranes are tight membranes that do not have definable pores. In the research community, some researchers classify Nanofiltration (NF) as a loose RO membrane that allows some ions to pass through, but others may refer to NF as a separate category. Both RO and NF utilize semi-permeable membranes which give them the ability to remove dissolved contaminants from water. They can hence be used in applications like softening or desalination of water.

Nano-filtration and RO are both pressure-driven process and can reject dissolved substances in water, and NF differs from RO only in terms of its lower rejection efficiencies of some dissolved ions especially monovalent ions like Na^+ and Cl^-. Take Cl^- for example, RO membranes can achieve over 99% rejection of Cl^-, while NF membranes achieve from 10 to 40% depending on the tightness of the membrane. But NF membranes can have a good rejection characteristic for removal of divalent ions and higher-molecule-weight organics that contribute to odor and taste of water, and rejection can be higher than 85%. This property of NF membranes makes them useful in removal of hardness ions like Ca^{2+}, Ba^{2+}, Mg^{2+} and $SO_4{}^{2-}$ from water at much lower driving pressures than required when using RO membranes. As a consequence, NF is sometimes also referred to as "membrane softening".

The molecular weight cut-off (MWCO) is a measure of the removal characteristics of a membrane, expressed in terms of Daltons (Da). RO membranes have a typical range of less than 100 Das, while for NF membranes, the MWCO range is approximately between 200 and 1000 Das. NF/RO membranes are designed to separate dissolved contaminants from water through diffusion rather than a sieving mechanism. When saltwater comes into contact with one side of the membrane under pressure, it in fact dissolves into the membrane material. This is followed by water desorption on the permeate side where the chemical potential is low. Due to this diffusion mechanism, the solubility of the species in the membrane material is of utmost importance in the rejection efficiency. For example, even though lithium ions (6.9 Da) are smaller, they are much more easily rejected by an RO membrane than ethanol (46.1 Da). This is mainly because charged lithium ions experience high repulsion by the membrane, whereas the uncharged ethanol molecules do not experience high repulsions when dissolving in the membrane material. Likewise, divalent ions experience stronger destabilization (compared to monovalent ions) and are hence more easily rejected [1]. In effect, many factors such as the (solute) concentration, composition, molecule shape and size, and operating conditions such as temperature, pressure, and cross-flow velocity affect the passage of solutes.

Osmosis is the natural flow of a solvent like water through a semi-permeable membrane due to the existence of a differential concentration

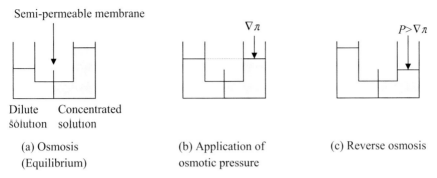

Figure 6.1   Conceptual diagram of osmosis, osmotic pressure and reverse osmosis (RO).

of a solute or solutes on the two sides. The flow direction is from the less concentrated solution to the higher concentrated solution and the flow stops when the chemical potentials on both sides of the semi-permeable membrane are equalized. To bring the columns on both sides to equal height, a pressure must be applied on the more concentrated solution side; the pressure applied is called osmotic pressure. If a pressure higher than the osmotic pressure is applied on the more concentrated solution side, the solvent will continue to flow from the more concentrated side to the less concentrated side, which is reverse to osmosis. Hence, this process is called reverse osmosis (Figure 6.1).

The value of osmotic pressure ($\nabla \pi$) which is a function of the concentration and temperature of the solution can be calculated by the Van 't Hoff equation, which will be discussed in the later section. The solvent that passes through from the more concentrated side to the less concentrated side of the membrane is called the permeate or filtrate, and the solution rejected by the membrane is referred to as the concentrate. In the case of desalination, the permeate will be freshwater and the concentrate with be a concentrated saline solution. The operational pressure required to separate the solvent varies depending on the $\nabla \pi$ of the feed solution, temperature, membrane property, and projected flux as well, and can range from about 100 psi for NF application in softening of brackish water to about 1000 psi for RO desalination of seawater.

6.2.2 Membrane Materials

The membrane materials refer to the substances from which the membrane layer itself is made. Normally, membrane materials are made from natural or synthetic polymers. As technology advances, new membrane materials are emerging these days. Some newer microfiltration (MF) membranes are ceramic membranes based on aluminum, and glass is also being used as a membrane material. The properties of membrane materials are directly reflected in their end applications. Some criteria for their selection are mechanical strength, temperature resistance, chemical compatibility, hydrophobicity, hydrophilicity, permeability, selectivity, and the cost of membrane material as well as manufacturing process [2]. The material properties of the membrane may significantly impact the design and operation of a membrane filtration system. For example, some membrane materials may not be tolerant with oxidants like free chlorine, so membranes made of these materials should not be used with chlorinated feed water. Some membranes are made of materials that have higher mechanical strength, and they can withstand higher transmembrane pressure, which can result in higher productivity of the filtration unit. Material properties influence the exclusion characteristic of a membrane as well. A membrane with a particular surface charge may achieve enhanced removal of particulate or microbial contaminants of the opposite surface charge due to electrostatic attraction. Additionally, some membrane materials are hydrophilic or hydrophobic. The hydrophilicity of material describes not only the ease that membranes can be wetted, but also the propensity of the material to resist fouling to some degree.

Nano-filtration and RO membranes are generally manufactured from cellulose acetate (CA) or polyamide (PA) materials and their respective derivatives. RO membranes are predominantly of two types, asymmetric membranes and thin film composite (TFC) membranes. The supporting material is commonly polysulfones, while the thin film is made from various types of polyamines and polyureas (discussed extensively in Chapter 11). There are advantages and disadvantages with CA and PA membranes. CA membranes are susceptible to biodegradation and hydrolysis, and they must be operated within a relatively narrow pH range and low temperature. But, CA membranes have relatively higher tolerance for continuous oxidant exposure. PA membranes, by contrast, are resistant to biodegradation and hydrolysis and can be operated in a wide range of pH and higher temperature conditions. While PA membranes have very limited tolerance for strong oxidants, they are compatible with weaker oxidants. So, for PA membranes, weaker disinfectants like chloramines must be used instead of free chlorine to control biofouling. Alternatively, chemicals should be used to ensure all the free chlorine is eliminated before the water reaches the membranes.

Table 6.1 Comparison of CA and PA membranes

Membrane Materials	Advantages	Disadvantages
CA	• Low purchase cost • High tolerance for chlorine	• Susceptible to hydrolysis and biodegradation • Inferior salt rejection • Low permeability and high-operating pressure required • Narrow pH range (4–8) • Low temperature limits (0–35°C)
PA	• Excellent hydrolytic resistance • Superior salt and organic rejection • Excellent water flux and low operating pressure • Wide operating pH range (2–11) • Wide operating temperature range (0–45°C) • Strong membrane structure and long membrane life	• Limited tolerance towards strong oxidants like chlorine • High purchase cost

For now, PA membranes require significantly lower pressure or energy to operate and have become predominant material used for NF and RO applications. Advantages and disadvantages of CA and PA membranes are summarized in Table 6.1.

6.2.3 Principles and Modeling of Membrane Systems

6.2.3.1 Membrane recovery

Membrane separation processes can be performed in either recovery mode or one-pass mode based on the nature and objective of tests. In recovery mode, the concentrate stream is recycled to the feed container while the permeate stream is abandoned. This mode is usually used in bench scale tests. In one pass mode, both concentrate and permeate stream are abandoned or collected in a separate container just as in most pilot and full scale plant operations. Membrane recovery may be calculated in different ways for the two different operation modes. For recovery mode, the membrane recovery is expressed as percent of permeate volume to the initial feed volume as the following:

$$R = \frac{\text{permeate volume collected}}{\text{initial feed volume}} \times 100 \qquad (6.1)$$

For one pass mode, the membrane recovery is defined as the percent of permeate flow rate to the feed flow rate as shown:

$$R = \frac{q}{Q} \times 100 \qquad (6.2)$$

where, q is the permeate flow rate (L/s) and Q is the initial feed flow rate (L/s).

6.2.3.2 Permeate flux

The permeate flux is the volumetric permeate flow rate passing through the unit area of the membrane and is defined by the following:

$$J = \frac{q}{a} \qquad (6.3)$$

Where J is the permeate water flux, L/(sm^2); q is the permeate flow rate, L/s; and a is the membrane area, m^2.

For RO/NF membranes, the permeate flux can also be calculated through the following equation:

$$J = A \times \text{NDP} \qquad (6.4)$$

Where, A is the water transport coefficient of the RO membrane L/s/m^2/psi; and NDP is the net driving pressure, psi.

For MF/UF membranes, the permeate flux can also be expressed in terms of the trans-membrane pressure and membrane resistance:

$$J = \frac{\text{TMP}}{R} \qquad (6.5)$$

Where, TMP is the trans-membrane pressure (psi); and R is the membrane resistance, (psi $\cdot$ s $\cdot$ m^2)/L.

6.2.3.3 Mass balance

Solute mass balances can be obtained for membrane tests. For different operation modes, the solute mass balance can be expressed by different equations. For recovery mode operation, it is defined as the following:

$$V_f C_f = V_p C_p + V_c C_c \qquad (6.6)$$

For one pass mode operation, the mass balance can be expressed in terms of the flow rates as shown below:

$$Q_f C_f = Q_p C_p + Q_c C_c \qquad (6.7)$$

where, V, Q, and C represent the volume, flow rate and solute concentration, with units of L, L/s, and mg/L, respectively. The subscripts f, p, and c respectively denote the feed, permeate, and concentrate stream.

6.2.3.4 Membrane permeation coefficient (A) and salt transport coefficient (B)

The membrane permeation coefficient A and salt transport coefficient B are important parameters to evaluate membrane fouling or scaling. When membrane fouling or scaling occurs, A will drop because of cake layer formation on the membrane surface and blocking of water passage through the membrane. B change is generally dependent on the fouling type of the membrane. When scaling occurs, B increases, while biological fouling or colloidal fouling of the membrane may cause a drop of B. A and B are mathematically defined as follows:

$$A = \frac{q}{a \times \text{NDP}} \tag{6.8}$$

$$B = \frac{J \times C_p}{C_f - C_p} \tag{6.9}$$

Where, q is the permeate flow rate (L/s); a is the membrane area (m^2); NDP is the net driving pressure (psi); J is the permeate water flux (L/(sm^2)); C_p is the salt concentration in the permeate (mg/L) and C_f is the salt concentration in the feed (mg/L).

6.2.3.5 Membrane rejection

Solute removal is defined as the percentage of a solute removed from the feed stream by the membrane and may be calculated by the formula shown below. Removal efficiency may be calculated for any parameter of interest (turbidity, total suspended solids, total organic carbon, etc.)

$$R_j = \frac{C_f - C_p}{C_f} \times 100\% = \left(1 - \frac{C_p}{C_f}\right) \times 100\% \tag{6.10}$$

Where, C_p is the salt concentration in the permeate (mg/L) and C_f is the salt concentration in the feed (mg/L).

6.2.3.6 Trans-Membrane Pressure (TMP)

The TMP is the difference between the average feed/concentrate pressure and the permeate pressure. For the crossflow mode of operation, the TMP can be calculated by the following expression:

$$\text{TMP} = \frac{P_f + P_C}{2} - P_p \tag{6.11}$$

where, P_f is the average feed pressure (psi); P_c is the concentrate pressure (psi); and P_p is the permeate pressure (psi).

For the dead end mode of operation (dead-end mode means that all the water reaching the membrane is pushed through it, while in crossflow mode, the water flows parallel to the membrane, and only a portion passes through):

$$TMP = P_f - P_P \qquad (6.12)$$

6.2.3.7 Net Driving Pressure (NDP)

The NDP is the pressure available to drive the feed water through the membrane minus the permeate and osmotic pressure:

$$NDP = \frac{P_f + P_c}{2} - P_p - \Delta\pi \qquad (6.13)$$

where $\Delta\pi$ is the differential osmotic pressure (psi).

Net driving pressure is the effective driving force for the flux, and it is an overall indication of the feed pressure requirement. It is used, with the flux, to assess membrane fouling.

6.2.3.8 Osmotic pressure

Osmotic pressure is the pressure required to balance the difference in chemical potential of a solute, and is denoted with the symbol π [3]. The osmotic pressure can be obtained through a formula derived from the Van 't Hoff equation:

$$\pi = i\emptyset CRT \qquad (6.14)$$

where i is the number of ions produced during the dissociation of the solute; $\emptyset$ is the osmotic coefficient, unitless; C is the concentration of all solutes (moles/L); R is the universal gas constant, $1.205935 \; L \cdot psi/(moles \cdot K)$; and T is the absolute temperature (K).

6.2.3.9 Langelier Saturation Index (LSI)

As seen is Chapter 4, the LSI is the most common method for determining the solubility of calcium carbonate in water. Waters that are negative on this index indicate an under-saturated condition with respect to $CaCO_3$ (tendency to dissolve $CaCO_3$), whereas waters that are positive indicate an oversaturated condition (tendency to precipitate $CaCO_3$). The LSI equation is as follows:

$$LSI = pH - pH_s \qquad (6.15)$$

In which the pH_s can be approximated as follows:

$$pH_s = (9.3 + A + B) - (C + D) \qquad (6.16)$$

Where, pH_s is the saturation pH for $CaCO_3$

$$A = \frac{\log_{10} TDS - 1}{10}$$

$$B = -13.12 \times \log_{10}(273 + {}^\circ C) + 35.55$$

$$C = \log_{10}\left[C_a^{2+} \text{ as } CaCO_3\right] - 0.4$$

$$D = \log_{10}[\text{Alkalinity as } CaCO_3]$$

An alternative approximation for the pH_s is presented in Chapter 4.

6.2.3.10 Silt Density Index (SDI)

The tendency for the water feed to foul a membrane can be evaluated with a filterability test called the silt density index (SDI). The SDI is primarily applicable in NF/RO. The test is very simple and consists of a 0.45-μm filter in a dead-end filtration cell (Figure 6.2). The time, in minutes, needed to collect the initial 500 mL of filtrate is recorded as T_i. The time needed to collect another 500 mL of filtrate after the filter has been online for 15 min is recorded as T_f. Standard conditions for the SDI determination call for a 47-mm diameter (1.85-inch diameter) filter, an applied pressure of 30 psi

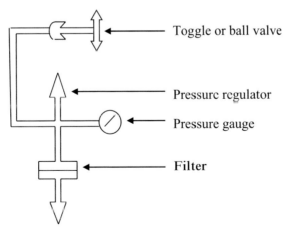

Figure 6.2   Hydraulic schematic of the typical SDI measurement.

Source: [4].

(± 1 psi) and a total test time of 15 minutes [5]. The temperature should be constant ($\leq \pm 1°C$) through the test. The SDI is calculated according to:

$$SDI_T = \frac{[1 - (^{t_i}/_{t_f})] \times 100}{T} \qquad (6.17)$$

where t_i is the initial time required to collect 500 mL of sample; t_f is the time required to collect 500 mL of sample after test time T (usually 15 minutes); and T is the total elapsed time, min (usually 15 min).

There is a chance that the water's fouling tendency is so high, that the filter will be blocked before 15 min is reached. Under such conditions the test time is reduced to before blockage occurs so that the final 500 mL can be collected. Care should be taken as this action reduces the accuracy of the test.

6.2.4 RO Separation System Design

The design of an RO separation system will need to take multiple factors into consideration, including feed water quality, flow rate, product water quality, RO trains required, etc. The RO train/array design shall also consider the membranes that will be used in the design. The maximum flux rate, applied pressure and rejection rate can be obtained from the membrane manufacturers. All of these factors or design criteria will dictate the RO system design and configurations including membrane trains, the number of stages, the number of pressure vessels in each stage, the number of membranes, etc.

In a typical RO setup, after pretreatment, the water passes through cartridge filters and the high-pressure feed pump sends it to the pressure vessels. Each pressure vessel contains several membranes placed one after another which desalinate the water (Figure 6.3). The cartridge filter, pumps, pressure vessels, RO membranes, various instruments, and the stages (as will be discussed later), all make up what is known as an RO train. An array is by definition: an arrangement of devices connected to common feed, product, and reject headers.

In some cases, a RO separation system has two or more membrane trains in parallel. Each train can work independently, irrespective of other trains. The majority of large RO facilities use membrane trains with a capacity equal to or less than 3 MGD (Million Gallons per Day), with 2 MGD being the most common size [6]. When deciding the number of trains, various factors are considered. Fewer membrane trains mean lower cost, less complex operation and maintenance, a smaller footprint, etc. More membrane trains have

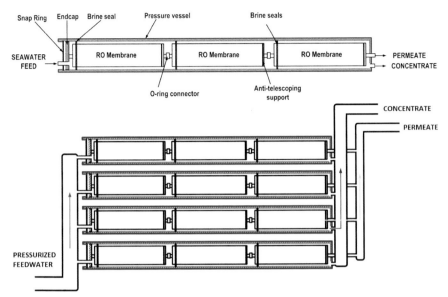

Figure 6.3 The setup of an RO array: 4 pressure vessels × 3 membrane elements.

Source: [7].

the advantage of continuous production of the permeate water even when membranes in one or more trains or cartridge filters require maintenance or cleaning, i.e., higher reliability. In the situation that two or more membrane trains are required, the capacity of parallel trains may be the same or different, and this is also true for the physical size of membrane trains.

A membrane train configuration can consist of one or more passes. A pass in an RO system consists of the treatment steps required for producing the permeate stream. For some applications, one-pass RO system may not produce the permeate quality that meets the treatment target or requirement, especially when the feed water has high total dissolved solids (TDS) or when ultra-low TDS in permeate water is required in some industrial applications. In these situations, a second pass may be required to further reduce the TDS in the permeate water. A schematic flow diagram is shown in Figure 6.4. In some applications, a full second pass RO may not be necessary but a partial second pass might be required. In a partial second pass installation, only a fraction of the first pass's permeate is treated by the second pass. The remaining fraction of the first pass is blended with the permeate of the second pass. This setup helps reduce the size of the second pass while meeting product water quality requirements.

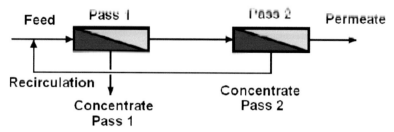

Figure 6.4   Schematic flow diagram of a two pass RO system.

Source: [8].

Each pass of the RO may have one or more stages. Single-stage RO systems are typically used in applications where high water recovery is not required. Normally, the recovery rate for a single stage seawater RO system is less than 50%. Multistage RO systems are used if a higher water recovery is required. In the multiple stage design, the concentrate from the first stage is used as the feed for the second stage RO vessels. Since the feed volume to the second stage is less than the first (some of the water has been extracted as permeate in the first stage), the number of pressure vessels in the second stage and the subsequent stages decreases to maintain the cross-flow velocity of the feed stream in each pressure vessel. In a two-stage RO system design, it is very common that the ratio of pressure vessels of first to second stage is 2:1. Figure 6.5 is a schematic flow diagram of a typical two-stage RO system.

In each stage, pressure vessels are arranged in parallel. In the pressure vessel, the cross-flow velocity decreases along the flow direction because of the loss of flow as membrane permeate. The TDS in the concentrate stream increases for the same reason. Within the same pressure vessel, the tailing

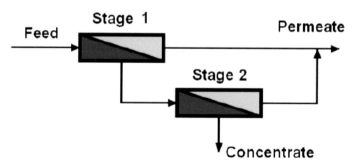

Figure 6.5   Schematic diagram of a typical two-stage RO system.

Source: [8].

membrane modules always have higher scaling tendency compared to the leading membrane modules, and permeate decline is always noticed from the tailing modules. The number of pressure vessels in each stage is determined by design capacity, number of stages, and required minimum cross flow velocity.

6.2.5 Restrictions of Membrane Application in Desalination

One factor that constrains membrane separation processes is membrane performance deterioration after a period of operation due to membrane fouling or scaling. Fouling and scaling may cause product water quality to deteriorate, the productivity to decline and the maintenance and energy cost to increase [9]. Usually, periodical chemical cleaning can help to recover all or the majority of membrane permeability. But when the membrane is fouled or scaled to a certain point where even chemical cleaning cannot help to recover its permeability, the membrane has to be replaced. Seawater RO desalination membranes have an average life of 3–5 years, but this very much depends on the conditions and the operation of the plant. The membrane cost usually constituents 20–30% of the capital cost [10] and membrane replacement constituents about 25–30% of operating cost [11] of a plant.

Fouling is a condition in which a membrane undergoes plugging or coating by some element in the stream being treated, in such a way that its output or flux is reduced [12]. Membrane foulants can be classified broadly into three categories: organic matter, colloidal and particulate matter, and biological growth. Dissolved organic matter, together with particulate matter, are the most serious foulant types, especially since they are the most difficult to remove via conventional treatment [13]. Colloidal clay fouling of RO membranes always causes headaches in surface water RO treatment plants (Figure 6.5). Biological activity can influence membrane operation in two ways: through direct attack resulting in membrane decomposition and through formation of flux inhibition layers, either on the surface or inside membrane pores [14]. CA membranes are particularly susceptible to biological attack. Organisms may be present in the feed or may come from other sources, i.e., airborne algae or fungi.

If solubility limits of salts are exceeded due to excess recovery rates, precipitation of salts occurs on the membrane surface, forming scale which impedes permeation of water. Some common inorganic scalants in the water include Ca^{2+}, Mg^{2+}, Ba^{2+}, CO_3^{2-}, and SO_4^{2-}. SEM images of RO membrane surfaces with and without fouling are shown in Figure 6.6.

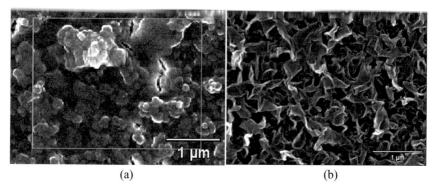

(a)   (b)

Figure 6.6   SEM image of clay fouled RO membrane (a) and virgin membrane (b).

Source: [4].

Membrane fouling and scaling can be mitigated by providing pretreatment or chemical treatment to the feed water. An example is slow sand filters. Precipitation with dolomitic lime and/or alum addition prior to sand filtration can be helpful to remove colloids and particulate matters [13]. In some feedwaters, where the hardness might be high and the water might impose high-scaling potential to membranes, ion exchange can be applied to remove

Figure 6.7   SEM image of scaled membrane surface.

Source: [4].

cations like calcium, barium, and magnesium or the anions like sulfate and carbonate. Antiscalants are usually used these days to delay scaling of RO/NF membranes, and chlorine and other disinfectants are very effective to control bio-fouling. At least one pretreatment method, and often a combination of multiple pretreatment methods are required to mitigate membrane fouling and scaling.

6.2.6 Concentration Polarization in Membrane Desalination

Membrane scaling is caused by mineral precipitation on the membrane surface, which can be significantly enhanced by concentration polarization (CP). During membrane separation, a concentration boundary layer forms near the membrane surface. Solutes tend to accumulate in the vicinity of the membrane surface due to convective flow toward the membrane and solute rejection. Under steady conditions, the solute concentration profile produces a diffusive transport away from the membrane surface, back into solution, that balances convective transport toward the membrane. In the boundary layer, local solute concentrations can be significantly higher than that of the bulk feed flow. When concentrations of solutes in the boundary layer exceed their respective solubility, precipitation reactions can occur on membrane surfaces, accelerating scale formation and prematurely lowering membrane permeability. Other adverse outcomes include reduction of membrane flux due to increased osmotic pressure at the membrane surface and enhanced trans-membrane solute transport (lower apparent rejection).

Various methods have been employed to minimize CP, normally by encouraging mixing near the membrane surface [15]. The objective of back-mixing enhancement is commonly pursued through the use of spacers in the feed solution channel. The feed channel spacer is an important component of membrane reactor design, primarily to create the feed channel itself by separating adjacent membranes. Feed spacers also increase the mass transfer coefficient in the concentration boundary layer [16], reducing both CP and membrane scaling. Spacer effects are, however, generally not uniform across the membrane surface. Direct inspection of scale deposits on membranes in spent spiral wound membrane modules showed that the feed channel spacer has a strong influence on the scaling patterns [17]. That is, CP effects tend to be heterogeneous and scale is most evident in stagnation regions introduced by the spacer [18].

The extent/distribution of stagnation regions is a function of feed spacer geometry, including the thickness of spacer filaments, filament angle, spacer porosity, and so forth.

Various models have been utilized to predict the CP index (β), which quantitatively describes the severity of CP. The index is commonly defined as the ratio of solute concentration at the membrane surface (C_m) to the bulk or feed solute concentration (C_f):

$$\beta = \frac{C_m}{C_f} \tag{6.18}$$

However, finding the concentration at the membrane surface is not easy, so predictive models are used. One such model which is independent of the solute concentration in the feed solution, water flux and the concentration boundary layer mass transport coefficient has been proposed by Yan [4].

The predicted CP index (β) using this model matched the values that were measured in the lab test using a spiral wound membrane for TDS removal from the Central Arizona Project water [4]. As expected, in all cases, β is greater than one; meaning by definition that the concentration at the membrane surface is higher than that of the bulk.

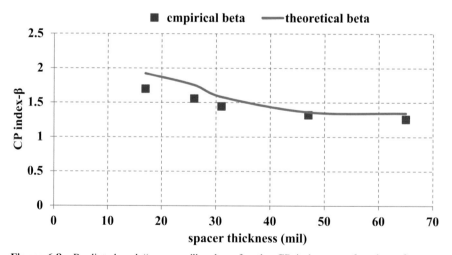

Figure 6.8 Predicted and "measured" values for the CP index as a function of spacer thickness using the developed model.

Source: [4].

6.2.7 RO Membrane Pretreatment

In most applications, pretreatment is required to protect RO membranes from fast fouling or scaling. Pretreatment, extensively discussed in the next chapter, can be any procedure or technology used to remove or reduce the substances that deposit on the membrane and cause membrane performance deterioration.

Selection of pretreatment methods is dependent on feed water quality. All the RO membrane manufacturers recommend the SDI of the feed water not to exceed 5. Screening, settling, coagulation, and filtration can help to remove the particulate matter and suspended solids from feed water and to reduce the SDI.

Some water supplies may have a high concentration of anions or cations that will precipitate out once super-saturation conditions are encountered in the desalination process. These ions include Ca^{2+}, Mg^{2+}, Ba^{2+}, CO_3^{2-}, SO_4^{2-}, and PO_4^{3-} which may form $CaCO_3$, $MgCO_3$ and $BaSO_4$, and deposit on the membrane surface. Adjustment of pH is a common and effective method to control $CaCO_3$ scaling. So, the pH of the feed water is normally reduced to 6.8 or lower by injection of acid to the feed stream. Anti-scalant injection is also a very popular chemical pretreatment to prevent membrane scaling. Anti-scalants are surface-active materials that interfere with precipitation reactions. In the last two decades, new generations of anti-scalants have emerged commercially. Active ingredients in these anti-scalants include polycarboxylic/carboxylic based polymers, polycarboxy-lates, polyphosphates/phosphates, phosphonates, phosphinocarboxylic acid, polyacrylic acid/polyacrylates, lignins, tannins, and other synthetic poly-mers. Commercial anti-scalants are usually proprietary solutions containing one or more active ingredients. Because of the complexity of the scale forming process, and lack of information regarding proprietary commercial anti-scalants, the precise mechanism of scale inhibition by anti-scalants are not clearly understood. It has been reported that anti-scalants work by three closely related mechanisms that interfere with one or more of the stages of crystal growth [19]. These three mechanisms are threshold effect, crystal distortion, and dispersancy effect. A single anti-scalant may inhibit or retard the scale by more than one mechanism. An example is polyacrylic acid, which is the most common inhibitor commercially avail-able. These polymers function as crystal distortion agents, but at higher molecular weight, they also exhibit dispersancy properties. Selection of the antiscalants should be based on the water quality, and in some applications, testing shall be done to determine the most effective antiscalant. Methods for comparing the efficiency of anti scalants have been developed. For example, by determining induction time by measuring conductivity and absorbance of the supersaturated solution [4].

In addition to chemical injection, ion exchange can also be used to remove scale-forming cations or anions. But ion exchange pretreatment is not very common and can be cost prohibitive. Another less common pretreatment is chemical precipitation. This method can be cost effective in certain applications when the concentration of certain cations or anions are high and they can be efficiently precipitated out of water by adding other chemicals. A good example is calcium (Ca^{2+}) in water. Chemical precipitation always requires settling or filtration (MF/UF, media) following chemical injection to remove the sludge from the RO feed stream.

When organic matter is present in the RO feed water, RO pretreatment can be selected from activated carbon absorption, coagulation, and slow sand filtration for surface water applications. For slow sand filtration, a thin layer of bio-film develops on the top skin of the filter, and this bio-film can effectively remove the natural organic matter that are present in the surface water.

Reverse osmosis membrane deterioration can also be caused by microbial growth on the membrane surface and channel spacers. It is very common that disinfectants are injected in the RO feed stream to control biofouling of the membrane. Selection of the disinfectants shall be determined by the material of the RO membranes and the tolerance of the material to disinfectants. CA membranes have relatively higher tolerance for continuous oxidant exposure, but polyamide membranes have very limited tolerance for strong oxidants, and are only compatible to weaker oxidants. Chloramines are commonly used for the RO systems using PA membranes instead of free chlorine. In the practical operations, sodium hypochlorite and ammonia chloride are used stoichiometrically to produce monochloramine. If free chlorine is used to inhibit biological growth, injection of chemicals such as sodium metabisulfite/sodium bisulfite prior to the membrane are essential to make sure all oxidants are reduced before reaching the membrane.

6.2.8 RO Membrane Chemical Cleaning

When RO membrane permeability has deteriorated, it must be cleaned with chemical reagents to recover its permeability. Usually, if an RO membrane shows 10–15% decrease in normalized permeate flow, it is an indication that chemical cleaning is needed. Cleaning reagents required for cleaning are usually dependent on the fouling characterization. Typically, calcium carbonate scale, metal oxides/hydroxides scale and inorganic colloidal foulants can be cleaned with citric acid solutions. Calcium, barium, or strontium sulfate scale and mixed inorganic/organic colloidal foulants can be effectively removed by a sodium tripolyphosphate (STPP) ($Na_5P_3O_{10}$) and Na-EDTA high pH solution. A high pH solution of STPP and Na-dodecylbenzene sulfonate (Na-DDBS) ($C_6H_5(CH_2)_{12}$-SO_3Na) can be used to clean biological matter and organic matter on the membranes. A proper cleaning solution should be selected for effective membrane permeability recovery. These and many other cleaning solutions are commercially available. For best cleaning efficiency, solutions are usually heated during cleaning, to as high as $40°C$.

Below is the typical setup for the RO membrane cleaning, recommended by a membrane vendor, Hydranautics.

The typical cleaning procedure is as follows:

1. Perform a low-pressure flush without running RO high pressure pump by pumping clean RO membrane product water through the pressure tubes to rinse out the concentrate water and particles in the pressure vessel. This clean water flush is about 10–15 min.
2. Circulate the cleaning solution through the pressure vessels for 20 min to 1 h depending on the situation. The cleaning solution may be heated for increased performance. At the start, send the displaced concentrate water to the drain. Then divert about 10% of the most highly fouled

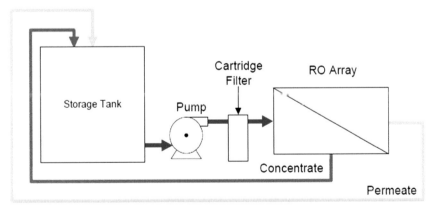

Figure 6.9 RO membrane cleaning setup.

Source: [20].

cleaning solution to the drain before circulating the remainder of cleaning solution. The flow rate during this stage is increased to the maximum design flow rate of the CIP pump.

3. Shut down the circulation pump and close all the valves and soak the membrane for at least one hour. If the results are satisfactory, we proceed to step 4; if not, it is possible to go back to step 2 and redo the circulation and soaking steps.

4. Rinse the pressure vessels with clean RO product water at low pressure to remove all traces of chemicals from the cleaning skid and the RO pressure vessels. This step may take about 30–60 min. Clean water rinse should continue until the pH and conductivity of the concentrate stream effluent of the rinse water is very close to the feed rinse water.

5. The cleaning solution is neutralized and drained. It is possible that a combination of cleaning solutions are used. In this case, return to step 2 and employ the next cleaning solution.

6.2.9 RO Projection Software

All the major RO membrane suppliers provide RO membrane users with projection software, which are useful tools for designing and simulating the operation of RO systems. Below is a list of the most popular membrane suppliers and the projection software they have released.

- GE Power and Water, Winflows
- Hydranautics, IMSDesign
- Dow Water and Process Solutions, ROSA
- Toray Membrane, TorayDS2

These software are available to users for free. Software from different manufacturers have different features, and instructions can be found from their respective download websites. Each RO projection software contains the database of the supplier's membrane specifications for use in the program. Some projection software like GE Winflows have their own features like three-pass systems, permeate split and recycle, anti-scalant dosing, energy recovery devices and the ability to combine stages. As an example, a review on how to use the IMSDesign 2016 projection software from Hydranautics is provided here. In real world applications, the software should be chosen depending on the membranes used in the desalination plant. Although the software are slightly different, they all have more-or-less the same function.

To use IMSDesign, there are five steps that need to be performed to create the RO projection:

Step 1: Analysis;
Step 2: Design;
Step 3: Calculation;
Step 4: Post-treatment; and
Step 5: Analysis save.

After the software is launched, the initial screen displayed is the setup screen (as shown in Figure 6.10(a)), where the default language, currency, and units are selected. After all of these are selected, click "Continue" to go to the next screen, where you can select from existing RO projects you have been working on, or start a new RO project. The screen is shown in Figure 6.10(b). Once the RO project is selected, it proceeds to the Analysis screen (Figure 6.10(c)). In this step, the project profile and RO feed water quality data will be input to the projection software. Water temperature and water source shall be specified here. The parameters

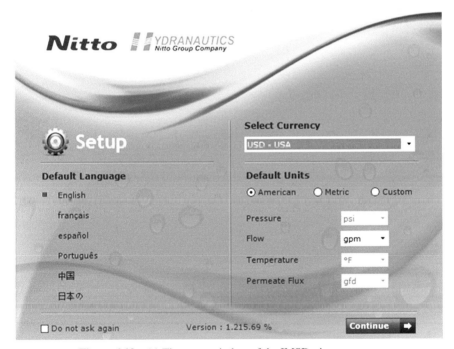

Figure 6.10   (a) The setup window of the IMSDesign program.

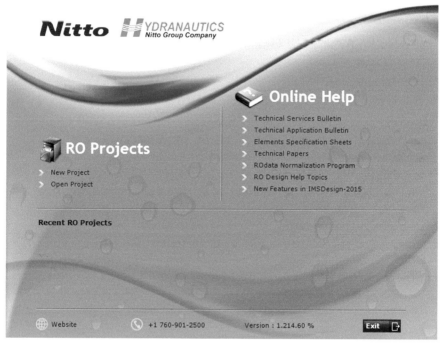

Figure 6.10   (b) The setup window of the IMSDesign program.

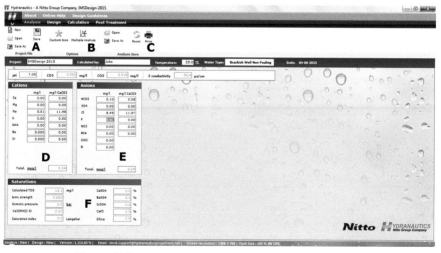

Figure 6.10   (c) Analysis screen in the IMSDesign program.

include pH, CO_3^{2-}, CO_2, conductivity of water, and cations and anions in the water. When the equivalent concentration of cations does not match anions, Na^+ or Cl^- shall be used to balance the cation and anion concentration. With the water quality data, IMSDesign will calculate the saturation index (Langelier), saturation levels of some sparsely soluble salts including $CaSO_4$, $BaSO_4$, $SrSO_4$, CaF_2, and silica. Some other parameters, including ionic strength and osmotic pressure are also calculated and shown here.

Once the water quality data input is completed, click on the "Design" function button to move to the RO system design screen. To do the RO membrane train design, the permeate recovery (%) and permeate flow rate need to be specified by the engineer. RO feed pH also needs to be specified, and depending on the pH of the water source, pretreatment chemicals (acid or caustic), and the dosing rate will be calculated and shown here. Passes and stages of the RO system shall be designed by the engineer depending on the RO application and job requirements. The software has built-in database of RO membranes that Hydranautics supplies. Once the element type is selected, the software will calculate the number of elements in each vessel and number of vessels required for the job. The membrane average flux and feed flow are calculated based on the input information by engineers (Figure 6.10(d)). If any design parameters (average flux, feed flow, etc.) are out of the design boundary, a warning window will pop out to remind the engineers that the RO system design must be changed to keep the design

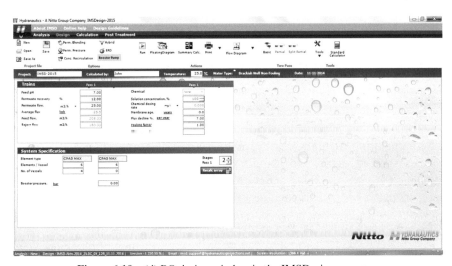

Figure 6.10 (d) RO design window in the IMSDesign program.

parameters within the design limits of the selected membrane type. After the RO design is completed, run the calculation, to allow IMSDesign to generate the calculation results, and report the design parameters including feed pressure and flux in each pass and stage of the unit.

The software provides additional design options that engineers can select for a specific RO application. The following additional design options can be selected:

- Permeate blending
- Permeate pressure
- Concentrate recirculation
- Hybrid (membranes)
- Energy recovery devices (ERD)
- Booster pump
- Custom element
- Recalculate array

If any specific design is required in the RO design, the above options can be selected from the "Options" group. After the options are performed, the design can be saved and printed.

If a two-pass RO design is required, click "Basic" in the "Two Pass" group, and the pass 2 options will appear under "System Specification". By default, the partial and split partial design functions are disabled in the program. To enable these two options, reduce the value of the Pass 2 "Permeate Flow" in the "Trains" section, and the program will select the "Partial Pass 2" option.

Following the completion of the RO system design, click "Run" in the "Actions" group to execute calculation.

In the "Calculation" phase, IMSDesign calculates power require-ment, chemical requirement and cost, and displays the calculation results (Figure 6.10(e)). IMSDesign uses the flow and pressure values that are entered at the time of creating a design to calculate the amount of power required for operating the high-pressure pump for the values that are specified for flow, pressure, recovery, and pump/motor efficiency data. The chemical requirement represents the three types of chemical dosing that are required while preparing the water treatment design. In IMSDesign, the dosing values are sodium-metabisulfite dosing feed, anti-scalant dosing feed, and NaOCl dosing feed. To calculate cost, click the "Costs" in the "Calculations" group, and the Cost Calculations table will appear showing cost fields, including

Figure 6.10 (e) Calculation window in the IMSDesign program.

capital cost, power cost, chemical cost, membrane replacement cost, mainte-nance cost, and total product water cost, and the appropriate value.

Post-treatment can be designed using IMSDesign by clicking "Post Treatment" on the secondary bar. A "Post Treatment" window appears and the engineers can add chemicals to the stream or treat the permeate (Figure 6.10(f)). Under "Dosing Rate of Chemicals", the concentration of the listed chemicals can be specified, and chemical consumption will be calculated. Also, if CO_2 reduction in the permeate is required, CO_2 removal can be simulated using the degasifiers.

After all the above steps are performed to create the RO project, the analysis and system design can be saved and printed. The project file can be saved by going to the "Project" file group and clicking "Save As". The file can be opened using IMSDesign in the future if engineers need to revisit the project.

6.3 Forward Osmosis (FO)

6.3.1 Introduction to FO Membrane Technology

Forward osmosis (FO) membrane separation technology is an osmotically driven membrane separation process that involves the diffusion of a solvent,

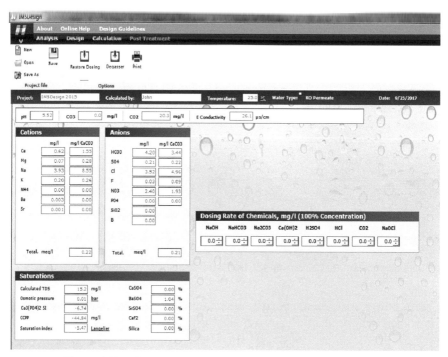

Figure 6.10   (f) Post-treatment window in the IMSDesign program.

water in most cases, through a semipermeable membrane and the rejection of most solute and all suspended particles in the solution. This process utilizes the natural osmotic pressure without applying hydraulic pressure to the feed water. Water spontaneously diffuses through the FO membranes from the feed stream of low osmotic pressure to a highly-concentrated draw solution stream of high osmotic pressure [21].

In the FO separation process, a low-pressure pump circulates the saline water to the feed water side of the FO membrane. A draw solution is circulated on the draw solution side, also known as the permeate side of the membrane. During this process, some of the feed water passes through the membrane due to the differential osmotic pressure on the two sides of the FO membrane. The majority of the TDS, all of the suspended solids, bacteria, and viruses are rejected and concentrated in the FO membrane concentrate stream. On the other side of the membrane, the draw solution is diluted due to the passage of permeate through the membrane and the volume increase of the draw solution. For continuous operation, an additional step is normally

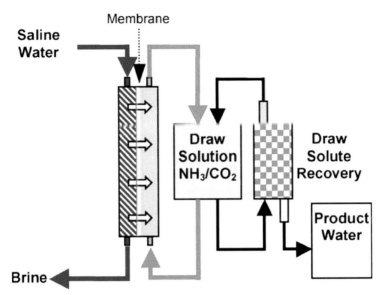

Figure 6.11 Schematic flow diagram of a typical forward osmosis process.
Source: [22].

required to separate the permeate from the draw solution (by using RO, thermal processes, etc.), to regenerate the draw solution so that the draw solution can be reused and circulated. After this permeate separation/draw solution regeneration step, the product water is collected, and the draw solution is concentrated and circulated back to the FO membrane for continuous operation.

Figure 6.11 shows a schematic of a typical FO process. In this example, an ammonia–carbon dioxide draw solution is used.

6.3.2 Forward Osmosis Membranes and Modules

Generally, any semipermeable membrane can be used for the FO process. Most researchers used RO membranes for testing during the early stages of FO studies. In all cases, researchers observed much lower flux and salt rejection than expected [23]. In the 1990s, Hydration Technologies Inc. (HTI), which was Osmotek Inc at the time, developed membranes specifically for the FO process. And since the 1990's for a long time, the FO membrane by HTI was the only available product for this technology on the market. The FO membranes by HTI are made of very similar materials that have been used for RO membranes. The membrane fabrication material has been proprietary, but it is thought that the HTI FO membranes are made of cellulose triacetate (CTA) [23]. Though the fabrication material is similar to RO membranes, the structure of the HTI FO membrane is very different. The HTI membrane consists of a dense cellulose triacetate polymer cast onto a polyester mesh for mechanical support. The membrane is very thin, having a total thickness in the order of 50 μm [24]. Typical asymmetric RO membranes consist of a thin dense active layer cast on a thick porous support layer. However, in forward osmosis, the porous support layer contributes to internal CP that reduces the performance of the process [25]. The CTA FO membrane lacks a thick support layer. Instead, the embedded polyester mesh provides mechanical support. These improved membranes showed improved performance when compared to RO membranes that were commercially available when operated in forward osmosis tests [26]. As a summary, the desired characteristics of membranes for FO would be having a dense active layer for high solute rejection, while having a thin membrane with minimum porosity as the support layer for low internal CP [23].

Today, in addition to FO membranes made of cellulose triacetate by HTI, membranes made of other materials are also commercially available. These materials include polysulfone and polyethersulfone, polybenzimidazole, poly(amide-imide), polyamide based TFCs, etc.

The FO membranes can be configured in different ways. Since in the FO process, both feed stream and the draw solution stream need to be circulated in the membrane modules. The design or configuration of the FO membrane modules is more complicated than the pressure-driven RO membranes. Spiral wound modules are the prevailing design for RO membranes, but the same design cannot be used in FO, because in the spiral wound design, the permeate side is sealed and no channel is available for circulation of the draw solution. For the bench-scale or pilot-scale tests, flat sheets and tubular configurations are available. Flat sheet membranes with plate-and-frame configurations are the prevailing design for full scale applications, because the flat sheet membranes are easy to pack and scale up. But the flat sheet membranes with plate-and-frame configurations have their limitations. These limitations include lack of adequate membrane support and low packing density of the membranes [23]. Lack of membrane support makes the membrane modules vulnerable and easily breakable under high hydraulic pressure, and for this reason, the membrane modules are limited to low hydraulic pressure operations. The low-packing density leads to a large footprint for the FO membrane separation system [23].

Although the spiral-wound configuration for FO membrane elements is not common, but due to the advantages of the spiral-wound configuration, it still attracts research and development attention. Back in the 80s, Mehta designed a unique FO membrane element advancing the design of the conventional RO membrane element design [27]. Below (Figure 6.12) is the demonstration of the flow pattern in the spiral-wound FO membrane element designed by Mehta. The draw solution flows through the membrane following the channel by the spacer similar to the feed water in the RO membrane design. But the central tube and the membrane is split into two sections forcing the feed water to flow through the membrane surface and to come out of the membrane from the opposite side of the water inlet.

Forward osmosis membrane elements in tube or hollow fiber are also available. These tubular membrane modules have several advantages over the flat sheet membranes. One of the advantages is that the tubular membranes are self-supported. The second advantage is that the tubular membranes are easier to fabricate, and high packing density can be achieved. Another advantage is that water can be circulated on both sides of the membrane, which meets requirements for the FO process (circulating feed water on one side of the membrane while the draw solution on the other side). But one disadvantage of hollow fiber type tubular membranes is that turbulent flow cannot be achieved

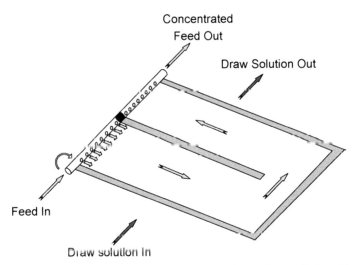

Figure 6.12   Flow patterns in the spiral wound FO membrane element designed by Mehta.

Source: [27].

within the hollow fiber, and so the CP index is high, which leads to increased fouling and scaling of the membranes.

6.3.3 Draw Solutions for the FO Process

The draw solution is the concentrated solution that is used in the FO process to provide the osmotic pressure on the permeate side of the FO membrane. As discussed, the difference between this osmotic pressure on the draw solution side and the feed water side is the driving force for the water to pass through the semi-permeable membrane. Other terms are also used in the literature to describe the draw solution including osmotic agent, osmotic media, driving solution, and osmotic engine [23].

Ideal draw solutions should be able to induce a high osmotic pressure that is higher than the feed water's osmotic pressure. For continuous operation, it is required that the draw solution be regenerated or re-concentrated after it is diluted by the permeate water. It is best that the regeneration/re-concentration process uses as little energy as possible. Another requirement for an ideal draw solution is that it is not toxic or not a source of contamination. In addition, it should not cause any fouling or scaling on the FO membranes. Being inexpensive, stable, highly soluble, and having a molecular size that is large enough to limit reverse draw solute flux through the FO membrane's active layer, yet small enough to be highly mobile and mitigate CP are other advantages [28].

Various chemicals have been suggested and tested as candidates for draw solutions by different research groups. The tested chemicals include sulfur dioxide solutions, aluminum sulfate, a mixture of glucose and fructose, potassium nitrate, and ammonia and carbon dioxide among many others [22]. As an example, the ammonia and carbon dioxide combination, in specific ratios, can create highly concentrated draw solutions inducing osmotic pressures in excess of 250 atm, allowing unprecedented high recoveries of potable water. The ammonia and the carbon dioxide in the solution can subsequently be thermally recovered and reused [29]. Some other chemicals for FO draw solutions are being tested and innovative draw solution recovery/regeneration methods are also under investigation.

6.3.4 CP in FO Processes and FO Membrane Fouling

Although to a lesser extent than RO, membrane fouling or scaling is still a potential challenge facing the FO membrane separation process. For this reason, depending on the specific application, proper pretreatment should be considered and designed to protect the FO membranes. Fouling in FO is structurally different from fouling in pressure-driven membrane processes such as RO. In RO, the structure of the cake layer formed is densely compacted. However, the cake layer in FO is less compact which means that 80–100% of the initial water flux can be recovered through simple rinsing of the membrane surface [28].

In RO desalination, a CP layer is formed on the membrane surface due to the convective flow of dissolved solids and the permeate water. For the same reason, a CP layer is formed on the feed water side of the FO membrane surface as well. The CP effect causes the actual concentration of dissolved solids within the layer to be higher than the concentration in the bulk flow. During the FO separation process, the draw solution is circulated on the permeate side of the FO membranes. As the permeate water passes through the membrane, the permeate water dilutes the concentration of the draw solution at the membrane surface, and another CP layer develops on this side. In this layer, the concentration of the draw solution solute is lower than in the bulk flow. To distinguish the CP layers on the two sides of the FO membranes, the layer on the feed water side is normally called concentrative external CP layer, while the layer on the draw solution side is normally called the dilutive external CP layer [23]. So, for membrane processes, there are 2 categories of concentration polarization: internal and external. External CP occurs at the interface of the membrane and the surrounding liquid, and can be either "concentrative" or "dilutive" as discussed. For asymmetric membranes (containing both a dense rejection layer and an underlying porous support) internal CP happens within the porous support layer. When the dense rejection layer faces the feed solution (which is the case for FO), the water permeating through the porous support layer dilutes the draw solutes inside the support, resulting in dilutive internal CP. Alternatively, if the dense rejection layer faces the draw solution, solutes inside the support are concentrated as water permeates through the membrane, resulting in concentrative internal CP.

In the calculation of the osmotic pressure and the membrane flux, the differential osmotic pressure at the membrane surfaces between two sides is the effective driving force that pushes water through the semi-permeable membrane. The concentrative external CP increases the osmotic pressure on the feed water side of the FO membranes, while the dilutive external and internal CP reduce the osmotic pressure on the draw solution side of the FO membranes. As a result of both the external and internal CP effects, the effective net osmotic pressure is much lower than the calculated value according to the theory, and this leads to a lower permeate flux of FO membranes compared to the expected value [25].

Models developed for RO membranes can still be used for the prediction of concentrative external CP for FO membranes, but, other models have also been developed by various research groups [25, 30, 31].

Membrane fouling has been studied in bench scale and pilot scale by researchers [24, 32]. Cath and his research group found that during short bench-scale FO experiments, water flux remained relatively constant and membrane fouling was minimal, and that the FO membrane fouling rate was substantially lower than the pressure driven RO/NF membranes tested under similar experimental conditions [24, 33]. The difference between membrane fouling of FO membranes and RO membranes is attributed to several factors including the low osmotic pressure of the feed solution, limited transport of foulants to the membrane due to lower permeate water flux, hydrophilicity of the FO membrane, low operating pressure process in the FO process, and low fouling potential in the feed water [24].

Long-term pilot-scale tests were conducted to investigate the impacts of feed water quality and operational conditions on FO membrane fouling. The pilot study results indicated that more rapid membrane fouling occurred when more suspended solids were present in the feed water and when the concentration of the draw solutions was higher. It was also observed that the orientation of the flow channel, whether the feed channel was facing up or facing down, changed the membrane fouling rate, and that the membrane spacer in the feed water channel changed the membrane fouling rate as well [24].

6.3.5 Advantages and Disadvantages of the FO Process

Due to the fact that the forward osmosis process utilizes osmotic pressure rather than hydraulic pressure for operation, FO membrane separation technology has many advantages over the pressure driven RO/NF membrane separation processes. These include high water recovery and salt injection, and potential utilization of renewable energy resources to drive the processes [34]. Particularly, lower sensitivity to membrane fouling and scaling allows FO to desalinate streams that would be too difficult for RO. Also, the FO membrane process requires relatively easier membrane and equipment design. Another advantage is the possibility of using a combination of electrical and thermal energy for desalination, as opposed to solely electrical energy in RO. All of these advantages make the FO separation process attractive for treating some very challenging streams.

While the FO technology has bright potential application for water desalination, there is more development work to be done on FO membranes, membrane modules and draw solutions to overcome some obstacles facing FO technology. Meanwhile, with elimination of hydraulic pressure, the [electrical] energy consumption of FO can be much lower than the conventional RO membrane process [35]. However as explained, FO technology normally requires a second process to separate the permeate from the draw solution. This second process for permeate extraction and recycling of the draw solution can be energy intensive (either in the form of electrical or thermal energy). This makes the process's overall energy consumption higher than RO. In other words, based on thermodynamic principles and kinetic requirements, the theoretical minimal energy of FO desalination is always higher than without FO, meaning that using FO cannot reduce the minimum energy of separation in filtration technology.

Overall, more research and development will be required before FO becomes mainstream desalination technology. So far, FO technology has not been commercialized on a wide scale. From the various prospective applications of FO, the following arouse particular interest [28]:

1. Leveraging low-cost thermal energy: If suitable thermolytic draw solutes are used, hybrid FO-thermal processes may use less energy than standalone thermal technologies. This is because relatively small amounts of draw solute must be vaporized and recovered from the draw solution, as opposed to vaporizing and recovering water in conventional thermal processes. Thus, low-cost energy (such as solar thermal energy, geothermal energy, and industrial waste heat) can be used. The higher the solute vapor pressure, the less total energy required for its recovery. As an example, in the ammonia–carbon dioxide system, ammonium bicarbonate and ammonium hydroxide salts, which have low molecular weights and high solubility, are dissolved in water to create a concentrated draw solution with high osmotic pressure. After the water flows across the FO membrane, the draw solution is removed and moderately heated. Thereupon, the ammonium salts are decomposed into ammonia and carbon dioxide gases, and freshwater is left behind. The gases can then be recovered and reconstituted into the draw solution again.

2. FO as pretreatment to RO: As explained above, when compared to RO, fouling and scaling are less problematic in FO processes. Hence for problematic feedwaters, FO can function as an advanced pretreatment

technology to remove suspended material, dissolved organic species and inorganic scalants. In this way, the coupled RO process for the draw solution downstream, is only exposed to engineered water that has negligible fouling and scaling potential. In addition, FO pretreatment can improve the permeate quality produced by RO.

3. Osmotic dilution: When a wastewater stream is to be treated, it can be used as feedwater, from which freshwater is removed via an FO process. Through natural osmosis, the wastewater becomes concentrated, and the draw solution is diluted. In this setup, seawater can be used as the draw solution, so that it is in this way diluted before being fed to an RO process. This results in not only a reduction in wastewater volume, but also reduced energy requirements for desalination by lowering the operating hydraulic pressure of the RO process. Alternatively, brine from a desalination plant can be used as the draw solution. As a result, osmotic dilution can be used to dilute the brine before it is discharged, in order to mitigate negative environmental impacts.

6.4 Electrodialysis (ED) and Electodialysis Reversal (EDR)

6.4.1 Introduction to ED and EDR Technologies

Electrodialysis (ED) is a voltage-driven membrane process, during which ions are transported through semi-permeable membranes under the driving force of an electric potential, and fresh water along with the uncharged fraction, is left behind as the product. The semi-permeable membranes are cation- or anion-selective. ED is primarily used for desalination of brackish water. Electric energy is consumed in proportion to the quantity of salts to be removed, and in most situations, the application of ED is limited to feed water of less than 10,000 mg/L TDS due to economics consideration arising from energy consumption.

The Electrodialysis Reversal (EDR) process is based on the same general principle as the ED process, except that both the product and concentrate channels are identical in construction and that the polarity is periodically alternated automatically. In the EDR operation, the polarity is normally reversed three to four times per hour to reverse the flow of ions passing through the semi-permeable membranes. This polarity reversal, along with the flushing step following the change, cleans the membranes and helps to reduce membrane fouling and scaling. This modification improves the tolerance of ED technology to operations treating water with high scaling or fouling potentials. For this reason, EDR technology has increased the utility of the ED process, and EDR technology has largely replaced ED technology in many markets.

6.4.2 ED/EDR Process Design

In the ED system, the basic unit is called a "cell pair", which consists of an anion selective membrane, a diluting spacer, a cation selective membrane, and a concentrating spacer. Below, in Figure 6.13, is a schematic several "cell pairs". Multiple cell pairs are normally packed between an anode and a cathode and the overall unit is called a "stack". The spacers are designed to create turbulent flow at the membrane surface. Repeating cell pairs are compressed between two end plates by tie rods and these cell pairs can be arranged either vertically or horizontally.

In the ED/EDR desalination system, the feed stream flows through the channel created by the desalinated flow spacer, while the concentrate stream flows in parallel through the channel created by the concentrate flow spacer. When water flows along the ionic-selective membranes, ions are

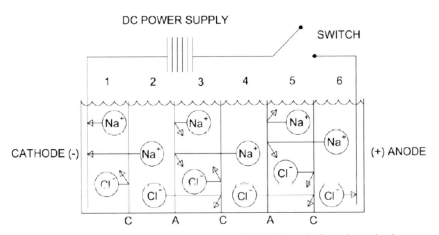

Figure 6.13 Schematic of several ED cell pairs [36]. "A" stands for anion-selective membrane, while "C" stands for cation-selective membrane. Channels 2, 4, and 6 are channels containing desalinated water. Channel 1, 3, and 5 are channels which contain the concentrate. As water flows through, the cations (such as sodium) will be pulled towards the cathode, yet stopped by the anion-selective membranes. Simultaneously, the anions (such as chloride) will be pulled towards the anode, yet stopped by the cation-selective membranes.

electrically transferred through membranes from the desalinated stream to the concentrate stream.

Depending on the system design and applications, the commercial ED systems can be operated in two different configurations. The first configuration is batch operation, in which the desalinated water is circulated back to the feed water storage tank and mixed with the feed water. The circulation will end when the treatment target is met and the water quality in the holding tank meets the requirements. During the process, the concentrate stream is also recirculated separately to reduce wastewater volume that requires disposal. The second configuration is the single pass continuous operation. In this operation, the feed water stream passes through the ED stacks once and flows to the product storage tank. In most cases, the ED stack contains two stages in series, and the water is desalinated in both stages. The concentrate stream is normally partially recycled to reduce waste volume. For both batch operation and continuous configurations, an acid solution is usually injected to the concentrate stream to prevent scaling.

The design of the ED/EDR system should be based on the characteristics of the feed water, flow rate, and the product water requirements. The efficiency of ED/EDR systems in a given application depends greatly on the process, design and mode of operation. Depending on the characteristics of the feed water, proper pretreatment units must be carefully selected. Generally, the feed water quality requirements are outlined by ED/EDR suppliers to guarantee system performance and to protect membranes or electrodes from fouling or scaling. Some contaminants need to be removed from ED/EDR feed water including suspended solids, iron, manganese, sulfides, residual chlorine, and sparsely soluble salts such as $CaSO_4$, $BaSO_4$, $SrSO_4$, and CaF_2. To remove alkaline scale or iron/manganese scale from the membrane, acid addition to the feed water is normally required for the ED system. Polarity reversal in the EDR system reduces or eliminates the scaling issue, and for this reason, acid addition is not required for EDR systems. Chemical injection systems must be specified for ED systems as part of the process design based on water quality and flow rate.

A chemical system is typically supplied with the ED/EDR systems to control the process conditions at the electrodes, where chemical reactions occur. As will be discussed, protons (H^+), chlorine gas, and oxygen can be formed at the anode. Hydrogen gas and hydroxyl ions can be formed at the cathode. So, the pH decreases at the anode and increases at the cathode [36]. A gas collection or disposal mechanism must be designed to

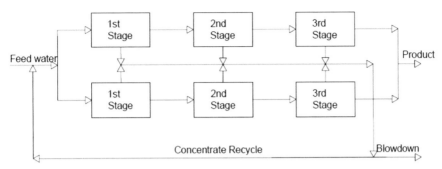

Figure 6.14 Example schematic of a three-stage, two-line ED/EDR system.
Source: [36].

remove the produced hydrogen, oxygen, or the chlorine from the electrode streams.

Similar to other membrane separation systems, chemical cleaning of ion selective membranes in ED/EDR systems is normally required after a certain amount of operation time. The cleaning frequency is dependent on feed water quality and operational conditions. Cleaning in place (CIP) systems must be provided for the ED/EDR system. In most cases, an acid solution can be used for mineral scale dissolution, and a sodium chloride solution with adjusted pH for organics removal [36]. Chemical injection equipment, chemical delivery, chemical disposal, and spacers required for the CIP system must all be considered within the process design stage.

In most cases, 25–60% of the TDS in the feed water can be removed by a single ED/EDR stack depending on the feed water characteristics [36]. If a higher TDS removal rate is desired, two or more stacks will be used in series. A separate power supply should be provided for each stage. The number of stages required in the design is normally determined by the economic analysis according to feed water quality and desired product water quality. To meet high flow rate requirements, two or more stack groups can be arranged in parallel, and in this configuration, each series of stacks is called a line. The number of stages improves the product quality while the number of lines increases the flow rate capacity. Figure 6.14 shows the schematics of a three-stage, two-line ED/EDR system.

6.4.3 ED/EDR Membranes

The ED/EDR process uses ion exchange membranes. These membranes transport dissolved ions across a conductive polymeric membrane, and they are produced in the form of foils composed of fine polymer particles with ion exchange groups anchored by a polymer matrix. Impermeable to water under pressure, the membranes are reinforced with synthetic fiber which improves the mechanical properties of the membrane [37].

There are two primary classes of iron exchange membranes: heterogeneous and homogeneous. Heterogeneous membranes are normally cheaper, have a thicker composition, a higher resistance and a rough surface which leads to higher membrane fouling potential. Homogeneous membranes normally have a thinner composition layer, with lower resistance, and a smooth surface which is less susceptible to membrane fouling. Because of their production process, the homogeneous membranes are more expensive than heterogeneous membranes to fabricate. For homogeneous membranes and heterogeneous membranes, the charged groups are attached to the membranes differently. For homogeneous membranes, charged groups are chemically bonded, while for heterogeneous membrane, the charge groups are physically mixed with the membrane matrix.

The ion exchange membranes are ion selective, and depending on the selectivity of the membranes, they can be classified as cation transfer membranes or anion transfer membranes. Both cation transfer membranes and anion transfer membranes are electrically conductive. Cation transfer membranes allow only positively charge ions to pass through while the anion transfer membranes allow only negatively charged ions to pass through. Cation transfer membranes generally consist of cross-linked polystyrene that has been sulfonated to produce $-SO_3H$ groups attached to the polymer. This $-SO_3H$ group ionizes in the water creating negatively charged sites for pairing of positively charged ions and releasing hydrogen ions (H^+) into the water. The anion transfer membrane matrix has fixed positive charges from quaternary ammonium groups ($-NR_3^+OH^-$) [38]. These positive charges provide pairing sites for the negative ions. In addition to charge of ions, the ion exchange membranes can be ion selective depending on the valence of the ions. For example, some ion exchange membranes may show high permeability for monovalent anions, such as Cl^- or NO_3^-, but have low permeability for divalent or trivalent ions, such as SO_4^{2-}, CO_3^{2-}, PO_4^{3-}, or other similar anions.

There are different ion exchange membranes commercially available on the market that are fabricated by various manufacturers. Membranes from each manufacturer may have certain advantages and disadvantages over other membranes, and they each may have different properties such as size, thickness, resistance, and composition, and may be a suitable fit for different applications. The membranes also share some common properties like low electrical resistance, insolubility in water, easiness in handling during fabrication and assembly, stability within a wide pH range from 1 to 10, resistance to osmotic swelling, resistance to membrane fouling, and so on. There are more than 10 major ion exchange membrane manufacturers globally, mainly from Japan, the USA, Germany, and China. Here are the names of some of the major players in the market.

- Asahi Chemical Industry Company (Japan, Aciplex),
- DuPont Company (USA, Nafion),
- FuMA-Tech GmbH (Germany, Fumasep),
- GE Water & Process (USA, AR/CR),
- MEGA a.s. (Czech Republic, Ralex),
- Tianwei Membrane Company Limited (China, TWAED).

6.4.4 Membrane Spacers

Similar to RO membranes, spacers are used between the ion exchange membranes. Spacers are used to create a flow channel for the water between membrane layers, to create turbulent flow to mitigate the CP in the boundary layer, and to improve the transport efficiency of ions from the fluid stream to the membrane surface. Membrane spacers are alternately positioned between membranes in the stack to create independent flow paths.

Ion exchange membrane spacers are normally made of polypropylene or low density polyethylene [38]. The spacers can be designed with different configurations, thickness, mesh sizes, and mesh shapes to meet specific applications. But one rule of thumb in the design is that the spacer shall promote the turbulence of the stream to mitigate the polarization without causing significant pressure drop throughout the ED/EDR stack. In the ED/EDR system, the flow velocity is normally low, in the range of 18–35 cm/s (0.6–1.2 ft/s) [38]. At low flow velocity, membrane spacers are a critical factor to reduce the polarization phenomenon. The ED/EDR system is a low-pressure system, and the feed pressure is generally limited to 50 psi which needs to be maintained throughout the system. For this reason, pressure drop through each stage must be minimized [38]. Hence, pressure drop through the flow channel within the ED/EDR system is one of the critical design parameters in spacer design.

There are typically two spacer designs: "tortuous path" and "sheet flow" design. In the tortuous path design, the spacer is folded back upon itself and

(a) Tortuous path spacers

(b) Sheet flow spacers

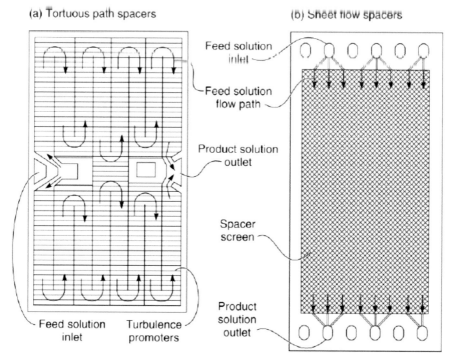

Figure 6.15 Two typical geometries of the spacers used in ED and EDR systems.

Source: [39].

the liquid flows through the path defined by the spacer. In this way, the path is longer than the linear dimension or the unit. The sheet flow design consists of an open frame with a plastic screen creating the flow channel between membranes. The typical geometries of the two types of spacers are illustrated in Figure 6.15.

6.4.5 ED/EDR Electrodes

Electrodes are a key component in ED/EDR systems, and many relevant aspects including electrochemical reactions, energy consumption, transport through the membranes, reversibility, etc., are important for the operation of the ED/EDR system [40]. Much research has been done to test and compare different electrodes for applications in the ED/EDR system. Burheim et al. [41] tested several different electrode materials and salts, and demonstrated that electrochemical reactions are controlled by mass transfer in the electrolyte rather than by the electrocatalytic properties of the electrode material.

Because of the corrosive nature of the anode compartments, electrodes are usually made of titanium and plated with platinum. Their life span is dependent on the ionic composition of the source water and the amperage applied to the electrodes. High concentration of chlorides in the source water and high amperages will reduce the service life of the electrodes [38]. Polarity reversal in EDR systems also results in a significantly shorter service life than when compared to non-reversing ED systems [37].

Oxidation and reduction reactions occur at the cathodes and anodes. The reactions transform ionic conduction to electron conduction, providing the required driving force for ion mitigation. Generally speaking, electrodes used in ED/EDR systems are inert and plate-shaped. For water treatment applications, the following reactions occur.

At the cathode, water molecules hydrolyze and the protons (H^+) receive electrons to form hydrogen gas (H_2) as a byproduct of hydroxide ions (OH^-).

$$2H_2O + 2e^- \rightarrow 2OH^-(aq) + H_2(g)$$

In addition to the reaction above, other cations may also receive electrons to form metal deposition on the surface of cathodes, creating concern of electrode scaling. The generic reaction is described as:

$$M^{n+} + ne^- \rightarrow M$$

The anions in the water, like OH^-, Cl^-, and SO_4^{2-}, diffuse through the anion exchange membranes and migrate to the anode. At the anode, the following reaction occurs to produce oxygen (O_2) and protons (H^+):

$$2H_2O \rightarrow 4e^- + 4H^+(aq) + O_2(g)$$

If chloride ions (Cl^-) are present in the water, the following reaction occurs to produce chlorine gas (Cl_2):

$$2Cl^- \rightarrow 2e^- + Cl_2(g)$$

Because of the reactions discussed above, the concentration of hydroxides may increase in the catholyte and protons may increase in the anolyte creating either scaling and/or corrosion.

6.4.6 Comparison between ED/EDR and RO

Today, RO is the prevailing technology for desalination. Compared to RO, ED/EDR has its advantages and drawbacks. In some cases, its advantages make ED/EDR an effective alternative to RO technology, but in most desalination applications, its drawbacks may outweigh its advantages over RO.

Since ED/EDR desalination is driven by electrical potential, only polar substances that can dissociate in the solute can be removed by this process; and nonpolar substances like silica, microorganisms, suspended solids, non-dissociated organics etc. cannot be removed by ED/EDR systems. One advantage of ED/EDR is that during its operation, it is the ions that migrate from the bulk solution past the membranes. Since there is no convection of water or nonpolar substances through the membrane, ED/EDR membranes are less subject to membrane fouling caused by nonpolar substances and can tolerate significant levels of these materials in the feed water if the reversal feature is incorporated. This advantage of ED/EDR over RO membrane technologies can become a disadvantage if the contaminants removal is evaluated. There is no barrier in the ED/EDR system to remove the mentioned nonpolar substances, and they will pass through the ED/EDR system. Therefore, ED/EDR cannot remove viruses and taste or odor compounds from drinking water to meet potable water standards.

Another advantage of the ED/EDR system is that higher water recovery can be achieved. As indicated in Figure 6.14, the concentrate is generally recycled to increase recovery. Water recovery of over 95% has been reported in some case studies [42]. In many cases, this advantage of ED/EDR may outweigh its disadvantages over RO membrane technologies, and make the ED/EDR process a valuable and effective alternative to RO. This is true for many brackish water desalination applications.

References

[1] Yoon S. (2015). *Membrane Bioreactor Processes: Principles and Applications*. Florida, FL: CRC Press.

[2] Chen, J. P., Mou, H., Wang, L. K., and Matsuura, T. (2006). *Membrane and Desalination Technologies*. New York, NY: The Human Press.

[3] MWH. (2005). *Water Treatment: Principles and Design, Second Edition*. Hoboken, NJ: John Wiley & Sons.

[4] Yan, D. (2011, May). *Study on Strategies to Reduce Membrane Scaling and Fouling in Drinking Water and Water Reuse Membrane Systems*. Retrieved from Proquest Website: http://search.proquest.com/docview/879043126

[5] ASTM International. (2006). *ASTM D4189-07 Standard Test Method for Silt Density Index (SDI) of Water*. West Conshohocken, PA: ASTM International.

[6] Robert, C. (2010). Reverse Osmosis Design and Concentrate Discharge Evolution in Florida the Past Three Decades. *Florida Water Resources Journal*, 19–31.

[7] Lenntech, Reverse Osmosis Desalination Process, Retrieved August 9, 2016, website: http://www.lenntech.com/processes/desalination/reverse-osmosis/general/reverse-osmosis-desalination-process.htm

[8] LANXESS Deutschland GmbH. (2012). *Guidelines for the Design of Reverse Osmosis Membrane Systems*. Technical Brochure, Leverkusen, Germany.

[9] Butt, F. H., Rahman, F., and Baduruthamal, U. (1997). Characterization of foulants by autopsy of RO desalination membrane. *Desalination*, 114, 51–64.

[10] Graham, S. I., Reitz, R. L., and Hickman, C. E. (1989). "Proc. 4th World Congress on Desalination and Water Reuse," in *4th World Congress on Desaliantion and Water Reuse* (p. 113). Kuwait: 4th World Congress on Desaliantion and Water Reuse.

[11] Leitner, G. F. (1987). *Water Desalination Report.* Tracys Landing, MD: Water Desalination Report & Global Water Intelligence.

[12] Slater, C. S., Ahlert, R. C., and Uchrin, C. G. (1983). Application of reverse osmosis to complex industrial wastewater treatment. *Desalination,* 48, 171–187.

[13] Potts, D. E., Ahlert, R. C., and Wang, S. S. (1981). A critical review of fouling of reverse osmosis membranes. *Desalination, 36*, 235–264.

[14] Wojslaw, J. A., Grantham, J. H., Arora, M. L., and Trompeter, K. M. (1983). Relationship between plugging index particle counts, and turbidity; commonly utilized parameters in the operation of RO facilities. *Desalination*, 48, 293–297.

[15] Shakaib, M., Hasani, S. M., and Mahmood, M. (2007). Study on the effects of spacer geometry in membrane feed channels using three-dimensional computational flow modeling. *J. Membr. Sci.*, 297, 74–89.

[16] Lipnizki, J., and Jonsson, G. (2002). Flow dynamics and concentration polarization in spacer-filled channels. *Desalination*, 146, 213–217.

[17] Schwinge, P., Wiley, D. E., and Fane, A. G. (2001). Simulation of the performance of spiral wound modules to assess fouling behavior. *Process Engineering with Membrane*, 205–209.

[18] Geraldes, V., Semiao, V., and Pinho, M. N. (2003). Hydrodynamics and concentration polarization in NF/RO spiral-wound modules with ladder-type spacers. *Desalination,* 157, 395–402.

[19] Dudley, L. Y., and Baker, J. S. (2006). *The Role of Anti-scalants and Cleaning Chemicals to Control Membrane Fouling.* Naperville, IL: Nalco Company.

[20] Hydranautics. (2010, July 26). *Hydranautics.* Retrieved July 26, 2010, from Hydranautics Service Bulletin: http://www.membranes.com/docs/tsb/tsb107.pdf

[21] Mulder, M. H. (2001). *Basic Principles of Membrane Technology.* The Netherlands: Kluwer Academic Publishers.

[22] McCutcheon, J. R., McGinnis, R. L., and Elimelech, M. (2005). A Novel Ammonia-Carbon Dioxide Forward (Direct) Osmosis Desalination Process. *Desalination*, 1–11.

[23] Cath, T. Y., Childress, A. E., and Elimelech, M. (2006). Forward osmosis: principles, applications, and recent developments. *J. Membr. Sci.* 281, 70–87.

[24] Cath, T. Y., Drewes, J. E., and Lundin, C. D. (2009). *A Novel Hybrid Forward Osmosis Process for Drinking Water Augmentation Using Impaired Water and Saline Water Sources.* Denver, CO: Water Research Foundation.

[25] McCutcheon, J. R., and Elimelech, M. (2006). Influence of concentrative and dilutive internal concentration polarization on flux behavior in forward osmosis. *J. Membr. Sci.* 237–247.

[26] Cath, T. Y., Gormly, S., Beaudry, E. G., Flynn, M. T., Adams, V. D., and Childress, A. E. (2005). Membrane contactor processes for wastewater reclamation in space part I. direct osmotic concentration as pretreatment for reverse osmosis. *J. Membr. Sci.* 257, 85–98.

[27] Mehta, G. D. (1982). Further results on the performance of present-day osmotic membranes in various osmotic regions. *J. Membr. Sci.* 10, 3–19.

[28] Shaffer, D. L., Werber, J. R., Jaramillo, H., Lin, S., Elimelech, M. (2015). Forward osmosis: Where are we now? *Desalination*, 356, 271–284. doi:10.1016/j.desal.2014.10.031

[29] McCutcheon, J. R., McGinnis, R. L., and Elimelech, M. (2006). Desalination by a novel ammonia-carbon dioxide forward osmosis process: influence of draw and feed solution concentrations on process performance. *J. Membr. Sci.* 278, 114–123.

[30] Lee, K. L., Baker, R. W., and Lonsdale, H. K. (1981). Membrane for power generation by pressure-retarded osmosis. *J. Membr. Sci.* 8, 141–171.

[31] Loeb, S., Titelman, L., Korngold, E., and Freiman, J. (1997). Effect of porous support fabric on osmosis through a loeb-sourirajan type asymmetric membrane. *J. Membr. Sci.* 243–249.

[32] Alsvik, I. L., and Hagg, M.-B. (2013). Pressure retarded osmosis and forward osmosis membranes: materials and methods. *Polymers*, 5, 303–327.

[33] Agenson, K. O., and Urase, T. (2007). Change in membrane performance due to organic fouling in nanofiltration (NF)/reverse osmosis (RO) applications. *Separat. Purif. Technol.* 55, 147–156.

[34] Cath, T. Y. (2010). Osmotically and thermally driven membrane processes for enhancement of water recovery in desalination processes. *Desalin. Water Treat.* 15, 279–286.

[35] *Modernwater.* (2010). Retrieved from www.modernwater.com: http://www.modernwater.com/assets/downloads/Factsheets/MW_Factsheet_Membrane_HIGHRES.pdf

[36] Watson, I. C., Morin, O. J., and Henthorne, L. (2003). *Desalting Handbook for Planners.* Tampa, FL: RosTek Associates, Inc.

[37] AWWA. (1995). *AWWA M38. Electrodialysis and Electrodialysis Reversal.* Denver, CO: American Water Works Association.

[38] Valero, F., Barcelo, A., and Arbos, R. (2011). "Electrodialysis Technology—Theory and Applications," in *Desalination, Trends and Technologies.* Spain: InTech.

[39] Baker, R. W. (2012). *Membrane Technology and Applications.* Newark, CA: John Wiley & Sons.

[40] Veerman, J., Saakes, M., Metz, S. J., and Harmsen, G. J. (2010). Reverse electrodialysis: evaluation of suitable electrode systems. *J. Appl. Electrochem.* 40, 1461–1474.

[41] Burheim, O. S., Seland, F., Pharoah, J. G., and Kjelstrup, S. (2012). Improved electrode systems for reverse electro-dialysis and electro-dialysis. *Desalination, 285,* 147–152.

[42] Allison, R. P. (1993). "High Water Recovery with Electricaldialysis Reversal," in *1993 AWWA Membrane Conference.* Baltimore: MD.

CHAPTER 7

Pretreatment

Nikolay Voutchkov[1]

[1]Managing Director, Water Globe Consultants, LLC, 824 Contravest
Lane, Winter Springs, FL 32708, USA
Phone: +1 203 253 1312
E-mail: nvoutchkov@water-g.com

7.1 Introduction

Seawater pretreatment is a key component of every membrane desalination plant. The main purpose of the pretreatment system is to remove particulate, colloidal, organic, mineral, and microbiological contaminants contained in the source seawater and to prevent their accumulation on the downstream seawater reverse osmosis (SWRO) membranes (i.e., to protect the membranes from fouling). The content and nature of foulants contained in the source seawater depend on the type and location of the desalination plant intake.

As compared to SWRO plants, thermal desalination facilities have much more simplified pretreatment because the evaporation process is not very sensitive to the content of suspended solids, algae, microorganisms, and colloidal substances in the seawater. The pretreatment of thermal desalination plants typically consists of the addition of antiscalants which aim to prevent the formation of mineral scale on the walls of the evaporation chambers. Such scale deposits reduce significantly the efficiency of the evaporation process over time. In addition, thermal desalination plants use antifoaming substances and corrosion protection chemicals. For thermal desalination facilities the pretreatment process must address:

- Scaling of the heat exchanger surfaces primarily from calcium and magnesium salts (acid treated plants);
- Corrosion of the plant components primarily from dissolved gases;
- Physical erosion by suspended solids;
- Effects of other constituents such as oil;
- Growth of aquatic organisms and content of heavy metals.

Thermal desalination systems are quite robust and normally do not include any physical treatment other than what is provided by the intake (i.e., no additional filters or screens). Chemical conditioning is utilized in thermal desalination in two treatment streams: the cooling water (which is the larger flow and generally returned to the feed source), and makeup water (used within the desalination process). Cooling water is normally treated to control fouling using an oxidizing agent or biocide. The makeup water is continuously treated with scale inhibitors (usually a polymer blend) and may be intermittently dosed with an anti-foam surfactant (typically during unusual feed water conditions).

The main focus of this chapter is seawater desalination by reverse osmosis. Therefore, pretreatment technologies used for thermal desalination plants are not discussed further.

Usually, subsurface intakes (wells and infiltration galleries) produce source water of an-order-of magnitude lower content of silt and particulates as compared to open-ocean intakes because the source seawater is naturally pre-filtered by the ocean bottom sediments. However, occasionally well source water may contain high levels of dissolved and colloidal foulants (iron, manganese, silica, etc.), and natural organic matter (NOM) that are difficult and costly to remove.

The fouling potential of source seawater collected through open ocean intakes also depends on the type and location of the plant intake. Deep off-shore intakes located away from underwater currents, wastewater discharges, industrial ports, ship traffic corridors, and river estuaries typically produce low-fouling seawater. The source water quality of on-shore intakes is often influenced by algal blooms, surface runoff and tidal movement, and could be quite variable and challenging.

Table 7.1 presents key source seawater quality parameters that have significant impact on the selection and configuration of the pretreatment system for a given SWRO desalination plant. This table also indicates threshold levels of significance of these parameters established based on full-scale experience.

Traditionally, seawater foulants are removed by a series of source water conditioning processes (coagulation, flocculation, and pH adjustment) followed by conventional granular media (anthracite or pumice and sand) filtration. Over the past 10 years, advances in microfiltration (MF) and ultra-filtration (UF) membrane technologies, and their successful application for water and wastewater treatment have created an impetus for using membrane pretreatment in seawater desalination plants [1, 17].

Table 7.1 Seawater quality characterization for pretreatment

Source Seawater Quality Parameter	Pretreatment Issues and Considerations
Turbidity (NTU)	Should be lowered to under 1 NTU (preferably under 0.5 NTU) before the water is allowed to reach the RO membrane. Spikes above 50 NTU for more than 1 h would require sedimentation or dissolved air flotation (DAF) treatment prior to filtration.
Total Organic Carbon (mg/L)	If below 0.5 mg/L—biofouling is unlikely. Above 2 mg/L—biofouling is very likely.
Silt Density Index (SDI_{15})	Source seawater levels consistently below 2 year around, typically indicate that no additional filtration pretreatment is needed. SDI > 4—pretreatment is necessary.
Total suspended solids (mg/L)	This parameter is needed to assess the amount of residuals generated during pretreatment. Does not correlate well with turbidity above 5 NTU.
Iron (mg/L)	If iron is in reduced form, SWRO membranes can tolerate up to 2 mg/L. If iron is in oxidized form, concentration >0.05 mg/L would cause accelerated fouling.
Manganese (mg/L)	If manganese is in reduced form, SWRO membranes can tolerate up to 0.1 mg/L. If manganese is in oxidized form, concentration >0.02 mg/L would cause accelerated fouling.
Silica (mg/L)	Concentrations higher than 20 mg/L may cause accelerated fouling. Analyze for colloidal silica if concentration >20 mg/L.
Free chlorine (mg/L)	Concentrations higher than 0.01 mg/L would cause RO membrane damage.
Temperature (°C)	$T \leq 12°C$ would cause significant increase in unit energy use. $T \geq 35°C$ may cause accelerated mineral scaling and biofouling. $T > 45°C$ would cause irreversible RO membrane damage.
Oil and grease (mg/L)	Concentrations higher than 0.02 mg/L would cause accelerated organic fouling.
pH (units)	Typical pH of seawater is 7.6–8.3. Long-term exposure to pH < 4 and pH > 11 may cause membrane damage.

Both granular media and membrane filtration pretreatment technologies may offer advantages or face challenges depending on the source seawater quality and origin. Therefore, selecting the most suitable pretreatment technology for a given project should be based on a comprehensive performance and life-cycle cost analysis, and whenever possible on side-by-side pilot testing. Pilot testing should target a time-frame encompassing events that could create pretreatment challenges such as: algal blooms, intake area dredging, heavy ship traffic, rainfall season, seasonal currents and winds etc.

7.2 Overview of Granular Media Filtration Technologies

At present, granular media (conventional) filtration is the most widely used seawater pretreatment technology [2]. A typical schematic of a desalination plant with granular media filters is shown on Figure 7.1.

This process includes source seawater conditioning by coagulation and flocculation followed by filtration through one or more layers of granular media (e.g., anthracite coal, pumice, silica sand, and garnet). Conventional filters used for seawater pretreatment are typically rapid single-stage dual-media (anthracite and sand) units. However, in some cases where the source seawater contains high levels of organics, algal content and suspended solids, two-stage filtration systems are used (Figure 7.2).

Under this configuration, the first filtration stage is mainly designed to remove coarse solids (i.e., particulates and large algae) and organics in suspended form. The second-stage filters are configured to retain fine solids and silt, and to remove a portion (20–40%) of the soluble organics contained in the seawater by biofiltration.

Depending on the driving force for seawater filtration, granular media filters are classified as gravity and pressure filters (Figures 7.3 and 7.4).

The two types of filters use the same type of media but have different head requirement to convey feed water through the media bed; operate at different filtration rate; and their media is installed in different types of vessels. Practical experience shows that pressure filters are more cost effective for small and medium size SWRO plants. Gravity pretreatment filters are used for all size SWRO desalination plants but have found wider application for large and medium size facilities.

Since the purpose of the pretreatment filters for SWRO plants is not only to remove over 99% of all suspended solids in the source seawater, but also to reduce the content of the much finer silt particles by several orders of magnitude, the design of these pretreatment facilities is usually governed by the filter effluent silt density index (SDI) target levels rather than by the target removal levels of turbidity and pathogens.

Granular media filters are typically backwashed using filtered seawater or concentrate from the SWRO membrane system. Filter cell backwash frequency is usually once every 24–48 h and spent (waste) backwash volume is 2–5% of the intake seawater (i.e., recovery rate of the filters is 95–98%). Use of SWRO concentrate instead of filtered effluent to backwash filter cells allows reducing backwash volume and energy needed to pump source seawater to the desalination plant.

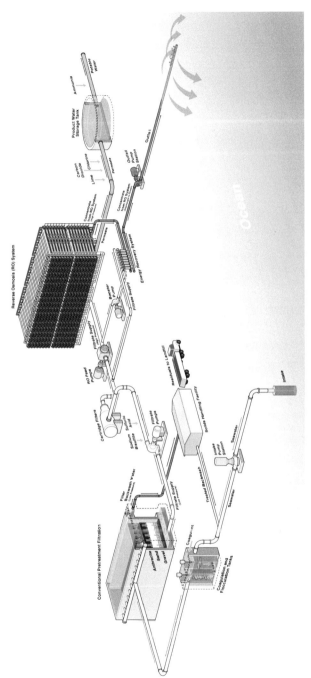

Figure 7.1 Typical schematic of SWRO desalination plant.

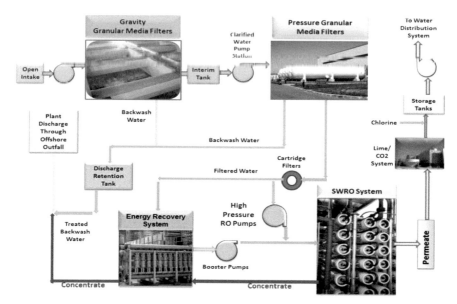

Figure 7.2 SWRO desalination plant with two-stage pretreatment filtration.

Figure 7.3 Gravity granular media filters.

Figure 7.4 Pressure granular media filters.

7.3 Seawater Conditioning Prior to Granular Filtration

Most seawater particles and microorganisms have a slightly negative charge which has to be neutralized by coagulation for efficient granular media filtration. In addition, these neutralized particles would need to be agglomerated in larger flocs that can be effectively retained within the filter media. Therefore, source seawater conditioning by coagulation (Figure 7.5) and subsequent flocculation (Figure 7.6) are necessary steps prior to granular media filtration.

Use of a coagulant is critical for the effective and consistent performance of granular media filtration systems. Without coagulation of the particulates in the source seawater, conventional granular media filters are likely to completely remove only particles which are larger than 50 μm. Coagulation allows granular filtration process to remove finer particles and micro-plankton from the source seawater. Well-operating filters can remove particles as small as 0.2 μm.

For comparison, membrane pretreatment systems (Figure 7.7) can remove particles as fine as 0.1 μm (MF membranes) or 0.01 μm (UF membranes) by direct physical separation. Therefore, for these systems coagulation of the source water particles is not as important and usually is only applied when the source seawater contains NOM that could be coagulated and removed via filtration.

The main purpose of the coagulation system is to achieve uniform mixing of the added coagulant with the source seawater and efficient neutralization of the charge of solid particles contained in the seawater. The two types of

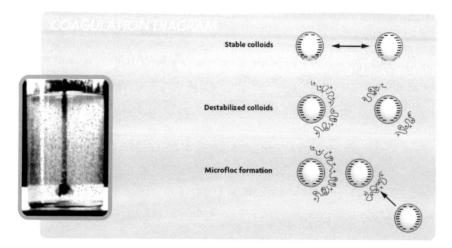

Figure 7.5   Coagulation process.

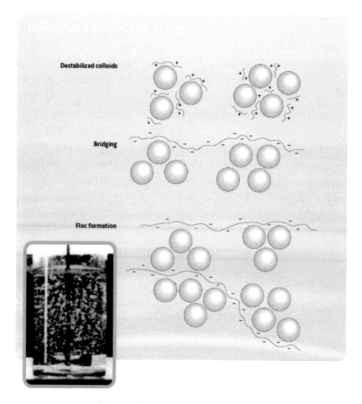

Figure 7.6   Flocculation process.

Figure 7.7 Typical membrane pretreatment system.

mixing systems most widely used in seawater desalination plants are in-line static mixers and mechanical flash mixers installed in coagulation tanks.

In-line static mixers and subsequent in-pipe flocculation are significantly less expensive than the construction of separate coagulation and flocculation tanks and therefore, they are often preferred by engineering contractors. However, when the source seawater originates from deep open-ocean intakes (i.e., intakes located deeper than 10 m below the ocean surface) very often the size and charge of the particles contained in this water are very small. Therefore, the complete coagulation and flocculation of such particles requires contact time that is much longer than that for particles contained in seawater collected within 5 m from the ocean surface. Adequate contact time and conditions for agglomerating fine particles originating from deep intakes can typically be achieved only by seawater conditioning in coagulation and flocculation chambers equipped with mechanical flash mixers. This is a common challenge in desalination plants with in-line coagulation and flocculation and source seawater originating from deep open intakes. This challenge is usually unsuccessfully attempted to be solved by increasing of the coagulant dosage which only worsens the performance of the pretreatment and SWRO systems.

Coagulants most frequently used for source seawater conditioning prior to sedimentation or filtration are ferric salts (ferric sulfate and ferric chloride). Aluminum salts (such as alum or polyaluminum chloride) are avoided because it is difficult to maintain aluminum concentrations at low levels in dissolved form since aluminum solubility is very pH dependent. Small amounts of aluminum may cause mineral fouling of the downstream SWRO membrane elements.

The optimum coagulant dosage is pH dependent and should be established based on an on-site jar or pilot testing for the site-specific conditions of a given application. In some cases, addition of sulfuric acid for source water pH reduction to 5.5–6.0 is practiced in order to optimize floc formation.

Overdosing of coagulants used for seawater pretreatment is one of the most frequent causes for SWRO membrane mineral fouling. When overdosed, coagulants accumulate on the downstream facilities and can cause accelerated fouling of the downstream cartridge filters and SWRO membranes. In such a situation, a significant improvement of source water SDI can be attained by reducing coagulant feed dosage or in case of poor mixing, by modifying the coagulant mixing and flow distribution system to minimize the content of unreacted chemical in the filtered seawater fed to the SWRO membrane system.

7.4 Seawater Pretreatment Prior to Filtration

Most of the existing pretreatment systems in operation today are single-stage/granular dual-media filters and are designed to operate without sedimentation or dissolved air flotation of the source seawater prior to filtration. However, source seawater may need to undergo additional pretreatment prior to filtration (sand removal, sedimentation, DAF, or initial filtration) depending on its quality.

7.4.1 Sedimentation

Sedimentation is typically used upstream of granular media and membrane filters when membrane plant source water has daily average turbidity higher than 30 NTU or experiences turbidity spikes of 50 NTU or more which continue for a period of over 1 h. If sedimentation basins are not provided, large turbidity spikes may cause the pretreatment filters to exceed their solids holding capacity (especially if pressure driven granular media filters are used), which in turn may impact filter pretreatment capacity. If the high solids load continues, the pretreatment filters would enter into a condition of continuous backwash, which in turn would render them out of service.

Sedimentation basins for seawater pretreatment are typically designed to produce settled water of less than 2.0 NTU and SDI_{15} below 6. To achieve this target level of turbidity removal, sedimentation basins are often equipped with both coagulant (most frequently iron salt) and flocculant (polymer) feed systems. The needed coagulant and flocculent dosages should be established based on jar and/or pilot testing.

If the source water turbidity exceeds 100 NTU, then conventional sedimentation basins are often inadequate to produce turbidity of the desired concentration of less than 2 NTU and of low silt and algal content. Under these conditions, sedimentation basins should be designed for enhanced solids removal by installing lamella plate modules or using sedimentation technologies that combine lamella and fine granular media for enhanced solids removal.

Typically, enhanced sedimentation technologies are used for treating source water from open ocean intakes which are under a strong influence of river water or of wastewater effluent of high turbidity. This condition could occur when the desalination plant intake is located in a river estuary area or is influenced by a seasonal surface water runoff. For example, during the rainy season, the intake of the Point Lisas seawater desalination plant in Trinidad is under the influence of the Orinoco River currents which carry a large amount of alluvial suspended solids. As a result, the desalination plant intake turbidity could exceed 200 NTU [3]. To handle this extremely high solids load, the plant source water is settled in a lamella sedimentation tanks prior to conventional single-stage dual-media filtration.

7.4.2 Dissolved Air Flotation

Dissolved air flotation (DAF) technology is suitable for removal of floating particulate foulants such as algal cells, oil, grease or other contaminants that cannot be effectively removed by sedimentation or filtration. DAF systems can typically produce effluent turbidity of <0.5 NTU and can be combined in one structure with dual-media gravity filters for sequential pretreatment of seawater.

Dissolved air flotation process uses very small air bubbles (resulting from pressurized dissolution of air in the water followed by pressure release) to float light particles (algae, fine silt, and debris) and organic substances (oil and grease) contained in the seawater. The floated solids are collected at the top of the DAF tank and skimmed off for disposal, while the low-turbidity seawater is collected near the bottom of the tank (Figure 7.8).

The time (and therefore, the size of the flocculation tank) needed for the light fine particulates contained in the seawater to form large flocs is usually 2 to 3 times shorter than that of conventional flocculation tanks, because the flocculation process is accelerated by the air bubbles released in the flocculation chamber of the DAF tanks. In addition, the hydraulic loading rate of the DAF system can be several times higher than conventional sedimentation Another benefit of DAF as compared to conventional sedimentation is the higher density of the formed residuals (sludge). While residuals collected at the bottom of the sedimentation basins typically have concentration of only 0.3–0.5% solids, DAF residuals (which are skimmed from the surface of the DAF tank) have solids concentration of 1–3%.

In some full-scale applications, the DAF process is combined with granular media filters to provide a compact and robust pretreatment of seawater with high algal and/or oil and grease content [4]. Although this combined DAF/filter configuration is very compact and cost-competitive, it has three key disadvantages: (i) complicates the design and operation of the pretreatment filters; (ii) DAF loading is controlled by the filter loading rate and therefore, DAF tanks are typically oversized; and (iii) flocculation tanks must be coupled with individual filter cells.

The feasibility of DAF use for seawater pretreatment is determined by seawater quality and governed by the source water turbidity, the size of the particles and the overall lifecycle pretreatment costs. The DAF process can handle source seawater with turbidity of up to 50 NTU. Therefore, if the source seawater is impacted by high turbidity spikes or heavy solids (usually related to seasonal river discharges or surface runoff), than DAF may not be a suitable pretreatment option.

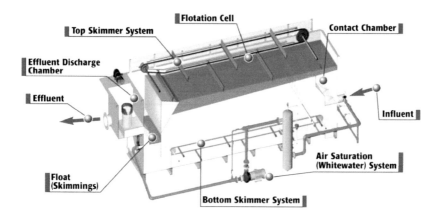

Figure 7.8 General schematic of the DAF clarifier.

In addition to source water turbidity concentration, another factor of key importance for the viability of DAF as pretreatment upstream of the filters is the type/size of plankton contained in the water. Often, plankton community in seawater is dominated by small algal cells with diameter of less than a few micrometers (μm) typically referred to as pico-plankton (0.2–2.0 μm [5]). As compared to fresh waters from lakes and rivers where pico-plankton usually does not dominate the algal community, in seawater pico-plankton is associated with more than 50% of the chlorophyll a (algal content) in most ocean waters [6] and in the cases of tropical or subtropical seawater, this content may reach 75% or more [7].

Bench testing experience at the Carlsbad seawater desalination demonstration project in California has indicated that pico-plankton may not be well removed by DAF even after elaborate coagulation and flocculation (observed removal efficiency of chlorophyll *a* and turbidity was typically less than 60%). In such cases, the use of conservatively designed deep single-stage dual media granular filters or two-stage dual media granular filters may render a better pretreatment than a combination of DAF and granular media or membrane filtration.

Taking into consideration the significant importance pico-plankton content may have on the selection of pretreatment system, it is essential to characterize both the algal concentration of the source water (measured as chlorophyll *a*) and the algal profile of the source water, which quantifies the total concentration of algal cells measured in cells/mL, as well as identifies the types of individual algae species contained in the sample, and determines the cell counts of these species in cells/mL. Algal profile should be completed under both normal and algal bloom conditions.

Although DAF systems have much smaller footprint than conventional flocculation and sedimentation facilities, they include a number of additional equipment associated with air saturation and diffusion, and with recirculation of a portion of the treated flow, and therefore, their construction costs are typically comparable to those of conventional sedimentation basins. Usually, the operation and maintenance (O&M) costs of DAF systems are higher than those of gravity sedimentation tanks due to the higher power use for the flocculation chamber mixers, air saturators, recycling pumps, and sludge skimmers. The total power use of DAF systems is usually 2.5–3.0 kWh/10,000 m^3/day of treated source seawater, which is significantly higher than that used by sedimentation systems (0.5–0.7 kWh/10,000 m^3/day of treated seawater).

Dissolved air flotation process with built-in filtration (DAFF) is used at the 136,000 m³/day Tuas seawater desalination plant in Singapore [8]. This pretreatment technology has been selected for this project to address the source water quality challenges associated with the location of the desalination plant's open intake in a large industrial port area (i.e., oil spills) and the frequent occurrence of algal blooms in the area of the intake. The source seawater has total suspended solids (TSS) concentration that can reach up to 60 mg/L at times and oil and grease levels in the seawater could be up to 10 mg/L. The facility uses 20 build-in filter DAF units, two of which are operated as standby. Plastic covers shield the surface of the tanks to prevent impact of rain and wind on DAF operation as well as to control algal growth. Each DAF unit is equipped with two mechanical flocculation tanks located within the same DAF vessel. Up to 12% of the filtered water is saturated with air and recirculated to the feed of the DAF units.

A combination of DAF followed by two-stage dual-media pressure filtration has been successfully used at the 45,400 m³/day El Coloso SWRO plant in Chile, which at present is the largest desalination plant in South America. The plant is located in the City of Antogofasta, where seawater is exposed to year-round red-tide events, which have the capacity to create frequent particulate fouling and biofouling of the SWRO membranes [9]. The DAF system at this plant is combined in one facility with a coagulation and flocculation chamber. The average and maximum flow rising velocities of the DAF system are 22 and 33 m³/m²h, respectively. This DAF system can be bypassed during normal operation and is typically used during red-tide events. The downstream granular media filters are designed for surface loading rate of 25 m³/m²h. Ferric chloride at a dosage of 10 mg/L is added ahead of the DAF system for source water coagulation. The DAF system reduces source seawater turbidity to between 0.5 and 1.5 NTU and removes approximately 30 to 40% of the source seawater organics.

Another example of a large seawater desalination plant incorporating DAF units for pretreatment is the 200,000 m³/day Barcelona facility in Spain [10]. The pretreatment system of this plant incorporates 10 high-rate AquaDAF units equipped with flocculation chambers; 20 first-stage gravity dual-media filters, and 24 second-stage pressurized dual-media filters. The purpose of the DAF system is mainly to remove algae during blooms and to reduce source water organic content. The intake of the desalination plant is located 2.2 km from the coast and 3 km away from the entrance of a large river (Llobregat River) to the ocean, which carries significant amounts of alluvial (NOM-rich) organics. After coagulation with ferric chloride and flocculation in flash-mixing chambers, over 30% of these organics are removed by the DAF system.

7.5 Selection of Granular Filter Media

Filter media type, uniformity, size and depth are of key importance for the performance of seawater pretreatment filters. Dual media filters have two layers of filtration media – typical design includes 1.0–1.4 m of anthracite over 0.6–0.8 m of sand (Figure 7.9).

Deep dual media filters are often used if the desalination plant filtration system is designed to achieve enhanced removal of soluble organics from seawater by biofiltration and/or to handle seawater with high picoplankton content. In this case, the depth of the anthracite level is enhanced to between 1.6 and 1.8 m.

If the source seawater is relatively cold (i.e., its average annual temperature is below 15°C), and at the same time is of high organic content, a layer of granular-activated carbon (GAC) of the same depth is used instead of a

Anthracite Coal

Silica Sand

Gravel

Figure 7.9   Typical configuration of granular filter media.

deeper layer of anthracite, because the biofiltration removal efficiency will be hindered by the low temperature. During biofiltration, a portion of the soluble organics in the seawater is metabolized by the microorganisms that grow on a thin biofilm formed on the granular filter media and the GAC media removes a portion of the seawater organics mainly by adsorption.

Tri-media filters have 0.45–0.60 m of anthracite as the top layer, 0.20–0.30 m of sand as a middle layer and 0.10–0.15 m of garnet or limonite as the bottom layer. These filters are used if the source seawater contains a large amount of fine silt or the seawater intake experiences algal blooms dominated by pico-plankton.

Typically, the depth of the filter bed is typically a function of the media size and follows the general rule-of-thumb that the ratio between the depth of the filter bed (l—in millimeters) and the effective size of the filter media (d_e—in millimeters), l/d_e, should be higher than 1500. For example, if the effective size of the anthracite media is selected to be 0.65 mm, the depth of the anthracite bed should be at least (0.65 mm $\times$ 1500 = 0.975 mm, i.e., approximately 1.0 m).

The depth of the GAC media is estimated based on the average contact time in this media, which is recommended to be 10 to 15 minutes. For example, if a filter is designed for a surface loading rate of 9 m^3/m^2.h, the depth of the GAC media should be at least 1.5 m (9 m^3/m^2h $\times$ 10 min/60 min/h = 1.5 m).

When each of the filter media layers are first placed in the filter cells, an additional 3–5 cm of media should be added to the design depth of the layer to account for the removal/loss of fine particles from the newly installed bed after backwashing. It should also be pointed out that if the filters are designed to achieve total organic carbon (TOC) removal by biofiltration, it would take at least 4 to 6 weeks for the filters to create sustainable biofilm on the surface of the filter media grains that can yield steady and consistent filter performance and TOC removal. If the seawater temperature is relatively low (i.e., below 20°C), then the biofilm formation process may take several weeks longer.

7.6 Selection of the Type of Granular Media Filter

Depending on the driving force for seawater filtration, granular media filters are classified as gravity and pressure filters. The main differences between the two types of filters are the head required to convey the water through the media bed, the filtration rate, and the type of vessel used to contain the filter media. Because of the high cost of constructing large pressure vessels with the proper wetted surfaces for corrosion resistance, pressure filters are typically used for small and medium size capacity SWRO plants.

Typically, gravity filters are reinforced concrete structures that operate at water pressure drop through the media of between 1.5 and 2.5 m. Dual media gravity filters are the predominant type of filtration pretreatment technology presently used in desalination plants of capacity higher than 40,000 m³/day. Downflow filters are preferred because they allow to retain algal biomass contained in the source seawater in the upper layer of the filter media and to minimize algal cell rupture which could cause release of soluble biodegradable organics in the filtered seawater which accelerate SWRO membrane biofouling.

Gravity pretreatment filters have found application for both small and large desalination plants worldwide. Most large seawater desalination plants in operation today have deep open-ocean intakes and use single-stage/dual media gravity filters (i.e., 320,000 m³/day Ashkelon SWRO plant in Israel; 250,000 m³/day Sydney Water and 125,000 m³/day Gold Coast desalination plants in Australia; and the 200,000 m³/day Hamma SWRO plant in Algeria).

Pressure filters have filter bed configuration similar to that of gravity filers, except that the filter media is contained in steel or plastic pressure vessels. They have found application mainly for small and medium size seawater desalination plants—usually with production capacity of less than 20,000 m³/day. One exception is Spain, where most of the pretreatment filters in seawater desalination plants are pressure filters.

In most cases for relatively good source seawater quality (TOC < 1 mg/L, SDI < 5, and turbidity <4 NTU) pressure filters are designed as single stage, dual media (anthracite and sand) units. Some plants with relatively poor water quality use two-stage pressure filtration systems. For small desalination plants, pressure filters are very cost-competitive, more space efficient and easier, and faster to install, and perform comparably to granular media gravity filters.

Often when the source seawater is collected via open intake, two-stage dual media (anthracite and sand) pressure filters are applied. One of the key cost disadvantages of these filters is that they operate under pressure and, therefore, use more energy than gravity filters. Single-stage pressure filters have found application for several large SWRO plants such as the 160,000 m³/day Kwinana plant in Perth, Australia (Figure 7.10) and the 120,000 m³/day Carboneras SWRO plant in Spain (Figure 7.11).

7.6.1 Removal of Algal Material from Seawater

Seawater always contains some amount of algae, whose concentration usually increases several times during the summer period and with several orders of magnitude during periods of algal blooms. When predominant phytoplankton species causing the algal bloom contain red pigment (peridinin) they cause reddish discoloration of the seawater which is the main reason why such algal blooms are referred to as "red tides". A number of algal species can cause red tides. In California, for example algal blooms of red-pigmented dinoflagellates are the most common cause of red tides in the summer.

Algal species in the seawater vary in type and size and often pico-plankton dominates the algal population in tropical and equatorial regions of the world such as the Middle East. Some plankton species that occur during algal bloom events, including many red-tide causing dinoflagellates, have cells that are relatively easy to break under pressures as low as 0.4–0.6 bars. As the algal cells break under the pressure (or vacuum) applied by the filtration pretreatment system, they release their cellular cytoplasm in the seawater. This cytoplasm has a very high content of easily biodegradable polysaccharides which serve as a food source of marine bacteria. When the amount of polysaccharides released by the broken algal cells exceeds a certain threshold in the filtered seawater, they would typically trigger accelerated microbiological growth and subsequent biofouling on the SWRO membranes.

Figure 7.10   The Kwinana desalination plant in Perth, Australia.

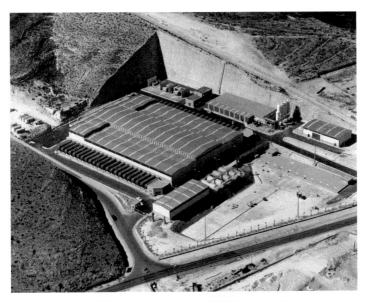

Figure 7.11 The Carboneras SWRO plant in Spain.

Because gravity filters operate at pressures below the algal-cell rupture threshold and pressure filters run at several times higher pressure than this threshold, the use of gravity filtration in source seawaters frequently exposed to heavy algal blooms (i.e., SWRO plants with on-shore or shallow off-shore intakes) is more advantageous. It is important to point out that this potential disadvantage of pressure filtration and the associated accelerated biofouling of the SWRO membranes, will manifest itself only during periods of algal blooms of measurable magnitude (typically when the total algal content in the source water is over 2,000 cells/mL and the TOC levels in seawater exceed 2 mg/L).

Overview of the performance of desalination installations in Spain and Australia, where pressure filters are successfully used for seawater pretreatment indicates that the source water quality of these plants is very good year-around (TOC <2 mg/L, SDI <5, and turbidity <5 NTU) and that the plants have deep open-ocean off-shore intakes or well intakes. At depths of 10 m and deeper, the concentration of algae entering the intake is significantly lower than at the ocean surface—and therefore, as long as the desalination plant intake is fairly deep, biofouling caused by rupture and decay of algal biomass may not be as problematic as it would be for shallow offshore intakes or intakes located at the ocean surface (i.e., near-shore open ocean intakes).

7.6.2 Useful Life of the Filter Structure

Typically, gravity filters are concrete structures that have useful lives of 50–100 years. Pressure filters are steel structures with a lifespan of 25 years or less. The internal surface of the pressure filters used for seawater desalination is typically lined with rubber or epoxy coating that may need to be replaced every 5–10 years and inspected occasionally.

7.6.3 Solids Retention Capacity and Handling of Turbidity Spikes

Gravity media filters have approximately two to three times larger volume of filtration media and retention time than pressure filters for the same water production capacity. Therefore, this type of filter can retain proportionally more solids and as a result, pretreatment filter performance is less sensitive to occasional source water turbidity spikes.

Pressure filters usually do not handle solids/turbidity spikes as well because of their smaller solids retention capacity (i.e., smaller volume of inter-media voids that can store solids before the filter needs to be back-washed). If source seawater is likely to experience occasional spikes of high turbidity (10 NTU or higher) due to rain events; algal blooms; naval traffic; ocean bottom dredging operations in the vicinity of the intake; seasonal changes in underwater current direction; or spring upwelling of water from the bottom to the surface; then pressure filters will produce effluent with inferior quality (SDI and turbidity) during such events. Therefore, the use of pressure filters would likely result in more frequent SWRO membrane cleaning.

Alternatively, in order to handle solids/turbidity spikes pressure filters could be designed as two-stage pretreatment systems with two sets of pressure filters operating in series. In similar source water quality conditions, comparable filtered water quality and performance reliability could be produced by a conservatively designed single-stage gravity-filtration system.

7.6.4 Costs

Pressure filters are prefabricated steel or plastic structures and their production costs per unit filtration capacity are 30–50% lower than those of concrete gravity filters. Since pressure filters are designed at approximately two to three times higher surface loading rates than gravity filters (20–25 m^3/m^2h vs. 6–10 m^3/m^2h), their volume and size are smaller and therefore, they usually are less costly to build and install. In addition, because pressure filter vessels are pre-fabricated, the installation time of this pretreatment system is approximately 20–30% shorter than that of gravity filters with concrete structures, which also has a positive impact on overall project construction costs and completion schedule.

On the other hand, the operation and maintenance (O&M) costs associated with pressure filters is usually higher than that of gravity filters. The useful life of the metal structures of the pressure filters is limited to 20–25 years while the concrete structures of the gravity filters can be used for 50–100 years. The internal surface of the pressure filters used for seawater desalination is typically lined with rubber coating that needs to be replaced every 5–10 years and inspected occasionally. Because pressure filters typically operate at several times higher feed pressures than gravity filters, the energy use for pressure filtration is proportionally higher.

In addition, contrary to pressure filters, the structure of the gravity filters could be designed to accommodate the installation of a vacuum-driven membrane pretreatment system and allow to take advantage of the advancement of vacuum-driven membrane pretreatment technologies as they reach maturity in the next 5–10 years.

It should be pointed out that because the site-specific conditions and priority of a given project often have a significant weight on the selection of the most suitable type of granular filtration system, worldwide experience shows that both gravity and pressure driven filters have found numerous applications in all sizes of projects. Table 7.2 shows the type and key design criteria of recent seawater desalination projects using granular media pretreatment.

7.7 Membrane Filtration Overview

Application of UF and MF membrane filtration for seawater pretreatment is relatively new. At present, approximately two-dozen operational full-scale seawater desalination plants worldwide are using membrane pretreatment [4]. However, many new plants with membrane pretreatment are currently under design and construction, and are expected to be in service by 2020.

Figure 7.12 depicts a general schematic of a seawater desalination plant with vacuum-driven membrane pretreatment. As indicated on this figure, seawater pretreatment includes several key components: (i) coarse and fine screens similar to those used for plants with conventional pretreatment; (ii) micro-screens to remove fine particulates and sharp objects from the seawater which could damage the membranes; and (iii) UF or MF membrane

Table 7.2   Examples of desalination plants with granular media pretreatment

Desalination Plant Location and Capacity	Pretreatment System Configuration	Average and Maximum Filter Loading Rates	Notes
Glen Rocky Plant, Gibraltar – 1,400 m³/day	Single-stage dual media vertical pressure filters	Filtration rate - 11 m³/m²h (avg.) and 16 m³/m²h (max)	Intake – 2/3 of volume from wells and 1/3 from off-shore open intake
Ashkelon, Israel – 325,000 m³/day	Single-stage dual media gravity filters	Filtration rate: 8 m³/m²h (avg.) and 12 m³/m²h (max)	Open intake 1,000 m from shore
Tuas, Singapore – 136,000 m³/day	Combined DAF and sand media filtration	DAF and filtration loading rates: 10 m³/m²h (avg.), 14 m³/m²h (max)	Near-shore open intake in industrial port
El Coloso, Chile – 45,400 m³/day	DAF followed by two-stage dual media horizontal pressure filters	DAF rate: 22–33 m³/m²h Filtration rate: 25 m³/m²h (avg.)	Open intake in industrial port frequently plagued by red tide events
Fujairah I, UAE – 170,500 m³/day	Single-stage, dual media gravity filters	Filtration rate: 8.5 m³/m²h (avg.) And 9.5 m³/m²h (max)	Open intake – high hydrocarbon content in seawater
Kwinana, Perth, Australia – 143,000 m³/day	Single-stage dual media pressure filters	Filtration rate: 14.0 m³/m²h (avg.) And 18.0 m³/m²h (max)	Offshore open intake
Carboneras, Spain – 120,000 m³/day	Single-stage dual media pressure filters	Filtration rate: 12.0 m³/m²h (avg.) 14.0 m³/m²h (max)	Offshore open intake
Barcelona, Spain – 200,000 m³/day	DAF, first-stage dual media gravity filters; second-stage dual media pressure filters	DAF rate – 25–40 m³/m²h Filtration rate (first stage): 10–15 m³/m²h Filtration rate (second stage): 20–25 m³/m²h	Open intake: 1,500 m from shore

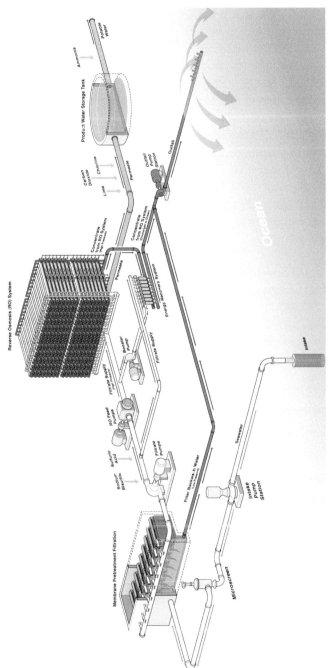

Figure 7.12 Schematic of SWRO plant with membrane pretreatment.

system. Often desalination plants with membrane pretreatment systems are constructed without the installation of cartridge filters between the pretreatment system and the SWRO membranes, taking under consideration the fact that the membrane filtration media size is an order-of-magnitude smaller than the size of the cartridge filters. However, some conservative designs incorporate cartridge filters to provide protection against the particulates that may be released into the filter effluent due to MF or UF fiber breakage.

Microfiltration and UF membrane systems have been shown to be very efficient in removing suspended solids, non-soluble and colloidal organics contained in the source seawater. Turbidity can be lowered consistently below 0.1 NTU (usually down to 0.04–0.08 NTU) and filter effluent SDI_{15} levels are usually below 3 over 90% of the time. Both MF and UF systems can remove four or more logs of pathogens such as *Giardia* and *Cryptosporidium*. In contrast to MF membranes, UF membranes can also effectively remove viruses.

It should be noted however, that membrane pretreatment does not remove significant amounts of dissolved organics and marine microorganisms, which typically cause SWRO membrane biofouling. Because of the very short seawater retention time in the membrane pretreatment systems, they do not provide measurable biofiltration effect, unless designed as membrane bioreactors. For comparison, granular media filters, depending on their configuration, loading rate and depth, could remove 20–60% of the soluble organics contained in the source seawater.

Depending on the type of the driving force, membrane pretreatment filters are divided into two categories – pressure and vacuum-driven. Pressure UF and MF systems consist of membrane elements installed in pressure vessels, which are grouped in racks (trains), similar to SWRO systems. Vacuum-driven pretreatment systems include series of filtration modules submerged into open tanks of size and configuration similar to that of gravity granular media filters.

The two most important parameters associated with the design of any membrane pretreatment system are the sustainable flux and feed water recovery. Membrane flux determines the amount of total membrane area and modules/elements needed to produce a certain volume of filtered seawater and is defined as the ratio between the total filtration area and the volume of water filtered through the membranes. This parameter is typically expressed either in cubic meters per square meter per hour ($m^3/m^2.h$) or as liters per square meter per hour (L/m^2h or lmh).

Feed water recovery indicates the fraction of the source seawater that is converted into filtrate suitable for seawater desalination. The recovery rate of membrane pretreatment systems is usually in the range of 88–94% and is typically lower than the recovery rate of granular media pretreatment filters for the same source water (95–98%). This is an important difference because it has an impact on the size of both plant intake and solids handling facilities.

Another important consideration is the operating pressure/vacuum of the pretreatment systems. Usually the vacuum driven membrane pretreatment systems operate in a trans-membrane pressure range of 0.2–0.8 bars. For comparison, pressure-driven systems typically run at pressures of 0.4–2.5 bars, which makes their pressure comparable to pressure-driven granular media filters.

As indicated previously, operating pressure may have a measurable impact on the rate of biofouling on the downstream SWRO membranes, if the source seawater is exposed to frequent algal blooms. In the case of algal bloom occurrence, both pressure and vacuum-driven membrane systems would typically operate at pressures higher than the threshold at which many algal cells break and would therefore release biodegradable organics in the filtered seawater. Unless the algae are gently removed by DAF or granular media filtration ahead of the membrane pretreatment system, the cell rupture caused by the membrane pretreatment system would accelerate SWRO membrane fouling. As a result, while membrane pretreatment systems may be able to produce better filtered water quality in terms of SDI and turbidity during heavy algal blooms they would often produce filter effluent of significantly higher SWRO biofouling potential than gravity granular media filters.

It is interesting to note that tests at the Carlsbad seawater desalination demonstration plant in 2005 when the plant was exposed to a series of heavy algal blooms (red tides) for a prolonged period of time (between May and September) clearly support the observation of the negative impact that membrane pretreatment may have on the biofouling potential of the source seawater [11]. Site-by-site testing of granular media filtration and vacuum-driven UF pretreatment systems followed by separate SWRO systems operated at the same performance conditions (flux, recovery, feed pressures, and product water quality) under these algal bloom conditions indicated that granular media filtration produced low-fouling effluent that required only one SWRO membrane cleaning during the entire algal-bloom period (May through September), while the SWRO membranes downstream of the UF system required four cleanings during the same period to maintain comparable performance.

When the effect of UF system operating pressure on algal cell rupture was studied again during the spring-summer algal bloom period of the following year (2006) it was found that if the maximum operating pressure of the UF system is limited down to the threshold for algal cell rupture of 0.4 bars (as compared to the usual maximum working vacuum of 0.8 bars), the biofouling effect of the UF pretreatment system on the downstream SWRO membranes was reduced dramatically.

While operating membrane pretreatment systems at lower maximum driving vacuum (0.4 vs. 0.8 bars) during heavy algal blooms is a workable solution for lowering the SWRO membrane biofouling rate, it reduces the length of the membrane filter cycle in half, which in turn practically doubles the number of filter backwashes. This in turn increases the total volume of backwash reject from pre-algal bloom levels of 6–8% up to approximately 12% of the total intake flow during algal bloom operations. Because of the significant impact of this algal-mitigation mode of operation of vacuum-driven membrane systems on intake and pretreatment system size capacities, its accommodation has to be reflected in the plant design. Alternatively, this algal bloom challenge could be addressed by installation of DAF or down-flow gravity granular media filters upstream of the membrane pretreatment to remove the algae cells without breaking them before they reach the membrane pretreatment filters.

7.7.1 Seawater Conditioning and Pretreatment Prior to Membrane Filtration

Micro- and ultrafiltration have a wider spectrum of particle removal capabilities than conventional granular media filtration. Single or dual granular-media filters usually have lesser removal efficiency in terms of source water organics in suspended form, disinfection byproduct precursors, fine particles, silt and pathogens. Membrane filtration technologies are also less prone to upsets caused by seasonal changes in source seawater temperature, pH, turbidity, color, pathogen contamination, and size and type of particles, because their primary treatment mechanism is mechanical removal through fine pores. Therefore, the upstream coagulation and flocculation of the seawater is of a lesser importance for their consistent and efficient performance. In contrast, the pretreatment efficiency of the conventional media filtration technologies is very dependent on how robust the chemical coagulation and flocculation of the seawater is ahead of the filtration process.

Seawater reverse osmosis desalination plants equipped with vacuum-driven membrane pretreatment systems are typically designed to operate without coagulant addition or at coagulant dosages which are several times smaller than those needed for granular media pretreatment. Pressure-driven pretreatment systems usually require source water coagulation prior to filtration during heavy algal bloom or rain events. In this case, the coagulation dosages are also smaller than those needed for source water conditioning for granular media pretreatment of the same water. Source seawater conditioning needs to address mineral and colloidal fouling.

Seawater pretreatment upstream of membrane filtration is similar to that needed upstream of granular media filtration with one exception. In order to protect the pretreatment membranes from physical damage, the source seawater has to be treated by micro-screening. Source seawater contains small sharp objects (such as shell particles), which can easily puncture the plastic pretreatment membranes and cause a rapid and irreversible loss of their integrity, unless the damaging particles are removed upstream of the membrane pretreatment system.

To remove sharp particles that can damage the pretreatment membranes, the RO plant intake system has to incorporate a micro-screening system of screen mesh size of 120 μm or less ahead of the membrane pretreatment system. Typically, previous experience with fresh surface source water (such as rivers or lakes) or wastewater indicates that 300–500 μm screens are adequate to protect the membranes from damage. However, numerous tests at the Carlsbad Seawater Desalination Demonstration Plant and West Basin Seawater Pilot Plant in California, indicate that screen size larger than 120 μm does not provide adequate protection of the pretreatment membranes against sharp particles in seawater.

In addition, seawater contains barnacles, which in their embryonic phase of development are 130–200 μm in size and can pass the screen openings unless they are 120 μm or smaller. If barnacle plankton passes the screens, it could attach to the walls of downstream pretreatment facilities, grow colonies, and ultimately interfere with pretreatment system operations. Once barnacles establish colonies in the pretreatment facilities and equipment, they are very difficult to remove and can withstand chlorination, which is otherwise a very effective biocide for most other marine organisms. Therefore, the use of fine micro-screens (80–120 μm) is essential for reliable operation of the entire seawater reverse osmosis desalination plant using membrane pretreatment. Micro-screens are not needed for pretreatment systems with granular media filtration because these systems effectively remove barnacles in all phases of their development and retain sharp debris.

The installation and operation of micro-screening system is only needed if membrane filtration is used for pretreatment and therefore, its additional capital and O&M costs have to be taken into consideration when comparing conventional and membrane filtration pretreatment. The performance and reliability of the conventional granular media pretreatment systems are not sensitive to the content of sharp objects in the seawater and do not require elaborate and costly micro-screening ahead of the filters. Usually, fine traveling screens of 3–4 mm openings provide adequate protection for conventional granular media pretreatment systems from coarse debris, large marine organisms, and particulates contained in the source water.

Typically, general selection choices of membrane pretreatment technology relate to the size of the membrane separation media (microfiltration or ultrafiltration) and the type of the driving filtration force (pressure or vacuum).

7.7.2 Considerations for Selecting Between UF and MF Pretreatment

To date, UF membranes have found wider application for seawater pretreatment than MF membranes mainly because they usually provide better removal of suspended organics, silt and pathogens from the source seawater. Results from a comparative MF/UF study [12] indicate that "tight" UF membranes (i.e., membranes that can reject compounds of size of 20 kilo-Dalton or kDa or smaller) can produce filtrate of lower SWRO membrane fouling potential than 0.1 μm MF membranes. For comparison, these MF membranes produce filter effluent of turbidity and SDI similar to that of 100 kDa UF membranes.

Under conditions where large amounts of silt particles of size comparable to the MF membrane pores are brought into suspension by naval ship traffic or ocean bottom dredging near the area of the intake, the silt particles may enter and lodge into the MF membrane pores during the filtration process and ultimately may cause irreversible MF membrane fouling. Since UF membrane pores are significantly smaller than those of the MF membranes, typically UF membrane pretreatment systems do not face this filtration pore plugging problem. Potential problem of incompatibility of source water particle size and membrane pore size can be identified by side-by-side pilot testing of MF and UF membrane systems during periods of elevated silt content.

7.8 Considerations for Selecting Between Pressure and Vacuum-Driven Membrane Filtration

Pressure and vacuum-driven membrane pretreatment systems mainly differ by their ability to handle source water quality variations, especially temperature and solids load, their operating pressure range and by their costs.

7.8.1 Source Water Quality Variations

Vacuum-driven membrane pretreatment systems are usually more advantageous for treating seawater of variable quality in terms of turbidity, such as waters collected by near-shore or shallow offshore intakes that experience frequent turbidity variations and spikes of 20 NTU or more. Pressure-driven membrane systems have limited capacity to retain solids due to the fact that the individual membrane elements are located in a tight membrane vessel of a very small retention volume. Therefore, if a pressure-driven system is exposed to large amount of solids, the membrane elements (modules) and vessels would fill up with solids very quickly, which in turn would trigger frequent membrane backwash and could result in destabilization of the membrane system performance. To address this deficiency of the pressure-driven membrane systems, some membrane manufacturers offer membrane modules with adjustable fiber density, which allows customizing membrane system design to the more challenging water quality.

Typically, the tanks in which the vacuum-driven membrane elements are installed provide a minimum hydraulic retention time of 10–15 min and have an order of magnitude higher volume and capacity available to handle source seawater of elevated turbidity and to temporarily store solids compared to pressure-driven systems. This renders the vacuum-driven membrane pretreatment systems more suitable for high-turbidity/high-variability sea waters. The air scouring applied by vacuum-driven systems typically apply for backwashing also improve their tolerance for high solid loads. In addition, since vacuum-driven systems usually operate at lower trans-membrane pressure, their rate of membrane fouling is lower and they have more stable operation during transient solids load conditions.

Pressure-driven membrane pretreatment systems however, are often more suitable and cost-competitive for cold source seawaters (i.e., for seawaters of minimum monthly average temperature of 15C or less) and/or for low turbidity and low organic content.

The productivity (sustainable flux) of vacuum driven systems is more sensitive to source water temperature/viscosity. The maximum transmembrane pressure available for vacuum-driven membrane systems is limited to approximately one atmosphere. This often makes pressure-driven systems the membrane pretreatment of choice for desalination plants with deep offshore open ocean intakes which yield low-temperature water in the winter.

7.8.2 Construction Costs and Energy Requirements

Depending on the size of the system and the intake seawater quality, the site-specific conditions may favor the use of either pressure or vacuum-driven type membrane pretreatment system. Pressure-driven systems are typically very cost competitive for small and medium-size plants because they can be manufactured and assembled in a factory off-site and shipped as packaged installations without the need for significant site preparation or construction of separate structures.

Equipment and construction costs of larger-size plants with more challenging source seawater quality are typically lower when using vacuum-driven membrane systems, especially for plant retrofits. An exception to this rule-of-thumb is the treatment of low-temperature seawater.

Because for the same water quality, vacuum-driven systems of the same type (MF or UF) usually operate at lower pressure, the total power use for these systems is lower. The vacuum-driven systems may use 10–30% less energy than pressure-driven systems for water sources of medium to high turbidity (5 to 20 NTU) and temperature between 18 and 35°C.

7.9 Lessons Learned from Existing MF/UF Systems

At present, the seawater desalination industry has limited experience with the use of membrane pretreatment. Over the past 10 years a total of about two dozen large full-scale SWRO plants with membrane pretreatment have been constructed worldwide. The majority of these plants have been in a continuous operation for less than five years and some of the plants have faced performance challenges and have undergone a number of modifications [13–16]. Most of the initial challenges have been addressed by the accelerated advancement of UF and MF membrane technology and by the development of new generations of membrane products tailored toward seawater pretreatment. Technological advancements have resulted in exponential growth of membrane pretreatment applications for seawater desalination over the past five years, with more than a dozen new large desalination plants planned to begin operation by 2020 [1, 4]. Table 7.3 lists key features for some of the largest SWRO plants using membrane pretreatment.

It is interesting to note that all desalination plants listed in Table 7.3 use UF membrane pretreatment systems. Very few plants have additional source water treatment (DAF, coarse media filtration) upstream of the membrane pretreatment system [17–19].

While UF and MF membrane experience for seawater applications is fairly limited at present, over 20 years of such full-scale experience exists for fresh water applications. A recent survey completed at 10 fresh drinking water treatment plants in the US [20] has identified a number of challenges these plants experienced during their startup, acceptance testing and full-scale operations (Table 7.4). Most of the lessons learned from these applications are relevant to seawater pretreatment as well.

Table 7.3 Large SWRO plants with membrane pretreatment

Desalination Plant Location and Capacity	Pretreatment System Type and Configuration	Hydraulic Loading Rate (Flux)	Notes
Adelaide, Australia – 300,000 m^3/day	Submersible UF membranes	Flux rate – 50–60 L/m^{2}h	Offshore open intake
Fukuoka, Japan – 96,000 m^3/day	Pressure UF membranes	Flux rate – 60–80 L/m^{2}h	Infiltration gallery
Kindasa, Saudi Arabia 90,000 m^3/day	Dual media granular pressure filtration followed by pressure UF membranes	Dual granular media filtration rate – 15–20 m^3/m^{2}h Flux rate – 75–100 L/m^{2}h	Near-shore open intake in industrial port
Palm Jumeirah, UAE – 64,000 m^3/day	Pressure UF system	Flux rate – 55–65 L/m^{2}h	Offshore open intake
Yu-Huan, China – 34,500 m^3/day	Submersible UF membranes	Flux rate – 30–50 L/m^{2}h	Offshore open intake
Colakoglu Steel Mill SWRO Plant, Turkey – 6,700 m^3/day	Pressure UF system	Flux rate – 50–60 L/m^{2}h	Offshore open intake

Table 7.4   UF/MF membrane system survey: lessons learned from ten US utilities [17]

Utility No.	Primary Membrane System Problems	Root Cause of Problem	Lessons Learned
1	Unable to meet design capacity	Lower achievable flux than projected Pilot testing did not address extreme water quality conditions	Pilot-test during extreme water quality conditions Use conservative safety factor when up-scaling pilot testing results
2	High CIP frequencies Excessive downtime and O&M costs.	Excessive membrane fouling Pilot testing did not address extreme water quality conditions.	Pilot-test during extreme water quality conditions Additional pretreatment may be needed to address extreme water quality conditions
3	Unable to meet design capacity	Undersized membrane ancillary support systems.	Ancillary support systems could be a significant bottleneck if undersized.
4	Higher membrane replacement costs	Lower than projected membrane life Potential membrane fouling & lack of previous data by suppliers	Additional pretreatment may be needed to obtain the useful membrane life indicated by membrane supplier
5	Excessive downtime & maintenance Lower than projected water quality	Excessive fiber breakage Fouling related to water quality deviations from specifications putting higher stress on the fibers than expected	Lack of experience with use of membranes for the given water quality may require change in membrane chemistry and durability
6	Unable to meet design capacity	Higher than anticipated downtime Manufacturer failed to include valve opening/closing time for integrity tests More membrane capacity needed to be installed	Complete thorough review of the downtime for all MF/UF system operational steps under worst-case operations scenario
7	Higher than Expected O&M costs	More frequent chemical cleaning needed than initially projected	Pilot test during extreme water quality conditions. Use conservative safety factor when up-scaling pilot testing results
8	Excessive system downtime	Failures in membrane potting System not handling water pressure/potting* materials not tested previously	Never use a membrane that has components or materials that have never been tested previously!

| 9 | Difficult system operation | Insufficient system training for staff | Plan for additional staff training beyond the minimum offered by the MF/UF manufacturer |
| 10 | Excessive downtime Failure to meet product water quality targets | Frequently failing membrane integrity testing Air leaking form gaskets and valves. | Make sure that replacement of failed gaskets, valves and seals is included in manufacturer membrane system warranty |

*Editor's note: potting means sealing the end of hollow fibers during module preparation. Typically, polyurethanes, silicone rubbers, and/or epoxy resins are used.

7.10 Concluding Remarks

Seawater pretreatment is a critical component of every reverse osmosis desalination plant. The main purpose of this plant component is to remove particulate solids, organic compounds and biofouling substances released by marine microorganisms, which can cause accelerated reduction of the productivity of the SWRO membranes. At present, the most commonly used technology for seawater pretreatment in SWRO desalination plant is granular media filtration. Over the last 10 years, membrane seawater pretreatment is becoming an attractive alternative to conventional granular media filtration and its use is gaining momentum. However, taking under consideration the numerous factors affecting the pretreatment costs of a full-scale seawater desalination plant, the selection of the most suitable pretreatment system for a given seawater desalination project has to be completed based on a thorough life-cycle cost analysis which accounts for all expenditures and actual costs associated with the installation and operation of the two systems.

As the desalination industry gains long-term experience with the operation of seawater membrane pretreatment systems and the existing membrane pretreatment technologies evolve and converge into compatible, standardized lower-cost modular products, the use of UF or MF membranes for seawater pretreatment is expected to become more competitive and attractive over time.

References

[1] Huehmer, R. P. (2009), "MF/UF Pretreatment in Seawater Desalination: Applications and Trends," in *Proceedings of World Congress in Desalination and Reuse, International Desalination Association, IDAWC/DB09-253*, Dubai, UAE.

[2] Voutchkov, N. (2008). *Pretreatment Technologies for Membrane Seawater Desalination*. Sydney: Australian Water Association.

[3] Irwin, K. J., and Thompson, J. D. (2006). Trinidad SWRO – Orinoco Fluctuations Fail to Make Filters Falter, Desalination and Water Reuse. IDA; Topsfield, MA November/December, 13/3 (2006), pp. 12–16.

[4] Bush, M., Chu, R., and Rosenberg, S. (2010). Novel trends in dual membrane systems for seawater desalination: minimum primary pretreatment and low environmental impact treatment schemes. *IDA J.* 2, 56–71.

[5] Sieburth, J., McN. Smetacek, V., Lenz, J. (1978). Pelagic ecosystem structure: heterotrophic compartments of the plankton and their relationship to plankton size fractions. *Lirnnol. Oceanogr.* 23, 1256–1263.

[6] Takahashi, M., and Bienfang, P. K. (1983). Size structure of phytoplankton biomass and photosynthesis in subtropical Hawaiian waters. *Mar. Biol.* 76, 203–211.

[7] Takahashi, M., and Hori, T. (1987). Abundance of picoplankton in the subsurface chlorophyll maximum layer in subtropical and tropical waters. *Mar. Biol.* 79, 177–186.

[8] Kiang, H. F., Young, W. W. L., Ratnayaka, D. D. (2007). The Singapore Solutions. *Civil Eng.* 77, 62–69.

[9] Petry, M., Sanz, M. A., Langlais, C., Bonnelye, V., Durand, J. P. Guevara, D., Nardes, W. M., Saemi, C. H. (2007). The El Coloso (Chile) reverse osmosis plant. *Desalination*, 203141–152.

[10] Vila, J., Compte, J., Cazurra, T. Ontanon, N. Sola, M., Urrutia, F. (2009). "Environmental Impact Reduction in Barcelona's Desalination and Brine Disposal," in *Proceedings of World Congress in Desalination and Reuse, International Desalination Association, IDAWC/DB09-309*, Dubai, UAE.

[11] Voutchkov, N., and Dietrich, J. (2005). "Pilot testing alternative pretreatment systems for seawater desalination in Carlsbad, California," in *Proceedings of World Congress in Desalination and Reuse, International Desalination Association, SP05-095*, Singapore.

[12] Kumar, S. A., Adam, S., and Pearce, W. (2006). Investigation of seawater reverse osmosis fouling and its relationship to pre-treatment type. *Environ. Sci. Technol.* 40, 2037–2044.

[13] Pearce, G., Allam, J., Chida, K. (2003). "Ultrafiltraton Pretreatment to RO: Trials at Kindasa Water Services, Jeddah, Saudi Arabia," in: *Proceedings of World Congress in Desalination and Reuse, International Desalination Association*, Paradise Island, Bahamas.

[14] Hashim, A., Arai, T., Tada, K., Iwahori, H., Tada, N., Ishihara, S., and Takata, M. (2009). "Spiral-wound UF-Membrane Pre-treatment Achievements for SWRO Application in the Arabian Gulf Region," in *Proceedings of World Congress in Desalination and Reuse, International Desalination Association, IDAWC/DB09-121*, Dubai, UAE.

[15] Choules, P., Boudinar, M. B., Mack, B. (2009). "Membrane pre-treatment design and operational experience from three SWRO plants," in *Proceedings of Annual Conference of the American Membrane Technology Association*, Austin, Texas.

[16] Knops, F., Dekker, R., Kolkman, R. (2009). "Ten Years of Ultrafiltration as Pretreatment to SWRO in the Arabian Gulf," in *Proceedings of World Congress in Desalination and Reuse, International Desalination Association*, IDAWC/DB09-071, Dubai, UAE.

[17] Jamaly, S., Darwish, N. N., Ahmed, I., Hasan, S. W. (2014). A short review on reverse osmosis pretreatment technologies. *Desalination*, 354, 30–38.

[18] Nadav, N., and Koutsakos, E. (2013). Innovative Design of the UF and SWRO Limassol Desalination Plant in Cyprus. *Desalinat. Water Treat.* 13, 1–3.

[19] Uemura, T., Kotera, K., Henmi, M., and Tomioka, H. (2011). Membrane technology in seawater desalination: history. *Recent Develop. Future Prospects* 33, 1–3.

[20] Atassi, A., White, M., and Rago, L. M. (2007). "Membrane failure headlines: facts behind the hype regarding initial problems experienced at low pressure membrane installations," in *Proceedings of Annual American Water Works Membrane Technology Conference*. Denver, Colo.: American Water Works Association.

CHAPTER 8

Post-Treatment

Alireza Bazargan[1,2], Reza Mokhtari[3]
and Mohammad Shamszadeh[4]

[1]Research and Development Manager and Business Development Advisor,
Noor Vijeh Company (NVCO), No. 1 Bahar Alley, Hedayat Street,
Darrous, Tehran, Iran
[2]Environmental Engineering group, Civil Engineering Department,
K.N. Toosi University of Technology, Tehran, Iran
E-mail: info@environ.ir
[3]Operations Manager, Noor Vijeh Company (NVCO), No. 1 Bahar Alley,
Hedayat Street, Darrous, Tehran, Iran
E-mail: r.mokhtari@nvco.org
[4]Member of the Board, Noor Vijeh Company (NVCO), No. 1 Bahar Alley,
Hedayat Street, Darrous, Tehran, Iran
E-mail: m.shamszadeh@nvco.org

8.1 Introduction

Modern desalination plants are very efficient at purifying water, but ironically, they are often "too efficient". As a result, the water lacks some required minerals and compounds, and hence without post-treatment, the desalinated water is too corrosive and cannot be supplied directly to the distribution network. The type and extent of the applied post-treatment process in a desalination plant should be selected based on local considerations and the potential applications of the water.

As stated in Chapter 4, the quality of unconditioned desalinated water varies considerably from plant to plant. This variation depends on the quality of the saline source water, the type of desalination technology used (MSF, MED, RO, etc.) as well as the specific design and operation of the desalination process. However, regardless of the desalination process, the desalinated water is slightly acidic, has a very low buffering capacity, and contains very little calcium and magnesium hardness [1]. Hence, post-treatment processes are required.

237

Table 8.1 Concentration of chemical constituents in the intake and permeate of NVCO's single-pass seawater reverse osmosis plant in Asaluyeh, Persian Gulf, Iran

Chemical Species	Seawater (mg/l)	Desalinated Water Prior to Post-treatment (mg/l)
Cl	21,924	150
Na	12,184	93
SO₄	3,060	3.1
Mg	1,459	2.9
Ca	500	1.4
K	438	5.4
HCO₃	157	1.9
Br	59	0.3
B	5.4	1.4
SiO₂	2.4	0
NO₃	2.2	0
Total dissolved solids (TDS)	39,809	260

The water entering a desalination plant can originate from various sources. The source water will have various unwanted substances, and by definition, the desalination process will aim to reduce the concentration of unwanted salts. As an example, the concentration of various chemical substances in seawater and the corresponding concentration after desalination via NVCo's single-pass reverse osmosis plant in Asaluyeh, Iran, are presented in Table 8.1. Keep in mind that the values for the desalinated water in Table 8.1 are reported prior to any post-treatment. It should be noted that thermal processes are more effective than membrane processes for reducing the total dissolved solids (TDS) concentration.

In a series of workshops held by the Water Research Foundation, desalination professionals and water purveyors identified the following five issues in post-treatment as their top priorities, and ranked them using the nominal group technique by importance [2]: identifying stabilization tools for producing consistently good water quality and ensuring effective results in the distribution system, determining effective permeate conditioning and corrosion control (design and operation of post-treatment systems), meeting the challenges of disinfection by-product formation during post-treatment, blending sources to achieve target water quality goals, and understanding the effects of blending permeate into an existing distribution system.

Although the desalination industry has significantly advanced since the 1980s, efforts on research and development have been asymmetrically concentrated. Comparatively, studies pertaining to post-treatment techniques are very rare. A rough survey of scholarly articles has shown that less than 1% of published studies by desalination experts focuses on post-treatment technologies [1]. In this chapter, post-treatment processes used for desalination water conditioning are first introduced, followed by a discussion on disinfection.

8.2 Post-Treatment Processes

As mentioned earlier, desalinated water has low buffering capacity and mineral content. In addition, desalinated water is "aggressive" to metallic and cement-based materials used in storage and distribution, and hence must be conditioned. Disinfection is also an indispensable part of post-treatment, particularly if the water is to be stored and distributed in a network. In this section, the most commonly used post-treatment processes are introduced.

Overall, there are various terms used when referring to post-treatment processes including terms such as re-carbonation (re-carbonization), re-mineralization, or potabilization. Although these terms may reflect different post-treatment objectives, there is often considerable overlap. In fact, the most popular remineralization processes combine re-carbonation coupled with increase in mineral content and the elevation of carbonate alkalinity. Potabilization is defined as the process of making the water suitable for drinking, including the addition of minerals, stabilization, re-carbonation (re-carbonization), and disinfection [1]. Suggested product water quality goals for post-treated water at its point of entry into the distribution network, are presented in Table 8.2. For review of these parameters, their definitions, and importance refer to Chapter 4.

8.2.1 Blending

The simplest post-treatment practice for improving water quality is to blend desalinated water with other water. As a minimum, the blending water is filtered through cartridge filters. Other treatment of the blending water, such as processing through activated carbon filters is also recommended.

If the blending water is of a high quality (saline water collected by beach wells, pure groundwater source, or good quality brackish water), blending can result in substantial savings in post treatment capital and operational costs (CAPEX and OPEX). However, blending carries the risk of contaminating the already desalinated water. Particularly, the microbial risk of the blending

Table 8.2 Suggested quality goals for desalinated water after post-treatment adopted from Duranceau et al. [2] with additions and modifications

Parameter	Target
pH	6.5–9.5
Alkalinity (mg/L as $CaCO_3$)	50–150
Hardness (mg/L as $CaCO_3$)	50–110
Calcium (mg/L as $CaCO_3$)	50–100
Sulfate (mg/L)	<250
Chloride (mg/L)	<250
Boron (mg/L)[a]	<2.4
Fluoride (mg/L)	<1.5
TDS (mg/L)	<300
Calcium Carbonate Precipitation Potential (mg/L)	4–10
Langelier Saturation Index	0–0.5
Sodium Adsorption Ratio (for agriculture)	<5
Turbidity (NTU)	<2

[a] Appropriate Boron concentrations are still debated.

water must be evaluated, in order to prevent pathogens and other undesirable organisms from entering the finished water. The amount of water used for blending may vary from 0 to 30% of the volume, depending on the quality/ salinity of the blending water [3]. In either case, blending alone may not suffice for meeting all water quality requirements and other processes may also be needed. Using blending as a post-treatment strategy is usually limited to thermal plants (in order to increase the very low TDS of the distillate) [4].

Generally, the NOM content in the desalinated water is very low, and the contribution of the blending water is not high enough to push the concentration of the finished water past the WHO guideline values. Nonetheless, if blending is to be incorporated, the potential problems caused by anthropogenic pollutants should not be overlooked.

Blending with 1% seawater (usually the maximum blending ratio used for seawater) adds about 15 mg/L of magnesium and about 5 mg/L of calcium to the finished water. If residual disinfectants are present during storage and distribution (which is most likely the case) they will react with the bromide in the blending water and possibly push the bromate (and other byproducts) concentration past the maximum WHO guideline threshold values. Careful consideration should be given to ensure that the final blended water does not exceed the guideline values for chlorides and bromates, is not corrosive, and does not have taste problems.

8.2.2 Direct Dosage of Chemicals

The most popular post-treatment process used for the re-mineralization of desalinated water is the sequential injection of carbon dioxide and lime slurry ($Ca(OH)_2$) into the desalinated water. The main advantage of this method is its relatively low-capital costs and small required space [1].

The engineering of lime slurry systems is challenging, due to scaling, difficulties in maintaining a consistent pH in the finished water, as well as keeping turbidity lower than 5 nephelometric turbidity units (NTU). If the system is not designed and operated correctly, a turbid finished water which exceeds regulatory limits will ensue. Nonetheless, due to its merits, the dosing of $Ca(OH)_2 + CO_2$ is *the* most popular post-treatment method used for re-mineralization, with over 90% of the world's desalination plants applying this technique [5]. Prior to lime dosing, carbon dioxide is injected into the unconditioned desalinated water. Carbon dioxide is typically delivered to the desalination plant in liquid form (although gaseous form is also possible) and is stored under pressure. By forming carbonic acid, the carbon dioxide helps decrease the pH and enhance the solubility of lime. The chemical reaction of this re-mineralization method is as follows:

$$Ca(OH)_2 + 2CO_2 \rightarrow Ca(HCO_3)_2 \leftrightarrows Ca^{2+} + 2HCO_3^-$$

Note that the molar masses of lime, carbon dioxide, calcium, bicarbonate, and calcium carbonate are 74.1, 44.01, 40.08, 61.02, and 100.09 mg/mmol, respectively. Hence, theoretically the introduction of 74 mg/L of $Ca(OH)_2$ and 88 mg/L of CO_2 in the desalinated water will lead to the addition of about 40 mg/L of Ca^{2+}, i.e., 100 mg/L calcium hardness as $CaCO_3$. In addition, the alkalinity increases by about 122 mg/L, equivalent to 100 mg/L as $CaCO_3$:

$$\left(40 \ \frac{mg}{L} Ca^{2+}\right) \left(\frac{100.09 \ mg/mmol \ CaCO_3}{40.08 \ mg/mmol \ Ca^{2+}}\right) \left|\frac{2}{2}\right| \approx 100 \ \frac{mg}{L} \text{ as } CaCO_3$$

$$\left(122 \ \frac{mg}{L} HCO_3^-\right) \left(\frac{100.09 \ mg/mmol \ CaCO_3}{61.02 \ mg/mmol \ HCO_3^-}\right) \left|\frac{-1}{2}\right| \approx 100 \ \frac{mg}{L} \text{ as } CaCO_3$$

An alternative process to the dosage of carbon dioxide and lime, is the direct dosage of $CaCl_2 + NaHCO_3$. The benefit of this method is that it does not require the handling of gases or slurries, and relies on simple dissolution of chemicals. The disadvantages of this option are that $CaCl_2$ is often more expensive than lime, and that the dosage of $CaCl_2$ and $NaHCO_3$ introduces unwanted sodium and chloride ions into the water, increasing the Larson Ratio [1]. The EPA has a secondary maximum chloride level of 250 mg/L [6] which should be respected. In addition, both chemicals are expensive and their storage under humid conditions (such as on the coast of the Persian Gulf) is challenging because they tend to cake in the presence of moisture. The final water may require pH adjustment [4].

Directly dosing $Na_2CO_3 + CO_2$ or $NaOH + CO_2$ is also an option. By considering the dosed species, it is obvious that this post-treatment process does not increase calcium concentration (and thus total hardness). However, this method can be used for increasing the carbonate alkalinity (buffering capacity) and the pH of the water to desired levels. Thus for corrosion-control purposes, the direct dosage of sodium hydroxide and carbon dioxide is possible. Various Swedish municipalities and the Massachusetts Water Resource Authority have reportedly applied this method [1]. The direct dosage of $Ca(OH)_2 + Na_2CO_3$ or $Ca(OH)_2 + NaHCO_3$ is also an option, which is rarely applied.

8.2.2.1 Design Considerations for $Ca(OH)_2 + CO_2$ Systems

One of the tallest structures in a typical desalination plant (probably *the* tallest structure in an RO plant) is the silo in which bulk powdered hydrated lime is stored. For large-size plants, these silos can be as large as 300 m^3, and are designed to store 15–60 days of lime. The silos are stainless steel structures with level sensors and fill lines with long-radius elbows.

The lime product is usually delivered to the plant as quicklime (CaO) which is mixed with water (slaked) in a slurry tank to generate hydrated lime. The hydrated lime slurry (milk of lime) is then fed into saturation tanks in which the slurry is mixed with unconditioned desalinated water to form saturated limewater. Usually, the removal of lime from the silo, its delivery to the slurry mixing tank, and its conveyance to the feed line of the lime saturators are automated and motorized. The saturator (thickener-clarifier) is a key feature of the process. It is designed to enhance lime particle dissolution by continuously feeding lime slurry into a reaction zone where it is mixed to form a homogenous saturated lime solution. The lime is diluted down to 0.1% in the saturator with unconditioned desalinated water. Polymer flocculants (0.5–1 mg/L) may also be added in order to prevent carryover of smaller slow-settling particles and reduce the turbidity [5]. Mixers or pumps are used to recirculate lime slurry from the tank bottom to the reaction zone to improve solubility. Higher temperatures result in a decrease of lime solubility, hence the need for CO_2 and adequate mixing is more pronounced. The design parameters include mixing energy, residence time, feed-well configuration, lime bed level control, and recycle flow. The bed level is important for optimizing solids dispersion by controlling the flocculation and distribution of the feed into the settling zone of the tank [6]. The saturators include a settling zone where the un-dissolved lime material and impurities are separated and removed with a bottom scraper and hopper. The removal of impurities is very important because assuming a lime purity of 90%, the addition of 100 mg/L of alkalinity and hardness as $CaCO_3$ would introduce 8.2 mg/L of impurities to the finished water, typically in the form of saturated solids [5]. Lime solids and impurities can accumulate in the lime feed system pipelines and equipment, and must be removed periodically. Finally, the saturated limewater is either directly injected into the desalinated water stream, or, is transferred to a limewater tank before its injection. Key design criteria for a typical system are presented in Table 8.3.

8.2.3 Limestone (Calcium Carbonate) Dissolution

Limestone is a naturally occurring calcium carbonate (calcite), which under favorable conditions, will dissolve and increase both calcium hardness and bicarbonate alkalinity of the water. This is usually accomplished by dosing desalinated water with carbon dioxide and flowing it through a limestone bed, bringing about the following chemical reaction:

$$CaCO_3 + H_2O + CO_2 \rightarrow Ca(HCO_3)_2 \leftrightarrows Ca^{2+} + 2HCO_3^{-}$$

Based on the reaction, the addition of 44 mg/L of carbon dioxide to the water will dissolve 100 mg/L of calcite, leading to a hardness and alkalinity increase of 100 mg/L as $CaCO_3$. A schematic of a typical limestone contactor process is provided in Figure 8.1. The ratio of the empty limestone contactor volume to the volumetric water flow rate is defined as the Empty Bet Contact Time (EBCT) and is usually in the range of 10–30 min [6]. This means that for a desalination plant with a production capacity 10 m^3/min, the contactor volume(s) will be 100–300 m^3. Depending on the design of the process and the desired output, it may be possible to feed only a portion of the desalinated water through the contactor. In order to adjust the final pH,

Table 8.3 Lime dosing system parameters [5]

Component/Parameter	Specifications and Comments
Lime Silo and Slurry System	
Lime dosage (assuming 100% pure lime)	0.74 mg/L per 1.0 mg/L of target alkalinity and hardness concentrations (as $CaCO_3$)
Lime consumption per 1000 m³/day of desalinated water (assuming 100% pure lime)	59.2–88.8 kg/day for addition of alkalinity and hardness in the range 80–120 mg/L (as $CaCO_3$)
Lime purity	85–94%
Storage time in silo	15–60 days
Silo vessel material type	Coated carbon or stainless steel
Silo vent filter type	Cartridge type with polyester felt cartridges
Lime flow facilitating equipment	Bin Vibrator Fluidized air pads
Typical lime slurry concentration	10%
Slurry tank retention time	3–5 h
Typical slurry pump	Progressing cavity—equipped with flushing system
Lime Saturator	
Saturator solids loading rate	1.5–4 kg lime/m²h
Saturator hydraulic loading rate	0.7–1 m³/m²h
Saturator retention time	1.5–2 h
Saturator height	Total height of 3.5–5 m, with 3–4.5 m of water
Retention time in reaction zone before recirculation	8–15 min
Final limewater concentration to be injected	900–1400 mg/L (no more than 1500 mg/L)
Turbidity of limewater	30–35 NTU
Possible retention time in the limewater storage tank	18–24 h
Target turbidity of final conditioned water after limewater injection	Preferably less than 2 NTU (no more than 5 NTU)

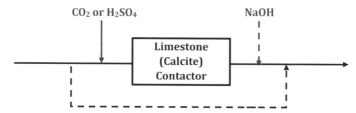

Figure 8.1 A typical limestone (calcite) contactor flow diagram.

a final step is usually incorporated. Most commonly, sodium hydroxide is added (although the addition of small amounts of $Ca(OH)_2$ may also be possible). Also, CO_2 desorption units can be implemented downstream in order to remove any excess CO_2 which may remain in the system. The difference between NaOH dosage and CO_2 stripping is as follows: NaOH dosage increases the alkalinity but does not affect the total inorganic carbon content (C_T value); whereas, CO_2 stripping does not affect the alkalinity and decreases the C_T.

There are a number of computer programs such as DESCON [7] and the "Limestone Bed Contactor: Corrosion Control and Treatment Process Analysis Program" [8], which can be used for calculating the height of the calcite bed required to reach the required water quality. Nevertheless, these programs should be used cautiously as real contactors may behave differently. The design parameters of the limestone contactor system include superficial flow velocity, carbon dioxide dosage (or acid concentration), contactor bed height, bed porosity, temperature, and particle size. The smaller the particle size, the more surface area will be available for contact, leading to less contact time required to reach a desired product water quality. In other words, smaller calcite particles allow for a smaller contactor size and hence reduced CAPEX. Particles are usually 1.5–8 mm in size [5,9], although more recently, a micronized calcite (0.2–25 μm) process has been proposed [10]. Hydrodynamic conditions are important as higher Reynolds numbers minimize the thickness of the diffusion boundary layer and increase the reaction rate. Secondary parameters include inner particle porosity, calcite crystal habit and purity (impurities reduce the dissolution rate and decrease water quality) [6]. Generally, as dissolution continues, the limestone is gradually depleted and the bed height decreases. Hence, the system requires regular reloading to maintain product water quality. Temporary turbidity spikes may occur after reloads.

The advantages of using the limestone dissolution method are the negligible addition of unwanted ions (such as sodium and chloride), the minimal increase in turbidity, the relatively high buffering capacity of the produced water, the cost effectiveness of the process, the easy operation and control, the minimal maintenance and operator skill requirement, and the improbability of overdose [9]. When comparing limestone contactors to the direct-dosage method (lime slurry), various factors should be considered:

- Capital costs (CAPEX): the CAPEX of limestone contactors is typically more than the direct-dosage method. In addition to more land required, the construction of large contactor tanks also incurs costs. It should be noted that the size of the contactors can be reduced with higher acid dosing (and subsequent higher use of base downstream for pH adjustment, entailing increased operation costs). Nevertheless, the equipment for handling limestone is significantly less costly than the system required for preparing and dosing lime.
- Operation costs (OPEX): the comparative OPEX costs are dependent on local lime, limestone, and carbon dioxide prices. Based on the chemical reactions, the use of limestone contactors theoretically requires 50% of the quantity of carbon dioxide compared to the direct dosing method with lime (although in practice it may be 65–85% [4]). For the same increase in alkalinity and hardness the consumption of lime is about three quarters of limestone; however, lime is usually more than twice as expensive. Hence, the OPEX costs of limestone contactors are typically less than the direct dosing of lime.
- Environmental impact: the carbon footprint in producing lime is several times higher than producing calcite; this is because of the high temperatures required, and the release of carbon dioxide in the lime production process.

In the design of limestone contactors, an alternative to using carbon dioxide is to use sulfuric acid, which ultimately forms carbon dioxide within the system. The produced carbon dioxide then reacts with the calcite as discussed before:

$$CaCO_3 + H_2SO_4 \rightarrow CaSO_4 + H_2O + CO_2$$
$$CaCO_3 + H_2O + CO_2 \rightarrow Ca(HCO_3)_2$$

In summary:

$$2CaCO_3 + H_2SO_4 \rightarrow CaSO_4 + Ca(HCO_3)_2$$
$$\leftrightarrows 2Ca^{2+} + 2HCO_3^- + SO_4^{2-}$$

Theoretically, when using sulfuric acid, twice as much limestone is dissolved (compared to when carbon dioxide is used). Meanwhile, half of the dissolved calcite is consumed in the formation of calcium sulfate. Also, when using sulfuric acid, the increase in calcium hardness is twice that of bicarbonate alkalinity increase (mg/L as $CaCO_3$). The advantages and disadvantages of

Table 8.4 The comparison of using H_2SO_4 instead of CO_2 in limestone contactors

Advantages of H_2SO_4	Disadvantages of H_2SO_4
• Can reduce desalinated water pH to any level required. • Much higher rate of calcite solubility, leading to smaller contactor sizes (lower CAPEX). • No need for storage of liquefied carbon dioxide under pressure.	• Twice as much calcite required for a given alkalinity increase. • Possibly higher chemical costs (depending on local prices). • Safety issues related to acid handling, transport, and storage. • Uneven increase of calcium hardness compared to alkalinity (2:1 ratio).

using sulfuric acid instead of carbon dioxide in limestone contactors are reviewed in Table 8.4.

It should be noted that the theoretical ratio of reactants involved in the dissolution of calcium carbonate is not always observed in practice. For example, modeling using the Minteq software has revealed that with a sulfuric acid concentration of 5 mmol/L, the actual reaction will deviate considerably to [6]:

$$1.55CaCO_3 + H_2SO_4 \rightarrow 1.55Ca^{2+} + 1.1HCO_3^- + 0.45H_2CO_3 + SO_4^{2-}$$

There are two configurations for limestone contactors: upflow and downflow. The difference being that in the upflow configuration water is pumped against the pull of gravity upwards through the bed of calcite particles, and in downflow contactors water is sent down through the bed in the same direction as gravitational force. Both upflow and downflow systems are typically pressurized. Downflow systems are usually smaller, while upflow contactors are favored in larger plants. The advantage of downflow contactors is that there is minimal loss of calcite particles from the bed (as opposed to upflow systems which may cause entrainment). On the other hand, upflow systems do not require backwashing, experience less head loss (because in downflow systems there is compaction in the calcite media), cause a smaller turbidity peak during start-up, and have lower costs [11]. An innovative upflow contactor designed by DrinTec^TM with automatic calcite feeding is depicted in Figure 8.2.

8.2.4 Dolomite Dissolution

A dolomite contactor is a lot like a calcite contactor, the difference being that dolomite is a naturally occurring substance that contains both magnesium and calcium carbonates. Hence, by using dolomite, it is possible to

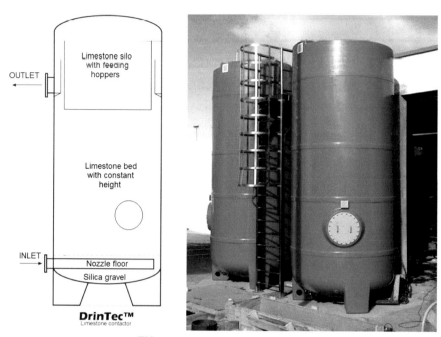

Figure 8.2 The DrinTec^{TM} upflow limestone (calcite) contactor with automatic feeding [12].

increase both the magnesium and calcium content of the water. Nevertheless, dolomite contactors have not found widespread appeal due to the following reasons [5]:

1. Naturally occurring dolomite is typically non-homogenous, and is inter-bedded with limestone. Hence, using dolomite contactors may lead to increased difficulty in controlling the exact quality of the finished water. In other words, dolomite-based systems must be designed with higher contingency.

2. In order to dissolve an appreciable degree of magnesium, the pH must be reduced to values less than 4.5. This, in turn, causes excessive dissolution of calcium, which will result in hardness levels above that which is necessary. Hence, the application of dolomite contactors is not practical in regions where a maximum limit is set for total hardness in water.

3. The limited availability of food-grade dolomite makes its application costlier than the traditional processes.

4. Dolomite dissolution is three times slower than limestone dissolution, leading to increased CAPEX costs resulting from increased land
   requirements and contactor size.

Alternatives to using dolomite contactors would be to directly inject
magnesium-containing chemicals (such as magnesium chloride or magnesium sulfate), MgO dissolution, or applying ion exchangers [6]. Other
innovative processes which employ ion exchange resins have also been
proposed.

8.3 Disinfection

Meeting the goal of safe potable water requires a multi-pronged approach.
First, the source water should be protected from contamination. Next, the
desalination process should be well designed and correctly operated. Finally,
the water should be appropriately disinfected and safely distributed.

The disinfection and safe distribution of drinking water is one of the
greatest achievements of the twentieth century which prevents countless
epidemics every year. During disinfection, a disinfectant (usually containing
chlorine) is added to the water. The chlorine-containing compound then forms
free available chlorine (also known as free residual chlorine) which destroys
pathogenic organisms. It is good to note that the disinfection task might not
be handled at the desalination plant but might be done at the receiving utility,
depending on the contract terms and length of product storage.

Nevertheless, almost all drinking water systems employ some type of
chlorine-based process, either alone or in combination with other disinfectants. If disinfection is to be carried out with chloramines, the amount of
the chlorine–ammonia compounds is referred to as the combined available
chlorine (or combined residual chlorine). The sum of the free available
chlorine and the combined available chlorine is called the 'total residual
chlorine' [13].

Free available chlorine is a combination of various chemical species
including Cl_2, $NaOCl$, $Ca(OCl)_2$, $HOCl$ and OCl^-. Typically, the main
species of free available chlorine are the electrically neutral hypochlorous
acid ($HOCl$) and the electrically negative hypochlorite ion (OCl^-); the former
being more reactive, a stronger oxidant, and having approximately 100-
times-higher disinfection ability than the later [13]. This is because pathogen
surfaces are naturally negatively charged, and hence the neutral hypochlorous

acid penetrates the coating of slime and cell walls more easily. The relative concentration of HOCl and OCl$^-$ is dependent on the pH (among other factors such as temperature). At higher pH (above 8) hypochlorite ions dominate, while at lower pH (lower than 5.5), hypochlorous acid is the dominating species [14]. This is evident from considering the following equation:

$$HOCl \leftrightharpoons H^+ + OCl^-$$

According to the Le Chatelier principle, when a chemical equilibrium is subjected to change (for example in the concentration of one of the species) then the system readjusts itself to counteract the effect of the applied change. This means that in a system with higher H$^+$ concentration (acidic, lower pH), the system will shift to the left and increase the relative abundance of HOCl. Nonetheless, bacteria and viruses are easy targets for free available chlorine over a wide range of pH [15].

In addition to destroying pathogens, chlorination reduces unpleasant taste and odors, eliminates slime and mold, helps remove any remaining iron and manganese from the water, and provides a residual disinfection which prevents microbial re-growth and protects the finished water throughout the distribution network [15].

The efficiency of the desalination process in terms of pathogen removal is dependent on the type of desalination process. If thermal desalination processes are used, the temperature and retention time are the controlling factors. Vegetative cells become inactive at a temperature of 50–60°C retained for 5–30 min. A possible approach is "flash" treatment which inactivates most vegetative pathogens by subjecting them to 72°C for 15 seconds. Higher temperatures and durations are required for endospores which must be inactivated at 70–100°C with longer retention times [16]. Water obtained from thermal desalination is unlikely to contain pathogens both due to the high temperatures involved as well as the unlikeliness of the pathogens to become entrained. Nonetheless, if vacuum thermal processes are employed (which require lower temperatures), care must be taken to ensure inactivation of all pathogens in the desalinated water.

Reverse osmosis is also very efficient for removing pathogens. If the membranes are correctly selected and maintained, they are excellent pathogen barriers. However, the integrity of the system (including seals, O-rings, etc.) is a critical factor in the pathogen removal efficiency. Pinholes in membranes are detrimental to pathogen removal. In addition to secondary disinfection (after desalination), primary disinfection (prior to desalination) is often applied to reduce the occurrence of biofouling and clogging of membranes. Ironically, biofouling could somewhat benefit pathogen removal due to cake formation and pinhole filling.

Care should be taken to remove free chlorine before the water reaches the membranes, because it will cause irreparable damage to their polymeric structure. Although conductivity measurements are usually carried out to ensure membrane integrity, their sensitivity typically corresponds to no more than 2 logs removal (99% removal) of pathogens. This means that if 4-log removal of pathogens is required, conductivity measurements will not provide a definite answer regarding membrane integrity. If ultrafiltration and/or nanofiltration is used as pretreatment prior to the RO process, practically no pathogens will remain in the desalinated water [16].

It is important to note that if pathogens are not completely rejected, and sterile conditions are not maintained, they can proliferate in product water conveyance lines and the distribution network. Hence, desalination processes are always succeeded by a disinfection step, as *secondary* disinfection. Secondary disinfection is defined as the disinfection used to maintain the water quality during storage and throughout the distribution system up to the point of use; whereas primary disinfection is a component of the pretreatment units aimed at destroying the source water pathogens [14].

An important issue associated with disinfection is the potential generation of unwanted disinfection byproducts. These are chemical compounds, such as trihalomethanes, which are formed unintentionally when chlorine and other disinfectants react with other constituents (such as natural organic material) in the water. Toxicology studies have reported that high concentrations of trihalomethanes can cause cancer in laboratory animals [15]. The presence of bromide can also lead to the formation of hypobromous acid and hypobromite. Hypobromous acid is considered a stronger substituting agent than hypochlorous acid and forms brominated byproducts (such as bromoform and bromodichloromethane) by reacting with organic matter [2]. However, the concentration of such chemicals in disinfected desalinated water is generally significantly lower than the confirmed hazardous thresholds. Nonetheless, the best approach of preventing such byproducts in the water is to design and operate desalination plants to effectively reduce the amount of natural organic material that exists in the water prior to disinfection. It should be noted that the health risks associated with inadequate disinfection are often much higher than those of the byproducts, and hence, disinfection should not be compromised while attempting to control byproduct formation.

8.3.1 The CT Value

When a disinfectant such as chlorine is added to water, it does not inactivate pathogens instantaneously. Rather, the rate of pathogen inactivation depends on five factors: the pathogen itself, the disinfectant concentration (C), the contact time (T), the temperature, and the pH.

The $C \times T$ value (also referred to as the CT value) is defined as the residual disinfectant concentration (mg/L) multiplied by contact times (minutes). For a given disinfectant, the higher the CT value, the more potent the disinfection. The CT concept offers water operators a structured operating criteria for water disinfection. For instance, if the operator decides to decrease the chlorine concentration while upholding the same disinfection potency, longer contact times are required. Obviously, the CT value required for 99.999% (5-log) inactivation of a particular pathogen will be higher than the CT value for its 99.99% (4-log) inactivation.

In order to properly use CT values the following steps are taken: first, the required CT for disinfection must be determined. Next, the actual applied CT from chlorination must be calculated. Finally, it should be ensured that the CT value achieved is higher than the value required. In determining the required CT, conservative calculations should be used. For example, since there is chlorine decay within the pipeline and the possibility of pathogen regrowth, it is important to calculate CT values at the point of water use. In using conservative calculations, we can ensure that the required disinfection is met, even under the most unfavorable conditions [17]. Hence, the following must be considered:

1. Minimum temperature: because chlorine's disinfection ability is reduced as the temperature is reduced.
2. Maximum pH: because as previously discussed, the higher the pH, the less potent the disinfection via free chlorine will be.
3. Minimum contact: because according to the definition of CT, lower contact times lead to decreased disinfection. The minimum contact time within the network should be considered, which occurs under peak flow conditions.
4. Minimum chlorine residual: since higher chlorine concentrations lead to better disinfection, we must ensure that the minimum residual during the day is still adequate.

For example, in order to achieve a 3-log reduction in *Giardia* (99.9% inactivation), at room temperature and pH of 8.0, with a chlorine concentration of approximately 2 mg/L, a contact time of roughly 1 h is required. If the water temperature drops to 10°C, the required contact time will approximately triple.

Since *Giardia cysts* are much more resistant to inactivation by chlorination than bacteria and viruses, if the target *CT* for Giardia is achieved, most other bacteria and viruses would certainly also be inactivated. The USEPA and other institutions have published extensive *CT* tables for most pathogens for different temperatures, pII, chlorine residuals and other factors. It should be noted that there are some pathogens such as *Cryptosporidium*, which are resistant to chlorination and are better removed with an alternative form of disinfection[1].

Care should be taken as there are other factors which can also affect the efficiency of the disinfection. These include attachment of pathogens to surfaces and particulates, encapsulation, and aggregation. For example the attachment of microorganisms to particles that cause turbidity, carbon fines, glass, and algae has been shown to decrease disinfection efficiency [14].

8.3.2 Comparison of Various Disinfection Methods

Although chlorine-based disinfectants are the most commonly used disinfectants by far, other alternatives also exist. No single disinfection method can be hailed as the "best" for all circumstances, and each system should be designed on a case by case basis. Depending on the pathogens in the water, a particular disinfection choice might not be suitable at all, while another may be excellent. Table 8.5, derived from World Health Organisation [14] and Chlorine Chemistry Council [15], provides a comparison of the most popular disinfection methods, including their advantages and disadvantages. It should be noted that possible disinfectants are not limited to those included in the table, and include other halogens (iodine, bromine), hydrogen peroxide, metal ions such as silver and copper, titanium dioxide, among others.

8.3.3 Residual Disinfection

It is common for desalinated water to be stored and pumped over long distances before reaching the end user. The challenges of maintaining

---

[1]Water treatment facilities should at least ensure 99%, 99.9% and 99.99% removal/ inactivation of Cryptosporidium oocysts, Giardia cysts, and viruses respectively. Newer guidelines in Europe and North America require less than 1 virus in every 1000 m^3 of water, less than 1 Giardia cyst in every 100 m^3 of water, and less than 1 Cryptosporidium oocyst per 10 m^3 of water. It should be noted that even such stringent guidelines will be easy to fulfill for properly operated desalination plants.

Table 8.5   Comparison of various disinfection methods

Disinfection Method	Advantages	Disadvantages	Comments
Elemental chlorine, Cl_2	• Provides residual protection • Highly reliable • Highly cost-effective • Easily applied and monitored • Unlimited shelf-life	• Unwanted halogenated byproduct formation • Oxidization of bromide • Not effective against Cryptosporidium • Hazardous gas requiring special handling and training (additional regulatory requirements)	• The most commonly used chlorine-based disinfectant • Transported and stored as a liquefied gas under pressure
Sodium hypochlorite, NaClO	• Provides residual protection • Highly reliable • Cost-effective • Easily applied and monitored • Less hazardous and easier handling than elemental chlorine	• Unwanted halogenated byproduct formation • Oxidization of bromide • Not effective against Cryptosporidium • Limited shelf-life • Corrosive and difficult to store • More expensive than elemental chlorine	• Produced by adding elemental chlorine to sodium hydroxide • Typical solution contains 5–15% chlorine
Calcium hypochlorite, $Ca(ClO)_2$	• Provides residual protection • Highly reliable • Cost-effective • Easily applied and monitored • Less hazardous and easier handling than elemental chlorine • Longer shelf-life than sodium hypochlorite	• Unwanted halogenated byproduct formation • Oxidization of bromide • Not effective against Cryptosporidium • Precipitated solids can form in solution and complicate feeding • Fire and explosion hazard	• Used in smaller plants • Commercially available in granular and tablet forms • Contains approximately 65% chlorine

	Advantages	Disadvantages	Notes
Onsite hypochlorite generation	• Provides residual protection • Highly reliable • Minimal chemical storage and transport compared to other chlorine-based methods • Easily applied and monitored • Less hazardous and easier handling than elemental chlorine	• Unwanted halogenated byproduct formation • Oxidization of bromide • Not effective against Cryptosporidium • Requiring higher maintenance and technical expertise than other chlorination methods • Higher costs • Requiring careful control of salt quality	• An electrolytic cell is used to extract weak hypochlorite solutions (typically less than 1%) from salt water. Due to the weak solution higher volumes are required.
Chloramines, NH_2Cl, $NHCl_2$, NCl_3	• Provides residual protection • Reduced formation of halogenated byproducts • No oxidation of bromide • More stable residual than free chlorine (excellent *secondary* disinfectant) • Lower taste and odor than free chlorine	• Weak disinfectant and oxidant • Requires shipment and handling of ammonia as well as chlorinating chemicals • Water cannot be used in aquariums or kidney dialysis (due to the presence of ammonia) • Poses the risk of nitrite formation in the distribution system, due to nitrification of excess ammonia	• Chloramines are derivatives of ammonia by substitution of one, two or three hydrogen atoms with chlorine atoms
Chlorine Dioxide ClO_2	• Effective against Cryptosporidium and other chlorine resistant organics • Several times faster than elemental chlorine at inactivating Giardia	• Inorganic byproduct formation (chlorite and chlorate) • Highly volatile residuals • Requires onsite generation and handling, entailing a high level of expertise, monitoring, and technical competence	• Various processes are possible, but usually sodium chlorite ($NaClO_2$) and elemental chlorine are mixed onsite to form chlorine dioxide

(Continued)

Table 8.5   Continued

Disinfection Method	Advantages	Disadvantages	Comments
	• Only moderately affected by pH (more effective at higher pH) • Will not form halogenated byproducts • Highly effective at removing foul taste and odor	• High operation costs (due to sodium chlorite) • Occasional creation of unique odors and tastes	• Characteristics different form chlorine • Volatile, gaseous, and easily stripped from solution
Ozone, O_3	• Strongest oxidant and disinfectant • Produces no chlorinated byproducts • Extremely effective against all pathogens including Cryptosporidium	• Difficult to apply and control due to high reactivity and low solubility • No residual disinfection • Forms brominated and other byproducts (such as ketenes and aldehydes) • High costs • Difficult to operate, maintain, control, and monitor • Breaks down organic matter into smaller compounds which can assist microbial regrowth	• Generated onsite by passing dry oxygen through high-voltage electrodes • More often applied for oxidation rather than disinfection purposes • More popular in the EU than other parts of the world
Ultraviolet radiation (UV)	• No chemical generation, storage, and handling • No known byproducts at levels of concern	• No residual disinfection • Not effective on some viruses • Difficult to monitor efficacy • If not applied adequately, some organisms can repair and reverse the destructive effects of UV radiation (photo-reactivation) • Does not provide taste or odor control	• Is a non-chemical disinfectant, generated by mercury arc lamps • UV radiation penetrates cell walls, damages genetic material, and prevents cell reproduction • Most effective in the 200–310 nm wavelength (maximum at around 265 nm)

water quality during storage and distribution are not specific to desalinated water, and can be faced in any water production plant. If adequate residual disinfectants are not present, microorganisms (both pathogenic and non-pathogenic) will have an opportunity to grow. Even if the microorganisms are non-pathogenic, they still contribute to an increase in the heterotrophic plate count and can render the water quality unacceptable [16]. Hence, "residual disinfection" is required. The maintenance of water quality during storage and distribution depends on the following:

- The amount of trace nutrients and biodegradable organic matter that can support bacterial growth; notably, the biodegradation of organic substances (such as humic acid) may increase due to oxidation/disinfection, which increases the nutrients available for pathogen growth.
- The condition of the surfaces on which the microorganisms can attach, for example the material of piping and reservoir surfaces; biofilm develops more quickly and diversely on iron pipe surfaces than polymers such as polyvinylchloride (PVC).
- The presence of corrosion and corrosion-sediments; free chlorine is more negatively affected than monochloramine, although the efficacy of both disinfectants is impaired.
- The integrity of the storage and distribution system; the re-introduction of microorganisms into the water supply may become problematic if the network is unsecured. As an example, transient pressure waves can create a short-lived negative pressure, leading to entry of microbes into the system at any point where water is leaking.
- The conditions which promote growth, such as favorable temperature, hydraulic conditions, and retention time;
- The amount and type of residual disinfectant in the system; the purpose of residual disinfection (also referred to as secondary disinfection) is to maintain the water quality during storage and distribution. Research has shown that various disinfectants differ in their interaction with pathogens. For example, although monochloramine is a much less reactive disinfectant than free chlorine, it is more effective for penetrating certain types of biofilm. This is because free chlorine is so reactive that it is essentially consumed before penetrating deep enough into the film; whereas chloramines are less reactive, and have time to diffuse into the biofilm and eventually inactivate the bacteria [14]. The amount of residual disinfectant required depends on various factors (as already discussed). However, as a rule of thumb, there should always be a

chlorine residual above 0.2 mg/L at all points of the water supply system. In order to achieve this target, the chlorine residual when water leaves the plant should at least 0.5   1 mg/L (depending on how far the point of use will be). Since chlorine residual can be tasted at high concentrations, it is advised to design the system in order to have less than 0.8 mg/L free chlorine at the point of use (unless higher concentrations are vital for health reasons) [18].

8.3.4 Disinfectant Decay Kinetics

Intuitively, when water is left for too long, some might believe that it is no longer clean. In reality, this sort of thinking does have scientific merit due to the decay of the disinfectant with time. Although there are complex models available for predicting how a disinfectant degrades, often, a simple first order model is used for bulk decay [19]:

$$\frac{dC}{dt} = -k_{d1}C$$

where dC/dt is a first order differential equation, C is the disinfectant concentration (mg/L), k_{d1} is the first order decay rate constant (1/time), and t is time (in complementary units to k_{d1}). The decay rate constant (which can be found in the literature) is dependent on many factors, including pH, temperature, the chemical constituents in the water, the piping and storage tank material, etc.

In other instances, instead of a first order a model, a second order model can be employed [19]:

$$\frac{dC}{dt} = -k_{d2}C^2$$

where k_{d2} is the second order decay rate constant (L/(mg.s)) and t is time (s). In addition to bulk decay discussed above, most modelling software also incorporate the effect of decay resulting from reactions with material at the pipe wall.

Here, an example will be presented for better understanding of how to use the above equations:

Assuming disinfection with chlorine gas with a first order decay rate constant of 1 (day^{-1}), and an average velocity of 1.6 m/s as the water is pumped through the network, would a customer living 9 km away receive adequately chlorinated water if the concentration of the disinfectant is 0.8 mg/L as the water is sent out from our desalination plant? Assume there is no wall decay.

In order to answer the question, we should first calculate how long it will take for the water to get to the customer:

$$t = \frac{distance}{velocity} = \frac{9000\ m}{1.6\ m/s} = 5625\ s = 0.065\ days$$

Now, we will integrate the first order model and rearrange, in order to isolate the term for concentration at the final point of use.

$$\frac{dC}{dt} = -k_{d1}C$$

$$\int \frac{dC}{C} = \int -k_{d1}dt$$

$$\ln\left(\frac{C_2}{C_1}\right) = -k_{d1}(t_2 - t_1)$$

$$\frac{C_2}{C_1} = e^{-k_{d1}(t_2-t_1)}$$

$$C_2 = C_1 \cdot e^{-k_{d1}(t_2-t_1)}$$

where C_2 is the concentration at the point of delivery (mg/L) and C_1 is the initial concentration. The difference between t_2 and t_1 is the time it takes for the water to reach the destination. So, by inserting the values from the example, the concentration at the point of use will be:

$$C_2 = 0.8 \cdot e^{-1(0.065)} = 0.8 \cdot 0.937 = 0.75\ mg/L$$

Since we had established that a concentration of at least 0.2 mg/L is desired at the point of use, then the disinfection in this example is adequate. However, would this level of disinfection still be adequate if due to varying conditions the value for the first order decay rate constant were to increase six-fold? Note that a several-fold increase of the rate constant is not too farfetched. In order to answer this question, we will once again use the derived formula:

$$C_2 = 0.8 \cdot e^{-6(0.065)} = 0.8 \cdot 0.677 = 0.54\ mg/L$$

Thus, the disinfection is still adequate. Note that in the above example, an average velocity of 1.6 m/s was assumed for the water in the network. However, in reality, in addition to the neglected wall decay, the water may remain stagnant in the network or in tanks for long durations of time, leading to further decay of the disinfectant.

8.4 Conclusion

Since modern desalination plants remove almost all minerals from water, it is often the case that desalinated water requires post-treatment for its proper storage, distribution, and utilization. This chapter has introduced various processes used for the conditioning of desalinated water. These processes are aimed at increasing buffering capacity, calcium (and magnesium) hardness, and the alkalinity of the water.

Although blending desalinated water with other water sources is the simplest method for its conditioning, it has its limitations and potential concerns. The most popular post-treatment process is the direct dosage of limewater (and carbon dioxide). It is estimated that more than 90% of desalination plants worldwide apply this method. Alternatively, flowing desalinated water through calcium carbonate beds is an alternate method for its post-treatment. Different post-treatment processes have been compared, indicating that although calcium carbonate contactors have slightly lower operating costs and considerably less environmental impacts, the capital costs and land requirements of the system for direct dosage of lime are significantly lower. It should be noted that the use of dolomite contactors and other innovative post-treatment methods have not gained much popularity.

In the second half of the chapter, the issue of disinfection has been discussed. The difference between primary and secondary disinfection has been defined, and the most popular disinfectants, namely elemental chlorine, sodium hypochlorite, calcium hypochlorite, onsite hypochlorite generation, chloramines, chloride dioxide, ozone, and UV radiation have been compared and contrasted. A chlorine (disinfectant) residual target above 0.2 mg/L at all points in the distribution network has been expressed as the guideline value. In order to have this concentration of chlorine at the point of use, the residual chlorine concentration as the desalinated water leaves the plant should be no less than 0.5–1 mg/L (depending on how far the point of use will be).

References

[1] Birnhack, L., Voutchkov, N., and Lahav, O. (2011). Fundamental chemistry and engineering aspects of post-treatment processes for desalinated water—a review. *Desalination* 273, 6–22.

[2] Duranceau, S., Wilder, R., and Douglas, S. (2012). Guidance and recommendations for posttreatment of desalinated water. *J. Am. Water Works Assoc.* 104, E510–E520.

[3] Cotruvo, J., Voutchkov, N., Fawell, J., Payment, P., Cunliffe, D., and Lattemann, S. (2010). *Desalination Technology: Health and Environmental Impacts*. Boca Raton, FL: IWA Publishing and CRC Press.

[4] Withers, A. (2005). Options for recarbonation, remineralisation and disinfection for desalination plants. *Desalination* 179, 11–24.

[5] Voutchkov, N. (2011). Re-mineralization of Desalinated Water. SunCam Inc.

[6] Shemer, H., Hasson, D., and Semiat, R. (2015). State-of-the-art review on post-treatment technologies. *Desalination* 356, 285–293.

[7] Letterman, R., and Kothari, S. (1995). "DESCON" http://rdletter.mysite. syr.edu/software.html (last accessed on August 1, 2017).

[8] Schott, G. (2003). *Limestone Bed Contactor: Corrosion Control and Treatment Process Analysis Program Version 1.02,* http://www.unh.edu/ wttac/WTTAC_Water_Tech_Guide_Vol2/limestone_design_criteria2.html (last accessed on August 1, 2017).

[9] Lehmann, O., Birnhack, L., and Lahav, O. (2013). Design aspects of calcite-dissolution reactors applied for post treatment of desalinated water. *Desalination* 314, 1–9.

[10] Blum, R. V., Buri, M., Poffet, M., and Skovby, M. (2010). *Micronized CaCO3 Slurry Injection System For The Remineralization of Desalinated and Fresh Water*. Patent No. WO 2013014026 A1. Washington, DC.

[11] Shih, W. Y., Sutherland, J., Sessions, B., Mackey, E., and Walker, W. S. (2012). *Upflow Calcite Contactor Study*. Austin, TX: Texas Water Development Board.

[12] DrinTec Solutions (2009). *Guidelines for the Operation and Management of DrinTec^{TM} Contactors*. St. Glasgow: DrinTec Solutions.

[13] The Dow Chemical Company, Tech Manual Excerpt. Filmtec Membranes. Water Chemistry and Pretreatment: Biological Fouling Prevention.

[14] World Health Organization (2004). "Inactivation (disinfection) processes," in *Water Treatment and Pathogen Control: Process Efficiency in Achieving Safe Drinking Water*, eds M. LeChevallier and K.-K. Au (London: IWA Publishing), 41–65.

[15] Chlorine Chemistry Council, and Canadian Chlorine Coordinating Committee (2003). *Drinking Water Chlorination: A Review of Disinfection Practices and Issues*.

[16] World Health Organization (2011). *Safe Drinking-Water from Desalination*. Rome: World Health Organization.

[17] Rush, B. (2002) *CT Disinfection Made Simple*. Alberta: Water Research Center.

[18] World Health Organization (2017). *Fact Sheets on Environmental Sanitation*. Guildford: Robens Institute, 181–186.

[19] Mackenzie, L. D. (2010). *Water and Wastewater Engineering Design: Principles and Practice*. McGraw-Hill Inc.

PART III

Science and Technology

CHAPTER 9

The Origins of Today's Desalination Technologies

Dr. Jim Birkett[1]

[1]West Neck Strategies, 556 W Neck Rd, Nobleboro, ME, USA
E-mail: westneck@aol.com

9.1 Background

Modern desalination technologies can be considered as commencing in the 1960s with the development and spread of multi-stage flash (MSF) evaporation, electrodialysis (ED) and, eventually, reverse osmosis (RO). But these technologies, and those which followed, did not just like Athena, "spring full-born from the brow of Zeus". Centuries of inquiry and experimentation preceded their appearance.

The realization of the need for desalination must precede written history. It is easy to imagine a primitive man kneeling beside a primitive sea, taking a drink from the water in his hands, spitting it out again, and then thinking in his primitive way "There must be a better way". However, we have no record of these encounters. We do know that the earliest writings which we can decipher refer to the importance of sweet water and the rejection of bitter waters.

Civilizations developed around sites with reliable water sources. If water was only seasonally available, nomadic tribes followed the water. If a region had no water, no one went there. Boats did not venture far from shore and kept in mind where water could be found. The world's population was low and life was sustainable by following these patterns.

But man is an ambitious and inquisitive creature. About 2000 years ago Aristotle wrote "*Salt water when it turns into vapor becomes sweet and the vapor does not form salt water again when it condenses*" [1]. This clearly is an understanding of the concept of distillation but just when it was reduced to practice is less clear. Certainly, the comments of Alexander of Aphrodisias (*c.* ad 200) that "*sailors at sea boil sea water and suspend large sponges from the mouth of a bronze vessel to imbibe what is evaporated. In drawing this off the sponge, they find it to be sweet water"* [1] are relevant. In the 1960s, the UK Atomic Energy Agency commissioned the now well-known depiction of this process [2] (Figure 9.1). And so we may state that the simplest of distillation technology was applied to desalination about 2000 years ago. But such single effect air-cooled batch stills were highly inefficient and were never widely used.

It is important to note that during the first and second centuries AD, Alexandria was the scene of experimentation on distillation. While seawater was not the specific subject of their investigations, the skills and concepts were passed on to subsequent generations. Particularly important was the design of equipment for evaporation and condensation, including the long-snouted alembic vessel (which has come to be one of the more popular "logos" of chemistry). At this time, the condensing portions of the apparatus were invariably air cooled. This was probably adequate for extraction and purification of relatively non-volatile distillates such as essential oils and heavier petroleum fractions, but water cooling would have been required for effective recovery of more volatile species.

Figure 9.1   Depiction of distillation with sponges [2].

9.2 The First Tentative Steps

About the same time another process was referred to by both Aristotle and Pliny the Elder. *"If one plunges a water-tight vessel of wax into the ocean, it will hold, after 24 hours, a certain quantity of water, that filtered into it through the waxen walls, and this water will be found to be potable, because the earthy and salty components have been sieved off"* [3]. Such a claim was to be reported repeatedly through the ages. Although it may suggest a RO type of mechanism at work, it is more likely just an attractive myth. Firstly, the wax would have to have had perm-selective properties. The simple animal, vegetable and mineral waxes available at the time did not. Second, to overcome the natural osmotic pressure of seawater, such a "waxen vessel" would need to withstand great pressures. Neither a waxen vessel nor an unglazed earthenware vessel, perhaps coated with such wax, would be likely to do so. Third, such a vessel would need to be lowered to and retrieved reliably from a depth of approximately 500 m. It is doubtful that the technology existed at the time for such an undertaking. Yet the concept was too attractive to be dropped and persisted for centuries.

Starting around 1500, new and vigorous explorations were made by European adventurers around and across the world. Their impact on the development of desalination is best phrased by Forbes: *"As the sea voyages became longer and longer in the course of the sixteenth and seventeenth centuries, and the ships no longer sailed along the coast but took to the open sea, it became an urgent thing to become independent on board ship of the water supplies of the coasts and islands on the route. Especially long voyages on the Pacific and other oceans where sailing ships might not be able to throw out their anchors for many a week, this was a most important question. The usual supplies of fresh water in wooden casks were often spoiled before the casks were opened and we need not wonder that the solution of this question occupied the minds of the Royal Society and other scientific bodies"* [1].

From this point onward there are numerous references [4] to the use of simple shipboard stills by European adventurers under emergency conditions. They were unreliable, fuel intensive and, as practiced at the time, produced a product of questionable quality. But as voyages of exploration and conquest ventured farther and for longer periods into the unknown, its potential became obvious. Hawkins [5] reported in 1622 that during his earlier voyages to the South Seas, he had been able to supply his men with fresh water by means of shipboard distillation. Sir Walter Raleigh reportedly experimented with the process while a prisoner in the Tower of London in the early 1600s [3].

Numerous other reports suggested that desalination by distillation was fairly well-known but usually dismissed as impractical. Indeed, one is struck by the off hand manner in which an explorer would refer to distillation, as though everybody knew about it but rarely bothered with it. For example, Hawkins in 1622 said: "... *with an invention I had in my shippe, I easily drew out of the water of the sea, sufficient quantitie of fresh water to sustaine my people with little expense of fewell; for with foure billets I stilled a hogshead of water, and therewith dressed the meat for the sicke and whole. The water so distilled, we found to be wholesome and nourishing*". Thus, Hawkins already had a still on board, ready for use when and if needed!

As discussed above, in simple stills the condenser was usually air-cooled. However, by the middle of the 17th century water cooled condensers began to appear as shown in the drawing by Le Febure [6] (Figure 9.2).

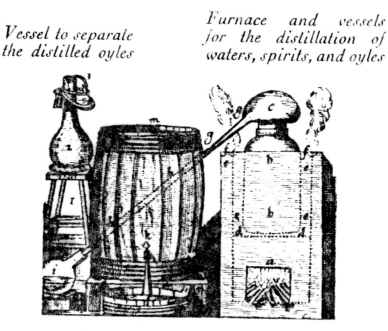

Figure 9.2   Distillation after Le Febure [6].

The disadvantage of such batch cooling of the condenser was that the water in the barrel would gradually become warmer and have to be replaced with new cold water. This problem was overcome by Houton [3] of Paris, France whose work was described in 1670 in the *Philosophical Transactions of The Royal Society of London*:

> "*for the cooling of which, he hath this new invention, that instead of making the Worm pass through a Vessell full of water (as is the ordinary practice) he maketh it pass through one hole made on purpose out of the Ship and so to enter again through another. So that the Water of the Sea performeth the cooling Part, by which means he saveth the room which the common Regrigerium would take up; also the labour of changing the Water, when the Worm hath heated it.*"

Up to this point these stills were hand made by individual craftsmen using materials at hand and to the simplest of plans. However as distillation became better known, business opportunities were anticipated. This led, in the late 17th century, to the first patent dispute involving desalination technology. Shortly, after Mr. Hauton's report, a Mr. William Walcot received an English patent (1675) for a device claiming the "*Art of Making Water Corrupted fit for Use, and Sea Water, Fresh, Cleare, and Wholesome in very large Quantities, by such Wayes and Meanes as are very Cheap and Easy.*" This patent [7] which granted the holder "*sole use and benefit of the said Invention for the space of fourteene yeares,*" offered no technical description of the device itself, only much praise for its virtues and legal language describing the extent of protection sought from possible infringers. Little is known about Walcot's origins or prior experience. Although the patent did not use the word "still" in the description of the device, we may presume that that is what it was.

The situation became complicated in 1683 when Fitzgerald et al., received a patent for "*A Way to render Salt or Brackish Water Sweet and Fitt for Drinking, Boyling Meate, Washing, or any other ordinary use, by the meanes of a certaine engine or engines not heretofore used or practised in our Dominions*". The Fitzgerald patent [8], like that of Walcot, was vague as to actual details of the device.

However Fitzgerald and his backers understood the power of publicity and importance of connections in high places. They issued numerous pamphlets and "press releases" lauding the claimed advances. Walcot claimed infringement and the dispute rose up to involve King Charles II himself. It certainly helped the Fitzgerald cause that his uncle, the scientist Robert Boyle, was well known in Court. Fitzgerald's backers also had more money and held higher positions in government and society than did Walcot. Under the direction of the king, Boyle tested the product water from the Fitzgerald device with a solution of silver nitrate. Finding no precipitate of silver chloride, he declared it pure.

By the end of 1683, the Privy Council declared the Walcot patent [9] to be null and void as it was *"inconvenient to His Majesty's subjects"*. This might be interpreted to mean that they felt that the Walcot device was unlikely to be reduced to practice, whereas the Fitzgerald device with stronger backing was more likely to be commercialized. However, as a sop to Walcot, they awarded him 1/6 of the proceeds that might eventually accrue to be earned by the Fitzgerald invention.

For all the legal excitement, there appeared to be no discernable technical advancement in either still. Evaporation was carried out in the batch mode (no continuous addition of seawater and/or discharge of concentrate). No successful business ventures were developed. However, it is worth noting that Boyle's testing of the purity of the Fitzgerald device was the first recorded use of analytical chemistry to judge the efficacy of a desalination system.

It is also interesting that both Walcot and Fitzgerald promoted the use of proprietary "ingredients" to improve performance and the quality of the product water. Thus, they saw the commercial advantages of the continuing sale of additives (long-term cash flow) as an adjunct to the one-time sale of the capital equipment. Fitzgerald even went so far as to stipulate that purchasers of his device must enter into a long-term contract to purchase a minimum quantity of "ingredient" in each 6-month period. The cost of this would be "one shilling for as much as makes about 90 gallons of water". (This was in addition to the first-cost of the still itself, 18 pounds for a 90 gallon per day unit.) Today the sale of chemicals (both proprietary and commodity) to desalination operations is a business worth several hundred millions of dollars annually, and based on very sound chemistry and scientific knowledge.

While Walcot and Fitzgerald may have had the correct entrepreneurial instincts, their timing was premature. There was not enough scientific understanding of the nature of water (and sea water) nor the technical tools to develop manufacturing facilities. These conditions were yet a century away.

9.3 The Appearance of an Industry in the 1800s

The "industrial revolution" of Europe, Great Britain, and America had no strict boundaries but occupied the period roughly from the late 1700s to 1900 (although some argue that it continues to this day). The period saw great changes in demographics, economics, commerce, culture and technology. The practice and technology of desalination were no exceptions; they leapt forward as well.

Until the early 1800s, desalination was practiced almost entirely in shipboard single stage stills, operated in batch mode, fired directly from the cookstove or furnace, and usually without recovery of the heat of condensation. This was bulky and inefficient. The product water quality was highly variable due to carry-over of mist and of decomposition products of impure feed and dirty boilers. Units were made individually by artisans to the vaguest of plans and specifications. Rarely were two alike.

However the groundwork had been laid for major changes. Lavoisier had discovered oxygen, Priestley had discovered oxygen's role in combustion, and Dalton had identified the formula of water to be H_2O. These three, together with Boyle, Cavendish and others had helped to establish the basis of chemistry and scientific thinking. Savery, Newcomen, Hornblower, and Watt had introduced and improved the steam engine to the point where the properties of steam itself became a subject worthy of study. Slowly, machine tools such as the machine lathe with compound rest were developed and fabrication skills increased.

In 1814 Smithson Tennant of England published the design of a simple but functionally correct two-effect still [10]. He even constructed a model demonstrating this for use in his teaching at Cambridge University. Unfortunately, he died in a riding accident shortly after his paper was published. He never had an opportunity either to teach this development to his students or to exploit it commercially, although he did speculate on its potential ship-board use. As Tennant himself put it, *"When water is deficient on board ship, it has been in some degree supplied by distillation from the ship's boiler, and if the steam from the boiler had been made to pass through the apparatus just described, the quantity would have been doubled."* His design is shown in Figure 9.3).

This was a very important advancement. Heretofore, the heat of condensation transferred to the condenser cooling water (or air) was lost to the environment. Tennant recovered this heat by using it to heat a secondary boiler and to generate additional water. To this day, all multi-effect distillation systems follow the same strategy, resulting in much higher fuel efficiency.

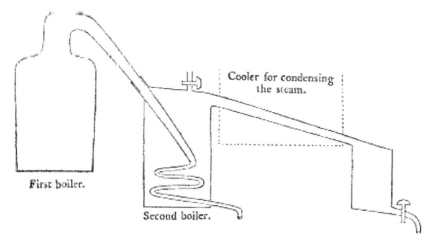

Figure 9.3   The Tennant two-effect still [10].

Interestingly, Rillieux [11] of New Orleans, USA, is generally considered to be the inventor of the multi-effect process in the 1840s. However he was seeking a more efficient means of obtaining sugar from dilute cane juices and focused on multi-effect *evaporation* (an efficient means of concentrating the retentate) rather than on multi-effect *distillation* (an efficient means of producing the distillate). These are fundamentally the same process, but in practice each design is optimized differently depending upon the desired product, retentate, or distillate.

Tennant's untimely death delayed the adoption of his technology by nearly 50 years. However, in the 1850s, Normandy (French-born but living and working in London) received patents [12] for multi-effect stills intended specifically for seawater conversion. Moreover, he established a company, Normandy's Patent Marine Aerated Fresh Water Company, which manufactured such equipment from 1858 until about 1910. Some 2,000 units were produced for both land-based and shipboard applications and were sold all over the world, including India, South America, Helgoland, Suez, Aden, and the United States. Figure 9.4 shows the remains of a Normandy unit placed at Key West in 1862.

Another design feature which distinguished Normandy's stills from his competitors was the capture of the dissolved air released during the heating of the sea water and its re-introduction to the steam prior to condensation. This led to the production of an aerated distillate which was then passed over bone charcoal for further removal of impurities. Today, we know that activated

Figure 9.4   The 1862 Normandy unit as discovered in 1977.

carbon can act as a catalyst for the reaction of dissolved oxygen with adsorbed compounds such as phenols on its surface, thus enhancing their removal.

Most important, Normandy's firm was the world's first successful international desalination company with patent protection, a factory, regular employees and a business structure including formal incorporation, investors and a Board of Directors. It exhibited its product line (and won prizes) at international exhibitions. It remained in business for about 60 years.

Normandy's success was in many ways the result of perfect timing. The Industrial Revolution of the late 18th century had brought with it not only the steam engine but knowledge of steam thermodynamics. It also furthered the development of machine tools such as the compound lathe, the milling machine, the screw-cutting lathe and more sophisticated drilling capabilities. Standardization of screw threads was initiated by Whitworth. Cast iron became more reliable. Normandy, a skilled chemist and ambitious entrepreneur, took full advantage of these new tools and the business opportunity.

Of course, Normandy's firm attracted many competitors from England, Scotland, Germany, France and probably elsewhere. At first, most offered simple single effect equipment which could not match Normandy's fuel efficiency. But by the late 1800s firms such as Mirlees–Watson of Glasgow and Lillie of Philadelphia entered the fray. They had been active in producing multi-effect evaporators for the sugar industry and readily made the conversion to multi-effect distillation.

It is interesting to note that early in the nineteenth century when sail propulsion still dominated the seas, ship-board stills were first accepted for emergency use only and simplicity of operation was more important than fuel efficiency. Indeed, in the early days of steam propulsion when stills provided boiler make-up water fuel efficiency was also of limited importance as there was usually excess steam available and even a single effect still would represent only a minor parasitic demand. Multi-effect stills did not become routine on steamships until the twentieth century. However, in the case of land-based applications for the provision of drinking water, where the still was fed from a dedicated boiler, multi-effect stills were more popular.

For several years England was involved in military activity in the Sudan, requiring the garrisoning of troops at Suokim. Water supplies were meager and in the 1890s two coal-fired six-effect distillers of 350 tons/day capacity were purchased from Mirlees–Watson of Glasgow [13]. Figure 9.5) gives a drawing of one such unit. Similar plants were installed at Quseir, Kamaran Island and Jeddah [14] (then part of the Ottoman Empire) on the Red Sea and at Mombasa. Toward the end of the century, Weir of Scotland also built a 100 m³/day plant on the Red Sea in Egypt to support a phosphate enterprise [13].

The Lillie Sugar Apparatus Manufacturing Company of Philadelphia developed the first horizontal tube spray-film distiller. The tubes were closed at one end and sloped slightly upwards so that the condensate could flow out [11] (Figure 9.6). At the closed ends were small vent holes to allow non-condensable gases to be drawn out. This may have been the first recognition of the need to remove such gases that would otherwise blanket the heat exchange surfaces and retard production. A triple-effect unit of the Lillie design was installed at Fort Jefferson on the island of Dry Tortugas in the Caribbean just prior to 1900 and produced 60,000 US gallons per day (225 m³/day) [15].

Perhaps, the best-known land-based desalter was the solar desalination system designed by Charles Wilson in 1872 and installed in Las Salinas,

Figure 9.5 The Mirlees–Watson unit at Suokim [13].

Chile [16]. This basin-type still with glass covers produced about 19 m^3 of fresh water daily, at a rate of 4.9 L/m^2/day. Lesser known was a similar but smaller solar system in German South West Africa (now Namibia), reported in 1887 [17].

The two solar installations noted above may have been based in part on the technology developed by Wheeler and Evans [4] and described in their 1870 patent. This design described a sloped basin still with double glazing, a "staircase" of feed water troughs on the darkened collector surface, heat exchange between feed and distillate, and a mechanism to rotate the apparatus to face the sun. Although the primary application was targeted to be alcohol production, the inventors specifically referred to usage *"when the object is to procure fresh and pure water from salt, alkaline, or otherwise impure sources..."*

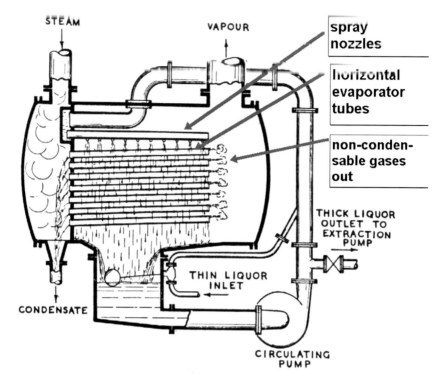

Figure 9.6   The Lillie horizontal tube spray-film evaporator [11].

Mention must also be made of the mammoth wood-fueled condenser constructed in Coolgardie, Australia in 1896, supplying water to a gold-mining operation [18].

Figure 9.7   The wood-burning still at Coolgardie [18].

Another development of note was the mechanical vapor compression evaporator invented by Piccard and Weibel in 1879 and patented by Weibel in 1881 [19]. The Piccard–Weibel system enjoyed some success, particularly in salt works, but was very large for its capacity and suffered from the immaturity and unreliability of reciprocating compressors of that period. Further vapor recompression improvements had to await the development of the steam jet nozzle compressor by Prache in 1908 [20].

Up to the end of the nineteenth century, there had been no investigation of membrane separations specifically applied to desalination. However mention should be made of the French Abbé Nollet who in 1748 discovered the semipermeability of animal membranes [21]. In France, in 1828 Dutrochet introduced the term *osmosis* while describing the passage of water through biological membranes [4]. But it was not until 1877 that van't Hoff first quantified it by the expression $\pi = kCT$ where π is osmotic pressure, k is a constant, C is concentration and T is temperature. In 1855, Fick published his laws of diffusion and also prepared and investigated early artificial (not plant or animal) semipermeable membranes [21].

As the century closed, the first hints of ED appeared with the work of Schwerin to purify sugar extracts in a three-chambered electrochemical membrane cell [22].

These efforts laid the groundwork for the membrane sciences we know today.

By the end of the century, all had changed in the desalination world. Multi-effect systems were commonplace, steam was the usual method for heat input and control, various designs were competing for the customers' attention, and there was an established pool of manufacturers eager and willing to produce the desired product. The work of Piccard and Weible had even explored evaporation separation through the use of mechanical rather than thermal energy!

9.4 The Early Twentieth Century – Evaporative Advancements

In 1908 Charles Prache, a resident of Paris, was awarded a patent [20] for a thermo-compressor nozzle design that paved the way for thermally-driven vapor compression desalting units, the most popular of which were known as Prache and Bouillon evaporators. A number of manufacturers including Aitens, Ltd. of England and SCAM (now SIDEM) of France licensed this technology and manufactured complete systems.

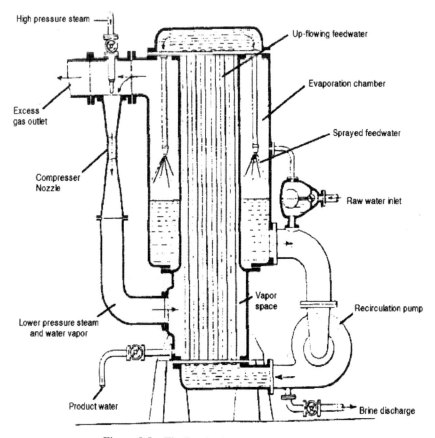

Figure 9.8   The Prache & Bouillon TVC design.

Despite the availability of distillation units from a variety of suppliers, production remained modest. For the first few decades of the century, most sales were to industrial users and ship-board applications. There was no market for municipal systems. Then, things changed rapidly in an unexpected location.

Until the 1920s, the Netherlands Antilles (Aruba, Bonaire, Curacao, St. Maarten, Saba, and St. Eustasius) were quiet islands under the Dutch flag. Populations were low and water supplies limited to rain water, ground water and occasional water importation. Then oil production began in nearby Venezuela and oil refineries established on Aruba and Curacao. The economies changed dramatically and immigration increased [23]. For example, on Aruba the population grew from about 8,000 in 1924 to 17,000 in 1932 and 50,000 in 1950. To this was added in later years, international tourism. The water deficit increased accordingly and in 1928 the government of Curacao acquired a small thermal vapor compression unit fabricated by Aiton, Ltd. of England under license from Prache and Bouillion of France. Its capacity was about 60 m³/day and its design similar to that shown above.

In practice, this unit was too small and unreliable but it served to establish the concept of municipal water supply through desalination. In 1929 Curacao purchased what would be the first of many multi-effect distillers from Weir of Scotland. One year later Aruba, convinced that this was the right approach to water security, followed suit. From 1929 until 1958, Aruba and Curacao would buy approximately 30 such Weir units, nominally rated at 200 m³/day. They were of a "double-deck" six effect configuration as shown below.

Figure 9.9   A completed 200 m³/day Weir unit installed at Aruba.

Vessels were of cast iron with heat exchange tubes fully submerged in the boiling seawater. The performance factor (gain output ratio) probably rarely exceeded 5 units water per unit of steam. These 200 ton/day units were built to last. The author saw such a unit on the Red Sea in Egypt in 1986, still operational after 50 years in service. Another such unit, installed in 1911 in Yemen, was reported still in service in 1964!

In the 1930s Kleinschmidt and others working at Little (USA) further refined the mechanical vapor compression system, including improvements in the compressor design [24]. One version was intended for use in submarines where its ability to run on electricity was an important feature. Subsequently the technology was licensed to firms such as Badger, MECO, Maxim Silencer, Cleaver-Brooks for engine-driven land based applications where it displayed an excellent production-to-fuel consumption ratio.

Klienschmidt's group even went so far as to build a prototype of a hand-cranked machine for emergency sea water conversion. One of the workers on the project reported that *"Under the conditions ….in the Harvard Fatigue Laboratory, where you had a 90°F temperature and 90% relative humidity, it was possible to operate this thing so you produced more water than you lost in weight by perspiration"*.

9.5 The Early Twentieth Century: Membrane Developments

In 1907 Bechold coined the term "ultrafiltration" as a specialized filtration tool for studying matter in colloidal suspension [25]. This coincided with much work by Zsigmondy, Bachman and others on microporous membranes [26]. Building upon this work, McBain and Kistler were able to report in 1931 that *"The writers have used even finer filters which hold back such molecules as sucrose and potassium chloride, while still allowing methyl alcohol and water to pass through freely"* [27]. Thus, their membranes were exhibiting the properties of RO membranes on dilute salt solutions although that term had not yet come into use.

Zsigmondy's continued work on cellulose acetate membranes led to the formation in 1929 of MembranFiltergesellschaft Sartorius Werke in Göttingen [21]. Thus was born the microfiltration industry. However, at that time the membranes were sold for research and analytical purposes, not for process separations. That would come much later.

By the 1930s, rudimentary ED had become a laboratory curiosity. Practical applications of it were stymied however by the lack of suitable membranes. Meyer and Strauss in 1940 published a design for a multi-compartment ED cell having alternating anion and cation permeable membranes [28]. They had the concept correct but unfortunately no such membranes were available at the time.

9.5.1 The Influence of World War II

The period of World War II saw dispersal of desalination technology around the world. Thousands of small units were manufactured by the belligerents and employed in the field. Both the United States and Germany used vapor compression systems on their submarines. Badger, MECO, Cleaver-Brooks (later to become Aqua Chem) and Maxim produced vapor compression units for the United States Army. Not all equipment performed well. Combat conditions included high levels of shock and vibration. Ship-board units additionally experienced instability due to pitch, roll and yaw. The war provided tough testing conditions that demonstrated the limitations and inadequacies of current desalination equipment.

After the war, much government and industry attention was refocused on the Middle East and its oil reserves. The first desalting plant in Kuwait was installed in 1946 by the Kuwait Oil Company, using the evaporator from an old World War I naval destroyer [29]. The unit was kept in operation off and on until 1950 when the company purchased their first commercial unit. The original unit was then borrowed by the Getty operation and continued in service for a number of additional years.

One interesting fall-out from the war was the transfer of membrane technology from Sartorius Werke in Germany to the United States where it ultimately led to the establishment of Millipore Corporation [21]. Continuing development of a wide range of cellulose acetate membranes spread to other companies including Gelman, Amicon, and Schleicher and Schuell.

Little noticed at the time was a theoretical paper *"Temperature and Pressure Phenomenon in the Flow of Saturated Liquids"* by Silver of Glasgow [30]. This would become key in the subsequent development of the MSF process.

9.6 The 1950s

The early 1950s was an interesting period in the development and application of desalination technology. Municipal desalination was largely limited to the Netherlands Antilles where it already had some 25 years of operating history. The largest production units at that point were 200 m^3/day, using multi-effect distillation (MED). World War II had led to refinements in mechanical vapor recompression (MVC) technology but unit sizes were small. Renewed interest in exploiting the oil fields of the arid Middle East was leading to the realization that reliable water supply would be an influencing factor, there and around the world.

In 1952, the United States government, under its Department of the Interior, founded the Office of Saline Water (OSW) to be its lead agency in supporting desalination research and to act as a clearinghouse for information related to the topic [31]. Similar initiatives were undertaken in Great Britain, Japan, the Netherlands and South Africa. Yet there was uncertainty as to how and where to proceed. Funding priorities were unclear.

The first "new" technology to make a successful appearance was ED. Although Meyer and Strauss had conceived the process earlier, suitable membranes were not then available. This changed with the development in the 1940s of synthetic ion exchange materials, based upon a polystyrene framework with attached functional groups. Juda and McRae of Ionics, Inc. (USA) found that these materials could be cast as thin sheets and make ED practical. Progress was rapid with the first publication in 1950 [32], first government funding in 1951, first public announcement in 1952, first patent in 1953 [33] and first commercial sales for brackish water desalination in 1954. Subsequent improvements including periodic reversal of polarity (ED reversal or EDR) made the systems even more rugged and resistant to scaling. Other work on ED took place at Permutit, Ltd. (UK), Rohm & Hass (USA) and TNO (Netherlands). There was similar activity in Japan, but with a focus on salt production rather than desalination.

In the Netherlands Antilles, water demand continued to rise and by the late 1950s, both Aruba and Curacao needed more capacity [23]. Rather than continue to buy more of the same Weir units, it was time to upgrade to something larger. Fortunately, Weir was able to respond with a unit offering a 10-fold increase in capacity. In 1959, Aruba purchased five such 2,000 m^3/day units and Curacao purchased two. They were so large that Weir had to outsource the casting to another foundry. Again, they were six-effect with three stacks of tube bundles per effect. Tubes were aluminum brass. From 1958 until 1963 ferric chloride was used for scale control but was then changed to sulfuric acid for environmental reasons. One such unit on Aruba is shown below.

*A view of the 2.240.000 l.g.p.d.
submerged tube evaporator plant at Aruba.*

Figure 9.10 A 2,000 m3/day MED unit on Aruba.

At the time of their installation, Aruba was the largest desalination facility in the world. Although it would eventually be surpassed by Kuwait and then other Middle Eastern countries, there was time enough to issue a commemorative postage stamp.

Figure 9.11 Commemorative postage stamp showing multi-effect units.

The multi-effect distillers which had been used on shipboard during the war were very prone to scaling on their heat exchanger tubes. In an effort to avoid this, attention turned to simple flash evaporation in which boiling would occur from the bulk liquid and not at a hot metal surface. The Bethlehem Steel Company, which had a long history of providing heavy equipment to the shipping industry, reported in 1955 and patented a 5-stage flash unit operating at a top brine temperature of 170°F (77°C) [4]. The unit boasted the same heat economy as a triple-effect distiller with lower weight and requiring less space. Such small flash units, usually with no more than five stages, achieved significant popularity, largely due to their resistance to scaling. Qualified manufacturers in the United States included Bethlehem, Griscom–Russell, Cleaver–Brooks (later to become Aqua Chem), Maxim Marine (later Riley-Beaird) and Badger Manufacturing.

Both Weir (Scotland) and Westinghouse (USA) were engaged in competition for contracts in Aruba and Curacao and now looked for new opportunities in Kuwait. To date, all large installations had been multi-effect submerged tube units of known design and operating limitations. Then in 1956, Westinghouse won a contract in Kuwait for 4 units, each with a capacity of 500,000 imperial gallons per day (2,250 m^3/day). They elected to build flash type units, each with four stages, stacked in a vertical configuration.

In Scotland, Weir assigned Robert Silver to analyze the Westinghouse design and to determine its competitive weaknesses. Silver had previously published on the topic of flashing flow and this led him to a new design in which the number of flash stages would be at least twice the performance ratio (number of units of product divided by the number of units of input steam). Although perhaps counter-intuitively, this design would significantly reduce the necessary area of heat transfer surface and thus the capital cost of the unit.

In the United States, completely independent research on the MSF process was underway. In early 1955 Reid Ewing of Cleaver Brooks (later Aqua Chem) prepared a patent disclosure for a MSF unit with many stages and having the heat exchange tubes parallel to the brine flow. This subsequently became known as the "long tube" arrangement, to distinguish it from the "cross tube" array with brine flow normal to the tubes. A 16 stage test unit was built and a patent application filed in 1958. In 1959, a full-sized (100,000 US gallons/day) unit was commissioned at the Mandalay Station of Southern California Edison Company. It had 26 stages and operated at a performance ratio of 7.7–7.9. It also incorporated brine recirculation, an advancement over the previous "once through" design.

Concurrently, and again independently, Adolph Frankel at Richardsons Westgarth (England) was developing proprietary technology on MSF designs with many stages.

As the decade drew to a close, two papers were published that would usher in another new and ultimately successful process, RO [34,35]. Reid and Breton at the University of Florida (USA) were able to demonstrate high salt rejection rates with synthetic membranes but were unable to achieve high enough fluxes to promise practicality.

9.7 The 1960s

This decade showed advancement on all fronts and ushered in world wide application of desalination. Of immediate commercial significance was the appearance and rapid acceptance of MSF technology for municipal supply. As an example we need to look no further than the Netherland Antilles again [23]. Within a few years, water demand in Aruba and Curacao was once again stressing supply. In addition, their small island neighbor, Bonaire, was experiencing shortages. The timing of their need coincided with the development and first commercialization of the MSF process.

In 1963 Curacao acquired two MSF units of 6,000 m^3/day each from Richardson Westgarth of England. That same year, Bonaire acquired a smaller 500 m^3/day unit from Weir of Scotland. In 1965, Aruba purchased a 6,000 m^3/day unit from Aqua Chem of the United States. It should be noted that these three manufacturers were the same three original developers of the MSF system! Thus the era of MSF began in the Netherlands Antilles with a full representation of available technology.

By the time of the first European Symposium on Fresh Water from the Sea, held in Athens in 1962 the three original key players in MSF had begun to understand what each of the others was doing [4]. Weir and Westgarth merged to become Weir–Westgarth. Bethlehem Steel and Griscom Russell lost whatever interest they may have had in land-based MSF units. The last MSF unit to be built by Westinghouse was for Key West, FL (USA) in 1965. However, for the record it should be remembered that the engineers of six firms, working concurrently and independently over a period of several years, ushered into the desalination field a novel seawater distillation process now used to provide massive sources of freshwater in many parts of the world. The question of who was first subsequently became moot as the continuing participants entered into royalty-free cross licensing agreements. Nonetheless, it is still argued about today.

Figure 9.12   The 1963 Richardsons Westgarth MSF on Curacao.

Figure 9.13   The original (1963) Weir unit on Bonaire.

There is one feature of the MSF design that is perhaps under-appreciated. Virtually all such plants are built and operated in conjunction with a steam turbine power station. This is known in the trade as a dual-purpose plant.

Figure 9.14 The 1965 Aqua Chem unit on Aruba.

At the steam station, only the hottest steam is used for power generation. Lower temperature steam is drawn off for heat supply to the MSF plant. For MSF, the size ratio at which it may be most efficiently coupled with a steam station is about one million gallons per day fresh water output to 10 MW of power output. It was fortuitous that this *production* ratio was a good match for the *demand* ratio in many Middle Eastern and island locations at that time.

If the 1960s was the period when MSF came of age, it was also the period when RO was truly born. Indeed, in 1960 the term "reverse osmosis" had yet to be introduced. The process was referred to as "hyperfiltration" and the separation mechanism unclear. Shortly after the papers of Reid and Breton in late 1950s, Loeb and Sourirajan at UCLA published (1962) [36] and patented (1964) [37] the results of their work on the casting of very thin asymmetric membranes of cellulose acetate. Not only did these membranes demonstrate fluxes an order of magnitude higher than those of Reid, their fabrication was amenable to high speed continuous casting, a necessity for inexpensive production. Their research also suggested the solution-diffusion theory of water transport through the membranes.

Once progress had been made toward a satisfactory membrane, it became important to package it into a modular form for use in a system. The first attempts were with plate-and-frame devices and tube-in-shell configurations.

These lacked compactness and reliability. Ultimately, the spiral design was demonstrated at General Atomic in 1964. Figure 9.15 shows the first functional spiral element [38]. *"This element was assembled to prove the concept and was made from materials found around the General Atomic laboratory. The permeate carrier was a sheet of felt with tiny glass beads to create an incompressible flow channel. The feed spacer was window screen. The glue was two part epoxy. The membrane was an asymmetric cellulose acetate. The assemble was held together with two rubber bands. When they laid out the materials they rolled them around two glass tubes which served as the feedwater inlet and permeate outlet...It worked and made about 90% rejection at 600 psi"* [39].

Figure 9.15   The first spiral element (1964) [38].

This design, perfected by Westmoreland [40] and Bray [41] to contain multiple leaves, became the preferred configuration. But it was not the only configuration.

While the above sheet membrane developments were taking place, scientists at Dow, DuPont and Monsanto were working with fine hollow fiber membranes for RO. Monsanto eventually chose to pursue its hollow fibers in the area of gas separations rather than desalination. Dow continued to develop hollow fine fibers of cellulose triacetate and enjoyed some commercial success in brackish water treatment and "membrane softening". Dow later abandoned its hollow fiber line with its acquisition of FilmTec in 1985. DuPont commercialized its aromatic polyamide hollow fiber line for both brackish and sea water applications, its first unit going to Long Boat Key (USA) in 1969. Although it was a dominant force for many years, it left the desalination arena in the 2000 due to stiffening competition from spiral-wound devices.

One other insight of the 1960s deserves mention. While developing and perfecting the spiral element, the team at General Atomic also explored the mechanism of water transport through the membrane and were able to show that it was a solution-diffusion mechanism rather than a simple size-exclusion one [42].

While most public attention was given in the 1960s to the bursting upon the scene of MSF and the emergence of RO from the laboratories, ED continued to establish its position as the technology of choice in treating brackish waters. The 1970 Desalting Plants Inventory (published by the US Department of the Interior) listed 44 ED plants worldwide as opposed to only 3 RO units. ED was strong in treating hot brackish waters with a high scaling propensity. If scale could not be prevented or treated chemically, it was even possible to dis-assemble the membrane stacks and scrub the membranes with a brush! Ionics Inc. (USA) remained the dominant player, due in part to a strong patent position and a willingness to defend it. Although ED ultimately lost most of its market share in brackish water treatment to RO, it established the acceptance of membrane separations for desalination.

Two other things of impact occurred during the 1960s. One was the formation of organized technical literature in the field. Previously, technical papers were published in various journals, chosen at the whim of the authors. There was no central place to look for information. This changed in 1962 with the introduction of the Elsevier journal *Desalination,* edited by Balaban. This has been of great benefit to the desalination community, user and supplier alike.

The second development to note was the appearance of well organized and attended international conferences and symposia devoted to the field. An example is the Symposium on Saline Water Conversion held in Washington, DC (USA) in 1960 under the auspices of the American Chemical Society and with the support of the Office of Saline Water (OSW). Among its many papers was one by Barnett Dodge of Yale University in which he made perfectly clear that there was indeed a minimum energy of separation for sea water, dependent solely on concentrations and temperatures of the incoming and outgoing streams but independent of process [43].

Another example mentioned earlier, was the 1962 first European Symposium on Fresh Water from the Sea held in Athens. This meeting (and subsequent such meetings) was organized by the Working Party on Fresh Water from the Sea, an *ad hoc* committee of the European Federation of Chemical Engineers. It did not have regular individual members such as today's International Desalination Association, European Desalination Society, etc. Such technical and trade organizations made up of individual members would have to wait until the 1970s.

9.8 Post 1960s

By 1970, desalination technologies, applications and supporting industry had acquired unstoppable momentum. There was no turning back. New developments in RO such as membranes of thin layer composite structure made by interfacial polymerization and advanced energy recovery systems still lay ahead. ED, while displaced by RO in many applications, would continue to evolve. MSF would continue to grow in unit size and reliability. Multi-effect systems would be reborn through their coupling with thermal vapor compression technology. But the underlying basic processes of 1970 would remain identifiable and continue to dominate the industry for decades to come.

References

[1] Forbes, R. J. (1948). *A Short History of the Art of Distillation*. White Mule Press: Leiden.
[2] Howarth, J. (1980). *Product Literature for P & B Evaporators*. Derby: Aiton, Ltd.
[3] Nebbia, G., and Menozzi, G. N. (1966). *Aspetti Storici Della Dissalazione*. Mangalore: Acqua Ind, (41–42).

[4] Birkett, J. D. (2000). *The History of Desalination before Large Scale Use*. Oxford: Desware Encyclopedia, EOLSS.

[5] Hawkins, R., (1622). *Observations of Sir Richard Hawkins, Knight, in his Voiage into The South Sea in the Year 1593*. London: John Jaggard.

[6] Gilliland, E. R. (1955). Fresh water for the future. *Ind. Eng. Chem.* 47, 2410–2422.

[7] Walcot, W. (1675). Purifying Water, English Patent No. 184.

[8] Fitzgerald, R., et al. (1683). Purifying Salt Water, English Patent No. 226.

[9] Birkett, J. D. (2011). Desalination Activities in Late 17th Century England. *IDA J. Desal. Water Reuse* 3, 14–20.

[10] Tennant, S. (1814). On the Means of producing a double Distillation by the same heat. *Philos. Trans. R. Soc. Lond.* 104, 587–589.

[11] Birkett, J. D. (2003). Advances in sea water desalination, 1800–1900. *Paper Presentation to International Water History Association*, Alexandria.

[12] Birkett, J. D. and Radcliffe, D. (2014). Normandy's patent marine aërated fresh water company; a family business for 60 years. *IDA J. Des. Water Reuse* 6, 24–32.

[13] Birkett, J. D. (1984). *A Brief Illustrated History of Desalination*. Vol. 50, Amsterdam: Elsevier Science Publishing, 17–62.

[14] Low, M. C. (2015). Ottoman infrastructures of the saudi hydro-state: the technopolitics of pilgrimage and potable water in the hijaz. *Comp. Stud. Soc. History* 57, 942–974.

[15] Crank, R. K. (1900). Six-day trial lillie dry tortugas. *Am. Soc. Naval Eng.* 3, 85–99.

[16] Harding, J. (1883). Apparatus for Solar Distillation. *Proc. Inst. Civ. Eng.* 73, 85–99.

[17] Huber. (1898). *Uber die Mittel fur Herstellung genussfahigen Wassers aus Meerwaser*. Marine: Rundschau, IX911, 1045–1057.

[18] Crisp, G. (2014). *Private Communication*. Availabie at: http://www.cosmos.esa.int/web/jwst/nirspec-pce2014

[19] Weibel, J. (1881). Concentrating Syrups and Other Liquids, US Patent No. 236, 657.

[20] Prache, C., *Thermo-Compressor*, France, Patent No. 904,276, (1908).

[21] Lonsdale, H. K. (1982). The Growth of Membrane Technology. *J. Mem. Sci.* 10, 81–181.

[22] Svedberg, T. (1928). *Colloid Chemistry*, 2nd ed. New York, NY: Chemical Catalog Company.

[23] Birkett, J. D. and Marchena, F. A. (2012). Early Desalination in the Caribbean and its Adoption in the Netherlands Antilles, *Paper Presentation to CaribDA 2012 Conference*, Aruba, 19–22.

[24] Latham, A. (1961). *Personal Experience with a Hand-Cranked Distiller*. London: Arthur D. Little Research.

[25] Cheryan, M. (1986). *The Ultrafiltration Handbook*. Basel: Technomics Publishing.

[26] Zsigmondy, R., and Bachmann, W. (1918). Uber Neue Filter. *Z. Anorg. Chem.* 103, 119–128.

[27] McBain, J. W. and Kistler, S. S. (1931). Ultrafiltration as a test for colloidal constituents in aqueous and non-aqueous systems. *J. Phys. Chem.* 35, 130.

[28] Meyer, K. H. and Strauss, W. (1940). Sur le passage du current membranes selectives. *Helv. Chim. Acta* 23, 795–800.

[29] Temperley, T. Personal communication, Okada: Temperley-Lieb, 25.

[30] Silver, R. S. (1947). Temperature and pressure phenomena in the flow of saturated liquids. *Proc. Roy. Soc.* 194, 1039.

[31] Ellis, C. B. (1954). *Fresh Water from the Ocean*, New York, NY: The Ronald Press Company.

[32] Juda, W. and McRae, W. (1950). Coherent ion-exchange gels and membranes, *J. AM. Chem. Soc.* 72, 1044.

[33] Juda, W. and McRae, W. (1953). Apparatus and Equipment for Electro-dialysis, US Patent No. 2,636,852.

[34] Reid, C. E., and Kuppers, J. R. (1959). Physical characteristics pf osmotic membranes of organic polymers. *J. Appl. Polym. Sci.* 2, 264.

[35] Breton, E. J. Jr., and Reid, C. E. (1959). *Filtration of Strong Electrolytes*, *AIChE Chem. Eng. Prog. Symp.* 24, 171–172.

[36] Loeb, S. and Sourirajin, S. (1962). Seawater demineralisation by means of an osmotic membrane. *Adv. Chem Ser.* 38, 117.

[37] Loeb, S. and Sourirajin, S. (1964). High Flow Porous Membranes for Separating Water from Saline Solutions, US Patent No. 3,133,132.

[38] Truby, R. L. (2015). "Development, Current Status and Future Potential of SWRO Membranes for Desalination," in *Proceedings of the AMTA Pre-Conference Workshop, IDA World Congress of Desalination and Water Reuse*, San Diego, CA.

[39] Truby, R. (2016). *Private Communication*. Available at: http://truby.com/truby-rates-the-oscar-hopefuls-2/

[40] Westmoreland, J. (1968). US Patent No. 3,367,504.

[41] Bray, D.T. (1968). Reverse Osmosis Purification Apparatus, US Patent No. 3,417,870.

[42] Lonsdale, H. K., Merton, U., and Riley, R. L. (1965). Transport properties of cellulose acetate osmotic membranes. *J. Appl. Polymer Sci.* 9, 1341–1362.

[43] Dodge, B. F. and Eshaya, A. M. (1960). "Thermodynamics of some desalting processes," in *Proceedings of the Symposium on Saline Water Conversion*, Washington, DC, 7–20.

CHAPTER 10

Research and Development Management

John Peichel[1] and Alireza Bazargan[2]

[1]Global Technology Leader, GE Water and Process Technologies,
5951 Clearwater Drive, Minnetonka, MN, USA
Phone: +1(952)457-2054
E-mail: john.peichel@ge.com
[2]Head of Research and Development, Noor Vijeh Co., No. 1 Bahar Street,
Hedayat Street, Darrous, Tehran, Iran
Phone: +98(21)22760822
E-mail: info@environ.ir

10.1 Introduction

The majority of companies today, regardless of size, recognize that to continue to be competitive, one must be engaged in research and development to some extent. Depending on the sector and many other variables, the amount of money spent on research and development (R&D) as a percentage of revenue can range from 0.5–15% and beyond. Software/internet, healthcare, electronics and automotive companies are the big spenders. Table 10.1 lists the top 20 spenders in R&D by sector and country. This shows the United States as an obvious powerhouse with 13 out of the top 20 spots. Of course, if the data were arranged differently, it would give another impression. For example, Figure 10.1 shows the gross domestic expenditures on R&D as a percentage of GDP on the x-axis, and the number of R&D researchers employed per 1000 employees on the y-axis.

Desalination is similar to other industries as in there is a fair number of standardized products and processes that are well-known in their strengths and weaknesses. In desalination, activities and emphasis may be slightly different from other types of development in that the combination of disciplines (chemistry, physics, biology, finance, sociology, etc.) all play an important role in the overall water picture. Meanwhile, the industrial, municipal and

Table 10.1   Top 20 R&D spenders of 2016 [1]

2016 Rank	2015 Rank	Company	Geography	Industry	R&D Spend ($Bn) *
1	1	Volkswagen	Germany	Automotive	13.2
2	2	Samsung	South Korea	Computing and electronics	12.7
3	7	Amazon	United States	Software and Internet	12.5
4	6	Alphabet	United States	Software and internet	12.3
5	3	Intel Co	United States	Computing and electronics	12.1
6	4	Microsoft	United States	Software and internet	12
7	5	Roche	Switzerland	Healthcare	10
8	9	Novartis	Switzerland	Healthcare	9.5
9	10	Johnson & Johnson	United States	Healthcare	9
10	8	Toyota	Japan	Automotive	8.8
11	18	Apple	United States	Computing and electronics	8.1
12	11	Pfizer	United States	Healthcare	7.7
13	13	General Motors	United States	Automotive	7.5
14	14	Merck	United States	Healthcare	6.7
15	15	Ford	United States	Automotive	6.7
16	12	Daimler	Germany	Automotive	6.6
17	17	Cisco	United States	Computing and electronics	6.2
18	20	AstraZeneca	Britain	Healthcare	6
19	32	Bristol-Myers Squibb	United States	Healthcare	5.9
20	22	Oracle	United States	Software and internet	5.8

* R&D spend data is based on the most recent full-year figures reported prior to July 1st.

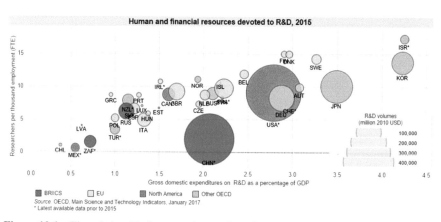

Figure 10.1   The relationship between the number of researchers per one thousand employees and the gross domestic expenditures on R&D as a percentage of GDP [2].

commercial plants have grown significant in size and scope which has created a larger water market but also poses the issue of entrenched familiarity for existing technology. Customers are less likely to try new products unless the cost, performance, and reliability are demonstrated to be much better than the status quo. This issue can raise a significant barrier to the research and development of new desalination technology. Of course, there are always companies who are trying to push the boundaries and develop new products and processes. As in most other areas of research, risk and reward usually have a direct relationship in desalination R&D. The more risk a person takes in developing new and improved products, the higher the potential rewards will ideally be. As evident from Figure 10.2, this is generally true for many different fields, including the relationship between cash and bonds (low risk) and domestic and foreign stocks (high risk) in investment banking.

For example, for processes which employ membranes, we recognize that there have been and continues to be two general areas of work: (i) the improvement of thin film composite and polysulfone support on the most common membrane types available on the market today, and (ii) a very diverse set of new processes and novel materials which are incorporating membranes to desalinate the water (e.g., forward osmosis, membrane distillation, application of carbon nanotubes and graphene oxide, etc.). The first area focuses on incremental improvement in the current commercialized membranes with the constraint of being completely backwards integrated into existing plants albeit with lower pressure and higher salt rejection. The second area has the goal of leap-frogging current desalination processes either with significantly lower capital expenditure (CAPEX) or operating expenditure (OPEX) and/or developing a process that is more robust and is more technically attractive compared to today's technology. The issues encountered in desalination R&D include identifying and understanding the customer, supporting the invention, deciding on patents versus trade secrets, creating mathematical models, collecting and analyzing data, understanding time requirements, and keeping a handle on costs.

For reverse osmosis membranes, the development of the technology can be linked back to the 1950s to the US Office of Saline Water of the US Department of Interior and their funding of reverse osmosis research using synthetic semi-permeable membranes. This funding resulted in key studies involving cellulose acetate material. Many inventors and inventions can be linked to membrane history, including the invention of thin film composite polyamide membrane by Cadotte at North Star Research in the late 1970s [3]. More information on the history and origins of desalination technology can be found in Chapter 9.

Figure 10.2 Higher risks usually entail the possibility of higher returns in case of success. The bars show the range of annual returns of each asset mix over the period of January 1926 to December 2015 [4].

10.2 The Customer

All R&D starts and ends with the customer. Fundamental knowledge of customer pain-points and how they are linked to market drivers is essential in deciding what research and development is most valuable. In the water treatment market, as specified in Chapter 1, water scarcity is the result of a number of common global trends including population growth, global warming, shrinking supply of potable water, and increased focus on environmental impact. The customer-base of the industrial and municipal segments has a unique linkage and sensitivity toward water of suitable quality for their processes and personal use. As a result, much of the research and development justification is based on satisfying both market segments through different channels or final product configurations.

In reverse osmosis desalination, we are almost always attempting to reduce energy requirements and risk of premature scale formation and/or fouling, in order to have higher recovery rates. Much interest in membrane-based solutions stems from the unique ability to remove inorganics and microorganisms in a robust, stable and almost chemical-free continuous process. Often times, a successful bid is based upon the total installed cost and expected cost of operations that utilizes a series of treatment steps. The more robust and higher performing the technology, the more likely the overall solution is deemed more cost effective than competing technologies. As a result, companies have been and continue to invest in research and development to address the competitive nature of the water treatment market.

As membrane technology improves, both the cost of desalination and cost of treating water for reuse will drop below the cost of "traditional water supplies". In many water-scarce regions of the world that drive demand for current and newer desalination technologies, this cross-over has already happened (Figure 10.3).

The gathering of customer feedback can be both straightforward and complicated at the same time. There are so many aspects to most customer's decision-making process including technology, operations, maintenance, footprint, complexity, ability to meet future needs, budget, etc. that true prioritization of research and development activities can be different based on who you ask and how much weight their opinion carries in the decision-making process. Ignoring any aspect, even though it may carry lower weight, can inhibit or restrict market adoption. Thus, the best customer feedback is one that includes all viewpoints that are part of the decision-making process,

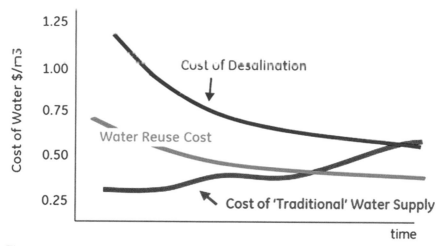

Figure 10.3   With time, desalination and water reuse have become cheaper options than conventional water supply. Prices depend on various factors. With current global conditions, water reuse is almost always cheaper. Graph courtesy of GE Water & Process Technologies.

and attempts to validate the new technology's ability to satisfy or at the very least, not detract from the decision makers' input on technology selection.

Once the customers and market are accurately assessed, we are then ready to dig deeper into what is the unmet need and what is required to fulfill it. Is it material invention, process development, path to the market, increased understanding of how to operate, or something else not available to the customers today? This understanding can then be evaluated against your company's/group's domain expertise of what is done today, some knowledge of the current and evolving IP landscape, along with the ability to focus individuals or teams by application or technology depending on the area of research and development to be focused on.

Review of current commercial activity, won and lost business, and customer inquiries can serve as excellent sources of information from which one can understand what fundamental water treatment and operational challenges customers are really facing. Sometimes customer forums, industry shows, focused working groups and conferences can highlight larger issues facing a region or industry but very rarely do they reveal the complicated decision making process contained within an industrial company or municipality. To understand the relative importance of varying factors, it helps to have a personal working relationship directly with customers and jointly operate small pilots on the customers' plant site. This wasy, one can analyze the performance of existing systems while paying special attention to the variety of stakeholders involved such as operators, managers, maintenance staff, financial and regulatory officers, etc.

For instance, let us consider a product that is designed to achieve very high water recoveries and close to zero liquid discharge (ZLD) where the salt in concentrated and removed from solution into a haulable slurry and/or salt. The resulting system could require several different technologies connected in series with various recycled flow streams. Each technology serves a unique and much needed purpose to the overall goal of extremely high water recovery. However, assume that although the resulting system is more effective than conventional evaporation/crystallization, it requires much more space than interested customers are generally used to allocating to such a system. Finally, the ongoing operation and maintenance requires time and expertise that will be different from the current staffing model in the target markets. These real requirements mean that the detractors, which should be small, create a disproportionally larger barrier to market adoption and successful commercialization than the cost model would predict. As a result, in addition to the technical solution, the final product must be evaluated on the merits of CAPEX, space and complexity which may suggest additional development and/or alternate paths to the market.

Another example would be, a customer starting with a very simple requirement to reduce the level of chloride contained in their wastewater discharge due to increasing concerns of salt levels in lakes or rivers. The customer may start by identifying sources of chloride such as water softeners that are regenerated with sodium chloride, or, the customer's process itself. In searching for a solution, the customer may discover that reverse osmosis is one technology capable of rejecting and concentrating chlorides. As a result, the customer may apply RO technology to the wastewater stream in which the permeate will meet the primary concern associated with wastewater discharge. But the customer will then have a smaller, even higher concentrated RO waste stream to manage! The better solution might be to replace the water softener with an alternative technology or try to treat or eliminate the source of chloride in their process. So, in order to develop the right technology, one must draw a complete water balance of the customer's entire process and evaluate at what point a solution may be most cost effective. An example of two options for treating water to be used in a boiler is shown in Figure 10.4. Generally, it is not possible to understand all the variables without close contact and in depth working conversation with the customer.

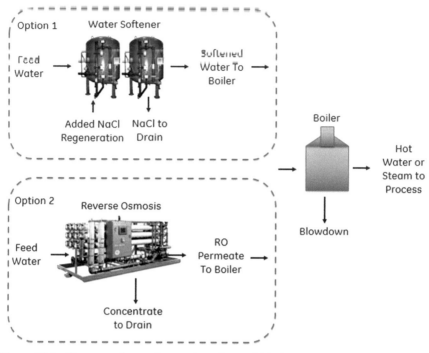

Figure 10.4   There are often various options, from which the more appropriate one should be chosen. Images courtesy of GE Water & Process Technologies.

10.3 The Invention

The process of invention requires a balance of individuals focused on a specific problem area, the support of a cross functional team and the corresponding tools for team members to use.

Usually the R&D department of a company is responsible for invention: coming up with new ideas and processes that are better than the competition. Without invention, the company's inability to provide more value to customers than the competition puts its long-term existence at risk. In many cases, the invention may be chemical or mechanical in nature, but the real significance of the invention lies in understanding how customers will use and derive value from it. In water treatment, this value will be different for large scale municipal plants versus small, industrial manufacturing. Yet, they often have the same water treatment problem that one is trying to solve through invention.

The need to successfully invent requires R&D to constantly revaluate people, skills, organizational structure, processes, tools & equipment, etc. and how well they are working to deliver the invention. Different approaches to team structure can help with the recruitment, assignment and engagement of employees. The team should contain people with a variety of educational backgrounds and levels of experience. An expense budget, a laboratory or shop, a computer database, statistical software tools, an ability to create quick prototypes, perform lab testing, and execute field testing are required tools. It helps to have both a small focused team that is loosely connected to a larger more structured organization that provides the resources necessary for prototyping, testing and customer engagement. In general, creative types handle the structure of commercialized products poorly and project managers and production employees are not able to problem solve and innovate very well. As it is, both need each other and need a way to interact in a mutually agreed manner.

Once people and tools are made available, it is important to somewhat define the area of innovation desired. This allows both the team and individuals to down-select what research to envisage, read about, and then attempt in the lab.

In desalination R&D, we have seen major themes of energy, improved performance and productivity, simplicity, robustness, and achieving a removal or separation that is difficult or impossible to achieve with today's technology.

For example, much of membrane desalination is evaluated on performance along a "goodness curve" shown in Figure 10.5; as you allow more water to flow through the membrane, there is a corresponding decrease in salt rejection. So, the invention of a new membrane material would be favorable if it increases flux with significantly less (or no) increase in salt passage, thus in effect meaning the development of lower energy, higher rejection membranes. A core team working on the novel membrane chemistry may primarily consist of PhD chemists with an emphasis in material science, a polymer chemist and a coatings expert. The supporting team would include equipment, controls, lab testing, and customer applications experience and expertise. The core team and supporting team would review progress on a regular (perhaps monthly) basis in order to share progress and get feedback on what challenges might the production process or the customers experience with the invented product. The team must collectively agree on the path forward otherwise it may be wise to stop or postpone additional work until the required (missing) technology or resources become available to overcome the key risk(s) to successful commercialization.

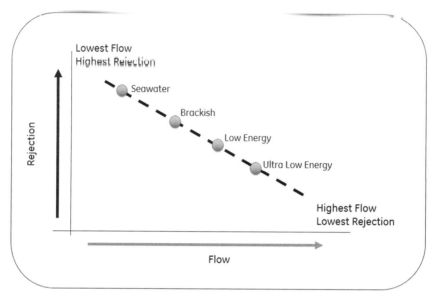

Figure 10.5   A typical membrane "goodness curve" illustrating the trade-off between salt rejection and flow for a series of membrane types identified by their application.

Many companies use a set of common tools for business leaders and project managers to evaluate and prioritize inventions. These tools include risk assessment, key milestones, clear goals, and objectives. Figure 10.6 shows the steps and bottlenecks that appear during the innovation process.

Another way to impact invention is through focus. Whether focus is on external companies and technologies, on material science, on innovation or improvements to existing water treatment processes. Experimentation in the lab versus traveling to meet other companies are choices that determine how time and money are spent. For larger organizations, it helps to have one team looking at the global landscape through patent searches, publications and products of competitors while another team is given a specific task to focus on in the lab relative to gaps in the current treatment capability. Endeavors can also be carried out in phases, as depicted in Figure 10.7.

10.4 Patents

Companies continue to be very focused on patents by measuring the number of patents and the strength of the patent portfolio, and by reviewing how long patents will still be in force, the commercial value protected by patents, etc.

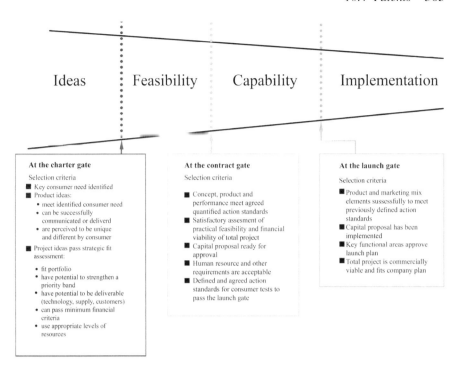

Figure 10.6 The process of innovation and project selection, courtesy Alireza Mohseni, inspired by [5].

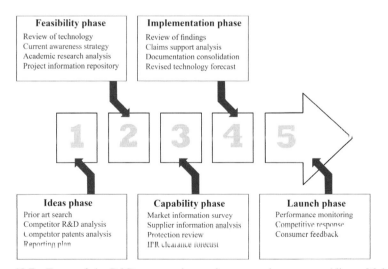

Figure 10.7 Focus of the R&D team as innovation proceeds, courtesy Alireza Mohseni, inspired by [5].

Even so, the decision of whether some invention is worth the cost of time and money to patent or to attempt to keep it as a trade secret must be considered. It is helpful to have some experience on the complete patent process from idea to issued (and possibly defended) patent. It is impossible to make good decisions without a person of adequately relevant legal background working through the decision with you.

As illustrated by the variety of costs in Table 10.2, the actual total costs of filing and maintaining patents can be quite large depending on the complexity and the number of countries in which patent protection is pursued and maintained. Decisions on what, when and where to patent are often supported based on the market and competitive landscape. Where will my customers be located? Will my competitors be producing or selling related technology in that country? How does protection work in that country and is it likely that having patent protection will be enforceable? All these questions are factors that help determine how much money should and is often spent on patents.

In the development of a technology and deciding whether to patent, some of the first key questions which must be asked are as follows: is the work novel/unique? Can you prove it? Is it easy to determine if others violate the patent? Is it financially worth filing the patent and maintaining it? Would licensing and purchasing the needed IP be more efficient? Which is better, enabling or blocking patents?

Enabling patents are those patents which prohibit competitors from copying your idea in the country where you have patent protection. An enabling

Table 10.2   How much does a patent cost over 20 years?

Stage	Activity	Cost
Ideation	Coming up with the patent idea, and carrying out experiments if required	$0–$200,000
Patent and Literature Search	Searching databases and publications; could include lawyer fees, consultant fees, and/or software	$0–$20,000
Writing Patent Application	Paying legal firm to write the patent	$5,000–50,000
Filing Fees for Initial Application	Fee paid to the legal entity that reviews and issues patents	$100–$7,000 per country or region
Maintenance and Renewal Fees	Fees paid to legal entity to maintain the patent	$0–varied depending on country
Patent protection/defense	Costs to monitor for patent infringement and litigate if required	$0–varied

patent supports the marketing plan as a company has a lower risk of competitors offering the same product at lower prices. Thus, the patent enables the company to proceed with all the investments required to commercialize a new or improved product that incorporates aspects covered in the patent.

Blocking patents are more strategic in that they are related to the markets and products you already have and so there is already or anticipated to be revenue and profits associated with those products. A blocking patent prevents competitors from patenting that idea which will compete with your product offering. In desalination, blocking patents are often used for ideas where either the market acceptance or the needed manufacturing capability is not ready or available to the company or person with the patent idea. For the purposes of this chapter, we are defining blocking patents as those owned by companies with no immediate plans to commercialize the technology covered by a patent. These companies usually produce their own products, and only make such patents to stop the competition. Such patents can also limit the area of new research in some countries which do not allow for you to even practice lab experiments for technology protected by a patent [6].

During the patent search, if a patent exists that is the same or contains claims that overlap with an idea or area of work to be done, a company can decide if that existing patent is worth licensing, or purchasing the patent outright. The cost and risk of trying to "invent around" an existing patent may be much higher than what the current patent owner is realizing; especially if the idea has not been or has no plans for commercialization. In some cases, the license can be for specific markets or applications thus allowing the same patent to be licensed to multiple companies simultaneously with clear rules about where the licensee is allowed to practice.

At the moment, key developments in desalination stay mostly as know–how or trade secrets. However, in terms of volume, patents, research papers, publications, and marketing literature are so abundant that skimming them would take more time than the amount of time spent in the lab. Particularly, many universities have membrane expertise and the amount of work on membranes happening globally is dizzying. The membrane industry tends to value membrane composition (materials and composition of matter) patents above all; followed by applications/process patents. The topic of how to manufacture membranes and/or devices is usually kept as a trade secret as this can be difficult to determine in the final product and thus impossible to determine infringement.

In order to address the issue, you must first define the market, application and product area of interest. Second, do a thorough search and review of

Table 10.3 Patents versus trade secrets

	Pros	Cons
Patent	Market protection; external validation of an idea's value and novelty; increased customer confidence	Time and cost to apply and maintain over the years
Trade Secret	Low Cost; immediate; flexible	Requires internal rigor; keeping information separated; if others patent the idea it may bar you from future practice

published patents. Third, develop a strategy of what inventions are beneficial to patent either to protect your products or prevent others from patenting and preventing you from future inventions. On the question of whether or not an invention would be a favorable potential patent, you will want to have a series of criteria by which to evaluate the idea, and set a patent strategy. Then, once it has been decided to patent, work with a knowledgeable legal counsel to prepare and file the patent.

In the United States the patenting law recently (as of 2013 [7] due to the America Invents Act) changed from first-to-invent to first-to-file. Care should be taken, and local conditions must be considered, because for example, the nature of the first-to-file rule at the European Patent Office (EPO) is different from the United States Patent and Trademark Office (USPTO). The American office allows for a "grace" period during which the inventor can patent the work which they have disclosed; but the EPO does not allow for any grace period and any prior disclosure of the work does not allow for it to patented at a later time [8]. The change from first-to-invent in the US to first-to-file may also be driving companies to choose trade secrets over filing patents due to costs. Unless the invention is clearly unique and has the potential to provide significant improvement in performance, companies may choose the trade secret route as a more expeditious and economical path to the market. A common difficulty may be for a researcher in the lab to unintentionally repeat someone else's invention and recognize the potential patentability of what they just discovered. Thus, companies and institutions that have a thorough and efficient understanding of the intellectual property space in which they are doing research and development are better positioned. Patentability is defined as being statutory, new, useful and non-obvious.

To be patentable, a patent must fall into one of the four statutory categories of invention which include machine, manufacture, composition of matter and process. An idea not falling into one of these four categories is not considered a patentable idea with two exceptions: abstract ideas and natural processes. To determine that an invention has passed the patentable subject matter test is the first step. The invention must also be novel (new to the area of invention) and non-obvious (often determined by one or more examples of prior art).

If the researcher is not well versed or does not have access to someone well versed in that area of research, much time and money could be wasted in creation of a patent filing that is easily rejected. In addition, unknowingly infringing someone else's patent could put at risk a very valuable future patent if patent infringement activity in the lab was involved in the invention process. Therefore, it is absolutely necessary to research existing patents, papers and publications in the area to avoid repeating someone else's lab work and/or patent invention.

If a person doing research realizes that their idea is already covered in an existing patent, the next step is to determine that the patent is still valid; meaning that the patent owner has continued to pay the maintenance fees required to keep it active and valid. If the patent is still valid and has a number of years of coverage remaining, the person in research has to come up with an improvement or a better idea in order to proceed with their research. Since the patents and rules on experimenting are country specific, it is also possible to proceed with lab research in other parts of the world, if that option is available to the researcher.

One of the landmark cases in patent law involved the use of membrane processes to desalinate dyes used in food, drug and cosmetic applications. Hilton Davis sued Warner-Jenkinson for patent infringement of its U.S. Patent No. 4,560,746 in 1991. An interesting aspect of the case is that in 1982, both Warner-Jenkinson and Hilton Davis had hired Osmonics (later part of GE Water & Process Technologies) under a secrecy agreement to perform lab test on a membrane separation process on a dye solution. Osmonics performed the Warner-Jenkinson test in August 1982, 1 week before it performed the first Hilton Davis test. Warner-Jenkinson did not learn of the Hilton Davis patent until October 1986, after it had begun commercial use of its ultra-filtration process to purify Red Dye #40. Hilton Davis learned of Warner-Jenkinson's process in 1989 and sued Warner-Jenkinson for patent infringement in 1991. Warner-Jenkinson versus Hilton Davis, 520 U.S. 17 (1997), was a US Supreme Court decision in the area of patent law, supporting the doctrine of equivalents while making some important enhancements to the law. This case and the decision serve as a very good example of why we patent and the potential commercial impact of patents when it comes to protecting commercial inventions.

In the decision of whether to patent or keep as a trade secret, companies must weigh the cost and protection offered by patents versus the speed and competitive advantage of trade secrets. For example, in the case of the Filmtec versus Hydranautics lawsuit, Hydranautics successfully argued that the Cadotte US Patent 4,277,344 on the invention of interfacially polymerized aromatic polyamide thin film membranes was actually owned by the public and so other companies such as Hydranautics had the right to produce such RO membranes. Reviewing the history of the invention will help better understand the ruling. In 1977, an inventor named John Cadotte, working for a not-for-profit research corporation called Midwest Research Institute (MRI) on a government-funded project, had discovered a reverse osmosis membrane that could be used for the desalination of water. Soon thereafter, Cadotte and others established a corporation, FilmTec, for the purpose of commercially manufacturing reverse osmosis membranes using the aforementioned technology that had been developed with government funds. Cadotte filed for a patent on his membrane, and assigned all his rights in his invention to FilmTec. United States Patent No. 4,277,344 (the 344 patent) was issued to Cadotte but the future litigations proved that it was not as clear cut as he had hoped [9].

10.5 Models

Mathematical models are becoming more and more of a necessity regardless of whether modeling a product or a process. A good, robust model includes technical relationships, human components of time and effort, along with financial quantities that help evaluate the probability of an outcome in terms of risk.

In water treatment, the complex interaction of chemistry and equipment and people make good, useful models very challenging to create if they are to deliver accurate predictions that allow product developers to make good tradeoff decisions. In reverse osmosis desalination, the longer a membrane runs over its useful life (which for seawater can be 3–5 years) the even more challenging it is to create a model that predicts and replicates actual real world conditions.

GE Water

Winflows Version 3.0.2
DataBase Version 2.04

Results Summary

Flow Data	gpm	Analytical Data	mg/L
RO Feed:	200.00	RO Feed TDS	579.70
Product:	150.04	Product TDS	7.91
Concentrate:	49.96	Concentrate TDS	2269.61

System Data		Single Pass Design
Temperature:	15.00 C	
System Recovery:	75.02 %	

Array Data

Pass 1

Recovery %: 75.00   Conc. TDS(mg/l):   2249.96   Conc. Flow: 49.84 gpm

Stage	Total			Flow, gpm		Pressure, psi		Perm TDS
	Housing	Element	Element Type	Feed	Perm	Feed	DP	mg/l
1	3	18 AG8040F		199.88	85.43	220.93	35.71	4.38
2	2	12 AG8040F		114.46	46.65	185.22	28.90	9.36
3	1	6 AG8040F		67.60	17.76	156.32	42.12	18.70
Total	6	36						

Analytical data

Cation	mg/l			Anion	mg/l		
	Perm	Feed	Conc		Perm	Feed	Conc
Ca	1.01	98.11	389.72	SO4	0.13	25.00	99.70
Mg	0.55	28.47	104.31	CI	0.27	31.79	126.48
Na	0.06	8.10	32.19	F	0.00	0.10	0.40
K	0.03	1.60	6.31	NO3	3.25	66.00	254.45
NH4	0.00	0.00	0.00	Br	0.00	0.00	0.00
Ba	0.00	0.08	0.32	PO4	0.00	0.00	0.00
Sr	0.00	0.11	0.44	B	0.00	0.00	0.00
Fe	0.00	0.07	0.28	SIO2	0.17	18.90	75.14
Mn	0.00	0.00	0.00	H2S	0.00	0.00	0.00
TDS mg/l	7.91	579.70	2269.61	HCO3	2.41	302.60	1169.63
pH	5.62	7.70	6.12	CO2	10.76	10.11	13.37
				CO3	0.00	0.76	10.05

Figure 10.8 A report output from Winflows, a membrane system design software by GE Water & Process Technologies.

In building a model for a process, equipment, or component, all the inputs and outputs and the initial volume and full scale volume need to be determined. Models need to represent fundamental relationships based on physics and chemistry. If the mechanism is understood and the model well designed, then it should be a good predictor of reality (with a confidence level above 0.95) when running Design of Experiments (DOEs) in the lab.

For example, some specific parts of the model in water treatment can be defined by basic water chemistry. One starts by describing the interaction with desalination technology in simple terms. Then, you proceed to the lab to test the model quickly realizing that you need to either very carefully avoid external factors from influencing your experiment, or to change the model to make it more complicated to account for environmental factors. One such factor is the impact of temperature on biological growth.

Biological growth is very common in all water treatment systems. It is difficult to model and almost impossible to simulate real world situations in a controlled lab environment. So, one has to develop a strategy for addressing the impact of biology in the model and determine an effective way to compare the model to real world applications. As another example, the prediction of energy consumption by various desalination processes is critical and models need to be calibrated with actual operation data. Modeling of treated water quality also needs to be addressed for post-treatment processes to achieve a target treated water quality goal.

Even back in 1950, researchers were doing side by side comparison of the cost to operate various technologies like evaporation, distillation, and freezing against ion exchange and reverse osmosis. This model-based approach was a necessary step to eliminating potential research options and focusing funding and resources on a few promising areas.

10.6 Data

Good, reliable data is both a blessing and a curse; why do we need data and why does data sometimes get in the way? When is there enough data? Of course the more data we have, the more potential knowledge we have gained from the extra data. But there is always the risk of drowning in the data that you have collected. Not to mention the extra time and money required for collection of the extra data. For scientists and engineers, new data may be fascinating and provide insight into various issues, but, some of this data may be irrelevant to developing a good product or service.

So you should always measure exactly what it is you need to know; how is the data collected; how accurate is the measurement; and whether or not the measurement is a direct measurement or indirect. A classic example is using conductivity as an indirect measurement of salt rejection when conductivity is influenced by the pH and the presence of carbonate/bicarbonate alkalinity.

Fortunately, there are many powerful data management and analysis tools. Developing a good database structure at the beginning of experimentation will often pay dividends in being able to analyze data in a multitude of ways to identify lurking variables and fundamental relationships. Good data is always based on exactly what it is you need to know and one must have an exact understanding of how accurate the measurement is. Always incorporate error of measurement and standard deviation into your results and mathematical

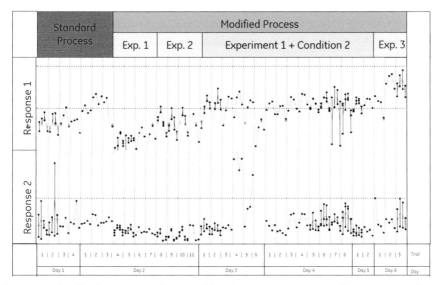

Figure 10.9 Good models can help elucidate impacts on responses. Data courtesy of GE Water & Process Technologies.

models to be able to determine if a new result is statistically different than your control.

Figure 10.9 shows an example of data from a Design of Experiment. At a quick glance, the researcher can compare various experiments in the modified process against the standard process, assess the impact on two important responses, and evaluate variation and response compared to specification limits. When data is noisy, this graphical representation becomes helpful to making good choices for down selection.

One must always formulate a hypothesis when creating experiments to generate data for the purpose of proving or disproving your hypothesis. In water treatment, a good hypothesis is one that can be validated or rejected based on experiments either in the lab or in the field. It is best if you can run replicates and compare results to a base case or control. This approach is necessary due to the wide variety of special cause factors that often come into play in water treatment applications.

Good, solid data collected on pilots which are operating on actual process streams that represent normal variation in water quality is essential for convincing markets and customers to move away from entrenched technology. In some cases, proof of product and/or application viability is what drives the additional investment in new technology for full commercialization.

We often encounter a lack of real operational data on existing technology due to age of equipment or system. Thus, there is very little baseline data by which to evaluate new technology. In desalination, environmental impact is often a key driver in moving customers from old, standard approaches to new technology. Here again, real data may not have been measured, collected or well understood thus putting the burden on the new technology offering to collect data from the existing system as well as the new product or service.

Finally, note that sampling methods and procedures for the collection of data are of importance. Data do not exist in a vacuum and are not separate entities that can be dealt with in isolation. The researcher must assume the responsibility for obtaining *representative samples* prior to analysis. Without representativity is the first stage in the entire data chain, there will be no way of evaluating the degree of sampling bias and errors embedded in the final results. It is not possible to distinguish if a sample is representative or not by analyzing the sample *itself*, but rather, the consideration of the sampling process is what can validate them. Correct samples, meaning those that are truly representative, originate only from fit sampling processes [10].

10.7 Time

How long does it take? What are the tradeoffs between speed, cost, and quality? What are the chances of success? Once an invention is realized in the lab; it can take 2–6 months of replication and optimization depending on equipment and resources available. Scaling is also highly dependent on the current state of commercial equipment. Pilot equipment is often used due to low cost, speed in customizing, and generation of directly scalable data. Depending on the performance improvement, the pilot phase can last 6–24 months. We are constantly in a search for applicable accelerated life testing using temperature, pH and pressure. In some cases, high levels of organic foulants are also found beneficial to compare two options side by side to see if one option (being developed) is better, worse or the same.

All research, including inventions, can be viewed as a project. As such, we can think about the trade-offs illustrated in the classic Project Management Triangle shown in Figure 10.10. Inventions that are simply the work of a single person doing paper studies may be low cost and done quickly, but very rarely are they robust and of the detail needed for full product development and commercialization in the water desalination area. The challenge is that

Figure 10.10 The Project Management Triangle also known as the "Triple Constraint". It is extremely unlikely (if not impossible) to have a project that is carried out with low cost, is done quickly, and has high quality [11].

whether research and development occurs as part of a large organization or a new venture capitalized startup, there is a limit to how much time and money can be spent on a new product or service before going to market and starting to return on that investment. Therefore, it is important to have a process to follow and assess your progress against time and money. The more honest and candid your assessment is of the development progress toward commercialization, the better decisions will be made.

The development of an invention from idea to commercial product follows the general path as follows: first the problem should be defined and inputs collected; then, the idea is formed; this is followed up by research regarding the existing relevant patents and papers; next comes designing an output and planning to build, test, learn, and repeat; then, the design should be verified with data that proves the idea can solve the problem; following the in-house lab testing, the design should be scaled up and validated with customer field tests; at this point it is proven that the product has solved the problem; finally, the invention should be controlled and reviewed for at least 1 year to successfully transit to commercial availability.

Companies or research groups will keep a list of ideas, projects, programs, and products as they progress through their own flowchart of commercialization. Often, the goal is to build a funnel or pipeline of many ideas from which one can down select which ideas to turn into programs that will hopefully turn into successful products. In the water desalination industry, this list is often segregated by the area of need being satisfied. Some common areas include: (i) increase salt rejection for higher water quality (ii) increase flow per unit pressure to reduce the cost of electricity, and (iii) increase robustness through broader pH ranges, lower fouling, and/or higher temperatures to reduce overall cost of operation. By coordinating research efforts in a given area, the amount of time required can be reduced because of the efficiency of learning from similar tasks and experiments. This focused approach can also reduce the time required to fully commercialize a product.

10.8 The Business Aspect: Costs, Revenue, Profit

But how does one evaluate the effectiveness of the R&D activity? A range of common measures related to technological innovation, as Jain et al. put it, "include inputs such as financial resources and people; processes such as resource efficiency, actual versus planned time to market, and milestone compliance; and output measures such as number of new products and services launched, market share growth, new product success rates, number of patents filed, and publications written. The choice of the measures most appropriate will vary depending on an organization's circumstances and goals (e.g., profit or not-for-profit status). It is wise to use a set of metrics versus only one. Collectively, these metrics should reflect not only research input, but its application, that is, innovation. For example, in a company, R&D as a percentage of sales alone is not a measure of innovation, as innovation implies that the R&D leads to a product that is marketed. Booz Allen Hamilton's annual survey of innovative companies distinguishes between those that dedicate the most funding to R&D and those that leverage it in the marketplace as measured by profitability per dollars of R&D invested…Similarly, in a government, not-for-profit, or university setting that emphasizes innovation, metrics should address not only investment in research, but also success in licensing and spin-off activity, or some other appropriate measure of how research findings are used for commercial or noncommercial purposes. For organizations that emphasize process innovation, appropriate measures include increased efficiency, cost reduction, and improvements in cycle time or resource use" [12]. Many companies track the percentage of revenue resulting from new products (products launched within the past 3 years). A sustainable company or business should have sustained or increasing percentage of revenue resulting from new products.

Other questions that one may ask are: how much does it cost? What are the advantages and disadvantages that big companies have vs. small start-ups? What is the relationship between big and small firms? What is the role of grants – whether governmental, from the university, or within companies?

A general rule of thumb is that piloting costs 3x the costs in the lab; and commercialization costs 3x the pilot cost. Of course, the closer the innovation is to the current commercial process the less the costs, and the further away, the more they will be. Costs include people, equipment, testing/analysis, and sample/trial runs generally in that order. As a result, we are investing in more and more trials, automatic analysis equipment and more sophisticated lab/pilot and production equipment to reduce the people cost portion of idea to commercialization. So, the motivation behind investing in equipment to reduce the time to prototype, test, gather and analyze data is to reduce both the overall time and number of people hours required to go from idea to commercialization. More time also entails the cost of overhead to support an employee working productively in a lab or research environment.

For an idea to be of commercial interest, we must prove to ourselves and others that (i) the idea works and (ii) the way it works delivers some additional value. The proof often comes in the form of data. Investment that speeds up prototyping, testing, data gathering and analysis typically allows us to evaluate and choose good ideas that become successful products. Conversely, the same data allows us to stop pursuing ideas that do not work in practice and have limited commercial value. The better the prototype and test reflect the operational reality of the idea, the easier and potentially less risky the decision to proceed or stop the steps taken toward commercial reality.

During initial project/product conception, it is common for a Product Manager to develop a business plan that outlines the potential customers, the markets those customers do business in, and why those customers/markets are compelling stories of future growth. If the new product is a simple extension of existing products or business, the Product Manager may use current business models and customer demand as a basis for outlining the new product business plan. If the new product is very different, the company will typically need to do market research, customer surveys, focus groups, test products, etc. to determine what features and benefits are required, how much value a customer will derive from such features and benefits, and thus forecast potential sales prices either directly or through distribution channels.

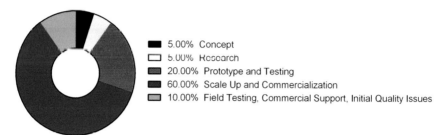

Figure 10.11   A general breakdown of costs from concept to commercialization.

The best business plans are those that have various "what if" scenarios: best case, worse case and most likely case. Often, the difference in scenarios is tied to the risk factors and what investment of time or money is required to mitigate those risks. During every stage of product development, the business plan is referred to in order to make decisions, set priorities and avoid scope creep during product development. In water desalination projects, the investment in capital ranges from small to very large with long cycle times between project initiation, design, installation, startup and commissioning. Thus, it can be very difficult for business plans to predict market situations, customer needs and timing for new products. Therefore, products are often directional in that they solve the continuous problems of removal or selectivity, lower energy and improved robustness. Figure 10.11 provides a general breakdown of costs from concept to commercialization.

10.9 Conclusion

As the world continues its attempts to meet the challenges of population growth and water scarcity, the technology of water desalination continues to be driven to reduce cost, improve energy efficiency and deliver operational robustness. In order for existing and new companies to provide products and services meeting the ever-increasing demands of quantity and quality, the companies and their staff must understand and organize their research and development activities in such a way as to address key areas that impact new technology. The areas include customer care approach to invention, value of patents, leveraging models, data-based decisions, time required and attention to business.

The customers' challenges continue to evolve and have raised the bar. A new technology must prove to deliver a better performance, at higher energy efficiency while maintaining robust and reliable operation. The water treatment facilities are so large that new technology that does not operate with a great degree of predictability will pose too large a risk for customers to adopt.

Our approach to invention must be driven by customer business models with a focus on the real pain points and opportunity to improve. People responsible for inventing must be engaged in the industry dialogue to hear what customers are saying, understand what current solutions are being employed, how competitors are addressing current problems and what value their company or organization can bring in the invention of new solutions.

Patents and trade secrets are an important strategy to protecting the value created by companies who invest in the process of invention and look for payback from the successful commercialization of new products and services. The patent process is complicated and expensive and thus requires thoughtful consideration on which approach is right for the idea, the market, and the area of the world to be commercialized.

The complicated integration of chemistry, mechanical engineering, and control of water desalination processes require the use of complicated models to evaluate the total cost of ownership and the risk trade-off profile when considering the application of new technology. These models have to be based on the understanding of the fundamentals which are reduced to mathematical quantification. Once a model is produced, value is derived from running many different scenarios thus providing a more data-driven decision-making approach to what prototypes, and lab and field pilots must prove. Ultimately, an invention must be tested as it would be used in the real world.

We live in a world full of data. Fortunately, computers can be used to produce a more visual representation of relationships contained in data. This can be very useful in providing insights. In research and development, we must remain ever vigilant in assessing the accuracy and resolution of measurement systems. We have to measure exactly what it is we need to know. When we cannot measure the most important values, we have to invent creative ways to measure useful factors in assessing a technology's effectiveness.

The time required to invent and prove the value of the invention is one of the most challenging aspects of research and development. We are constantly evolving organizational and decision-making models to support rapid learning and constant iteration to test ideas and show success or failure with increased speed. The organizational structure that supports the speed of development the best is often the winner on first to file patents and commercializing new products.

The barriers of research and development are completely intertwined with the aspects of running a business. Decisions based on the value of an invention are always framed in terms of potential market size, the cost of production, the sale price and the long term operating cost to be borne by the customer. Inability to understand and quantify the business aspects of an invention are often met with few resources to be invested which greatly hinders the process of development and commercialization.

In today's ever changing, dynamic world, the research and development of any organization is a key part of the strategy to develop and maintain its competitive advantage. If persons involved in the leadership and execution of technology development can address the important areas, we believe that the organization will have a higher chance of success; compared to an organization that proceeds without taking these aspects into account.

The desalination industry has enjoyed a period of sustained growth primarily due to the improvement of commercialized ideas that started with focused research and development activities. Likewise, the future of the industry will continue to depend heavily on a well-thought, balanced and all-encompassing approach to new technology and invention.

References

[1] PwC's Strategy&. (2017). The Global Innovation 1000: The top innovators and spenders [Internet]. [cited 7 June 2017]. Available from: https://www.strategyand.pwc.com/innovation1000

[2] OECD. (2017). Research and Development Statistics (RDS) [Internet]. OECD Better Policies for Better Lives. [cited 7 June 2017]. Available from: http://www.oecd.org/innovation/inno/researchanddevelopmentstatisticsrds.htm

[3] Kucera, J. (2015). Reverse Osmosis: Design, Processes, and Applications for Engineers. 2nd ed. Wiley.

[4] Diversify Your Portfolio [Internet]. Fidelity.com [cited 7 June 2017]. Available from: https://www.fidelity.com/learning-center/investment-products/fixed-income-bonds/diversify-your-portfolio

[5] Ganguly, A. (1999). Business-Driven Research & Development: Managing Knowledge to Create Wealth. Palgrave Macmillan

[6] Clarkson, K., Miller, R., and Cross, F. (2016). Business Law: Text and Cases. 14th ed. Cengage Learning.

[7] America Invents Act: Effective Dates. United States Patent and Trademark Office; 5 October 2011, pp. 1–7. Available from: https://www.uspto.gov/sites/default/files/aia_implementation/aia-effective-dates.pdf

[8] Kravets, L. (2013). First-To-File Patent Law Is Imminent, But What Will It Mean? [Internet]. TechCrunch. [cited 7 June 2017]. Available from: https://techcrunch.com/2013/02/16/first-to-file-a-primer/

[9] FilmTec Corporation v. Hydranautics. 1993. United States Court of Appeals, Federal Circuit. 94-1034. Available from: http://caselaw.findlaw.com/us-federal-circuit/1287586.html

[10] Petersen, L., Minkkinen, P., and Esbensen, K. (2005). Representative sampling for reliable data analysis: Theory of Sampling. *Chemometrics and Intelligent Laboratory Systems*. 77 (1–2):261–277.

[11] Trombo, M. (2016). Project management tips and techniques for your small business [Internet]. National Leasing. [cited 7 June 2017]. Available from: http://www.nationalleasing.com/en/blog/entry/project-management-tips-and-techniques-for-your-small-business

[12] Jain, R., Triandis, H., and Weick, C., (2010). Managing Research, Development and Innovation: Managing the Unmanageable. 3rd ed. Wiley.

CHAPTER 11

Membrane Chemistry and Engineering

Steven Jons[1], Abhishek Shrivastava[2], Ian A. Tomlinson[3], Mou Paul[4] and Abhishek Roy[5]

[1]The Dow Chemical Company, 5400 Dewey Hill Road, Edina, MN 55439
E-mail: stevejons@dow.com
Phone: +1(952)-897-4249
[2]The Dow Chemical Company, 5400 Dewey Hill Road, Edina, MN 55439
E-mail: ashrivastava@dow.com
Phone: +1(952)-897-4395
[3]The Dow Chemical Company, 1821 Larkin Center Drive,
Midland, MI 48674
E-mail: iatomlinson@dow.com
Phone: +1(989)-636-4967
[4]The Dow Chemical Company, 5400 Dewey Hill Road, Edina, MN 55439
E-mail: mpaul@dow.com
Phone: +1(952)-914-1025
[5]The Dow Chemical Company, 5400 Dewey Hill Road, Edina, MN 55439
E-mail: alroy@dow.com
Phone: +1(952)-914-1026

11.1 Introduction

Compared to thermal methods, desalination by membranes is a relatively new approach to water purification, resulting in large part from innovations during a particularly productive period in the latter half of the 20th century [1–3]. The first synthetic ion-exchange membranes were made in 1950, and commercial electrodialysis (ED) installations for reducing the saline content of water followed shortly [4]. Reverse osmosis (RO) was born in the same time period. Osmotic transport through natural membranes had been previously observed, but the use of membranes for passing water and removing salt was only first proposed by Hassler in 1950. Subsequent studies used synthetic membranes to show NaCl rejection by RO, but fluxes were much too low at acceptable pressures for practical desalination. In the late 1950s, Loeb and Sourirajan [5] discovered a breakthrough process for producing highly asymmetric cellulose acetate membranes that resulted in an order of magnitude increase in permeability. A commercial desalination membrane made by this process was introduced in 1965 [6]. The 1970s saw the development of interfacially polymerized membranes with even higher permeability, and a preferred chemical reaction identified by Cadotte [7] is still used in most RO membranes today. In terms of membrane modules, Loeb and Sourirajan's [8] cellulose acetate RO membranes were originally implemented with a tubular geometry. The hollow fiber module was invented by Mahon [9], and DuPont, Dow Chemical, and Toyobo subsequently commercialized hollow fiber RO modules between 1969 and 1979. The currently dominant membrane configuration, the spiral wound module, was invented by Westmoreland and Bray [10, 11]. Today's combination of flat sheet polyamide membrane in a spiral wound configuration was first sold by FilmTec Corporation in 1980. Subsequent to these seminal advances, enhanced manufacturing and more incremental improvements to the membrane, module, and system have resulted in a reliable and economical means for purifying water.

The relative importance of RO has especially grown in the last few decades [12]. By the mid-2000s, approximately 60% of the total installed desalination capacity globally was accounted for by thermal desalination and the remaining 40% by RO. With continuing advancements in membrane technologies and improved implementation (especially high-pressure pumps, energy recovery devices, and smarter system designs) the adoption rate of RO continued to rise. After 2000, 70% of the desalination plants installed globally were membrane desalination plants. Today, membrane-based desalination accounts for more than 60% of installed desalination capacity worldwide. Outside the Middle East, adoption of membrane-based technology has been much more rapid.

The acceptance of RO can principally be understood in terms of the lower cost of water compared to thermal desalination. An important point is that no phase change is required for RO, and so no energy inefficiencies for that conversion are experienced. Energy consumption accounts for a significant portion of the cost of water [13, 14], and membrane based desalination has been seen to result in lower total cost of water for most systems. For example, the current specific energy consumption (energy required to produce a unit volume of pure water from a given feed water) of large scale seawater desalination plants with reverse osmosis ranges between 2 and 4 kWh/m^3, whereas for large-scale thermal seawater desalination plants it is reported to be between 5 and 16 kWh/m^3 [14–16]. Specific energy consumption may be separated into thermal and electrical components, and just the auxiliary electricity (primarily for brine recirculation pumps) in MSF systems is often comparable to the total energy required for RO desalination plants [17]. Reported desalination costs [14] for large-scale desalination plants range between US $0.2 and US $1.2/m^3 for reverse osmosis desalination and between US $0.5 and US $1.5/m^3 for thermal desalination plants. The largest reverse osmosis desalination plant in the world recently started at Sorek, and will provide 627,000 m^3 of treated water per day by treatment of Mediterranean seawater, profitably selling water for $0.58/m^3.

Beyond RO, other processes for membrane-based desalination have also been proposed and advanced over the same time period. This chapter introduces nanofiltration (NF), forward osmosis (FO), membrane distillation (MD), and ED. Of these, NF and ED are particularly important and fairly mature commercial process for desalination, both being used separately for selectively removing ions, but also in combination with other membrane and thermal desalination unit operations.

Commercial membranes are generally sold within modules (also referred to as elements) that define the flow paths around membranes, and several (or several thousand) modules are frequently combined to form a system. These membrane modules provide high active area and flow patterns that minimize fouling, polarization, and energy losses. This chapter also discusses the most common module configurations for membrane-based processes.

11.2 Membrane Processes

A diverse set of membrane-based processes have the potential to be used in separating salts from water. An applied pressure difference causes separation of salts from water in RO and NF. While the membranes for FO are similar to those of RO, the transport of water in the FO process is driven by a concentration gradient. Salt separations in ED (and related electrodialysis reversal (EDR) and electrodeionization (EDI) processes) result from ion transport associated with current induced by an applied electrical potential. And, transport of water vapor through a MD membrane is caused by differences in vapor pressure across the membrane, such as induced by temperature or pressure differences. This section focuses on the membranes used in each of these processes.

Reverse osmosis is the most widely practiced commercial process for membrane based desalination. The key component in the RO process is the membrane. This semipermeable membrane selectively allows solvents (most commonly water) to pass through it, but it rejects most of the solutes such as ions and organic macromolecules. The process differs from "natural" osmosis, in that pressure in excess of the osmotic pressure is applied to the high concentration solution. The pressure drives water through the membrane in the direction opposite to "natural" osmosis and results in a purified permeate stream. This section describes the chemistry and fabrication of RO membranes, although many aspects are also relevant to NF and FO membranes as well.

A critical aspect of commercially successful reverse osmosis membranes is the thin barrier layer, since resistance to flow is approximately proportional to this thickness. In the 1950s, Mylar® used for tape recordings was the thinnest (about 6 μm) commercially available integral polymer film [6]. The ability to create membranes having a barrier layer of less than 1 m thickness was first accomplished in that decade by the phase inversion technique [5]. It was subsequently discovered in the 1970s that even thinner RO barrier layers could be formed by interfacial polymerization. Both approaches are further described below.

A phase inversion takes place when a thermodynamically stable solution comprising of a homogeneous mixture of polymer and solvent (and sometimes other additives) is caused to separate into a polymer-rich and polymer-lean phase, eventually resulting in a bicontinuous porous solid [18, 19]. The earliest RO membranes with practical flux for commercial desalination were made by Loeb and Sourirajan [20, 21] using a non-solvent induced phase separation (NIPS). In the original recipe, a homogenous mixture of cellulose acetate, acetone, and a perchlorate salt was immersed in water. Many subsequent studies on cellulose-based membranes have since focused on identifying recipes enabling improved performance and on better understanding the process by which these highly asymmetric, integrally skinned membranes were formed [22–24].

Figure 11.1(a) illustrates a process suitable for continuous manufacturing of membrane by NIPS. A thin layer of polymer solution is cast onto a non-woven web, such as by extrusion die or doctor blade. After a suitable time,

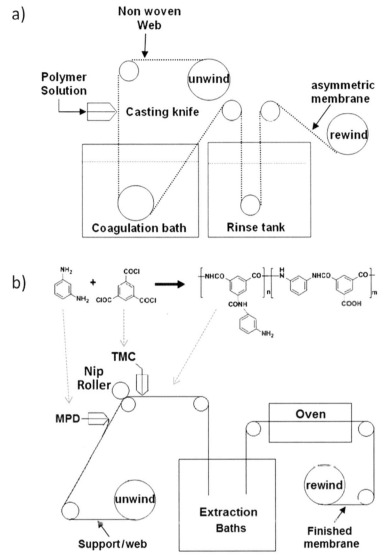

Figure 11.1 Typical asymmetric membrane production processes via (a) non-solvent-induced phase separation and (b) interfacial polymerization.

the coated web is immersed in a coagulation bath containing a non solvent (typically water) to produce the asymmetric membrane via phase inversion. The membrane is then rinsed (mainly with water) to remove residual solvent.

Several process and chemistry variables such as concentration of polymer, additives in the polymer solution, time, and temperature in the coagulation bath, moisture content in the air, etc. can be varied to impact structural parameters. For instance, partial evaporation of the solvent before contact with the coagulation bath can increase polymer concentration at the surface to create a tighter and thicker barrier layer. The NIPS process can result in an asymmetric structure with a dense, thin barrier layer at the top surface and a more open porous support layer underneath. This structure enabled the Loeb-Sourirajan membranes to achieve a previously unprecedented combination of mechanical stability, high flux, and high salt and neutral rejection.

Although the initial performance of asymmetric cellulose acetate membranes excited scientists and demonstrated the feasibility of seawater desalination, the cellulose acetate membrane is now rarely used. These membranes were unstable to moderate pH extremes, often resulting in decreased rejection over time. Moreover, as compared to subsequently developed interfacially polymerized membranes, they had generally less advantageous performance (flux and salt passage), requiring more energy but resulting in lesser separation quality at the same flux. Nonetheless, the abovementioned phase inversion process is still used with other polymers to form highly asymmetric support layers for interfacially polymerized membranes.

11.2.1.2 RO membranes made by interfacial polymerization

Almost all newly installed commercial RO membranes are now prepared by an interfacial polymerization [25, 26]. As typically practiced, an aqueous amine solution is imbibed in a porous support and reacted at an oil/water interface with a solution containing acid chlorides. Cadotte discovered that interfacial polymerization between an aromatic diamine, *m*-phenylenediamine (MPD), and a trifunctional acid chloride, trimesoyl chloride (TMC), resulted in a membrane with particularly good flux and rejection [7, 27]. This combination of monomers (MPD+TMC) is still used in producing most RO membranes.

Commercial RO membranes made by this method are typically a three-layer composite structure (Figure 11.2). A web layer provides mechanical strength and is commonly a non-woven polyester sheet. A porous support layer, usually a phase inverted asymmetric ultrafiltration (UF) membrane, is formed on the web to create a smooth surface for the discriminating layer and to provide a reservoir for amine during the interfacial polymerization process. The porous support layer is typically cast from polysulfone or

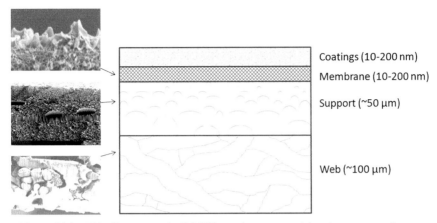

Coatings (10-200 nm)
Membrane (10-200 nm)

Support (~50 µm)

Web (~100 µm)

Figure 11.2 Thin film composite RO/NF membrane comprising three or more layers.

polyethersulfone in a suitable solvent (e.g., DMF and NMP), but several other polymers have been used for specialty RO/NF applications, such as those requiring greater solvent resistance [28]. The interfacially polymerized polyamide membrane is formed on the support layer, and this discriminating layer is typically less than 200 nm thick. Components of the thin film composite (TFC) are created individually and are of different thickness and chemical composition. Thus, the flat sheet composite structure allows for optimization of the separate parts for their respective purposes.

Figure 11.1(b) illustrates a conventional RO membrane production process. The unwind roll initially contains a composite sheet comprising a UF support layer coated on a non-woven web, such as may be formed by the process illustrated in Figure 11.1(a). The UF support layer is coated with an aqueous MPD solution, and then the excess MPD solution is removed by air or nip roller. Next, TMC solution in an organic solvent such as dodecane is applied onto the MPD saturated support, and interfacial polymerization between MPD and TMC takes place to produce a thin layer of polyamide on the support layer. Excess TMC is removed, and the membrane is rinsed with water to extract residual solvents and monomers. Finally, the membrane may be dried in an oven to generate the finished product. Additional polymeric layers contacting the discriminating thin film (below or above it) may further be present to enhance adhesion, molecular separations, or fouling resistance.

One advantage of interfacial polymerization is that a strong and cross-linked polymer film can be produced in a very short time and at low temperature. Also, formation of the barrier layer at the interface is to a large extent a self-limiting process, and so a very thin, defect-free discriminating layer can be made. Factors that influence the process include the monomers (e.g., different amine and acid chlorides), additives (e.g., surfactants, acid scavengers, and phase transfer catalysts), and polymerization conditions (e.g., reaction time, temperature, and concentrations). Separation properties of the resulting membrane depend on both its structure and composition. A thinner barrier layer allows greater permeability, whereas increasing crosslinking density generally decreases permeability of water and various solutes (both ionic and neutral species). Despite the high crosslinking ratio, there remain the amine and carboxylic acid end groups, the latter resulting from hydrolysis of unreacted acid chlorides. Carboxylic acid end groups, which are de-protonated at typical operating pH conditions, give commercial RO membranes a generally negative charge. This facilitates rejection of negatively charged ions through electrostatic interactions with the membrane (e.g. Donnan exclusion, reduced ion solubility, and/or reduced mobility of these ions within the polymer).

11.3 Nanofiltration

Reverse Osmosis and NF membranes lie on a continuum. The term "Nanofiltration" was originally used at FilmTec Corporation to describe "loose RO" membranes with pores greater than about 1 nm in diameter. While today's definition of NF is not universally agreed upon, the membranes are commonly described as having pore size less than about 2 nm, and their separations are characterized by low passage of divalent ions, substantial passage of monovalent ions, and a molecular weight cutoff (MWCO) for neutral molecules between about 150–2000 Da. Rejection of specific solutes during an NF separation is commonly impacted by both size and charge, as well as other ions in the feed solution, and operating conditions (e.g., flux). NF membranes typically require much lower operating pressure (and less energy) than RO.

As compared to RO, which has high rejection of almost all salts, an important advantage of NF is frequently the ability to selectively separate one solute over another. Many municipalities choose NF membranes to remove a large portion of the divalent cations associated with hardness to prevent corrosion of pipes while passing a sufficient amount of other salts to enhance taste. Offshore oil production platforms treat water prior to injection into the reservoir (to increase the recovery of oil) with NF to remove sulfate ions and prevent barium sulfate scaling and reservoir souring (H_2S production by microorganisms in the reservoir), while passing other ions that help maintain a productive reservoir. It is also important that, because of substantial passage of monovalents into the permeate stream, there is not a large difference in osmotic pressure between feed and permeate streams, and so the applied pressure can be relatively low. NF can also be used to further treat the reject stream from an RO system to enable higher overall recovery of water. An NF reject stream that contains almost all of the calcium and sulfate ions may be appropriate to send to a crystallizer. By contrast, when an NF permeate stream has high concentrations of monovalents but few scale-forming species, it may be possible to treat the NF permeate again with RO and extract additional water. While NF is usually implemented as a single pass, it has also been demonstrated that two-passes of a tight NF membrane can be sufficient to desalinate seawater [29]. Further, a recent review article has documented how NF may be combined with other membrane types (RO, FO, and ED) and/or thermal processes [Multi-stage flash (MSF), Multi-effect distillation (MED)] to desalinate seawater at lower cost [30].

Like RO, the majority of commercial NF membranes are made by inter-facial polymerization. Instead of reacting TMC with the aromatic amine MPD, as is done to make RO membranes, most commercial NF membranes today are formed by reacting TMC with a cyclo-aliphatic amine, piperazine. As with RO, the unreacted acid chloride groups of TMC hydrolyze into carboxylic acids during the membrane fabrication process and these negative groups impact performance by both increasing hydrophilicity and promoting rejection of negative ions. Common constituents that may also be present in these interfacial polymerizations include co-monomers, co-solvents, sur-factants, and phase transfer catalysts. Especially in the open literature, many alternative monomers have been explored to make membranes with improved performance (flux, rejection, and selectivity) or enhanced stability (chlorine, pH, and temperature). Among these, different positively charged membranes have been created using interfacial polymerizations involving polyethyleneimine (PEI).

Beyond interfacial polymerization, there are a number of other means that are used to form polymeric NF membranes. A phase inversion pro-cess can be economically used to form an asymmetric membrane with a porous support and relatively thin top skin layer. Preparing a phase inversion membrane with a charged polymer such as sulfonated polyethersulfone can result in enhanced rejection through the electrostatic interactions described previously. NF membranes may also be made through post-treatment of a microfiltration (MF) or UF support layer. A polymer barrier layer can be applied via different coating methods (e.g., dip coating, slot die, doctor blade, and spraying) and this allows a wide variety of polymers (e.g., polyvinyl alcohol, chitosan, dopamine, and poly(dimethyl siloxane) to be used. In the layer-by-layer technique, alternating applications of cationic and anionic polyelectrolytes can be used to create a thin and relatively stable coating. MF and UF supports may also be modified to make NF membranes by covalently bonding charged polymers to the surface (e.g., with reactive groups or by grafting with UV radiation, low temperature plasma, or redox reactions). The wide variety of materials in these processes allows different needs and selectivity to be addressed beyond water permeability and rejection. Current research efforts focus primarily on anti-fouling characteristics and resistance to high temperature, pH, and solvents. Beyond polymers, ceramic membranes are another important class of commercial NF membranes. These can often be advantaged for specialty separations with challenging feeds (especially with extremes of temperature, chlorine, or pH), but their greater costs make ceramic membranes less applicable to desalination.

11.3.1 Forward Osmosis (FO)

In the FO process, a feed solution passes across one side of the membrane and a draw solution with higher osmotic pressure passes across the other. Due to the difference in osmotic pressure across the membrane, water from the feed solution permeates through the membrane and into the draw solution. In a few cases, such as diluting concentrated beverages or fertilizer, the diluted draw solution may be the final product. However, in most FO applications, it is then needed to separate the permeated water from the draw solution in a regeneration step.

Membranes for FO are very similar to those used in RO (The asymmetric membrane is oriented so that the selective barrier layer surface faces the feed solution and away from the draw solution). Currently, both TFC membranes formed by interfacial polymerization (MPD + TMC) and cellulose-based membranes formed by phase inversion are used. As compared to RO, where the porous support layer is much less important, the FO support should ideally be thin, highly porous, and have low tortuosity. Early studies used a wide variety of biological and commercial membranes, and found lower flows than expected [31]. These relatively thick composite membranes resulted in a water flux decline over time, and this was attributed to internal concentration polarization within the web and support layer. Back diffusion of draw solutes was insufficient to compensate for dilution by permeate, causing the concentration of ions inside the membrane to be different from the bulk draw solution. Studies found that, despite their inherently lower water permeability than commercial TFC membranes, greater net-driving-pressure and improved flux could be obtained in FO by using a thinner (<50 m) CTA membrane having an embedded mesh [32]. In 2012, MHI commercialized another FO membrane with better performance that also had an embedded mesh, but further incorporated an interfacially polymerized TFC membrane on a support that had been optimized for low resistance [33]. Experiments have also interfacially polymerized (MPD + TMC) to form a membrane on a porous electrospun matrix of nanofibers [34]. The resulting composite membrane had five times higher flux than observed for a commercial cellulose acetate FO membrane. Between 2010 and 2015, the number of publications per year focused on FO membranes increased about fivefold.

FO is in an early stage of piloting and demonstration, exploring new membranes and potential applications in wastewater treatment and desalination. It will be a natural fit in a few niche applications where the product is a diluted draw solution that requires no further treatment. It may be shown that FO also has particular advantage in de-watering difficult waste waters, where water may be extracted by a draw solution and then the purified water is subsequently separated from the diluted draw solution (such as by RO). For large-scale desalination, a critical limitation remains the economics of regenerating the draw solution [35].

11.3.2 Electrodialysis (ED) and Related Processes

In ED, an electric potential difference induces the flow of current between two electrodes, and this current is supported by the movement of ions. Anions and cations are transported through their respective ion-exchange membranes. Electrodialysis reversal (EDR) and electrodeionization (EDI) are related processes that also rely on electrically driven transport of ions through ion-permeable membranes, and the primary differences between system geometries for these three very similar processes will be noted in Section 11.3. As compared to previously described methods of desalination, where water is purified by passing it through a membrane, the ions in these processes are instead removed from a feed solution by transporting them through ion exchange membranes. As such, these processes do not provide a barrier to viruses or bacteria and they cannot separate neutral species from feed water. However, they are effectively used to remove salts from water. Both ED and EDR are employed to desalinate brackish water and seawater, but because the energy required to transport ions increases with the salinity of the feed water, these processes are typically more cost competitive with RO for brackish water than seawater. These methods can also be useful when very high recovery is required (e.g., zero liquid discharge) because ED and EDR can further concentrate an RO reject stream after use of RO becomes limited by high operating pressures [36–38]. By contrast, EDI is used to remove ions at much lower concentrations (including weakly ionized species such as NH_4^+, $B(OH)_4^-$, $HSiO_3^-$, and HCO_3^-) from a feed solution. EDI is principally used to make ultrapure water by further treating RO permeate, especially for electronics and pharmaceutical applications.

For efficient use in ED and other processes the membranes must have a high perm-selectivity (highly permeable to counter ions but impermeable to co-ions), low electrical resistance, and good mechanical and chemical stability. Cation exchange membranes are functionalized with negatively charged ionic groups (e.g. $-SO_3^-$, $-COO^-$, $-PO_3^{2-}$, $-PO_3H^-$, $-C_6H_4O^-$) to have selective permeability for cations. Similarly, anion exchange membranes contain positively charged groups (e.g. $-NH_3^+$, $-NRH_2^+$, $-NR_2H^+$, $-NR_3^+$, $-PR_3^+$, $-SR_2^+$) and are selectively permeable to anions. In both cases, the electrostatic interactions decrease the concentration and mobility of co-ions (having the same charge as the fixed charge of the polymer) within the membrane, whereas counter ions (having the opposite charge) more readily enter and are exchanged within the polymer. As these membranes can dimensionally change with electrolyte concentration or pH, a support matrix may increase stability.

A variety of charged membranes may be made through use of monomers having moieties that either are charged or can be made charged by subsequent treatment, through attachments of charged groups to the membrane (e.g., grafting), or through inclusion of charged polymers [39–41]. Nafion[TM] polymer is a sulfonated tetrafluoroethylene commonly used to form cation exchange membranes. It is resistant to high pH and is the most common membrane used in the ED method (chlor-alkali process) to produce caustic soda and chlorine. Anion and cation exchange membrane are also made from other polyelectrolyte polymers, either self-supported or by incorporation on a support. For instance, sulfonated polysulfone and sulfonated poly(ether ether ketone) (PEEK) are other charged polymers often mentioned in the literature for use in making cation exchange membranes [39, 42]. The chemical structure of some representative ion exchange membranes are shown in Figure 11.3.

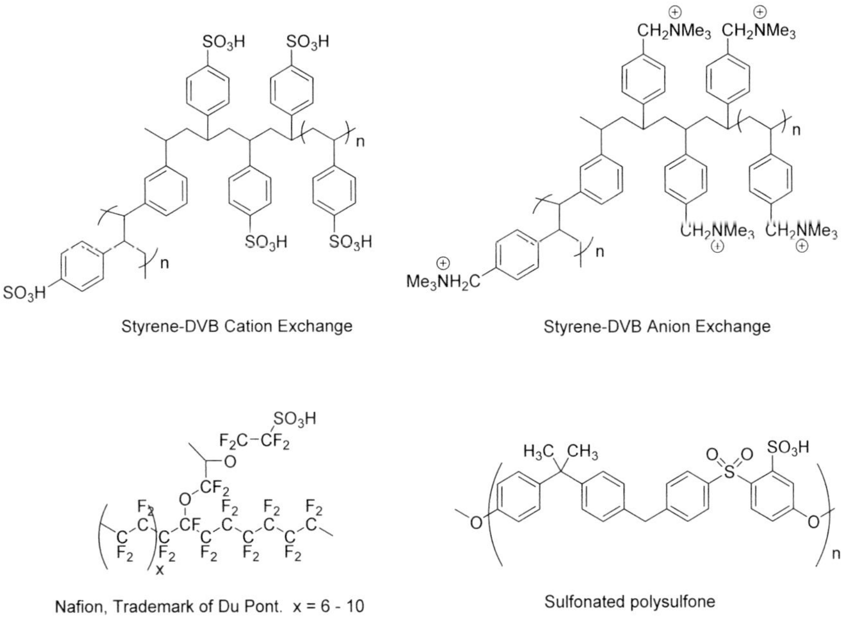

Figure 11.3   Selected ion exchange membrane materials.

Low-cost ion-exchange membranes are often prepared by mixing ground ion exchange resins, potentially with plasticizers, into a thermoplastic binder (e.g., polyethylene, polypropylene, or polyvinyl chloride) and molding or extruding the membrane film with an evaporating solvent or at a temperature above the melting point of the binder [39, 43]. Cation exchange resins are frequently based on sulfonated styrene-divinylbenzene, made by sulfonation using oleum and chlorosulfonic acid. Anion exchange resins are most commonly prepared by chloromethylation of the styrene-divinylbenzene core followed by reaction with a tertiary amine (e.g., trimethylamine). Higher charge density within the membrane increases selectivity, but it promotes swelling. Increased crosslinking can improve membrane stability by reducing swelling, and it typically increases selectivity, but it also increases electrical resistance.

11.3.3 Membrane Distillation (MD)

Membrane distillation is another often mentioned approach to desalination by membranes [44, 45]. In this thermally driven process, a difference in vapor pressure between opposing sides of the membrane causes water molecules (vapor) to transport across the membrane. Liquid water and salts are retained.

Membranes appropriate to MD are porous and hydrophobic, so that pores are not wettable by liquid water but vapor may permeate through the pores. Consequently, the most commonly employed membranes are commercial ultrafiltration membranes made from PTFE or PVDF with pore sizes between about 0.02 and 5 microns. The membrane's thermal conductivity is also important, since this can be responsible for important efficiency losses.

Even more so than FO, the MD separation process is still in its infancy, having been used on the lab scale and in a few larger demonstration systems. MD does not require the high pressures of RO and may operate at a temperature less than the liquid (e.g., water) boiling point. As compared to reverse osmosis, MD is not substantially impacted by osmotic pressure of the feed solution, so there is potential to generate even ultrapure water from vary saline feeds. However, for common concentrations, both capital and energy costs for MD are substantially higher than for RO. At present, there are also significant practical challenges (e.g., thermal losses, scale accumulation, fouling, and wetting) requiring further research, and so it is not yet clear if this will emerge as an important commercial technology.

11.3.4 Microfiltration and Ultrafiltration

MF membranes typically remove suspended particles of greater than about 0.1 microns. UF membranes are frequently characterized as excluding molecules greater than about 5000 molecular weight. Both of these membranes separate by size and they are not suitable for separating salt and water. (An exception may be the above-mentioned MD process, which can use a sufficiently hydrophobic MF or UF membrane, such that liquid water does not pass through small pores.)

Although porous MF and UF membranes are not responsible for desalination, they are important in pre-treating the feed to RO and NF systems. By preventing passage of particles, bacteria, and much organic matter, fouling of downstream RO and NF membranes can often be prevented. However, it should be noted that small molecules (nutrients) supporting biogrowth will still pass, and so biofouling is not necessarily eliminated. Also, the MF/UF systems in plants require frequent cleanings, such that the operational problem is effectively moved upstream from the RO.

MF and UF membranes may be inorganic (e.g., ceramic) or polymeric. Ceramic membranes can be more stable to heat and chemical exposure during extreme applications, but these are usually much more expensive than the commercial polymeric membranes typically applied in pre-treatment for desalination. Polyethylene and polypropylene are especially cheap among polymers used for MF and UF. However, it is more often that polymers (e.g., polyvinylidene fluoride, polysulfone, polyether sulfone, and polyacrylonitrile) are selected for preferred chemical and mechanical properties. The previously described non-solvent induced phase separation (NIPS) is one of the more typical methods for membrane manufacture, resulting in an asymmetric structure that is oriented in operation so that small surface pores face the feed side solution. Alternatively, a thermally induced phase separation generally results in a structure with more evenly-sized pores throughout the membrane depth. Commercially relevant microporous membranes (esp. polypropylene) may also be formed by a fabrication process that creates pores through a stretching an extruded film or capillary.

11.4 Configurations

The above section has described various approaches to desalinate water using membranes and provided an overview of the chemistry of those membranes. However, there are several compatible form factors (e.g., flat sheet, hollow fiber, capillary, and tubular) and associated element designs. This section provides an overview of the important configurations (membrane and modules) typically implemented for desalination.

11.4.1 Membrane Configurations

The flat-sheet thin film composite membrane was previously illustrated in Figure 11.2. Several other membrane geometries are identified in Figure 11.4. Commercial membranes used for desalination take many forms. Ion-exchange membranes for ED, EDR, and EDI processes are used as flat sheet. Commercial membranes for RO, NF, and FO are also predominantly based on flat sheet, although hollow fiber and tubular geometries are available. MD is still in early stages of development and an optimal membrane and module configuration has yet to be determined, but experiments have been performed with each of P&F, hollow fiber, tubular, and spiral wound modules [45].

Membranes may be symmetric or asymmetric. For a symmetric membrane, the structure and transport properties are uniform throughout the membrane and flux is inversely proportional to thickness. By contrast, an asymmetric structure will have a tight discriminating layer on one surface and a more open (and less resistive to flow) supporting structure below it. It was previously noted that desalination was first made practical by the discovery of highly asymmetric, phase inverted cellulose acetate (CA) membranes.

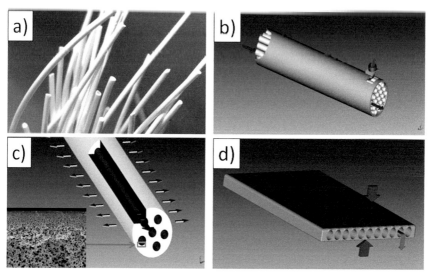

Figure 11.4 Membrane geometries including (a) hollow fiber, (b) tubular, (c) multi-capillary fibers (with inset illustrating asymmetric surface structure), and (d) capillary sheet.

Hollow fiber (Figure 11.4(a)) is the most common configuration for MF and UF membranes in RO pretreatment. The hollow fiber configuration is also employed currently with cellulose triacetate (CTA) membranes for desalination by RO. Since CTA membranes have lower intrinsic permeability than existing interfacially polymerized membranes with similar salt passage, these CTA membranes are mostly used in the Middle East and are rarely specified for new installations. Hollow fibers may be configured to generate a permeate stream with flow from inside to outside or vice versa. In either case, the discriminating layer of an asymmetric membrane is adjacent to the feed stream. In an outside to inside configuration, as is used with current commercial hollow fiber RO, the feed stream flows across a barrier layer on the peripheral surface (outside) and permeate is produced into the fiber lumen (inside). This configuration typically has greater total active area for the same fiber volume, but the alternative geometry (inside to outside) may have greater mixing of feed solution to reduce fouling or polarization.

There are important trade-offs associated with hollow fiber dimensions. As compared to flat sheet geometry, sufficiently small hollow fibers would provide an increase in active membrane area per volume. However, smaller fibers are associated with an increased potential for fiber breakage and greater pressure drops down both the lumen and in the spaces between fibers. Early commercial RO modules from DuPont and Dow used fine hollow fibers that often had an inner diameter of about 40 microns (even 20 microns for DuPont's B10 permeator used for seawater desalination), but pressure drops in the modules were still considered acceptable because of the comparatively high resistance of early RO barrier layers to water permeation. By contrast, water permeability of MF and UF barrier layers is much higher, and the resistance to flow in and around the fibers needs to be less. It is common that commercial UF hollow fibers have a lumen of about 0.7–0.9 mm diameter.

Tubular membranes (Figure 11.4(b)) are typically used because their larger inner diameters can be advantaged in environments with minimal pre-treatment, a great number of suspended solids, or highly viscous feeds. While the defining dimensions are only approximate, increasing diameter is the principal difference between hollow fiber (<0.5 mm), capillary membranes (0.5–10 mm), and tubular membranes (>10 mm). The flow in a tubular membrane is usually inside to outside, with the inner surface of the tube supporting a discriminating surface.

Alternative membrane geometries combine multiple elongated bores into a single structure. In one approach, several hollow capillaries having a dis-criminating layer on their inner surface are simultaneously extruded within a single more porous fiber (Figure 11.4(c)). In this way, high membrane active area can be provided within a structure that has greater physical strength than a single hollow fiber. Similarly, a capillary sheet geometry (Figure 11.4(d)) can provide an array of hollow capillaries within an extruded flat sheet, with the discriminating surfaces on either the inside (capillary wall) or outside (flat) surfaces. Both orientations have been proposed for RO [46]. Tubular modules may similarly be formed as a single monolithic porous structure containing several hollow tubular channels. As is common with individual tubular membranes, the inner surfaces of hollow cavities may be post-treated to form a discriminating layer.

11.4.2 Flow Path Considerations

Independent of membrane geometry, the velocity and orientation of feed flow relative to the surface is important. Figure 11.5 illustrates dead-end and cross flow configurations for flat sheet membrane. In a simplest configuration, membranes could operate in dead-end (or "frontal") filtration. In this case, feed flow would be directed at (i.e., perpendicular to) the filter surface. This dead-end filtration is relatively common for UF and MF. Retained particles can rapidly build-up on the filter surface, forming what is called

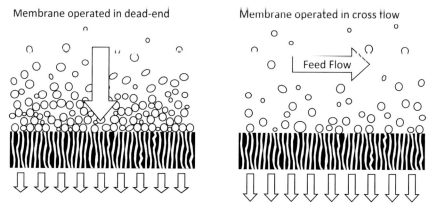

Figure 11.5   Dead-end flow versus cross flow membrane operation.

a cake, before periodic removal by cleaning, scouring with air bubbles, or backflushing. However, dead-end filtration is generally inappropriate for RO and NF membranes, as salts and small molecules rejected by the membrane are often unable to sufficiently diffuse away from the surface and increased concentrations near the membrane results. (When the surface concentration of a free and diffusive solute is higher near the membrane than in the bulk, the situation is referred to as concentration polarization and the effect causes both increased apparent passage of the solute and increased osmotic pressure). Both surface fouling and polarization layers can result in rapidly decreased flux.

The most common configuration for RO and NF is the cross-flow configuration (Figure 11.5(b)), also referred to as tangential filtration, which can be used to reduce fouling and polarization. The feed stream passes across (i.e., parallel to) the surface of a membrane, and two exiting streams are generated. The permeate stream is the portion of fluid that passes through the membrane. The retentate (or reject) stream is the remainder of the feed stream. As with dead-end operation, increased permeate flux tends to concentrate material near the membrane surface. However, the tangential velocity in cross flow configuration can reduce both polarization and fouling near the membrane surface. In practice, additional features on the membrane surface or within the feed channel (e.g., spacers) contribute to further mixing. For a given system and feed water, a balance may be found between permeate flux and feed flow that allows continuous operation.

For FO membranes, it was mentioned that polarization also exists on the draw side of the membrane, within the web and support. To maintain high osmotic pressure across the barrier layer, cross flow is desirable on both opposing surfaces of the membrane.

"Recovery" is defined as the ratio of permeate volume to feed volume. This term may be applied to a membrane cell, module, or system of modules. Because the osmotic pressure of feed water increases with recovery, flux is decreased at the tail end of a system operating at constant pressure. Most seawater desalination plants operate with recoveries between 40 and 60% as this often provides the most economical operating point. Higher values (e.g., 60–80% recovery) are more common for BW separations. In either case, individual RO and NF spiral wound modules are typically operated with recovery less than 15%. Velocity in the feed channel spacer is commonly between about 8–22 cm/sec, and flux through the membrane is usually 5–20 gfd.

11.4.3 Module Configurations

While often omitted from a discussion of modules, practitioners may occasionally have need for simple small-scale laboratory tests using membranes and a challenge feed of interest. For desalination, by RO or NF, a few cells are commonly mentioned in the literature. Among these, the classic Amicon dead-end stirred cell now sold by Millipore comes in different configurations and is suitable for low pressures filtrations (MF, UF, NF). It can operate at pressures up to 5 bar, and a magnetically coupled rotating stirrer reduces the cake and/or polarization layers. POROMETER nv has a system of multiple (8 or 16) dead-end cells that allow several membrane coupons to be simultaneously tested at pressures up to 100 bar with a common feed and pressure. Sterlitech sells a variety of high- and low-pressure test equipment, including both dead-end stirred and cross flow geometries. Some of these cells allow for cross flow at both surfaces of the membrane, as is preferred for FO.

For larger-scale separations, the plate and frame (P&F) is among the simplest of configurations. A pair of flat sheet membranes are oppositely oriented and arranged to sandwich a common spacer material there between. Multiple sets of these paired sheets are stacked, with a second spacer material type separating each pair. The stack further comprising end plates which are compressible against the stack, so that pressurized liquid may be circulated within the stack. Various seals provide at least two sets of isolated channels (e.g., feed and permeate) within the stack.

Figure 11.6(a) shows a common P&F configuration that comprises several stacks of alternating layers: feed-side channel plate, membrane sheet, permeate-side support plate, and membrane sheet. Using a pressure greater

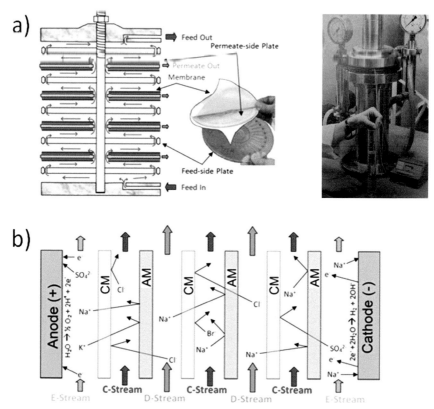

Figure 11.6   Illustration of typical flow paths used in (a) membrane-testing systems and (b) electrodialysis (ED), and photograph of assembled P&F unit (stack of only two sets of coupons).

than the applied feed pressure, the stack of plates is compressed and the compression creates seals against the membrane surfaces. The feed-side plates have radial channels to guide feed fluid back and forth between the center and periphery of the stack, over the front surface of each membrane. Permeate from the back surfaces of adjacent membrane sheets exits the permeate-side support plates. Flows from each pair of membranes sheets may be separately collected.

As compared to flat cells, the P&F configuration can provide a relatively large area of flat sheet membrane. Membrane loading is also easy. For these reasons, several manufacturers, including Dow Chemical, use P&F units as part of the membrane quality control program. A P&F configuration is sometimes also advantageous for operating at high temperature or high pressure.

Disc-Tube RO systems, as sold by Rochem and others, are a variant of the P&F within a pressurizable shell. Such units may operate at high pressures (up to about 150 bar) to overcome an osmotic feed pressure limitation and can include wide open channels to limit particulate fouling [47]. However, capital costs are larger than for more common configurations (esp. spiral wound elements). Also, while membrane sections may be replaced in Disc-Tube units, the cutting of circular flat sheet coupons is an inherently inefficient use of membrane area. Another P&F variant (VSEP – Vibratory Shear-Enhanced Process) reduces polarization and fouling by providing rapid oscillations of the unit. While also very capital intensive, it has shown advantages in highly fouling process waters and high recovery separations prone to scaling [48].

A P&F configuration is also common for ED. As also illustrated in Figure 11.6(b), anode and cathode electrodes can bound opposite ends of a membrane stack that comprises alternating planar anion (*AM*) and cation (*CM*) selective membrane sheets. The channels between adjacent membrane sheets define a *dilute* stream (D-stream) and a *concentrate* stream (C-stream), the former reduced in concentrations from the original feed by passing through the ED stack. In operation, the electric potential drives cations within the dilute stream through the CM towards the cathode, but the AM prevents cations from further passing again into the next dilute stream. Similarly, anions are removed from the dilute stream and accumulated within the proximate concentrate stream (nearer the anode). Additional *electrode* streams (E-stream) may be present between the membranes and the anode or cathode. In some ED embodiments, net-like spacers define flow channels and create mixing within the stream. In EDR, periodic switching of both flow streams (dilute vs. concentrate) and electrodes (anode vs. cathode) can remove scale and enable operation in higher salt concentration. In electrodeionization (EDI) an ultrapure dilute stream is produced and ion-exchange particles within the dilute stream allow current to pass through the low conductivity water. In each of these related electromembrane processes, system throughput may be scaled by increasing number of paired membranes within the stack.

The spiral wound module is by far the most common geometry for RO and NF, and it is used for other membrane applications as well. The spiral wound module provides a large amount of active membrane area in a small volume. Figure 11.7(a) illustrates a conventional module formed by concentrically winding membrane envelopes and adjacent feed spacer sheets about a central permeate collection tube. Each membrane envelope comprises two rectangular sections of membrane that sandwich a permeate spacer. This sandwich

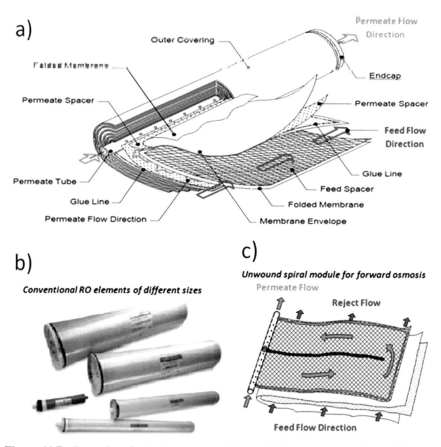

Figure 11.7 Examples of spiral wound modules used for reverse osmosis and forward osmosis.

structure is sealed along three edges to form an envelope while a fourth edge of the permeate spacer abuts the permeate collection tube. In operation, pressurized feed fluid enters the module from one end, flows across the front surfaces of the membrane, and exits the module at an opposing end. A portion of the feed fluid passes through the membrane as permeate and spirals along the permeate spacer toward holes in the permeate tube. By forming a module from multiple membrane envelopes, instead of a single very long envelope, resistance of the permeate flow to the tube is decreased. The most common spiral wound modules are 8-inches in diameter, 40-inch long, and comprise about 20–30 membrane individual envelopes. Figure 11.7(b) depicts some of the commercial modules that are commonly available in diameters between

1.8 inches to 8 inches. Dow Chemical and Hydranautics have each supplied 16" diameter elements to the previously mentioned Sorek desalination plant.

Although the spiral wound module in Figure 11.7(a) illustrates the most typical geometry, additional ports and alternative flow paths are sometimes used. For instance, some less common configurations use a radial feed flow, such as where feed is supplied near the permeate tube and concentrate is removed at the module's periphery. When the radial flow path is longer than the axial length of the module, greater feed velocity (and reduced propensity to scale) may be attained under conditions of similar operating flux and recovery.

In the case of FO, cross flow is also desirable on the permeate side of the membrane envelope so that osmotic pressure from the draw solution remains high. Figure 11.7(c) illustrates a simplified (single-leaf) unrolled module that accomplishes this. In the illustrated spiral wound module, both the permeate tube and permeate spacer are divided in two, so that a permeate flow path may be established that enters and leaves at opposite ends of the permeate tube.

Spiral wound electrodeionization (EDI) modules have also been sold that include multiple concentric windings of anion exchange membrane, dilute channel, cation exchange membrane, and concentrate channel. With electrodes positioned at the center and periphery, a partial cross section of the module from central tube to periphery would be similar to the ED stack of Figure 11.6(b). Fabrication costs are reduced and external leaks are prevented because the spiral module fits within a cylindrical pressure vessel. Ion removal efficiency varies across the stack due to the changing diameter and current densities. Omexell Inc. had made use of the variable current density to improve ion removal, manufacturing a spiral wound EDI module that included two passes of the dilute stream; a first pass was near the module's periphery and a second pass was near the central electrode.

Hollow fiber modules are currently especially important for MF and UF, but they are also being used for RO. Mahon of Dow Chemical invented the hollow fiber module (Figure 11.8(a)) and proposed its use for RO [9]. DuPont commercialized the first hollow fiber RO module in 1969, using a membrane of linear aramid polyamide. This was followed by offerings from Dow Chemical and Toyobo, using cellulose triacetate membranes. DuPont has now exited the business and Dow Chemical has since purchased FilmTec Corp. and bet on interfacial polymerization and spiral wound modules. One can still purchase hollow fiber modules made with CTA fibers from Toyobo, and these have now replaced many in the original DuPont systems. To reduce pressure drop, Toyobo hollow fibers have somewhat larger dimensions

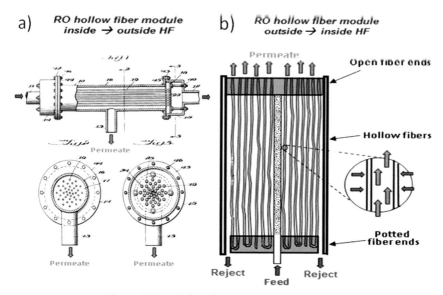

Figure 11.8 Hollow fiber module embodiments.

(e.g., 70 micron ID, 160 micron OD) than the earlier DuPont hollow fine fibers. Today, the Toyobo franchise is partially protected from spiral elements of higher permeation (but standardized sizes) by differences in vessel size and porting arrangements. Another important tradeoff is that CTA membrane are sensitive to pH (which frequently causes reduced salt rejection over time), but the membranes are cleanable with chlorine. This leads to standardization around different operating practices.

The primary cause for the transition from hollow fiber to spiral modules has been the better performance (and associated reduced energy) from inter-facially polymerized flat sheet membranes. However, efforts are continually underway to improve the hollow fiber barrier layer as well. Independent of the type of membrane, hollow fiber modules can have several advantages compared to spiral wound modules [49]. Construction of hollow fiber modules is simplified because they are self-supported; they don't require spacers. They have an advantage in surface area to volume compared to current spiral wound modules, based on commercial fiber sizes and common flat sheet spacer thicknesses. This high surface area and tight packing of fibers also can allow greater water recovery within the hollow fiber module at the same flux. (A greater feed water volume is present within the typical spiral wound feed spacer.) With this tight fiber packing, a problem with hollow fiber modules has been fouling between fibers.

A wide variety of pressurized hollow fiber module configurations are practiced, especially in UF where we see many different non-standardized approaches to connect feed and permeate ports to flow paths within the element. The following description (and Figure 11.8(b)) is consistent with a hollow fiber RO module from DuPont. A bundle of parallel hollow fibers is contained in a shell and the fibers extend between two opposing ends. Each end of the fiber is potted (e.g., in epoxy or urethane), and the potting of at least one end is cut to create a cross section of exposed and open lumens from which to remove permeate. (Lumens on the opposite end are sometimes also exposed, so that permeate can be removed at both ends to reduce pressure drop within the lumen.) In RO modules from DuPont and Toyobo, feed was introduced through a central tube that extended through the potting on at least one end. Feed propagated radially from the center to the inner periphery of the module's shell and exited the module from one end as reject.

The tubular module illustrated in Figure 11.4(b) is similar to a shell and tube heat exchanger. Multiple hollow membrane tubes are commonly located in and supported by a common outer cylinder. This tubular geometry is now most commonly used for MF and UF, but it is also employed for specialty applications using RO and NF when minimal pre-treatment is desired but the separation still requires a tight barrier. Utilization of tubular membranes for desalination is uncommon due to much greater costs compared to traditional spiral elements.

11.5 Future Developments

This chapter has mentioned both established (RO, NF, ED) and developing (FO, MD) approaches to use membranes for separating salt from water. While the degree to which each method will be relevant and successful in the future is presently unknown, there is potentially an optimal niche for each of these. Nonetheless, the current dominant approach of using pressure-driven membrane separation of ions from water is already successful and widely implemented, and there are several potential directions for further advancement in that space.

Membrane research for RO/NF desalination often falls into one of three major areas: (i) improved performance properties,(ii) reduced susceptibility to fouling and scaling, and (iii) increased stability in challenging environments. Improving conventional polymeric RO membranes to provide higher rejection and/or lower energy consumption has been and will continue to be an active area of exploration. A substantial savings in capital would also be available if high permeability RO modules could reliably operate at higher flux, without the concomitant accumulation of scale and foulants. Several efforts have focused on modifying membranes or modules so as to make them either less prone to fouling/scaling or easier to clean (Improved feed water pre-treatment, membrane cleaning procedures, and/or module designs may also enable higher flux). There is also need for membranes that sustain their performance under challenging process conditions (e.g., exposure to extreme temperatures and pH, oxidizing environments, or incompatible solvents) and research continues to increase lifespans through modification of the polymeric membrane and/or their supports.

Various new membranes are also being currently evaluated as future solutions to the above needs. Among these, graphene, graphene oxide, carbon nanotubes, and zeolite membranes can be categorized as nanoporous material membranes. Nanoporous materials may provide significant advantage on energy consumption and solute rejection compared to conventional RO and NF membranes. RO and NF membranes desalinate water by a pressure mediated, solution-diffusion process in which ions enter and diffuse across the membrane at a lower rate than water due to stochastic combination of steric and charge interactions. In order to drastically reduce the energy requirements, the transport of salt and/or water through the membrane has to be fundamentally altered. Nanoporous materials provide such a separation mechanism which is to a greater degree based on entrance pore size and molecular sieving. Charges induced at the entrances of nanopores may further increase rejection of ions beyond that expected based on just the size of hydrated species. However, as elaborated upon in a recent review [50] of simulated transport studies, large deviations for graphene and carbon nanotubes from macroscopic transport theories have been observed, and these may be ascribed to a variety of identified interactions between water, solutes, and pore walls.

Graphene membranes comprise a two-dimensional hexagonal lattice of covalently bonded carbon atoms having natural or induced defects that create pores for water permeation. (Defects may be induced by ion or electron beams or by plasma or chemical etching.) Graphene membranes can be thinner, stronger, and more oxidatively stable than conventional TFC RO membranes. Computational research suggests that thin monolayer graphene sheets with pores can increase membrane water permeability at high solute rejections by more than two orders of magnitude [51]. Practical issues for full scale commercialization of graphene membranes include manufacturability of large areas of defect free graphene sheet and the difficulty in obtaining a high density of nanopores with monodispersity [52]. In an alternative approach, membranes have been made from partially overlapping two-dimensional sheets of graphene oxide or chemically converted graphene [52, 53]. Transport through the nanometer-sized confined channels between adjacent layers contributes to rejection of solutes by both size and electrostatic interactions. High rates of water transport have been hypothesized based on both low friction of water with walls and coupled movement of interacting water molecules. NF membranes have been demonstrated by using these partially stacked layers. However, in light of the longer path length with this approach, it may be difficult to similarly achieve an RO membrane with both high water permeability and good rejection.

Carbon nanotubes (CNT) can be envisioned as rolled graphene sheets in tubular form with nanometer scale diameters. Researchers [54, 55] have claimed that hydrophobicity and atomic smoothness of the inner surfaces of the CNTs provide extremely low resistance to water flow, and that tube diameter may be selected to reject solute species based on size. Simulations also suggest that connected chains of aligned water molecules facility water transport within CNTs. Efforts [55] have been made to manufacture CNT membranes where CNTs were embedded in a polymer matrix (mixed matrix membranes) but the inability to manufacture defect free carbon nanotubes of small diameters and align them uniformly in an economical fashion has prevented large scale manufacturing for CNT membranes.

Zeolites are solid crystalline structures made of aluminum, silicon, and oxygen that form a framework with defined channels inside. Zeolite membranes are commonly formed as polycrystalline thin films on a macroporous ceramic or metal substrate by *in-situ* crystallization or by a seeded secondary crystal growth method. They can have ordered pores in the 4–8 Angstrom size, large enough for water to pass but sufficiently small to reject most hydrated ions. Ion passage has also been found to be impacted by a charged double layer from ions adsorbed on pore walls. Early zeolite membranes demonstrated moderate rejection of salts, but structures have been much thicker than those resulting from interfacial polymerization and water flux has so far been very low [56, 57]. While economics for desalination may not prove competitive, it has been noted that these membranes would have advantage over polymeric membranes in cleaning (e.g., with chlorine).

Another class of membranes gaining widespread attention is biomimetic desalination membranes. Biomimetic membranes are synthetic membranes that include or are based upon a biological component. One of the most widely researched biomimetic membranes currently are aquaporin membranes. Aquaporins are a class of water channel proteins found in cell walls that are responsible for transport of water to cells while rejecting solutes and toxins. It has been estimated that membranes incorporating aquaporins could have good selectivity and permeabilities of about 600 L/m^2/h/bar, many times higher than even todays polymeric NF membranes [58]. Several approaches have been used to incorporate aquaporins within a barrier layer [59], and one of the more facile methods used in several reports has been to add liposomes containing aquaporins directly to the amine (MPD) solution in an interfacial polymerization. Beyond the degree of improved performance available, an important question will be the ability of these aquaporin proteins to withstand extreme temperatures and pH conditions.

Reflecting on past breakthroughs from Loeb-Sourirajan and Cadotte, new membranes may provide a leap in performance again. However, although permeabilities might be dramatically improved, operational flux will still be limited by concentration polarization and fouling. Increased permeability can also be implemented to reduce energy use, but it is important to realize that membranes are only a part of the energy required for desalination [16]. Further, while implementation could potentially be as facile as adding a component (e.g., aquaporin containing proteoliposomes) to the amine within current manufacturing processes, it is more likely that considerable research will be needed before new membranes are commercialized. After demonstrating improved performance, significant focus will need to be targeted to issues of manufacturing.

During the last four decades, the energy required to desalinate seawater using membranes has decreased by almost an order of magnitude [15, 16]. While continued improved performance of the membrane has been critical, there have also been important advancements to membrane modules, pumps, energy recovery devices, and smarter system designs. As we strive to both make water more affordable and decrease our carbon footprint, improved membranes and membrane processes will all have to be a part of the solution.

References

[1] Lonsdale, H. (1982). The growth of membrane technology. *J. Membr. Sci.* 10, 81–181.

[2] Bennett, A. (2013). 50 th anniversary: desalination: 50 years of progress. *Filtration Sep.* 50, 32–39.

[3] Glater, J. (1998). The early history of reverse osmosis membrane development. *Desalination* 117, 297–309.

[4] Grebenyuk, V., and Grebenyuk, O. (2002). Electrodialysis: from an idea to realization. *Russ. J. Electrochem.* 38, 806–809.

[5] Sidney, L., and Srinivasa, S. (1964). High flow porous membranes for separating water from saline solutions. U.S. Patent No. 3133132. Washington, DC: U.S. Patent and Trademark Office.

[6] Lonsdale, H. K. (1987). The evolution of ultrathin synthetic membranes. *J. Membr. Sci.* 33, 121–136.

[7] Cadotte, J. E. (1981). Interfacially synthesized reverse osmosis membrane. U.S. Patent No. 4277344. Washington, DC: U.S. Patent and Trademark Office.

[8] Balla, A., et al. (1969). Desalination assembly and its method of manufacture. U.S. Patent No. 3446359. Washington, DC: U.S. Patent and Trademark Office.

[9] Mahon, H. I. (1966). Permeability separatory apparatus, permeability separatory membranc clement, method of making the same and process utilizing the same. U.S. Patent No. 3228876. Washington, DC: U.S. Patent and Trademark Office.

[10] Bray, D. T. (1968). Reverse osmosis purification apparatus. U.S. Patent No. 3417870. Washington, DC: U.S. Patent and Trademark Office.

[11] Westmoreland, J. C. (1968). Spirally wrapped reverse osmosis membrane cell. U.S. Patent No. 3367504. Washington, DC: U.S. Patent and Trademark Office.

[12] Gude, V. G. (2016). Desalination and sustainability—An appraisal and current perspective. *Water Res.* 89, 87–106.

[13] Fritzmann, C., et al. (2007). State-of-the-art of reverse osmosis desalination. *Desalination* 216, 1–76.

[14] Ghaffour, N., Missimer, T. M., and Amy, G. L. (2013). Technical review and evaluation of the economics of water desalination: Current and future challenges for better water supply sustainability. *Desalination* 309, 197–207.

[15] Elimelech, M., and Phillip, W. A. (2011). The future of seawater desalination: energy, technology, and the environment. *Science* 333, 712–717.

[16] Shrivastava, A., Rosenberg, S., and Peery, M. (2015). Energy efficiency breakdown of reverse osmosis and its implications on future innovation roadmap for desalination. *Desalination* 368, 181–192.

[17] Ghiazza, E., Borsani, R., and Alt, F. (2013). "Innovation in multi-stage flash evaporator design for reduced energy consumption and low installation cost," in *Proceedings of the International Desalination Association (IDA), World Congress on Desalination and Water Reuse*, Tianjin.

[18] Mulder, J., (2013). *Basic Principles of Membrane Technology*. Dordrecht: Springer.

[19] Lalia, B. S., et al. (2013). A review on membrane fabrication: Structure, properties and performance relationship. *Desalination* 326, 77–95.

[20] Loeb, S., and Sourirajan, S. (1960). *UCLA Dept. of Engineering Report*, 60.

[21] Sourirajan, S., and Govindan, T. (1967). "Membrane separation of some inorganic salts in aqueous solutions," in *Proceedings of the First International Symposium on Water Desalination* (Washington, DC: U.S. Government Printing Office).

[22] Strathmann, H., and Kock, K. (1977). The formation mechanism of phase inversion membranes. *Desalination* 21, 241–255.

[23] Fahey, P. M., and Grethlein, H. E. (1971). Improved cellulose acetate membranes for reverse osmosis. *Desalination* 9, 297–313.

[24] Merten, U., et al. (1967). Performance of cellulose acetate membranes in sea water desalination. *Desalination* 3, 353–358.

[25] Wittbecker, E. L. (1996). Reflections on "Interfacial polycondensation. I," by Emerson L. Wittbecker and Paul W. Morgan, J. Polym. Sci., XL, 289 (1959). *J. Polym. Sci. A Polym. Chem.* 34, 515–16.

[26] Petersen, R. J. (1993). Composite reverse osmosis and nanofiltration membranes. *J. Membr. Sci.* 83, 81–150.

[27] Cadotte, J. E. (1980). Interfacially synthesized reverse osmosis membrane and its use in removing solute from solute-containing water. E.P. Patent No. 15149A1. Washington, DC: U.S. Patent and Trademark Office.

[28] Marchetti, P., et al. (2014). Molecular separation with organic solvent nanofiltration: a critical review. *Chem. Rev.* 114, 10735–10806.

[29] Adham, S., et al. (2003). Long Beach's dual-stage NF beats single-stage SWRO. *Int. Desalination Water Reuse Q.* 13, 18–21.

[30] Zhou, D., et al. (2015). Development of lower cost seawater desalination processes using nanofiltration technologies—A review. *Desalination* 376, 109–116.

[31] Cath, T. Y., Childress, A. E., and Elimelech, M. (2006) Forward osmosis: principles, applications, and recent developments. *J. Membr. Sci.* 281, 70–87.

[32] McCutcheon, J. R., McGinnis, R. L., and Elimelech, M. (2005). A novel ammonia—carbon dioxide forward (direct) osmosis desalination process. *Desalination* 174, 1–11.

[33] Ren, J., and McCutcheon, J. R. (2014). A new commercial thin film composite membrane for forward osmosis. *Desalination* 343, 187–193.

[34] Song, X., Liu, Z., and Sun, D. D. (2011). Nano gives the answer: breaking the bottleneck of internal concentration polarization with a nanofiber composite forward osmosis membrane for a high water production rate. *Adv. Mater.* 23, 3256–3260.

[35] Mazlan, N. M., Peshev, D., and Livingston, A. G. (2016). Energy consumption for desalination—A comparison of forward osmosis with reverse osmosis, and the potential for perfect membranes. *Desalination* 377, 138–151.

[36] Walker, W. S., Kim, Y., and Lawler, D. F. (2014). Treatment of model inland brackish groundwater reverse osmosis concentrate with electrodialysis—Part III: Sensitivity to composition and hydraulic recovery. *Desalination* 347, 158–164.

[37] Zhang, Y., et al. (2012). RO concentrate minimization by electrodialysis: techno-economic analysis and environmental concerns. *J. Environ. Manage.* 107, 28–36.

[38] Oren, Y., et al. (2010). Pilot studies on high recovery BWRO-EDR for near zero liquid discharge approach. *Desalination* 261, 321–330.

[39] Nagarale, R., Gohil, G., and Shahi, V. K. (2006). Recent developments on ion-exchange membranes and electro-membrane processes. *Adv. Colloid Interface Sci.* 119, 97–130.

[40] Mizutani, Y. (1990). Structure of ion exchange membranes. *J. Membr. Sci.* 49, 121–144.

[41] Xu, T. (2005). Ion exchange membranes: state of their development and perspective. *J. Membr. Sci.* 263, 1–29.

[42] Tang, Y., et al. (2014). Novel sulfonated polysulfone ion exchange membranes for ionic polymer–metal composite actuators. *Sens. Actuators B Chem.* 202, 1164–1174.

[43] Vyas, P. V., et al. (2001). Characterization of heterogeneous anion-exchange membrane. *J. Membr. Sci.* 187, 39–46.

[44] Drioli, E., Ali, A., and Macedonio, F. (2015). Membrane distillation: Recent developments and perspectives. *Desalination* 356, 56–84.

[45] Alkhudhiri, A., Darwish, N., and Hilal, N. (2012). Membrane distillation: a comprehensive review. *Desalination* 287, 2–18.

[46] Billovits, G. E., et al. (2014). Spiral wound module including membrane sheet with capillary channels. U.S. Patent No. 8911625. Washington, DC: U.S. Patent and Trademark Office.

[47] Peters, T. A. (1999). Desalination of seawater and brackish water with reverse osmosis and the disc tube module DT. *Desalination* 123, 149–155.

[48] Shi, W., and Benjamin, M. M. (2011). Effect of shear rate on fouling in a Vibratory Shear Enhanced Processing (VSEP) RO system. *J. Membr. Sci.* 366, 148–157.

[49] Moch, I. (1989). A twenty year case history: B-9 Hollow fiber permeator. *Desalination* 74, 171–181.

[50] Thomas, M., Corry, B., and Hilder, T. A. (2014). What have we learnt about the mechanisms of rapid water transport, ion rejection and selectivity in nanopores from molecular simulation? *Small* 10, 1453–1465.

[51] Cohen-Tanugi, D., and Grossman, J. C. (2015). Nanoporous graphene as a reverse osmosis membrane: Recent insights from theory and simulation. *Desalination* 366, 59–70.

[52] Hu, M., and Mi, B. (2013). Enabling graphene oxide nanosheets as water separation membranes. *Environ. Sci. Technol.* 47, 3715–3723.

[53] Han, Y., Xu, Z., and Gao, C. (2013). Ultrathin graphene nanofiltration membrane for water purification. *Adv. Funct. Mater.* 23, 3693–3700.

[54] Goh, P.S., Ismail, A. F., and Ng, B. C. (2013). Carbon nanotubes for desalination: Performance evaluation and current hurdles. *Desalination* 308, 2–14.

[55] Ahn, C. H., et al. (2012). Carbon nanotube-based membranes: Fabrication and application to desalination. *J. Ind. Eng. Chem.* 18, 1551–1559.

[56] Li, L., Dong, J., and Nenoff, T. M. (2007). Transport of water and alkali metal ions through MFI zeolite membranes during reverse osmosis. *Sep. Purif. Technol.* 53, 42–48.

[57] Zhu, B., et al. (2015). Application of robust MFI-type zeolite membrane for desalination of saline wastewater. *J. Membr. Sci.* 475, 167–174.

[58] Kumar, M., et al. (2007). Highly permeable polymeric membranes based on the incorporation of the functional water channel protein Aquaporin Z. *Proc. Natl. Acad. Sci. U.S.A.* 104, 20719–20724.

[59] Tang, C. Y. et al. (2013). Desalination by biomimetic aquaporin membranes: review of status and prospects. *Desalination* 308, 34–40.

CHAPTER 12

State-of-the-Art Desalination Research

Seyed Hamed Aboutalebi[1], Alexandros Yfantis[2] and Nikolaos Yfantis[2]

[1]Condensed Matter National Laboratory, Institute for Research
in Fundamental Sciences, 19395-5531, Tehran, Iran
E-mail: hamedaboutalebi@ipm.ir
[2]Sychem Advanced Water Technologies, 518 Mesogeion Av., 153 42 Agia
Paraskevi, Athens, Greece
Phone: +30 210 6084940
E-mail: info@sychem.gr

12.1 Introduction

Water scarcity, along with the rapid development of urban areas is boosting the ever-increasing demand for safe and clean water [1, 2]. The urgent adoption of new water recycling technologies and the development of alternative potable water sources are necessary to address the ongoing drought experienced by almost 2.6 billion people around the world [2–4]. The United Nations Department of Economic and Social Affairs estimates that by 2025, two-thirds of the world's population will be living under water stressed conditions [5–7]. Hence, the development of novel desalination[1] technologies, and not incremental changes, is crucial to the realization of energy efficient and reliable desalination plants. This new generation of desalination plants is expected to provide low cost and high quality potable water to deal with the vast demand over the next decades.

This chapter will provide a review for some of the most recent research trends with promising laboratory results. Of course, the chapter is not exhaustive and some technologies have not been discussed.

A categorization of several desalination processes is shown in Figure 12.1.

---

[1]Desalination is a process to remove salt from *saline* (the Greek word *alas* to Latin *sal*).

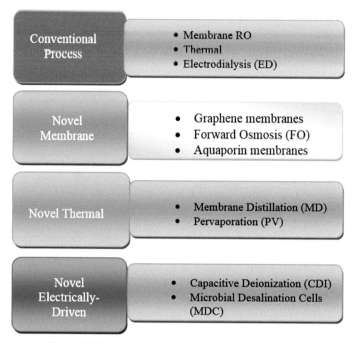

Figure 12.1   Categorization of desalination processes.

There are currently three methods of choice for sea water desalination; thermal distillation, reverse osmosis (RO), and to a much lesser extent electro-dialysis (ED). Also, the high capital and energy costs of thermal desalination combined with corrosion problems often experienced in the plants have made them less attractive for large scale application in countries where the energy price is on the rise [4, 6]. However, solar thermal distillation and humid-air distillation technologies are finding an increased market-share for small communities in remote locations with access to solar energy [4, 8, 9].

12.2 RO Technologies

RO, which accounts for more than half of the world's installed capacity, although quite practical from the energy efficiency point of view ($\sim$5 Wh L^{-1} which can theoretically be decreased down to 2.5 Wh L^{-1}), suffers particu-larly from the high cost of electrical energy, corrosion and fouling problems. Moreover, although RO is being used as a state-of-the-art desalination tech-nology, it still faces several challenges that should be addressed to realize a reduction in the total cost of water production. The first and foremost problem is fouling. To overcome such an issue, either membranes with low propensity to fouling should be developed or pretreatments with oxidants (e.g., chlorination with sodium hypochlorite) for biofouling control should be exercised. However, the current membrane technology implemented in RO systems has a low tolerance of such oxidants [9]. In order to avoid membrane scaling, besides antiscalant chemicals, strong acids (sulfuric or hydrochloric acid) can also be used; but their storage and transport contributes to the complexity of the overall processes leading to hidden risk factors [10]. Another ignored aspect of the process is the direct discharge of RO concentrates into seawater which may cause risk for the marine ecosystem and the environment depending on the coast morphology and size of application [11]. As the coast morphology can differ significantly in combination with a very big range of installation capacities, it is hard to generalize the environmental impact of RO concentrates; and a case by case evaluation is required.

Although many conventional and emerging technologies are being applied increasingly to decrease the environmental problems associated with RO plants, most emerging technologies are yet at their preliminary stage; having high capital costs while still encountering technical obstacles, renders them quite impractical for large-scale applications. As such, the direct discharge into seawater through proper diffusion mechanisms is still regarded as the most viable economic choice for RO technologies.

The cost of water produced by RO units can be categorized into three inter and intra-related components, namely: capital costs, energy costs and other operation costs. These are the driving force behind developing alternative technologies. Consequently, to reduce the total practical cost of RO units, progress in all three components should be made. Although recent works suggest that the theoretical minimum energy for water desalination can be approached, this seems more like wishful thinking rather than a practical approach. To illustrate this case, we should have a look at the principle operation of asymmetric RO membranes. It should be noted that for an effective desalination process, the diameter of the pores in the desalination membrane should be smaller than the hydration diameter of ions to effactually exclude them. Having larger pore diameters, typically results in a thick separation layer to guarantee acceptable salt rejection consequently leading to a reduced flux rate. Therefore, reaching the minimum theoretical energy required for the desalination process is impractical in real-world RO applications as the inherent energy losses from diffusion, viscous dissipations and other thermodynamic aspects play a huge factor governing the process. To overcome such an issue, however, hydrophobic membranes can be employed, in which the water flux can be more than three orders of magnitude higher than fluxes predicted by continuum hydrodynamic models [12–14]. Membranes fabricated from this method can exceed the salt rejection efficiency of currently used membranes with much higher flux (~4 times higher) [15]. However, the hydrophobic nature of such membranes make them prone to biofouling [3]. Moreover, the high cost of the manufacturing process combined with the difficulty in the functionalization of the pore entrance, required to reduce hydrophobicity and increased selectivity, make these membranes challenging to fabricate, to say the least. Even if such a membrane can be fabricated, the increase in the flux rate is eventually ruled by the concentration polarization layer at the membrane (Figure 12.2). This ultimately means that the theoretical efficiency of RO systems cannot be achieved necessitating the need for the development of new technologies for desalination as alternatives or in combination with RO.

12.3 Current State-of-the-Art Materials for Novel Membrane-Based Processes

To develop alternative solutions in water desalination processes requires significant advances in materials currently employed to achieve benchmark performance. The goal is, therefore, to improve on the positive aspects of the industry-standard materials or develop new materials of choice based on the required criteria while avoiding the negative effects [3]. The ultimate membrane should exhibit a set of specific characteristics including minimum thickness to allow for the highest flux possible, high mechanical strength to maximize its lifetime and well-defined pore sizes for increased selectivity [17–21].

To this aim, recently developed carbon-based materials can be regarded as potential materials of choice that might be used as an enabling platform for the development of membrane-based water desalination technology [22]. However, in order to achieve superb selectivity for various ions, controlled nanopores with precise functionalities should be introduced on the surface of the almost impermeable single-layer atom thick graphene [17, 23].

Graphene is defined as a two-dimensional material that consists of a hexagonal (i.e., honeycomb structure) lattice of covalently bonded sp^2 carbon atoms (Figure 12.3).

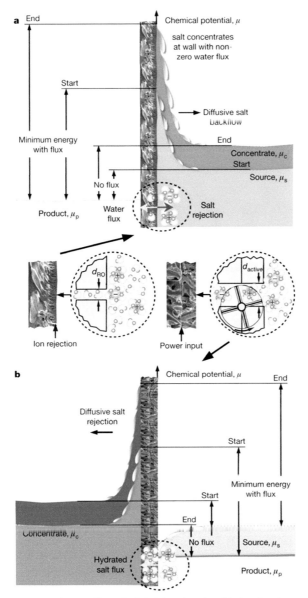

Figure 12.2 Concentration gradients in RO (a) and active (b) desalination membranes. The energy levels marked with 'start' and 'end' correspond to the evolution of each process. The darker blue color denotes higher concentration. Insets depict different mechanisms of salt ion separation. The active process with energy input shows a conceptual strategy for overcoming the Born barrier with fixed charges. Reproduced from Reference [3] with permission from Nature Publishing Group.

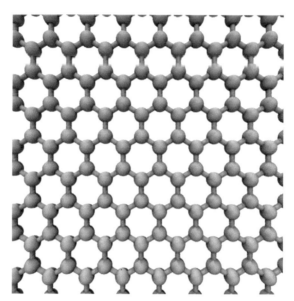

Figure 12.3 Crystal lattice of graphene [24].

Graphene has a large theoretical specific surface 2630 m^2g^{-1} and high electrical and thermal conductivity (ca. 5000 $Wm^{-1}K^{-1}$). Graphene in its free standing 2D form has been isolated in the past decade [25, 26].

Theoretical studies have suggested the possibility of the superior performance of graphene-based membranes to the current state-of-the-art polymer-based filtration membranes [25–30]. In contrast to polymeric membranes, in which the kinetics of water transport is slow, molecular dynamics studies show that the water permeability of graphene membranes is some orders of magnitude higher [27, 29]. This, combined with more than 99% salt rejection puts graphene at the forefront of research for developing new membrane technologies. At the same rejection rate, graphene membranes can provide water transport of 66 $L/cm^2 \cdot$ day $\cdot$ MPa in contrast to RO membranes that can only provide about 0.01–0.05 $L/cm^2 \cdot$ day $\cdot$ MPa [27, 31]. A schematic of water transport through a graphene membrane is depicted in Figure 12.4.

Following the lead of theoretical predictions, experimental research activities have begun to explore the possibility of the use of graphene monolayers as membranes with promising results [17, 32, 33]. Pore size tuning of graphene membranes for ion-selectivity can be performed employing ion bombardment and oxidative etching or plasma etching (Figure 12.5) [25, 32].

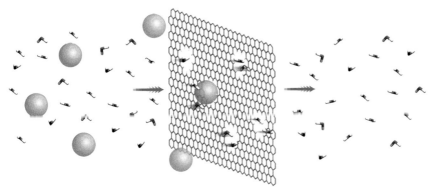

Figure 12.4 High pressure applied to the salt water (*left*) drives water molecules (red and white) across the graphene membrane (*right*), while salt ions (spheres) are blocked. Chemical functionalization of the pores with hydrogen (white) increases water selectivity, whereas functionalization with hydroxyl groups (not shown) increases the speed of water transport. Reproduced from Reference [27] with permission from Nature Publishing Group.

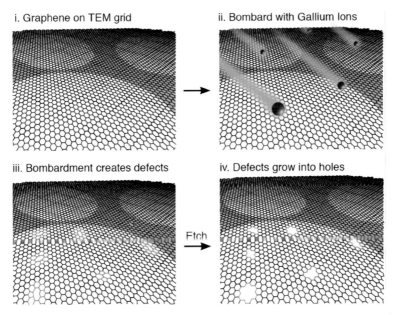

Figure 12.5 Process to create controlled pores in graphene membranes. Controlled sub-nanometer pores in graphene are created by ion bombardment followed by chemical oxidation. Ion bombardment generates reactive defect sites in the graphene lattice that preferentially etch during exposure to acidic potassium permanganate etchant. Reprinted with permission from Reference [32]. Copyright (2014) American Chemical Society.

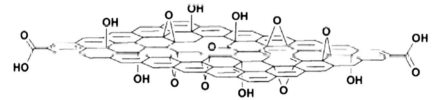

Figure 12.6   Graphene oxide structure [36].

However, the problem that should be addressed in all these systems is the cost-effective scalable manufacturing of large membranes with controllable pores and narrow size distribution while maintaining the structural integrity of the whole system; which is impractical with the current technology unless breakthroughs in the processing of graphene are achieved [27].

As such, current efforts are shifting towards using graphene oxide (GO) instead of graphene for practical applications [20–22, 34] (Figure 12.6).

The much lower scalable production cost of GO combined with its ease of processing into different shapes such as free-standing papers [18] fibers [35], etc. [37, 38] with sufficient mechanical strength can act as an enabling platform to process this material into different shapes desired for practical water desalination purposes based on their rheological properties (Figure 12.7) [22, 26, 38–40].

Although, the exact atomic structure of GO is still under debate, there are various oxygen functional groups that render GO a material of choice for these applications [19]. As an example, the presence of hydroxyl groups on the surface can increase water transport speed (Figure 12.2). Moreover, the structure of GO and its functional groups can be tuned and controlled based on the application, from a completely dense to a highly porous architecture (Figure 12.8).

Furthermore, the presence of multi-functionalities on the surface of GO can enable the easy decoration and hybridization of GO with different materials, resulting in an enhancement in ion selectivity and adsorption [19, 21, 22, 36, 40–47].

Although graphene-based membrane technologies are exciting for water desalination, the challenge of adding precise functional groups on the surface has served as a motivation to look for other candidates in the class of 2D materials that might act better than graphene. One of these recently explored candidates is the single atom-sheet molybdenum disulfide (MoS_2) [48, 49]. However, all the results are based on theoretical calculations and no experimental results are available, as of yet. Nevertheless, the theoretical calculations show that through mechanical induced stretching, the size of the nanopores in MoS$_2$ single layers can be adjusted leading to a biomimetic system with acceptable performance [49].

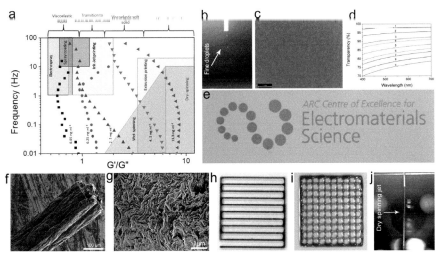

Figure 12.7 A correlation between rheological properties and the key prerequisites for various manufacturing techniques have enabled us to process and fabricate GO via a wide range of industrial techniques. **(a)** Ratio of elastic and storage moduli for various GO concentrations measured over a range of testing frequencies. Overlaid are the approximate processing regimes for a number of industrial fabrication techniques. When the viscous modulus (G'') dominates, the GO dispersion is suitable for high rate processing methods where the dispersion must spread on contact with the substrate. However, when the elastic modulus (G') is high the rheological properties suit fabrication methods requiring the dispersion to keep its given shape, such as extrusion printing and fiber spinning. **(b)** Photograph of electro-spraying of a viscoelastic liquid of a GO dispersion at a concentration of 0.05 mg ml^{-1}. **(c)** Photograph of a GO thin film that was spray coated and thermally reduced (overnight at 220°C) utilizing a transitional state to a viscoelastic liquid GO dispersion of 0.25 mg ml^{-1}. **(d)** Transparency of the spray coated reduced GO thin films as a function of coating layers; the numbers show the number of coating layers. **(e)** Ink-jet printed ACES logo using an LC GO viscoelastic soft solid at a concentration of 0.75 mg ml^{-1}. **(f)** As-prepared wet-spun fibers from an LC GO viscoelastic soft solid at a concentration of 2.5 mg ml^{-1}. **(g)** Cross-section of the wet-spun LC GO fiber, showing that GO sheets are stacked in layers with some degree of folding and are ordered due to the formation of nematic liquid crystals. **(h)** Extrusion printed pattern using an LC GO viscoelastic gel of 4.5 mg ml^{-1}. **(i)** Multi-level extrusion printed 3D architecture using an LC GO viscoelastic gel of 13.3 mg ml^{-1}. **(j)** Dry-spinning of LC GO fibers utilizing an LC GO viscoelastic gel of 13.3 mg ml^{-1}. Reproduced from Reference [38] with permission from Royal Society of Chemistry.

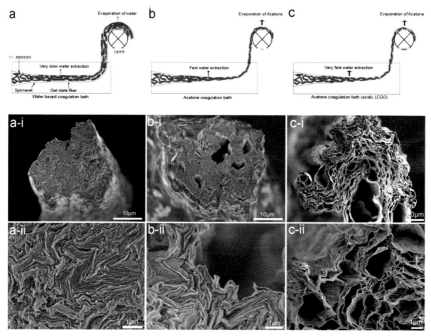

Figure 12.8 Proposed strategies for the evolution of structure in different coagulation baths from highly dense (left) to highly porous architectures (right). **(a)** Employing a water-based coagulation bath results in slow expulsion of water from the as-injected gel-state fiber-like structure. Therefore, to be able to pick up such fibers, the length of the bath should be optimized to enable the formation of a solid-like sheath around the core of the fiber. The fiber can then be taken from the bath and transferred on a spool for the evaporation of the water from the fiber, resulting in a highly dense structure **(a-i and a-ii)**. **(b)** Using an acetone coagulation bath results in the high rate of water extraction from the surface as a result of the difference in inhibition rate consequently leading to higher rate of solidification and porous fiber structure **(b-i and b-ii)**. **(c)** The slight acidic condition of LC GO dopants (pH ∼3) further changes the difference in inhibition rate, resulting in much higher water extraction rate and consequently more porous geometry **(c-i and c-ii)**. Reprinted with permission from Reference [20]. Copyright (2014) American Chemical Society.

12.4 Forward Osmosis (FO)

Forward osmosis[2] or direct osmosis is a separation/desalination method with remarkable and versatile industrial application in the field of water and wastewater treatment [50–52]. In FO, we exploit the osmotic pressure between two solutions of different concentrations. The solutions are separated

---

[2]From the Greek word osmos that means impel, push.

by a semi-permeable membrane. One solution is concentrated (draw) and the other is more dilute (feed). The difference of osmotic pressure forces water (without the salts) to permeate across the membrane from the dilute side to the concentrated side. After the permeation of water, it is separated from the draw solution to yield desalinated water. A mixture of ammonia and carbon dioxide has been used as the predominant draw solution which ensures a high pressure. An advantage of this draw solution is the ability to be regenerated by thermal energy [53].

A remarkable advantage of the FO process is the low proneness to fouling because the structure of the foulant layer formed is different from the layer in RO. The cake layer of RO is compact, whereas the FO cake is thicker but less compact. In some cases, high salinity waters have high fouling potential. FO is more effective in treatment of these difficult waters. These waters include landfill leachate, industrial waste water and flue gas desulphurization wastewater for which the concentrate is treated as hazardous waste and must be minimized.

The ability of FO as pre-treatment to remove dissolved constituents from the feedwater can also help conventional RO desalination to satisfy strict water quality requirements. Generally, FO is not proposed to replace RO, but rather to be used for feed waters that cannot be treated by RO efficiently.

12.5 Aquaporin Membranes

Aquaporins are protein channels that control water flux across biological membranes. They are found in human tissues with the purpose of rapid transport of water molecules across cell membranes. Due to osmotic gradients only water molecules are transported by diffusion while charges are rejected. Kumar et al. (2007) have used the protein Aquaporin-Z from *Escherichia coli* bacterial cells in a polymeric membrane [54].

Aquaporin membranes have shown promise for desalination processes. Due to the absence of applied pressure, it is expected that the energy consumption will be lower in comparison to RO membranes. However, there are not enough available reports of experimental data using real feed sources. Membranes based on aquaporins are not yet fully commercialized [55]. A short review article by Tang et al. (2013) has summarized the status and prospects of biomimetic aquaporin membranes concluding that: "based on their unique combination of offering high water permeability and high solute rejection, aquaporin proteins have attracted considerable interest over the last years as functional building blocks of biomimetic membranes for water desalination and reuse. Several design approaches have been pursued in facing the challenge of making the biomimetic membranes as stable, robust, scalable, and cost-effective as their polymeric counterparts in the form of existing technologies such as RO membranes. One type of approach aims at making ultrathin <10 nm supported films with incorporated aquaporins. While attractive in terms of flux potential this approach rests on the ability to form large defect free membranes—a technical challenge yet to be met— even at the square micron scale. Other designs use aquaporins stabilized in vesicular structures as a structural and functional element. Recent progress with this type of design, involving interfacial polymerization, has led to large (>400 cm^2) robust membranes (with lifetime of months) with water permeabilities >4 L m^{-2} h^{-1} bar^{-1} and salt rejection values >96%". [72]

12.6 Thermal-Based Processes

The most important conventional thermal distillation processes are multistage flash (MSF), multi-effect distillation (MED), and vapor compression (VC). In the past decades, modifications to thermal-based desalination processes have been studied expansively in order to increase the process efficiency.

Recent research and development have also focused on a combination of phase change and membrane utilization. Two methods of this category are discussed briefly here: **membrane distillation** and **pervaporation**. Other methods based on pure thermal principles have also been developed aiming at reducing the energy consumption. This category includes technologies such as **humidification–dehumidification** and **adsorption distillation**.

12.6.1 Membrane Distillation

Membrane distillation (MD) is a thermal-based desalination process that is being considered as a possible alternative to conventional desalination technologies such as MSF and RO [56, 57]. In MD water vapor molecules evaporate from the feed solution and are transported through the pores of hydrophobic membranes as distillate (the size of the pores is 0.1–1μm). The MD membranes are fabricated mainly from polymeric materials such as PTFE (polytetrafluoroethylene), PP (polypropylene), PE (polyethylene) and PVDF (polyvinylidene fluoride). The driving force in MD is the vapor pressure difference across the membrane. The operating temperature of MD is low, ranging from 50–90°C and the operating pressures are lower than conventional pressure—driven membrane processes such as RO.

Moreover, MD has the theoretical ability to achieve 100% salt rejection. The low operating temperature permits the utilization of low grade heat such as conventional solar energy, geothermal and waste heat. When energy sources of this type are available, MD might be competitive with RO. A number of configurations have been reported.

In MD, the role of the membrane is crucial since the selected membrane must satisfy strict requirements: to have suitable porosity, to be hydrophobic (should not be wetted by liquids), to allow only the vapour to transport through the pores, to have no capillary condensation inside the pores etc. It has been observed that wetting of the membrane surface during prolonged use leads to accelerated deposition of organics. Wetting and fouling of the membrane require pre-treatment of the feed water source [58].

Research regarding MD has been growing exponentially in recent years. Meanwhile, pilot plant demonstrations are also underway. In the past couple of years, several review papers on MD have been published. Three recommended reviews published within the past couple of years, each covering an aspect of MD desalination, are as follows:

- A review on fouling of membrane distillation
  Naidu, G., Jeong, S., Vigneswaran, S., Hwang, T. M., Choi, Y. J., and Kim, S. H. (2016).
  Desalination and Water Treatment 57 (22), 10052–10076.
- Critical review of membrane distillation performance criteria
  Luo, A., and Lior, N. (2016).
  Desalination and Water Treatment 57 (43), 20093–20140.
- Membrane synthesis for membrane distillation: A review
  Eykensa, L., De Sittera, K., Dotremonta, C., Pinoyc, L., and Van der Bruggen, B. (2017).
  Separation and Purification Technology 182, 36–51.

12.6.2 Pervaporation (PV)

A well-known application of pervaporation is the separation of liquid mixtures like dehydration of organic solvents and evaporation of volatile compounds from aqueous solutions (commercialized process). Nowadays, pervaporation is also being considered as an attractive and promising method for water desalination although is not yet commercialized. In PV, generally certain components of the feed solution permeate preferentially through a dense porous membrane acting as a molecular sieve and evaporate downstream as shown in Figure 12.9.

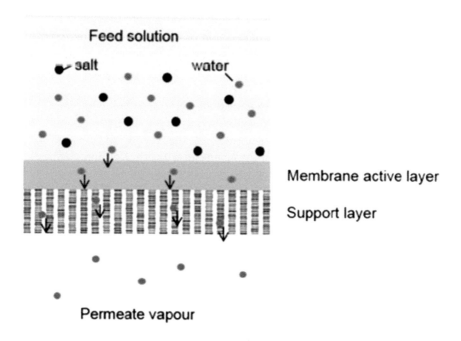

Figure 12.9   Desalination by the pervaporation process as water passes through a dense pervaporation membrane [59].

The driving force in PV for the mass transfer of permeate is the chemical potential gradient between the feed side and the permeate side of the membrane. For desalination purposes, PV includes a combination of diffusion of water through a membrane and its evaporation into a vapor phase (freshwater) on the other side. PV and MD are often confused because they both have a membrane with an upstream side in contact with hot feed liquid, and a freshwater permeate vapor that emerges on the downstream side of the membrane. However, the two processes have at least three distinct differences [59], the most fundamental of which is as follows: the membrane employed in the two processes is inherently different. In MD, the membrane does nothing more than act as a support medium for the vapor–liquid interface and does not actually contribute to the separation performance. However, in PV, a dense molecular sieve membrane is required with high selectivity to allow only certain components in the feed solution to be transported through. In other words, in PV, the membrane functions as a selective barrier for separation between the liquid feed and vapor phase. Under similar conditions,

MD has significantly higher fluxes than PV, whereas the selectivity of PV is considerably higher.

Many kinds of membranes can be used for PV such as polymer-based (sulfonated polyethylene, cellulose acetate), inorganic (zeolite, amorphous microporous silica), hybrid materials, etc. The main advantage of PV over RO is the ability to handle a wide feed concentrate range in contrast to RO (because the phase change process need not overcome the osmotic pressure of the feed). Bench-scale or pilot-scale studies have been performed however a lot of research should be done to improve the efficiency of the PV process before it will become commercially attractive [59]. PV has displayed very high salt rejection, generally above 99.9%. Favourably, the salt rejection can remain more or less independent of any variation in operating conditions.

12.7 Novel Electrically-Driven Processes

12.7.1 Capacitive Deionization (CDI)

Capacitive deionization is an alternative desalination method being developed, probably more efficient for low-salinity feed water sources (total dissolved solids, TDS <15,000 mg/L). In CDI technology, a saline solution flows through an unrestricted capacitor-type module that consists of many electrodes having high-surface areas. In general, a carbon aerogel of high specific surface area (400–1100 m^2/g) and a very low resistivity <40 mΩcm is utilized. A power source creates an electric field that provokes adsorption of anions and cations of the solution on each electrode. After adsorption of ions, the saturated electrode is regenerated by desorption of the ions under zero potential or reverse electric field (potential reversal). Consequently, the absorption ability of the electrode is the most important parameter of this method [60].

The main advantages of CDI include: the absence of external pressure, high rejection of salts, and the possibility of electrode self cleaning by polarity reversal. However limited data regarding desalination of seawater using CDI is available. Several modifications to CDI have been reported in order to increase efficiency, including:

1. M-CDI (membrane-CDI)
   Recently a modification to CDI in which ion–exchange membranes are utilized for selective transport of ions to the electrodes has been proposed [61]. This innovation has led to higher efficiency and lower energy consumption.

2. Flow through electrode CD

   In this case the feed water flows through the electrodes, instead of flowing between them. And instead of carbon aerogels, a new material called hierarchical carbon aerogel monoliths (HCAMs) is used as the electrode. The porous carbon permits less separation between electrodes leading to better performance of the system. This method is of course faster than conventional CDI, is more energy efficient and is operated without membrane components. Also, it has been reported that the constant current (CC) operation mode consumes much less energy than the constant voltage (CV) mode [62].

3. ED (electrodialysis) combined with CDI

   Another system has been proposed utilizing ED (electro dialysis) combined with CDI [63]. The system includes a three-step process: the water flows in a cell with positively and negatively charged electrodes. The electrode surfaces are covered with ion–selective membranes, and ions in the feed water are attracted to the opposite charged electrodes, passing through the selective membrane and finally accumulating on the electrodes in the third step, due to its porous structure. When the electrodes have been saturated, their polarity is reversed (potential reversal).

12.7.2 Microbial Desalination Cell (MDC)

Since 1972 fuel cells based on biological activity have been demonstrated for electricity production. The first direct biochemical fuel cell is the well-known sulfate reduction cell. A smooth platinum sheet covered with species of *Desulfo vibrio desulfuricans* can serve as the biocathode in the microbial fuel cell (MFC). Seawater is the electrolyte that contains *Al*gae for nutrition and sulfate anions for metabolism. A magnesium sheet serves as a counter electrode. On the biocathode the following reaction takes place consuming electrons:

$$SO_4{}^{2-} + 4H_2O + 8e^- \xrightarrow{\text{bacteria}} S^{2-} + 8OH^-.$$

A potential difference develops between the cathode and anode that can generate electricity. In the same way, when the bacterial metabolism liberates electrons the electrode acts as the bioanode [64].

It is noteworthy to mention that the above reaction explains the biological corrosion[3] of iron materials in the soil due to the presence of sulfate reducing

---

[3]Biocorrosion, microbiologically induced corrosion (MIC).

bacteria using sulphate anions for their metabolism. Under such conditions, the corrosion rate (anodic reaction) accelerates [65]. It is known that the corrosion phenomenon can be interpreted as a short-circuited galvanic or corrosion cell [66].

The recently developed microbial desalination cells (MDCs) are based on the same principle and can be considered as an extension of the microbial fuel cell [67, 68]. An MDC can treat wastewater containing bacteria with simultaneous production of electricity [69].

As shown in Figure 12.10, an MDC consists of an anode and cathode chambers and in the middle a chamber of desalination constructed by an anion-exchange membrane (AEM) and a cation-exchange membrane (CEM) on either side. An external wire is used for closing the circuit. The middle chamber is filled with the water to be desalinated. In the anode chamber, oxidation (degradation) of organic matter by bacteria producing CO_2 and H^+ takes place. In the cathode chamber, usually O_2 uses the electrons from the anode undergoing reduction and producing water. Consequently, a potential difference develops between the cathode and anode. In the middle chamber, the anions (such as Cl^- and SO_4^{2-}) migrate from the saltwater across the AEM into the anode, while the cations (such as Na^+ and Ca^+) move across the CEM into the cathode chamber. This electrochemical mechanism can

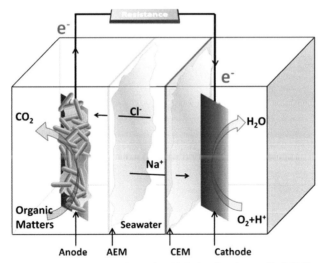

Figure 12.10 A typical scheme of a microbial desalination cell (MDC). AEM: anion exchange membrane; CEM: cation exchange membrane [68].

remove more than 99% of the salt from the saline water and simultaneously produce enough energy for the operation of the system.

The crucial parameter is the formation of the biofilm by accumulation of bacteria adhered to the anode. It is noteworthy that biofilms are created in the human body as well for example sulfate reducing bacteria in the oral cavity [70].

As stated, due to electricity production of the microbial cells, an external energy source is absent and therefore energy consumption is minimal. This may be the biggest advantage of the MDC. The desalination performance of conventional MDC can be increased by using multiple pairs of ion-exchange membranes (IEMs) inserted between the anode and cathode chambers as shown in Figure 12.11. This is called the stack structure MDC. Stacked MDCs achieve more efficient desalination when compared to single chamber MDCs [71] due to improvement of charge transfer.

In a conventional MDCs, the different chambers are separated by ion exchange membranes. In the anode, the oxidation of the organic material releases protons (H^+) which are not allowed to diffuse to the cathode where hydroxyls (OH^-) are produced through the reduction reaction according to the following equations:

$$n(CH_2O) + n(H_2O) \ nCO_2 + 4ne^- + 4NH^+(\text{anode})$$

$$O_2 + 2H_2O + 4e^- + 4OH^- \ (\text{cathode})$$

This may create an imbalance of pH inside the cell because pH increases in the anode and decreases in the cathode. The pH changes can negatively

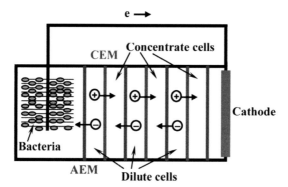

Figure 12.11   Stack structure microbial desalination cell (MDC) modified from reference [71].

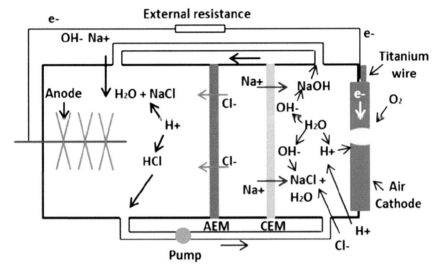

Figure 12.12 Recirculation microbial desalination cell (r –MDC) [69].

influence the metabolism of the bacteria. In order to tackle this problem a recirculation MDC has been developed as shown in Figure 12.12.

It is reported that the main advantage of the r-MDC is the increase of desalination efficiency [69]. In general, microbial desalination technology is still at lab-scale and pilot-scale evaluation. Intensive research is necessary in order to elucidate the details.

Acknowledgements

The authors would like to thank Professor Dr rer.nat.TUM D.K.Yfantis for his helpful suggestions.

Assessment of New Desalination Technologies Using the Coefficient of Desalination Reality (CDR)

Tom Scotney, Deputy Editor, Global Water Intelligence, UK

Although the desalination market may be dominated by a few established technologies, a diverse and changing landscape of technological innovation is constantly at work. Below a variety of emerging technologies in desalination that seek to address desalination's biggest obstacles, such as energy use, volume reduction, and membrane improvements, among others are outlined. Bridging the gap between the lab and the field is a notoriously difficult process, particularly in the water market. While a huge amount of innovation goes on at a technological level, it can often be decades before even a tried and tested technology becomes standard for installations. The slowness of adoption is a result of a number of factors, including:

- The difficulty of replicating field conditions—which often include low or variable water quality—in the lab.
- Problems of scale that often emerge when looking to take a test-level process up to a working plant size.
- Conservative attitudes to new technology in a sector with a high level of exposure to public opinion.
- The fact that water utilities—which are often supported by government subsidy instead of income from their customers—have less incentive to reduce their operating costs.

This makes it very difficult to assess the likelihood of success for new desalination technologies going forward, even if they offer strong benefits on paper.

Table 12.1   Assessment of new desalination technologies by CDR rating

Technology	Developer	CDR
Cavitation with coagulation	CaviGulation	1
Electromagnetic frequency technology	Paramount discoveries	1
Novel multi-stage flash technology	Centriforce	1
Magnetic desal	Blue Fin Water	1.4
Biomimetic ultra-thin membrane	AguaVia	2
Boron nitride nanotube technology	Australian National University	3
Sonic desalination	Globe Protech	3.3
AirDrop irrigation desalination	Swinburne University of Technology	3.4
Aquaporin nano-polymer membrane vesicles	Danfoss AquaZ	3.5
Microbial desalination fuel cell	Penn state/Tsinghua University/ KAUST	3.5
Multi-effect drying and condensation	3MW	3.5
Graphene-based membrane permeation	University of Manchester	3.7

Carbon nanotubes	Porifera/NanOasis	3.9
Decanoic acid solvent extraction process	MIT	3.9
Radial de-ionization	Atlantis Technologies	3.9
Solvent-extraction desalination	MIT	3.9
Autonomous desalination module	Aquamodule	4
Biomimetic aquaporins	Aquaporin A/S	4
Solar sponge	MIT	4.3
Locally powered water distillation system	Dean Kamen	4.5
Carbon nanotube membrane distillation	New Jersey Institute of Technology	4.6
Membrane distillation + vapor compression	Dais Analytic	4.8
Capacitive deionization	Atlantis	4.9
Graphene membrane	Lockheed	4.9
Clathrate separation	Aqueous Logic	5
Desal with deep sea pressure differentials	Econopure	5
Electrochemical desalination	Evoqua Water Technologies	5.2
Microbial desalination	Microbial Desal Cell	5.2
Microbial electrolysis and desalination cell	University of Colorado	5.2
Upflow microbial desalination fuel cell	University of Wisconsin–Milwaukee	5.2
High-res. Electrical impedance spectroscopy	INPHAZE	5.4
Multifunctional TiO_2 nanocomposite membrane	NanoSun/Nanyang Technological University	5.6
DNA UF membrane	Cerahelix	5.9
Membranes without feed spacer mesh	Aqua Membranes	6.3
Hydrophilic graft of RO membranes	Samueli School, UCLA	6.4
Ion Concentration polarization	MIT and Pohang University of Science and Technology	6.5
Membrane distillation with solar pond	University of Nevada-Reno and Col. Sch. Of Mines	6.5
Vacuum-based evaporative process	New Mexico state university/ Sterling Water	6.5
Electrochemically mediated desalination	Okeanos Technologies	6.6
FO thermolytic draw solution	Trevi Systems	6.6

(Continued)

Table 12.1 Continued

Technology	Developer	CDR
Brine squeezer (High recovery RO)	Osmoflo	6.9
Low-temperature evaporator	Hittite	7
Humidification-dehumidification process	MIT/King Fahd University of Petroleum	7.1
Dissolved air filtration and microfiltration	Technical University of Berlin	7.4
Pervaporation for subsurface irrigation	DTI-r	7.5
UFO-MBR	Colorado School of Mines	7.5
Nonmetallic tubes	Technoform	7.8
Indirect fired steam generation	Hipvap	7.9
Zero discharge desalination	University of Texas-El Paso	7.9
Polymeric membrane (PolyCera)	Water Planet Engineering	8.3
Multi-stage flash fluidized bed evaporator	Klaren BV	8.4
Chemical crystallizer	Advanced Water Recovery	8.6
Ion exchange with bridges and solar ponds	Saltworks	8.6
Vacuum multi-effect membrane distillation	Memsys clearwater	8.7
Membrane capacitive deionization	Voltea	8.8
Wind-aided intensified evaporation	Ben Gurion University	8.8
Manipulated osmosis (forward osmosis)	Modern Water	8.9
Nano-engineered composite membranes	Nano H_2O	8.9
Anti-biofilm selenium compound	Selenium, ltd	9.1
Closed-circuit desalination	Desalitech	Commercialized
Dewvaporation	Altela	Commercialized
Flat sheet UF	Microdyn-Nadir	Commercialized
FO membranes (particularly for pre-treatment)	Porifera	Commercialized
Humidification-Dehumidification	Gradiant	Commercialized
Continuous Ion exchange	Clean TeQ	Commercialized
Membrane distillation	Memsys	Commercialized
MVC crystallizer	SaltTech	Commercialized
Nano composite membrane	NanoH$_2$O (LG Chem)	Commercialized
Package plant	Water Planet Engineering	Commercialized
SDI/MFI monitor	Convergence	Commercialized
Thermolytic FO	Oasys	Commercialized
Fibracast UF membrane	Anergia	Commercialized

Source: GWI, 2016.

One gauge of a technology's potential is the Coefficient of Desalination Reality (CDR). Developed by the Water Desalination Report (WDR) publication, it serves as an indicator of WDR's confidence that a product or process is based on sound scientific principles, is commercially viable in the market for which it is intended, and will be available within a reasonable time frame.

It is determined using a 1- to 10-scale that considers three separate aspects of a particular process:

- Laws of Nature Assessment: Gauges a process' apparent adherence to the basic laws of physics.
- Commercial Assessment: Predicts the commercial and competitive viability of the technology.
- Deployment Assessment: Assesses the plans and schedule for commercialization.

The sum of the three scores is averaged, and false, misleading or ridiculous claims made about competing technologies by the technology's proponents may result in point reductions.

A 1.0 is the lowest possible score and indicates a high degree of skepticism based on the available information. It does not necessarily mean a process is without merit, however, and may only indicate there is not enough information for a proper evaluation. A perfect score of 10.0 indicates that a technology has been successfully commercialized.

Admittedly, a CDR is somewhat subjective and it should not be considered to be WDR's final assessment of a process' viability. As new information is made available for review, the CDR may be revised.

References

[1] Lafforgue, M., and Lenouvel, V. (2015). Closing the urban water loop: lessons from Singapore and Windhoek. *Environ. Sci.* 1, 622–631.
[2] Radcliffe, J. C. (2015). Water recycling in Australia—during and after the drought. *Environ. Sci.* 1, 554–562.
[3] Shannon, M. A., et al. (2008). Science and technology for water purification in the coming decades. *Nature* 452, 301–310.
[4] Montgomery, M. A., and Elimelech, M. (2007). Water And sanitation in developing countries: Including health in the equation. *Environ. Sci. Technol.* 41, 17–24.

[5] Subramani, A., and Jacangelo, J. G. (2015). Emerging desalination technologies for water treatment: a critical review. *Water Res.* 75, 164–187.

[6] Kim, S. J., et al. (2010). Direct seawater desalination by ion concentration polarization. *Nat. Nano* 5, 297–301.

[7] Qiu, J. (2011). China to spend billions cleaning up groundwater. *Science* 334, 745–745.

[8] Fath, H. E. S. (1998). Solar distillation: a promising alternative for water provision with free energy, simple technology and a clean environment. *Desalination* 116, 45–56.

[9] Bourouni, K., Chaibi, M. T., and Tadrist, L. (2001). Water desalination by humidification and dehumidification of air: State of the art. *Desalination* 137, 167–176.

[10] Yang, Y., et al. (2014). An innovative beneficial reuse of seawater concentrate using bipolar membrane electrodialysis. *J. Membr. Sci.* 449, 119–126.

[11] Pérez-González, A., et al. (2012). State of the art and review on the treatment technologies of water reverse osmosis concentrates. *Water Res.* 46, 267–283.

[12] Hummer, G., Rasaiah, J. C., and Noworyta, J. P. (2001). Water conduction through the hydrophobic channel of a carbon nanotube. *Nature* 414, 188–190.

[13] Hinds, B. J., et al. (2004). Aligned Multiwalled Carbon Nanotube Membranes. *Science* 303, 62–65.

[14] Holt, J. K., et al. (2006). Fast mass transport through sub-2-nanometer carbon nanotubes. *Science* 312, 1034–1037.

[15] Fornasiero, F., et al. (2008). Ion exclusion by sub-2-nm carbon nanotube pores. *Proc. Natl. Acad. Sci. U.S.A.* 105, 17250–17255.

[16] Li, W.-W., Yu, H.-Q., and He, Z. (2014). Towards sustainable wastewater treatment by using microbial fuel cells-centered technologies. *Energy Environ. Sci.* 7, 911–924.

[17] Koenig, S. P., et al. (2012). Selective molecular sieving through porous graphene. *Nat. Nano* 7, 728–732.

[18] Aboutalebi, S. H., et al. (2011). Spontaneous formation of liquid crystals in ultralarge graphene oxide dispersions. *Adv. Funct. Mater.* 21, 2978–2988.

[19] Aboutalebi, S. H., et al. (2012). Enhanced hydrogen storage in graphene oxide-MWCNTs composite at room temperature. *Adv. Energy Mater.* 2, 1439–1446.

[20] Aboutalebi, S. H., et al. (2014). High-performance multifunctional graphene yarns: toward wearable all-carbon energy storage textiles. *ACS Nano* 8, 2456–2466.

[21] Aboutalebi, S. H., et al. (2011). Comparison of GO, GO/MWCNTs composite and MWCNTs as potential electrode materials for supercapacitors. *Energy Environ. Sci.* 4, 1855–1865.

[22] Sun, P., et al. (2015). Highly efficient quasi-static water desalination using monolayer graphene oxide/titania hybrid laminates. *NPG Asia Mater.* 7: e162.

[23] Bunch, J. S., et al. (2008). Impermeable atomic membranes from graphene sheets. *Nano Lett.* 8, 2458–2462.

[24] Cohen-Tanugi, D., and Grossman, J. C. (2015). Nanoporous graphene as a reverse osmosis membrane: recent insights from theory and simulation. *Desalination* 366, 59–70.

[25] Surwade, S. P., et al. (2015). Water desalination using nanoporous single-layer graphene. *Nat Nano* 10, 459–464.

[26] Nair, R. R., et al. (2012). Unimpeded permeation of water through helium-leak–tight graphene-based membranes. *Science* 335, 442–444.

[27] Wang, E. N., and Karnik, R. (2012). Water desalination: graphene cleans up water. *Nat Nano* 7, 552–554.

[28] Cicero, G., et al. (2008). Water confined in nanotubes and between graphene sheets: a first principle study. *J. Am. Chem. Soc.* 130, 1871–1878.

[29] Cohen-Tanugi, D., and Grossman, J. C. (2012). Water desalination across nanoporous graphene. *Nano Lett.* 12, 3602–3608.

[30] Lin, L.-C., and Grossman, J. C. (2015). Atomistic understandings of reduced graphene oxide as an ultrathin-film nanoporous membrane for separations. *Nat. Commun.* 6: 8335.

[31] Lee, K. P., Arnot, T. C., and Mattia, D. (2011). A review of reverse osmosis membrane materials for desalination—Development to date and future potential. *J. Membr. Sci.* 370, 1–22.

[32] O'Hern, S. C., et al. (2014). Selective ionic transport through tunable subnanometer pores in single-layer graphene membranes. *Nano Lett.* 14, 1234–1241.

[33] O'Hern, S. C., et al. (2012). Selective molecular transport through intrinsic defects in a single layer of CVD graphene. *ACS Nano* 6, 10130–10138.

[34] Yousefi, N., et al. (2012). Self-alignment and high electrical conductivity of ultralarge graphene oxide-polyurethane nanocomposites. *J. Mater. Chem.* 22, 12709–12717.

[35] Jalili, R., et al. (2013). Scalable one-step wet-spinning of graphene fibers and yarns from liquid crystalline dispersions of graphene oxide: towards multifunctional textiles. *Adv. Funct. Mater.* 23, 5345–5354.

[36] Mahmoud, K. et al. (2015). Functional graphene nanosheets: The next generation membranes for water desalination. *Desalination* 356, 208–225.

[37] Chidembo, A., et al. (2012). Globular reduced graphene oxide-metal oxide structures for energy storage applications. *Energy Environ. Sci.* 5, 5236–5240.

[38] Naficy, S., et al. (2014). Graphene oxide dispersions: tuning rheology to enable fabrication. *Mater. Horiz.* 1, 326–331.

[39] Jalili, R., et al. (2013). Organic solvent-based graphene oxide liquid crystals: a facile route toward the next generation of self-assembled layer-by-layer multifunctional 3D architectures. *ACS Nano* 7, 3981–3990.

[40] Aboutalebi, S. H. (2014). *Processing Graphene Oxide and Carbon Nanotubes: Routes to Self-Assembly of Designed Architectures for Energy Storage Applications.* Doctor of Philosophy thesis, University of Wollongong, Wollongong, NSW.

[41] Islam, M. M., et al. (2016). Liquid-crystal-mediated self-assembly of porous α-Fe2O3 nanorods on PEDOT: PSS-functionalized graphene as a flexible ternary architecture for capacitive energy storage. *Part. Part. Syst. Charact.* 33, 27–37.

[42] Islam, M. M., et al. (2015). Self-assembled multifunctional hybrids: toward developing high-performance graphene-based architectures for energy storage devices. *ACS Cent. Sci.* 1, 206–216.

[43] Chidembo, A. T., et al. (2014). Liquid crystalline dispersions of graphene-oxide-based hybrids: a practical approach towards the next generation of 3D isotropic architectures for energy storage applications. *Part. Part. Syst. Charact.* 31, 465–473.

[44] Chidembo, A. T., et al. (2014). *In situ* engineering of urchin-like reduced graphene oxide-Mn2O3-Mn3O4 nanostructures for supercapacitors. *RSC Adv.* 4, 886–892.

[45] Islam, M., et al. (2014). Liquid crystalline graphene oxide/PEDOT: PSS self-assembled 3D architecture for binder-free supercapacitor electrodes. *Front. Energy Res.* 2.

[46] Gudarzi, M. M., et al. (2011). Self-aligned graphene sheets-polyurethane nanocomposites. *MRS Proceed.* 1344:mrss11-1344-y02-06.

[47] Mustapić, M., et al. (2015). Improvements in the dispersion of nanosilver in a MgB2 matrix through a graphene oxide net. *J. Phys. Chem. C* 119, 10631–10640.

[48] Heiranian, M., Farimani, A. B., and Aluru, N. R. (2015). Water desalination with a single-layer MoS2 nanopore. *Nat. Commun.* 6. 8616.

[49] Li, W., et al. (2016). Tunable, strain-controlled nanoporous MoS2 filter for water desalination. *ACS Nano* 10, 1829–1835.

[50] Shaffer, D. L., et al. (2015). Forward osmosis: Where are we now? *Desalination* 356, 271–284.

[51] Klaysom, C., et al. (2013). Forward and pressure retarded osmosis: potential solutions for global challenges in energy and water supply. *Chem. Soc. Rev.* 42, 6959–6989.

[52] Cath, T. Y., Childress, A. E., and Elimelech, M. (2006). Forward osmosis: Principles, applications, and recent developments. *J. Membr. Sci.* 281, 70–87.

[53] McCutcheon, J. R., et al. (2005). A novel ammonia-carbondioxide forward (direct) osmosis desalination process. *Desalination* 174, 1–11.

[54] Kumar, et al. (2007). Highly permeable polymeric membranes based on the incorporation of the functional water channel protein Aquaporin Z. *Proc. Natl. Acad. Sci. U.S.A.* 104, 20719–20724.

[55] Subramani, A., Jacangelo, J. (2015). Emerging desalination technologies for water treatment : a critical review. *Water Res.* 75, 164–187.

[56] Yfantis, N. D. (2011). *Design and operation of desalination units by reverse osmosis—Case studies*. Dissertation, National Technical University of Athens, Athens.

[57] Yfantis, N. D., and Yfantis, A. D. (2009). "Design of a UF-RO combined system for wastewater reuse in Crete – A case study," in *Proceedings of the 11th International Conference on Environmental Science and Technology – CEST 2009, Chania 3–5 September*, Crete, 1569–1578.

[58] Warsinger, D. M., et al. (2015). Scaling and fouling in membrane distillation for desalination applications: a review. *Desalination* 356, 215–244.

[59] Wang, Q., et al. (2016). Desalination by pervaporation: a review. *Desalination* 387, 46 60.

[60] Seo, S. H., et al. (2012). Investigation on removal of hardness ions by capacitive deionization (CDI) for water softening applications. *Water Res.* 44, 2267–2275.

[61] Kim, Y. J., Choi, J. H. (2010). Enhanced desalination efficiency in capacitive deionization with an ion selective membrane. *Sep. Purif. Technol.* 71, 70–75.

[62] Qu, Y. et al. (2016). Energy consumption analysis of constant voltage and constant current operations in capacitive deionization. *Desalination* 400, 18–24.

[63] Voltea (2016). *Capacitive Deionization.* Available at: http://voltea.com/ [accessed November 5, 2016].

[64] Ebert, H. (1972). *Elektrochemie (Electrochemistry).* German: VOGEL VERLAG.

[65] Yfantis, D. K., Yfantis, A. D., and Anastassopoulou, I. (1998). Biological corrosion of metallic parts in an underground irrigation system – a study of alternative materials. *Br. Corros. J.* 33, 237–240.

[66] Yfantis, D. K. (2008). *Corrosion and Protection of Materials.* Athens: National Technical University of Athens.

[67] Mercer, J. (2014). *Microbial microbial fuel Cells: Generating Power from Waste.* Available at: illumin.usc.edu [accessed April 26, 2014].

[68] Ping, Q., et al. (2013). Long term investigation of fouling of cation and anion exchange membranes in microbial desalination cells. *Desalination* 325, 48–55.

[69] Saeed H. M., et al. (2015). Microbial desalination cell technology: a review and a case study. *Desalination* 359, 1–5.

[70] Anastassopoulou, I., Theofanides, T., Yfantis, D., Yfantis, K. (2016). *Biomaterials-Applications.* Available at: www.kalippos.gr

[71] Kim, Y., and Logan, B. E. (2013). Microbial desalination cells for energy production and desalination. *Desalination* 308, 122–130.

[72] Tang, C. Y., Zha, Y., Wang, R., Hélix-Nielsen, C., and Fane, A. G. (2013). Desalination by biomimetic aquaporin membranes: Review of status and prospects. *Desalination* 308, 34–40.

PART IV

Energy and Environment

CHAPTER 13

Desalination Powered by Renewable and Nuclear Energy Sources

Veera Gnaneswar Gude[1]

[1]Department of Civil and Environmental Engineering,
Mississippi State University, Mississippi State, MS 39762, USA
E-mail: gude@cee.msstate.edu

13.1 Desalination Technologies and Renewable Energy Coupling Schemes

Currently available desalination technologies can be categorized as phase change (thermal) and non-phase change (membrane) processes. Phase change processes involve heating the feed water (seawater or brackish water) to "boiling point" at the operating pressure to produce steam which is condensed in a condenser unit to produce freshwater. Desalination technologies based on this principle include solar distillation (SD) such as solar stills and active and passive solar desalination systems; multi-effect evaporation/distillation (MED); multi-stage flash distillation (MSF); thermal vapor compression (TVC) and mechanical vapor compression (MVC). Non-phase change processes involve separation of dissolved salts from the feed waters by mechanical or chemical/electrical means using a membrane separator between the feed (seawater or brackish water) and product (potable water). Desalination technologies based on this principle include electrodialysis (ED) and reverse osmosis (RO). Other processes that involve a combination of the two principles in a single unit or in sequential steps to produce pure or potable water also exist and have been reviewed elsewhere in the book (see Chapters 3, 5, 6, and 12).

Various renewable energy sources are available to provide for energy needs in desalination processes [1]. These include solar energy, wind energy, geothermal energy, wave energy and nuclear energy. Solar energy can be harvested as heat or electricity or both while wind energy can be used to produce electricity. Geothermal energy source can be used to produce both process heat and electricity that could be used for desalination.

Wave energy can also be used to produce electricity for use in desalination. Potential renewable energy-desalination technology combinations are shown in Figure 13.1. The renewable energy sources (RES) should be integrated with the relevant desalination technology that has the capability to utilize the supplied energy in the most effective manner. Some renewable energy source dependent desalination technologies must be placed on the same site (co-location) while some do not have this requirement. Accordingly, the following thermal desalination-renewable energy combinations require co-location (located on same site): (i) wind-shaft-MVC; (ii) solar thermal-TVC; (iii) solar thermal-MSF; (iv) solar thermal-MED; (v) solar thermal-SD; (vi) geothermal-TVC; (vii) geothermal-MSF or MED. The other electricity-driven combinations that do not require co-location are: (a) Wind-MVC; (b) Wind-RO; (c) Solar PV-RO; (d) Solar PV-MVC; (e) Geothermal-MVC; and (f) Geothermal-RO.

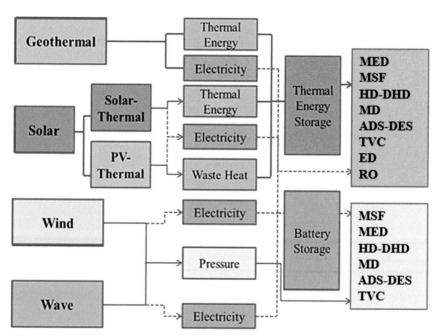

Figure 13.1 Desalination technologies coupled with renewable energy and storage systems (PV-Thermal, photovoltaic thermal; MED, multi-effect distillation; MSF, multi-stage flash distillation; HD–DHD, humidification-dehumidification; MD, membrane distillation; ADS-DES, adsorption desalination; TVC, thermal vapor compression; ED, electrodialysis; RO, reverse osmosis (adapted from [1]).

13.2 Global Overview

Availability of renewable energy sources varies around different parts of the world. Some water scarce regions are rich in solar energy sources, while others have ample wind energy and geothermal sources, or are situated along coastal lines. Figure 13.2 shows desalination technology and renewable energy combinations used in different installations around the world [2]. PV-RO has been the most commonly used technology due to its inherent simplicity and ease of operation and elegant nature. Its share is about 31% followed by other combinations of wind-RO (12%), solar MD (11%), solar MED (9%), solar MEH (multi-effect humidification and dehumidification) (9%), and solar MSF (7%) and other combinations. Photovoltaic (PV) energy can be used to power electrodialysis and electrodialysis-reversal and hybrid combinations including membrane and thermal processes. Solar stills are included under "others" which are known as the simplest small-scale technology used broadly worldwide. Figure 13.3 shows the development stage and capacity range for various renewable energy driven desalination technologies. Some of the renewable energy desalination schemes have not yet been well-studied. It can be noted that small and novel applications for geothermal

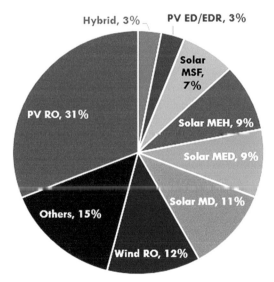

Figure 13.2 Renewable energy utilization for desalination by relative share of all renewable energy driven desalination units around the world (data taken from [2]). It should be noted that overall, all renewable energy driven desalination units account for a tiny percentage of the total world desalination capacity.

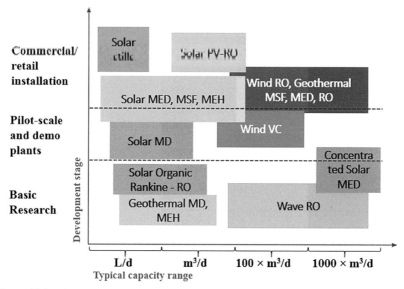

Figure 13.3 Comparison of different renewable-energy-driven desalination technologies.

and wave energy technologies are in the basic research phase while the solar energy based desalination schemes have passed this stage and are now industrialized. Wind and geothermal energy sources have the potential for large scale applications. Desalination processes such as membrane distillation can also utilize solar and geothermal energy.

13.3 Solar Energy for Desalination

Water scarce regions around the world are commonly enriched with abundant solar insolation throughout the year. Solar energy can be harvested as either thermal energy or electrical energy to support desalination processes. Solar desalination has a history of several centuries dating back to the work by Muslim alchemists [3] and possibly even earlier. A solar distillation apparatus by Della Porta was available in the 15th century [4]. Solar energy can be used for seawater desalination either by producing thermal energy required to drive the phase change processes or by producing electricity required to drive the membrane processes.

Solar energy is harvested in devices called solar collectors which are essentially heat exchangers that extract and transfer the heat to the load. Several types of solar collectors are available depending on the application.

A comparison of solar collector devices and the representative temperature ranges for their suitability in desalination and power generation applications are shown in Table 13.1 [3].

Solar radiation is concentrated by reflecting or refracting the incident light on the aperture area onto a smaller receiver/absorber area. There are two definitions for the "concentration ratio". An optical concentration ratio is defined as the ratio of the solar flux on the receiver to the flux on the aperture, whereas the geometric concentration ratio is defined as the ratio between the concentrator area and the aperture area that receives the concentrated radiation. A higher concentration ratio indicates a higher energy harvesting potential in the solar collector, usually reflected by higher temperatures [3].

Solar collector systems that concentrate solar energy to high temperatures can be used to produce electricity and or heat. This configuration is usually termed as "cogeneration" and allows for simultaneous power and water production. In a cogeneration scheme, a steam turbine is included which converts the steam produced from the solar energy collector system into electricity and condenses the steam in a condenser. The low energy steam

Table 13.1 Potential solar collector applications for desalination and power production

Collector Type	Absorber Type	Concentration Ratio	Indicative Temperature Range (°C)	Potential Application
Flat plate collector (FPC)	Flat	1	30–80	Desalination
Evacuated tube collector (ETC)	Flat	1	50–200	Desalination
Compound parabolic collector (CPC)	Tubular	1–5	60–240	Desalination and power production
Compound parabolic collector (CPC)	Tubular	5–15	60–300	Desalination and power production
Linear Fresnel reflector (LFR)	Tubular	10–40	60–250	Desalination and power production
Parabolic trough collector (PTC)	Tubular	15–45	60–300	Desalination and power production
Cylindrical trough collector (CTC)	Tubular	10–50	60–300	Desalination and power production
Heliostat field collector (HFC)	Point	100–1500	150–2000	Desalination and power production
Parabolic dish reflector (PDR)	Point	100–1000	100–500	Desalination and power production

could then be used as heat source for MSF or MED plants depending on the temperatures. Electricity produced from this unit can be fed to the grid for end-users. Auxiliary electrical energy requirements for the desalination system could also be supplied from this system as well. In a stand-alone desalination system, solar energy collector systems are solely dedicated to the desalination system.

For efficient utilization of solar energy harvested in the solar collection systems, a thermal energy storage unit can be included in the process configuration [1]. Energy storage allows for capture, storage and release of thermal energy throughout the day. Since solar energy is intermittent in nature and fluctuates with time, a thermal energy storage system can help match the difference between the energy demand and supply. Thermal energy storage systems require a medium for heat accumulation, storage and transfer. Water is a commonly available, low cost medium suitable for use in thermal energy storage systems.

Solar desalination systems are classified into two categories, i.e., direct and indirect collection systems [5]. Direct collection systems use solar energy to produce distillate directly in the solar collector, whereas indirect collection systems employ two sub-systems (one for solar energy collection and one for desalination). The use of direct or indirect systems depends on the type of solar energy harvesting technology and the related suitable desalination mechanism.

Conventional desalination systems are similar to solar systems since the same type of equipment is applied. The prime difference is that in the former, either a conventional boiler is used to provide the required heat or grid electricity is used to provide the required electric power, whereas in the latter, solar energy is applied.

13.3.1 Direct Solar Desalination: Solar Stills

The first desalination process "solar still" was based on direct solar energy utilization. Solar stills imitate the natural process of water evaporation from the oceans and their transport and condensation in the cooler regions to precipitate as freshwater (rain). This simple natural principle is the key behind thermal desalination processes. Solar stills are the simplest and cheapest direct solar harvesting desalination units (Figure 13.4). These units utilize direct solar energy to evaporate the freshwater from salt water leaving behind the concentrated brines. Solar stills incorporate evaporating and condensing units into a single chamber. Often, the glass roof of the unit serves as the

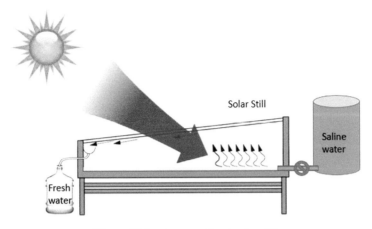

Figure 13.4 A conventional solar still.

condensing surface which rejects the latent heat to the ambient environment. The energy efficiency of typical solar stills varies between 30 and 40% [6]. The stills with an external condensing surface maintained in the shade or at lower temperatures have shown higher energy efficiencies up to 70%. The energy efficiency and the distillate production can be improved by adding multiple effects in the same unit albeit with added cost and complexity. The energy efficiency is calculated as the ratio of the amount of distillate produced to the amount of solar energy received by the solar still unit.

13.3.2 Indirect Solar Desalination Using Solar Collectors

Figure 13.5 shows the schematics of a solar powered thermal desalination systems. In Figure 13.5(a) MSF technology is used whereas as MED technology can be seen in Figure 13.5(b) [7]. Solar energy collectors are a special kind of heat exchanger that transforms solar radiation to internal energy of the transport medium. This is a device which absorbs the incoming solar radiation, converts it into heat, and transfers this heat to a fluid (usually air, water, or oil) flowing through the collector. The solar energy thusly collected is carried from the circulating fluid either directly to the hot water or space conditioning equipment, or to a thermal energy storage tank from which it can be drawn for use at night and/or cloudy days. There are basically two types of solar collectors: non-concentrating or stationary and concentrating. A non-concentrating collector has the same area for intercepting and for absorbing

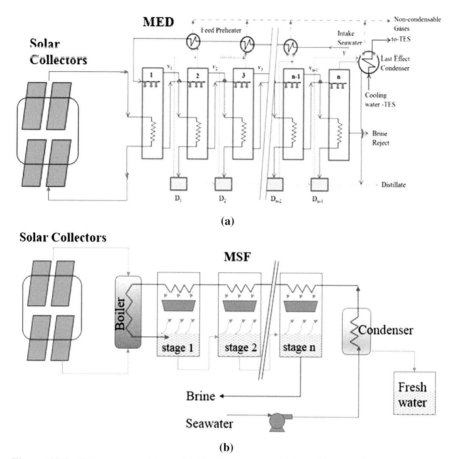

Figure 13.5   Solar energy driven desalination systems: **(a)** multi-stage flash desalination system and **(b)** multi-effect distillation system [7].

solar radiation, whereas a sun-tracking concentrating solar collector usually has concave reflecting surfaces to intercept and focus the sun's beam radiation to a smaller receiving area, thereby increasing the radiation flux.

Other indirect solar desalination systems include low temperature multi-effect desalination, humidification – dehumidification, and membrane distillation systems [1]. A comparison of solar powered desalination systems is shown in Table 13.2. Each has its own advantages and disadvantages. Hence the process should be selected based on site-specific constraints and desalination requirements [7, 8].

Table 13.2 Advantages and limitations of indirect solar desalination systems (non-membrane processes)

Process	Multi-Stage Flash Desalination	Multi-Effect Distillation	Vapor Compression Desalination
Advantages	• The plant is more reliable • Suitable for large scale production of distilled water • Plant can tolerate feed water of any quality • Minimal or no feed water pre-treatment is required	• The system is reliable • System could be operated even at temperature below 70°C • Low thermal energy consumption • The produced distillate is of high quality • Feed water does not need pre-treatment • CO_2 emissions are lower than MSF desalination system	• High efficiency • Does not require additional cooling medium • Low energy consumption • Suitable for low capacity applications • Feed requires little pre-treatment • Product water is of high quality
Disadvantages	• High energy consumption • Corrosion takes place due to high temperature operation • The plant is heavy and costly	• It consumes electrical power for vacuum pump • The system is heavy and costly • Suffers from corrosion problems	• Size of the plant is limited by the size of compressor • Corrosion of compressor • High initial cost • Higher water production cost

13.3.3 Solar PV-RO

Reverse osmosis systems driven by PV electricity are more suitable for small communities that are remote with brackish water and abundant solar energy sources [1]. These systems are especially favorable in locations without access to the power grid. PV systems can be designed with and without a backup storage unit. PV systems with battery storage allow for continuous operation. This provides a simple solution to the intermittent nature of the solar energy. Because the energy storage systems are expensive, a fossil fuel back-up system is desirable (grid-connected systems) or the system can be operated intermittently to match the availability of solar energy. Systems without a grid-connection are generally described as standalone or autonomous systems.

A typical solar-PV powered RO desalination system is shown in Figure 13.6. It consists of an array of PV modules which convert the solar energy into electricity to be used in the desalination process. The PV system components may include a battery charger, an inverter and a control circuit. The desalination system includes a pre-treatment filter system followed by a high-pressure pump and RO membrane unit.

Solar PV-based RO technology is simple and elegant. Solar PV-battery-operated RO systems of small to medium capacity (0.5 to 50 m^3/day) have been demonstrated in many parts of the world. For example, Herold and Neskakis [9] presented a small PV-driven RO desalination plant on the island of Gran Canaria with an average daily drinking water production of 0.8–3 m^3/day. The plant was supplied by a stand-alone 4.8 kW PV system with additional battery storage of 60 kWh. The nominal production was 1 m^3/day. The Energy Research Institute of King Abdulaziz City for Science and Technology (KACST) conducted extensive research on a PV-battery-inverter RO system in Sadous, Saudi Arabia. The RO system produced on average 5.7 m^3/day, converting brackish water from 5700 ppm TDS to 170 ppm TDS with an average 30% recovery rate [10].

Another PV-RO unit was developed at Murdoch University, Australia to produce 100 gallons per day of water from feed water containing up to

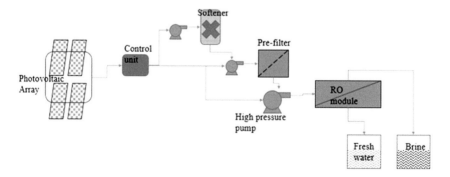

Figure 13.6   Solar PV-powered RO desalination system.

Table 13.3 The advantages and disadvantages of Solar PV powered RO (modified from [7])

Process	Solar PV Powered RO
Advantages	<ul><li>Flexibility in capacity expansion</li><li>Operation in ambient temperature</li><li>Can be built as portable or compact unit</li><li>Highly suitable for treating brackish water and ground water</li><li>Lower power consumption compared to thermal processes</li></ul>
Disadvantages	<ul><li>Pre-treatment of feed water is required</li><li>Biological fouling of membrane is possible</li><li>Membranes usually need replacement within a couple of years</li><li>The requirement of large surface areas for PV cells. For example, with 10 m^2 of PV area no more than 4 m^3 of freshwater per day can be produced from seawater</li></ul>

5000 ppm TDS [11]. The system was designed for 15–20% water recovery to avoid the problems with scaling. Carvalho et al. [12] presented the cost of a PV-RO desalination plant with batteries installed in the community of Ceara, of Brazil. The specific energy consumption of produced water was around 3.03 kWh/m^3. Battery-less small-scale PV-RO plants for stand-alone applications have also been developed [13].

13.4 Wind Energy for Desalination

Wind turbines operate on a simple principle. The energy in the wind turns two or three propeller-like blades around a rotor. The rotor is connected to the main shaft, which spins a generator to create electricity. Wind energy is the fastest growing source of electricity in the world. In 2012, nearly 45,000 megawatts (MW) of new capacity were installed worldwide. This stands as a 10 percent increase in annual additions compared with 2011 [14]. Small wind turbines are generally used for providing power off the grid, ranging from very small, 250 watt turbines designed for charging up batteries on a sailboat, to 50-kilowatt turbines that power dairy farms and remote villages. Like old farm windmills, these small wind turbines often have tail fans that keep them oriented into the wind. Large wind turbines, most often used by utilities to provide power to the grid, range from 250 kilowatts up to the enormous 3.5 to 5 MW machines that are being used offshore. In 2009, the average land based wind turbines had a capacity of 1.75 MW [15] Utility-scale turbines are usually placed in groups or rows to take advantage of prime windy spots. Wind "farms" like these can consist of a few or hundreds of turbines, providing enough power for tens of thousands of homes.

Wind-power technology is mature and increasingly competitive with conventional energy sources and considered one of the most promising renewable energy sources. Wind resources are often available in areas where problems of water scarcity are pronounced, such as coastal areas with a relatively warm climate and a dense population [16]. In addition to its technological readiness, wind-power is well suited for desalination since the freshwater output can be easily stored. Therefore, it is possible to produce drinking water when the power supply is available and store it for when it is not, allowing the system to be adapted to the intermittent nature of the wind. This alleviates the need for expensive back-up systems. In addition, wind power, while in operation, has a low impact on the environment.

Wind power is more suitable for RO (Figure 13.7) and mechanical vapor compression (Figure 13.8) among all the desalination processes. The MVC process is more tolerant of intermittent operation than RO but it is not traditionally used with a variable power supply. Several research installations have been tested to determine the feasibility of reverse-osmosis and vapor compression units with wind-power [17, 18]. Two autonomous wind-driven MVC desalination plants have been built to operate with variable power. In both plants a variable speed compressor and a resistive heating element in the brine tank allow a variable amount of power to be absorbed by the unit [19, 20]. The plant on Borkum Island uses a 60 kW wind turbine, an MVC unit with a 4–36 kW compressor and a 0–15 kW resistive heater. It produces distillate at 0.3–2 m^3/h with a specific consumption of 16–20 kWh/m^3 [21, 22]. Another wind powered MVC desalination plant was demonstrated on the Island of Rugen, in the Baltic Sea with a capacity

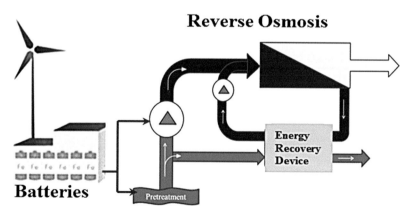

Figure 13.7 Wind powered RO desalination system (adapted from [1]).

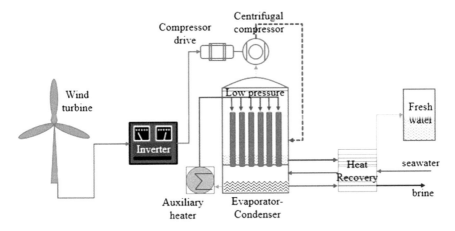

Figure 13.8 Wind-powered MVC desalination system.

of 360 m³/day [19]. The wind energy production capacity for this application was 300 kW. The distillate production varied between 2 and 15 m³/h depending on the wind speed conditions. Advantages of this system were: low-energy demand due to efficient heat recovery; variable rotor speed for higher energy gains; complete utilization of wind power for desalination; low process temperatures; and only wind energy used for operation. The economics of a wind-powered desalination system differ from conventional plant economics since it is almost entirely based on the fixed costs of the system. There are no fuel costs for the system since the wind turbine, being largely a capital expenditure, replaces fuel costs. Therefore, energy efficiency is not the main determining factor, but rather the economics of the process [16].

To understand the feasibility of wind-powered desalination, RO and vapor compression processes were analyzed and the cost of water was determined [16]. The dynamic effects of the processes, also taking into account the variability of the wind, have been studied, and the feasibility of using wind as the power source for desalination has been proven. The RO system is a better choice if there is a weak grid connection. It can produce water at slightly lower costs, mainly due to the lower specific energy consumption. For stand-alone applications in remote locations, the MVC is a suitable process for desalination due to its variability in operation. Also, MVC produces a much purer product. As the economics of the wind-powered desalination process are strongly site-dependent, a thorough analysis of local conditions is indispensable. In general, it was concluded that wind-powered desalination can be competitive with other desalination systems while providing safe and clean drinking water efficiently in an environmentally responsible manner offering a sustainable solution to growing freshwater needs.

Wind-powered RO systems can be operated in different configurations: with an alternative electrical supply (weak grid or diesel generator); matching the available wind energy to the load; and the operational characteristic of RO membranes such as operating at constant conditions and/or with storage devices [23]. Due to intermittence in the production of wind energy, suitable combinations of other renewable energy sources can be employed to provide smooth operating conditions. For example, a combination of wind power and PV cells can drive the desalination process round-the-clock with a battery bank system [24, 25]. Combining these two renewable energy sources with desalination may have several inherent advantages. However, maintenance of the battery bank system can be a major concern. Another noteworthy fact is that freshwater can be stored in vast quantities for extended periods of time. Therefore, it is possible to produce water and store it when the power supply is available, for use when the power supply drops. This alleviates the need for expensive back-up systems.

13.5 Geothermal Energy for Desalination

Geothermal energy is a proven and well-established source for electricity production, district heating and cooling and industrial process applications. Geothermal energy can be used directly in combination with MED, MEH, TVC, and MD (low temperature) or with MSF (medium temperature). Moreover, thermal energy conversion into shaft power or electricity would permit coupling with such desalination systems as RO, ED, and MVC.

Geothermal energy has the potential for desalination due to the following advantages [26, 27]: (i) it provides a stable and reliable energy supply; (ii) geothermal production technology is mature and its performance is unaffected by seasonal changes and weather fluctuations; (iii) it is cost-effective with simultaneous electricity production; (iv) it saves imported fossil fuels improving local energy security and environmental sustainability; and (v) geothermal sources have the lowest surface area or land requirements per unit (megawatts, MW) of all renewable energy sources and energy demand can be matched from the smallest to the largest energy-consuming utilities.

Geothermal energy also has the following disadvantages: geothermal brines have high-salt concentration that may create operational problems; hard-scale formation; and problems of disposal, if not near the sea. The scale formation could be mainly due to the hydrogen sulfide and ammonia in geothermal heat sources [28, 29].

Depending on the source temperature, various desalination technologies can be integrated with the geothermal sources. For temperatures in the range of 40 and 70°C, desalination applications which include simple evaporation basins, membrane distillation or low temperature multi-effect distillation units (LTMED) can be used. Higher heat sources (>70°C) can be used for either MED or MSF desalination processes. Much higher source temperatures between 120 and 200°C are suitable for cogeneration schemes (Figure 13.9). The application of geothermal water in desalination is relatively an unexplored technical concept. A limited number of studies evaluating the potential of geothermal water as a heat source for desalination are available [30–34].

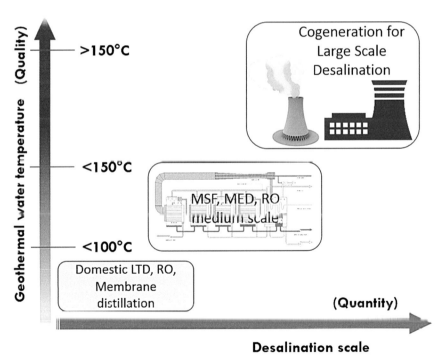

Figure 13.9 Geothermal energy applications in desalination.

The first study of geothermal desalination was proposed and analyzed by Awerbuch et al. [35] to produce power and water from geothermal brines in a novel process. In this process, a separator, steam turbine and an MSF unit were included. The separator was used to make sure that the steam flashed from the hot brine extracted from the geothermal well was sent to the steam turbine while the non-evaporated hot brine was used as the feed water to the MSF unit to produce freshwater.

Trivedi [36] described a case study of a low-enthalpy geothermal energy driven seawater desalination plant on the Milos Island in Greece. The proposed design consists of coupling MED units to a geothermal groundwater source with temperatures ranging from 75 to 90°C. The study showed that the exploitation of the low-enthalpy geothermal energy would help save the equivalent of 5000 TOE/year for a proposed plant capacity of 600–800 m^3/day of freshwater. Even in the case of limited geothermal energy, thermal desalination processes such as MED, TVC, SF and MSF can benefit greatly when coupled to geothermal sources by economizing considerable amounts of energy needed for pre-heating.

An integrated configuration including a multi-effect boiling unit and a MSF unit was evaluated in Baja California, Mexico [37] with a geothermal source at 80°C. An optimum desalinated water (freshwater) to geothermal source (heat source water) was found to be 1:14. In another study, coastal geothermal desalination including MSF–MED configuration was studied. This study reported a much smaller ratio of 1:5.9 for the freshwater to geothermal source. About 20 m^3/day of freshwater was produced from a geothermal heat flow of 118 m^3/day [38]. Elsewhere, Sephton Water Technology developed a pilot project for simulating process conditions for a 15-effect vertical tube evaporator MED plant which could produce up to 14 pounds of distilled water for each pound of geothermal steam used.

Kamal [39] reported that an enhancement of feed water temperature for seawater RO plants located in southern California induced a substantial reduction in the cost of potable water. The membrane productivity increase is about 2–3% per 1°C increase of the feeding temperature. Most of the membranes commercialized for RO desalination processes can tolerate temperatures up to 40°C. Increasing the feed water temperature up to 40°C will increase its productivity by about 20–30%. Meanwhile, few membrane suppliers such as Backpulseable^{TM} membranes (polypropylene tubular membranes) can tolerate temperatures up to 60°C which could entail ever higher productivities.

Geothermal source utilization around the world is expected to increase significantly in direct heat utilization and electricity production and industrial application uses. In conjunction with this development, geothermal source utilization in desalination applications can be used to combat the negative environmental impacts by the non-renewable energy powered desalination. However, not all geothermal sites are suitable for desalination application which needs to be carefully evaluated case by case. Uncertainties related to resource characteristics and its availability (reservoir volume and other flow, pressure, and temperature characteristics) should be determined in advance to match the life span of the desalination plants. Due to inefficiencies in the power generation scheme, cogeneration schemes may not be suitable in all locations.

13.6 Wave Energy for Desalination

Wave energy originates from the sun. Of the 173,000 TW of solar power arriving at the earth's atmosphere, 114,000 TW is absorbed in the atmosphere, oceans and the earth's surface. About 1200 TW of this thermal energy is then converted into the kinetic energy of the wind [40]. The shearing action of the wind on the surface of the ocean generates currents and waves, involving energy transfer at a rate of around 3 TW [41]. Thus, only a tiny fraction of solar radiation is eventually converted into ocean-wave energy. Nevertheless, ocean waves represent a clean renewable energy resource.

Ocean waves are mathematically complex; consequently, their complete description requires several parameters. An additional difficulty is related to the conception of the power take-off mechanism (air turbine, power hydraulics, electrical generator or other) which should allow the production of usable energy. The problem here lies in the variability of the energy flux absorbed from the waves, in several time-scales: wave-to-wave (a few seconds), sea states (hours or days) and seasonable variations. Naturally, the survivability in extreme conditions is another major issue [42]. The wave energy level is usually expressed as power per unit length (along the wave crest or along the shoreline direction); typical values for "good" offshore locations (annual average) range between 20 and 70 kW/m and occur mostly in moderate to high latitudes. Seasonal variations are in general considerably larger in the northern than in the southern hemisphere [43], which makes the southern coasts of South America, Africa and Australia particularly attractive for wave energy exploitation.

Wave energy harvested as electricity can be utilized to power a membrane desalination process such as an RO process. It can also be used to run a mechanical vapor compression unit. A generic wave powered desalination scheme is shown in Figure 13.10(a). It is comprised of an impulse turbine with an alternator and rectifier, a possible energy storage system (battery) and an inverter to convert the current. An RO system designed for energy recovery with a pressure exchanger scheme is shown in Figure 13.10(b). An offshore RO desalination system powered by wave energy is shown in Figure 13.10(c).

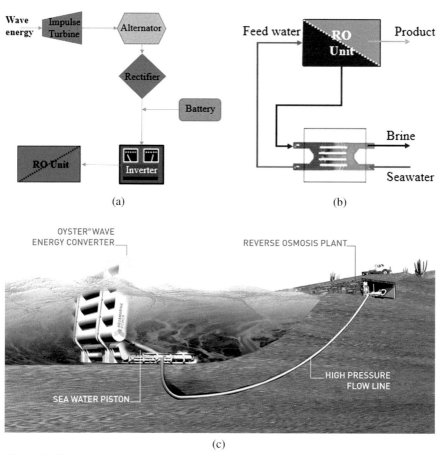

(a)   (b)

(c)

Figure 13.10   Examples of wave energy applications in desalination Aquamarine Power and the oyster wave energy harvesting system.

Buoy (oscillating water body device) and oscillating water column technologies have been successful at RO and vapor compression desalination [40]. The first reported desalination process powered by wave energy was designed using oscillating buoys to drive piston pumps anchored to the seabed. These pumps fed seawater to submerged RO modules. A simple linear reciprocating pump driven by the motion of the waves connected via non-return valves produced flow in one direction to a RO membrane. This unit produced an average of 2 m^3/day. This system operated successfully for over 18 months. An energy recovery device was not reported in this case.

An autonomous wave-powered desalination plant has been demonstrated in Kerala, India [44]. A 10 m^3/day RO desalination plant was developed based on the oscillating water column principle. This arrangement encloses a column of air on top of a column of water. The wave action causes the water column to rise and fall, which alternately compresses and depressurizes the air column. Energy is extracted from the system and used to generate electricity by allowing the trapped air to flow through a turbine. This electricity was used to drive the RO unit at a feed flow rate of 30–40 L/min and the product water recovery was 10–15 L/min. This is an indirect process, similar to the schemes that involve other renewable energy sources, with one main advantage when compared to other resources, which is the distance to the seawater. In another study [45], the potential for autonomous wave-powered desalination by RO utilizing a pressure exchanger–intensifier for energy recovery was considered. A numerical model of the RO plant showed that the specific energy consumption would be less than 2.0 kWh/m^3 over a wide range of seawater feed conditions, making it particularly attractive. The specific hydraulic energy consumption of the desalination plant is estimated to be 1.85 kWh/m^3, while maintaining a recovery-ratio of less than 25–35% to avoid the need for chemical pre-treatment to eliminate scaling problems [46].

Overall, wave energy is a clean energy which reduces fossil fuel dependence and is environmentally friendly. The electricity costs have improved in recent years and the energy conversion efficiency is also very high. The main disadvantage of wave power, as with the wind from which is originates, is its (largely random) variability in several time-scales: from wave to wave, with sea state, and from month to month (although patterns of seasonal variation can be recognized). The assessment of the wave energy resource is a basic prerequisite for the strategic planning of its utilization and for the design of wave energy devices. Other issues include construction of strong, cheap and efficient conversion devices, ecological impacts related to the alteration of tides and waves, highly location dependent energy source, diffusivity of energy source, irregularity in direction, durability and size, and extreme weather. There are only around 20 sites in the world that have been identified as possible tidal power stations, and more research and pilot-scale experience is required.

13.7 Nuclear Energy for Desalination

In nuclear reactors, the fission energy generated is released as heat which is suitable for power generation and other process heat applications [47, 48]. Most of the plants in operation are "light water" reactors, meaning they use normal water in the core of the reactor, as opposed to "heavy water" reactors. U.S. reactors account for more than one-fourth of nuclear power capacity in the world. In the United States, two-thirds of the reactors are pressurized water reactors (PWR) and the rest are boiling water reactors (BWR). In a boiling water reactor, the water is allowed to boil into steam, and is then sent through a turbine to produce electricity. In pressurized water reactors, the core water is held under pressure and not allowed to boil. The heat is transferred to water outside the core with a heat exchanger (also called a steam generator), boiling the outside water, generating steam, and powering a turbine. In pressurized water reactors, the water that is boiled is separate from the fission process, and so does not become radioactive. After the steam is used to power the turbine, it is cooled off to make it condense back into water. Some plants use water from rivers, lakes or the ocean to cool the steam, while others use tall cooling towers. The hourglass-shaped cooling towers are the familiar landmark of many nuclear plants. For every unit of electricity produced by a nuclear power plant, about two units of waste heat are rejected to the environment [49].

Commercial nuclear power plants range in size from about 60 MW for the first generation of plants in the early 1960s, to over 1000 MW [50]. Many plants contain more than one reactor. The Palo Verde plant in Arizona, for example, is made up of three separate reactors, each with a capacity of 1,334 MW_e. Some reactor designs use coolants other than water to carry the heat of fission away from the core. Canadian reactors use water loaded with deuterium (called "heavy water"), while others are gas cooled. Some plants have used helium gas as a coolant (called a High Temperature Gas Cooled Reactor) and others use liquid metal or sodium.

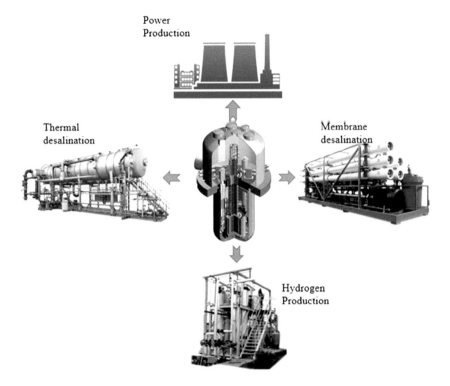

Figure 13.11 High-temperature gas nuclear reactor for power production combined with desalination and hydrogen production [51].

In nuclear power reactors, heat can be extracted at various temperature levels, both in the form of hot gas or steam. High pressure steam can be used to generate electricity as shown in Figure 13.11. Cogeneration (simultaneous power and water production) or poly-generation (power, water and other industrial process heating and cooling) schemes can be developed depending on the size and working principle of the nuclear power reactors. Low pressure and temperature steam may be used to drive MSF desalination or any other distillation units. Electricity can be generated from the nuclear power reactor to drive the high-pressure pumps of RO desalination plants [52]. Coupling MSF and RO with nuclear reactors will yield some economic and technical advantages. A comparison between nuclear powered MSF and RO desalination plants is shown in Table 13.4.

Table 13.4   Advantages and consideration for nuclear powered MSF and RO processes [52]

Item	Nuclear-MSF	Nuclear-RO
Availability	about 0.90%	about 0.90%
Capital cost	High	Low
Operation interface	Yes	No
Energy consumption	Steam and power	Power only
Safety consideration	Radiation contamination	Nuclear plant only
Load regulation	Flexible	No load regulation
Coupling	Require extensive care	Direct coupling
Thermal pollution	High	Medium
Manpower	Highly qualified	Qualified in nuclear
Plant site	On the same site	Anywhere
Transportation cost	Required	Not required

13.7.1 Experience with Nuclear Desalination

Small- and medium-sized nuclear reactors (SMRs) are identified to be suitable for desalination, often with cogeneration of electricity using low-pressure steam from the turbine and hot seawater feed from the final cooling system [53]. Currently SMRs are defined as power reactors that have power outputs in the range of 100–400 MW_e. The development of cost-competitive small- and medium-sized nuclear power reactors are suitable for developing countries as well as for coupling with desalination plants of 10,000 m^3/day (2.64 mgd) to 100,000 m^3/day (26.4 mgd) capacities.

The feasibility of nuclear desalination has been proven with over 150 reactor-years of experience, mainly in Kazakhstan, India and Japan. Large-scale deployment of nuclear desalination on a commercial basis will depend primarily on economic feasibility [54]. Desalination costs are estimated to be similar to fossil-fueled plants. The BN-350 fast reactor at Aktau, in Kazakhstan, supplied up to 135 MW_e of electric power while producing 80,000 m^3/day of potable water over 27 years through 10 MED trains, about 60% of its power being used for heat and desalination. The plant was designed as 1000 MW_{th}. If oil/gas boilers are used in conjunction with this plant, total desalination capacity can reach 120,000 m^3/day. To date, the BN-350 is the biggest example of nuclear desalination in history. The construction of the plant began in 1964 and it began operation in 1973. By 1995, the plant's operating license had expired and the facility continued to operate at a low capacity until it stopped operation in 1999. Table 13.5 provides a summary of production capacities and locations of some nuclear desalination plants around the world.

Table 13.5   Examples of some nuclear desalination plants around the world [55]

Plant Name	Location	Gross Power [MW(e)]	Water Capacity [m³/day]	Reactor Type/ Desal. Process
Shevchenko	Aktau, Kazakhstan	150	80000–145000	FBR/MSF&MED
Ikata-1,2	Ehime, Japan	566	2000	LWR/MSF
Ikata-3	Ehime, Japan	890	2000	LWR/RO
Ohi-1,2	Fukui, Japan	2 × 1175	3900	LWR/MSF
Ohi-3,4	Fukui, Japan	1 × 1100	2600	LWR/RO
Genkai-4	Fukuoka, Japan	1180	1000	LWR/RO
Genkai-3,4	Fukuoka, Japan	2 × 1180	1000	LWR/MED
Takahama-3,4	Fukui, Japan	2 × 870	1000	LWR/RO
NDDP	Kalpakkam, India	2 × 170	1800	PHWR/RO
Diablo Canyon	San Luis Obispo, USA	2 × 1100	2180	LWR/RO

LWR: light water reactor;

PHWR: Pressurized heavy water reactor.

Cogeneration or dual-purpose power plants have received much attention due to increasing water demands and electricity needs. Although the units are rather small, in Japan, 10 desalination facilities linked to pressurized water reactors operating for electricity production produce about 14,000 m³/day of potable water, with over 100 reactor-years of experience. MSF was initially employed, but MED and RO have been found more efficient in this system.

India investigated the possibility of nuclear desalination research since the 1970s. A demonstration plant coupled to twin 170 MW_e nuclear power reactors (PHWR) was set up at the Madras Atomic Power Station, Kalpakkam, in southeast India in 2002. This hybrid nuclear desalination demonstration plant comprises of an RO unit with 1800 m³/day capacity, an MSF unit of 4500 m³/day and a recently-added barge-mounted RO unit. This is the largest nuclear desalination plant based on hybrid MSF-RO technology using low-pressure steam and seawater from a nuclear power station.

A 10,200 m³/day MVC plant was set up at Kudankulam, India, to supply freshwater for the new plant. It has four stages in each of the four streams. An RO plant supplied the plant's township initially. However, a full MVC plant was commissioned in 2012, with a capacity of 7200 m³/day to supply the plant's primary and secondary coolant and the local town. A low temperature (LTE) nuclear desalination plant was operated by waste heat from the nuclear research reactor at Trombay since 2004 to supply make-up water in the reactor.

Pakistan commissioned a 4800 m³/day MED desalination plant in 2010, coupled to the Karachi Nuclear Power Plant (KANUPP, a 125 MWe PHWR) in 2014. It was quoted as 1600 m³/day and has been operating a 454 m³/day RO plant for its own use. China General Nuclear Power (CGN) has commissioned a 10,080 m³/day seawater desalination plant using waste heat to provide cooling water at its new Hongyanhe project at Dalian in the northeast Liaoning province. Other countries such as Russia, Eastern Europe and Canada have also gained experience with nuclear reactors.

Large-scale deployment of nuclear desalination on a commercial basis will depend primarily on economic factors. Some issues related to environmental pollution and operational safety due to radiative activity are impeding the development of nuclear desalination. With improved process safety procedures and technology developments in reactor design, nuclear reactor based desalination is expected to become a viable alternative for freshwater production throughout the world.

13.8 Conclusion and Selection Criteria

Considering the energy requirements for desalination processes, renewable energy sources present exciting opportunities to produce freshwater in an environmentally friendly manner. While some of the renewable energy technologies such as PV and wind are mature, their application in desalination are still economically debatable.

It should also be noted that desalination unit processes do not care from where their energy comes. Heat is heat, electricity is electricity. Unless there is some unique symbiosis and integration, in the end it will depend upon the final energy production cost and public acceptance (particularly in the case of nuclear desalination). Where renewable energies are competitive with other production methods, they will prevail, regardless of whether they are powering a desalination plant or any other operation. At the end of the day, the question is not "Will it work?" but rather, "Will it compete?". The end-user must be convinced that he is getting the best buy for his prevailing circumstances.

The selection of appropriate renewable energy-powered desalination process requires complete analysis of the plant from start to end. Irrespective of the production volume, the following factors need to be considered when selecting a desalination process [26]:

1. Need for desalination:

   a. Capacity
   b. Quality

2. Energy source, infrastructure, hardware:

   a. Solar
   b. Wind
   c. Wave
   d. Geothermal
   e. Nuclear

3. Feed water source:

   a. Seawater
   b. Brackish water
   c. Produced water (wastewater reuse)

4. Process:

   a. Type of process
   b. Specific energy requirements
   c. Characteristics

5. Economic feasibility:

   a. Financing
   b. Capital costs
   c. Operating costs

6. Environment:

   a. Brine disposal
   b. Intake structures
   c. Use of chemicals

Acknowledgements

This work was supported by the Civil and Environmental Engineering Department, the Bagley College of Engineering, and the Office of Research and Economic Development at Mississippi State University. This work was also partially supported by the research grant SU836130 from the United States Environmental Protection Agency (USEPA).

References

[1] Gude, V. G. (2015). Energy storage for desalination processes powered by renewable energy and waste heat sources. *Appl. Energy* 137, 877–898.

[2] Papapetrou, M., Biercamp, C., Wieghaus, M. (2010). *Roadmap for the Development of Desalination Powered by Renewable Energy: Promotion for Renewable Energy for Water Production through Desalination.* Stuttgart: Fraunhofer Verlag.

[3] Kalogirou, S. A. (2005). Seawater desalination using renewable energy sources. *Progr. Energy Combust. Sci.* 31, 242–281.

[4] Nebbia, G., and Menozzi, G. N. (1966). *Aspetti Storici della Dissalazione delle Acque Salmastre* Vol. 41. Italiaans: Gedrukt boek, 3–20.

[5] Kalogirou, S. (1997). Survey of solar desalination systems and system selection. *Energy* 22, 69–81.

[6] Sampathkumar, K., Arjunan, T. V., Pitchandi, P., Senthilkumar, P. (2010). Active solar distillation—A detailed review. *Renew. Sustain. Energy Rev.* 14, 1503–1526.

[7] Sharon, H., and Reddy, K. S. (2015). A review of solar energy driven desalination technologies. *Renew. Sustain. Energy Rev.* 41, 1080–1118.

[8] Mezher, T., Fath, H., Abbas, Z., and Khaled, A. (2011). Techno-economic assessment and environmental impacts of desalination technologies. *Desalination* 266, 263–273.

[9] Herold, K., and Neskakis, A. (2001). A Small PV-Driven Reverse Osmosis Desalination Plant on the Island of Gran Canaria. *Desalination* 137, 285–292.

[10] Banat, F., Qiblawey, H., and Al-Nasser, Q. (2012). Design and Operation of Small-Scale Photovoltaic-Driven Reverse Osmosis (PV-RO) Desalination Plant for Water Supply in Rural Areas. *Comput. Water Energy Environ. Eng.* 1, 31–36.

[11] Matthew, K., Dallas, S., Ho G., and Anda, M. (2001). "Innovative Solar Powered Village Potable Water Supply," in *Proceedings of Women Leaders on the Uptake of Renewable Energy Seminar*, Perth.

[12] Carvalho, P., Riffel, D., Freire, C., and Montenegro, F. (2004). The Brazilian experience with a photovoltaic powered re-verse osmosis plant. *Progr. Photovolt.* 12, 373–385.

[13] Mohamed, E., Papadakis, G., Mathioulakis, E., and Belessiotis, V. (2005). The effect of hydraulic energy recovery in a small sea water

reverse osmosis desalination system; experimental and economical evaluation. *Desalination* 184, 241–246.

[14] Global Wind Energy Council (GWEC). (2012). *Global Wind Report 2012*. Brussels: Global Wind Energy Council.

[15] American Wind Energy Association (AWEA). (2009). Anatomy of a Wind Turbine.

[16] Forstmeier, M., Mannerheim, F., D'Amato, F., Shah, M., Liu, Y., Baldea, M., and Stella, A. (2007). Feasibility study on wind-powered desalination. *Desalination* 203, 463–470.

[17] Carta, J. A., Gonzalez, J., Subiela, V. (2003). Operational analysis of an innovative wind powered reverse osmosis system installed in Canary Island. *Solar Energy* 75, 153–68.

[18] Pestana, I. N., Latorre, F. J. G., Espinoza, C. A., Gotor, A. G. (2004). Optimization of RO desalination systems powered by renewable energies. Part I: wind energy. *Desalination* 160, 293–299.

[19] Plantikow, U. (1999). Wind-powered MVC seawater desalination-operational results. *Desalination* 122, 291–299.

[20] Plantikow, U. (2003). "Production of potable water by a wind-driven seawater desalination plant—experience with the operation," in *Proceedings of the World Wind Energy Conference*, Madrid.

[21] Coutelle, R., Kowalczyk, D., and Plantikow, U. (1991). "Seawater desalination powered by wind-powered mechanical vapour compression plants," in *Seminar on New Technologies for the Use of Renewable Energies in Water Desalination*, 49–64.

[22] Henderson, C. R., McGowan, J. G., Manwell, J. F. (2004). "A feasibility study for a wind/diesel hybrid system with desalination for star island, NH," in *Proceedings of the Global WINDPOWER Conference and Exhibition*, Chicago.

[23] Miranda, M. S., and Infield, D. (2003). A wind-powered seawater reverse-osmosis system without batteries. *Desalination* 153, 9–16.

[24] Tzen, E., Morris, R. (2003). Renewable energy sources for desalination. *Solar Energy* 75, 375–9.

[25] Tzen, E., Theofilloyianakos, D., and Kologios, Z. (2008). Autonomous reverse osmosis units driven by RE sources experiences and lessons learned. *Desalination* 221, 29–36.

[26] Gude, V. G., Nirmalakhandan, N., and Deng, S. (2010). Renewable and sustainable approaches for desalination. *Renew. Sustain. Energy Rev.* 14, 2641–2654.

[27] Gude, V. G. (2016). Geothermal source potential for water desalination-Current status and future perspective. *Renew. Sustain. Energy Rev.* 57, 1038–1065.

[28] Boegli, W. J., Suemoto, S. H., and Trompeter, K. M. (1977). Geothermal desalting at the East Mesa test site. *Desalination* 22(1–3), 77–90.

[29] Bourouni, K., Deronzier, J. C., and Tadrist, L. (1999). Experimentation and modelling of an innovative geothermal desalination unit. *Desalination* 125(1–3), 147–153.

[30] Bouchekima, B. (2003). Solar desalination plant for small size use in remote arid areas of South Algeria for the production of drinking water. *Desalination* 156, 353–354.

[31] Bourouni, K., Deronzier, J. C., and Tadrist, L. (1999a). Experimentation and modelling of an innovative geothermal desalination unit. *Desalination* 125, 147–153.

[32] Bourouni, K., Martin, R., and Tadrist, L. (1999b). Analysis of heat transfer and evaporation in geothermal desalination units. *Desalination* 122, 301–313.

[33] Mohamed, A. M. I., and El-Minshawy, N. A. S. (2009). Humidification–dehumidification desalination system driven by geothermal energy. *Desalination* 249, 602–608.

[34] Bouguecha, S., and Dhahbi, M. (2003). Fluidised bed crystalliser and air gap membrane distillation as a solution to geothermal water desalination. *Desalination* 152, 237–244.

[35] Awerbuch, L., Lindemuth, T. E., May, S. C., and Rogers, A. N. (1976). Geothermal energy recovery process. *Desalination 19*, 325–336.

[36] Trivedi, K Karytsas. (1996). *Mediterranean Conference on Renewable Energy Sources for Water Production, European Commission, EURORED Network*. Santorini: CRES, 128–31.

[37] Rodrıguez, G., Rodrıguez, M., Perez, J., Veza, J. (1996). "A systematic approach to desalination powered by solar, wind and geothermal energy sources," in *Proceedings of the Mediterranean Conference on Renewable Energy Sources for Water Production*, Santorini, 20–25.

[38] Gutiérrez, H., Espíndola, S. (2010). *Using Low Enthalpy Geothermal Resources to Desalinate Seawater and Electricity Production on Desert Areas in Mexico, GHC Bull*. Available at: http://geoheat.oit.edu/bulletin/bull29-2/art5.pdf

[39] Kamal, I. (1992). Cogeneration desalination with reverse osmosis: a means of augmenting water supplies in Southern California. *Desalination*, 88, 355–369.

[40] Davies, P. A. (2005). Wave-powered desalination: resource assessment and review of technology. *Desalination* 186, 97–109.

[41] Gregg, M. (1973). *Scientific American,* 65–77.

[42] Antonio, F. D. O. (2010). Wave energy utilization: a review of the technologies. *Renew. Sustain. Energy Rev. 14*(3), 899–918.

[43] Barstow, S., Gunnar, M., Mollison, D., and Cruz, J. (2008). "The wave energy resource," in *Ocean Wave Energy*, ed. J. Cruz (Berlin: Springer), 93–132.

[44] Sharmila, N., Jalihal, P., Swamy, A. K., Ravindran, M. (2004). Wave powered desalination system. *Energy* 9, 1659–1672.

[45] Folley, M., Suarez, B. P., and Whittaker, T. (2008). An autonomous wave-powered desalination system. *Desalination* 220, 412–21.

[46] Folley, M., Whittaker, T. (2009). The cost of water from an autonomous wave-powered desalination plant. *Renew. Energy* 34, 75–81.

[47] IAEA (1997). *Nuclear Desalination of Sea Water, Proceedings of 1997 Symposium.* Vienna: International Atomic Energy Agency.

[48] IET (2008). *Principles of Nuclear Power*, 3rd Edn. Stevenage: Institute of Engineering and Technology.

[49] IAEA (1998). Nuclear heat applications: design aspects and operating experience, IAEA-TECDOC-1056.

[50] Thomas, S., Bradford, P., Froggatt, A., and Milborrow, D. (2007). The economics of nuclear power.

[51] IAEA (2015). New technologies for seawater desalination using Nuclear energy. Iaea-tecdoc-1753.

[52] Al-Mutaz, I. S. (2003). Hybrid RO MSF: a practical option for nuclear desalination. *Int. J. Nucl. Desalination* 1, 47–57.

[53] ANS (2010). *Interim Report of the American Nuclear Society President's Special Committee on Small and Medium Sized Reactor (SMR) Licensing Issues.* La Grange Park, IL: American Nuclear Society.

[54] Seneviratne, G (2007). *Research Projects Show Nuclear Desalination Economical, Nuclear News.* London: World Nuclear Association.

[55] Khamis, I, (2010). Nuclear Desalination New Technologies for Seawater Desalination Using Nuclear Energy, TecDoc 1753. Vienna: International Atomic Energy Agency (January 2015).

CHAPTER 14

Energy Consumption and Minimization

Konstantinos Plakas[1], Dimitrios Sioutopoulos[1] and Anastasios Karabelas[1]

[1]Laboratory of Natural Resources and Renewable Energies,
Chemical Process and Energy Resources Institute, Centre for Research
and Technology – Hellas, 6th km Charilaou-Thermi, P.O. Box 60361,
GR-57001, Thermi-Thessaloniki, Greece
Phone: +30 2310 498181
E-mail: kplakas@cperi.certh.gr; sioutop@cperi.certh.gr;
karabaj@cperi.certh.gr

14.1 Introduction

Considering that the economics of desalination are tied to the cost and quantity of energy used for the process, the primary goal of this chapter is to present a review of the energy requirements and the energy minimization approaches implemented in all types of desalination processes (thermal and membrane-based), alone or in hybrid configurations. Particular attention is paid to the design and operating strategies, aiming at reducing energy usage, including the integration of renewable energies as a source of sustainable clean energy supply.

14.2 Energy Issues in Desalination

14.2.1 Interrelation between Water and Energy

According to the European Environment Agency (EEA), more than 44% of freshwater abstraction in Europe is carried out to cool thermal power plants [1]. At the same time, water is also necessary for alternative or renewable forms of energy such as hydropower, concentrated solar energy and biofuels. On the other hand, a significant amount of energy is used to collect, treat, distribute, and heat water supplies and to collect, treat, and discharge wastewater. Despite the interdependencies of energy and water, historically the two resources have been managed and regulated independently of each other. However, failure to consider the interrelationships of energy and water introduces vulnerabilities such that constraints in one resource can introduce constraints in the other. For example, electricity grid outages or other failures in the energy system can create constraints in the water and wastewater sectors. Droughts and tight water regulations constitute water constraints that can seriously impact the energy sector. Understanding and accounting for this inter-dependency, often referred to as the 'water–energy nexus' is becoming increasingly important to ensure sustainable and cost-effective resource (water and energy) utilization, as many areas around the world are at high risk of water stress, if they are not already facing serious water shortages.

As we plan a sustainable global future, this water–energy nexus, is becoming increasingly important in desalination, where a sizeable energy expenditure is required for the production of drinking water. At present, desalination consumes at least 75.2 TWh of electricity per year, equivalent to around 0.4% of global electricity needs [2, 3]. Most of the energy required for desalination presently comes from fossil fuels, with less than 1% of the capacity being dependent on renewable sources [3]. However, as the number of desalination plants increases, dependence on fossil fuels will no longer be sustainable from an economic and environmental perspective. In this context, it is important to maximize the energy efficiency of desalination through energy minimization strategies, innovation, and introduction of new technologies, keeping this nexus in mind. Moreover, synergies between water and energy systems must be pursued. Such opportunities include the use of waste heat from electricity generation for desalination (*cogeneration*) and the implementation of renewable energy technologies to decouple water production from fossil fuel supply and meet the growing energy demand.

The intensity of the water–energy nexus in desalination is shaped by conditions in the particular geographic location considered (i.e., regional and national), depending on the energy mix, demand characteristics, resource availability and accessibility, as well as political, regulatory, economic, environmental, and social factors, in addition to available technologies. Therefore, regarding the water-energy interrelationship, the landscape is fragmented, complex, and evolving. Moreover, there is also variation across the globe in relevant energy policies, including renewable portfolio standards, regulation of oil and gas development activities, and regulation of thermoelectric-plant water intake and discharge. For example, the co-location of power plants and desalination facilities is not always feasible. In California (US), federal utility laws prohibit existing power plants, which are co-located with other facilities, from selling power at a preferential rate to those facilities [4].

An extensive analysis of the water-energy nexus is presented in several reports prepared by the UN [5], the International Energy Agency [6], the Stockholm Environment Institute [7], the World Energy Council [8], and other organizations. The United Nations 'World Water Development Report (WWDR) 2014' provides probably the most extensive analysis of the nexus in the literature to date, drawing upon information, data and analyses from a broad range of literature on the subject. The report deals with water demands, energy requirements for water provision, water availability, and the water requirements for power generation. It also expands the energy–water links to include issues related to food and agriculture, thus broadening the scope of the nexus. In all reports it is stressed that, as easily accessible freshwater resources are depleted, the use of energy-intensive technologies, such as desalination or more powerful groundwater pumps, is expected to expand rapidly. In the Middle East and North Africa (MENA) region for instance, the shortage of water will be met mostly through desalination by 2050 [8]. The expansion of desalination will require a careful consideration of its social, economic and environmental impacts as well as its associated energy demand. Along these lines, the issue of energy demand and minimization for desalination is described in the next sections, with emphasis given to the *Specific Energy Consumption* (*SEC*), as a key figure of merit for the desalination processes.

14.2.2 Energy Demand for Desalination

For desalination, the total energy used by the process (or plant) divided by the amount of water produced, gives a value for the energy consumed per unit volume of desalinated water, and is generally used as a metric to quantify energy consumption. This metric is termed SEC and normally has the units of KWh/m^3. The quality of energy used to drive the separation, characterizes the type of desalination process. For example, the distillation-based methods [e.g., Multi-Stage Flash (MSF), Multi-Effect Distillation (MED), Vapor compression (VC)] use thermal energy to heat and evaporate seawater, and the vapors generated are condensed to produce freshwater. On the other hand, reverse osmosis (RO) and electrodialysis (ED) use electrical energy to separate the ions from water by applying a pressure-difference or an electrical field, respectively.

Specific energy for a given desalination process is a function of the thermodynamic properties of the feed (feed composition and concentration, temperature), the recovery, and the inherent efficiency of the technology used for desalination. To assess the energy efficiency of a given process, one must consider the theoretical minimum amount of energy required to separate pure water from the salts, which corresponds to separation occurring as a reversible thermodynamic process [9]. Specifically, the theoretically *minimum energy consumption* (referred to as *SEC-min*), for separating a saline solution into pure water and concentrated brine under ideal conditions, can be calculated by applying the second law of thermodynamics to separations; i.e., from Gibbs Free Energy of non-mixing for salt–water mixtures [9–11]. Spiegler and El-Sayed [9] have presented an analysis leading to an estimation of the SEC-*min,* for a recovery ratio approaching zero as well as for finite recovery. It is understood that the SEC-*min* is independent of the desalination method. In other words, the minimum energy requirement for the separation process is common for all desalination systems.

As indicated [9], the SEC-*min* depends on the salt content of the saline solution (Table 14.1) and the water recovery. For modest recovery, SEC-*min* is approx. 1.1 kWh/m^3 for typical seawater (35 g/L) and 0.12 kWh/m^3 for typical brackish water (4 g/L); however, for a recovery rate of 50%, the respective SEC-*min* values increase to 1.6 and 0.17 kWh/m^3, whereas for a recovery rate of 70%, these values are 2.0 and 0.21 kWh/m^3. At a recovery rate approaching zero the SEC-*min* would be near 0.8 kWh/m^3 for seawater with 35 g/L salt concentration (Figure 14.1).

In practice, however, the energy requirements in all desalination processes are considerably higher than those computed for the ideal reversible separation. According to Table 14.2, for seawater desalination the electrical energy demand using thermal-based technologies is in the range 6.5–28 kWh/m^3 compared to 2–6 kWh/m^3 for membrane-based technologies. This difference

Table 14.1   Salt concentration of different water sources [12]

Water Source or Type	Approximate Salt Concentration (g/L)
Brackish waters	0.5–3
North Sea (near estuaries)	21
Gulf of Mexico and coastal waters	23–33
Atlantic Ocean	35
Pacific Ocean	38
Persian Gulf	44
Dead Sea	~300

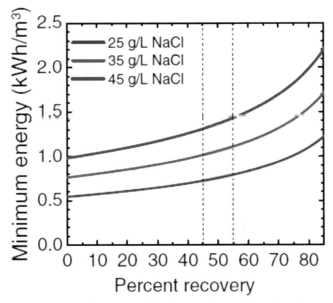

Figure 14.1 Theoretical minimum SEC for desalination as a function of percent recovery for common waters: 25 g/L is typical of the less saline open waters (e.g., from the Tampa Bay, Florida estuary), 35 g/L is typical for seawater, and 45 g/L is characteristic of the Persian Gulf. Minimum energies for recoveries between 45 and 55%, the range in which modern SWRO plants operate, are highlighted [11].

is attributed to process irreversibilities, i.e., due to friction losses, non-equilibrium, and other thermal losses, including boiling point elevation, flow resistance through membranes, and pump efficiencies. Hence, the deviation of the actual energy required in any given desalination system depends on the system design and engineering characteristics as well as its principle of operation, which affect the quantity and type of losses encountered during separation.

Considering that energy costs account for the largest percentage of a water utility operating budget—as much as 55% by some estimates—it is evident why saving energy is at the top of the desalination industry priorities. The objective of saving energy has dominated the focus of technological innovation and research in this sector, with special attention given to the adoption of best practices by the desalination plant operators, as well as to the prioritization of related R&D. These efforts have been directed to three general areas:

1. Improvements in the established desalination technologies as already widely applied in large-scale plants through, e.g., enhanced pre- and post-treatment as well as new stage designs and improved operational strategies.
2. Development of novel technologies (i.e., Membrane Distillation (MD), Forward Osmosis (FO), Capacitive Deionization, alone or hybrids).
3. Utilization/integration of degraded/waste heat sources and of renewable energies in developing desalination technologies.

Over the last few decades, the above efforts have resulted in a reduction in the SEC for water desalination. For reverse osmosis in particular, which has a low carbon footprint, the increase in energy efficiency and reduction in cost has been due to higher efficiency pumps, higher efficiency energy-recovery devices, lower energy and higher salt rejection membranes, high-efficiency membrane modules, more active surface area in each membrane module, optimized feed spacers and more efficient system designs [10, 11, 27]. In the case of the thermal processes, efforts are concentrated on multi-objective optimizations such as the minimization of specific heat transfer areas and the maximization of distillate production or exergy efficiency [28–30]. Other than the stand-alone optimization of thermal-based configurations, the water industry has turned its interest also to alternative/hybrid schemes, such as MSF–MED [31], or reverse osmosis (RO)–thermal technologies in which the permeate from a RO desalination component is mixed with distillate from thermal desalination [32]. Such facilities have already been implemented in the Persian Gulf countries (i.e., Saudi Arabia at Jeddah and Yanbu-Medina) with the aim of ensuring, as economically as possible, uniform water supplies under the specific, greatly varying load conditions in these countries [33].

14.3 SEC in Membrane-Based Processes

14.3.1 Energy Consumption in RO Desalination

Reverse osmosis desalination is established as a mature technology serving as the benchmark for comparison with other emerging desalination technologies. Numerous large-scale seawater reverse osmosis (SWRO) desalination plants have been built mostly in Mediterranean countries, the Middle East and Australia, and construction of new plants is expected to increase. Notable examples are the large-scale SWRO plants recently constructed in Algeria (Magtaa plant, 500,000 m^3/day of drinking water), in Spain (the largest RO facility in Europe completed in 2009, covering 20% of Barcelona's water needs [200,000 m³/day]) and in Israel (the world's largest RO desalination plant [624,000 m³/day] in Sorek).

This rapid growth in global RO desalination capacity is related to the dramatic decrease of the SEC over time (Figure 14.2). From approximately 16 KWh/m³ in the early 70s, the SEC has dropped to 3–6 kWh/m³ (including pre and post treatment). The lowest energy consumption reported for an RO system is 1.58 kWh/m³ with a feed water recovery of 42.5% and a flux of 10.2 LMH (Lm⁻²h⁻¹) [34]. This tremendous reduction in the SEC is attributed to continuous technological improvements including enhanced system design and operation, high efficiency pumping, energy recovery devices and/or systems (such as Combined Heat and Power) and improved membrane materials [20, 27, 35, 36].

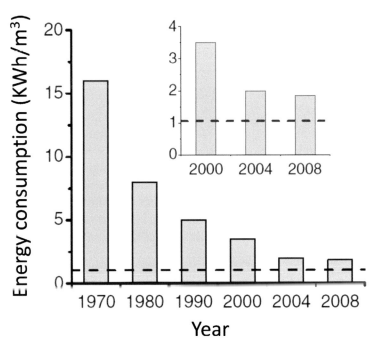

Figure 14.2 The change in energy consumption for the RO unit in SWRO plants from the 1970s to 2008. The horizontal dashed line corresponds to the theoretical minimum energy required for desalination of 35 g/L seawater at 50% recovery (1.06 kWh/m³). The data is only for the desalination unit process, and excludes energy required for intake, pretreatment, post-treatment, brine discharge, etc. [11].

Several reviews on emerging desalination technologies and methods to reduce energy consumption have been recently published [11, 36–38]. Subramani et al. [36] presented a detailed review of energy optimization techniques available for seawater desalination along with a review of renewable energy resource utilization. Elimelech and Phillip [11] reviewed strategies to reduce energy consumption for seawater desalination and concluded that optimization of pretreatment and post-treatment is the best method for energy reduction. Peñate and García-Rodríguez [37] reviewed the current trends and future prospects for seawater RO desalination to improve process performance and obtain high productivity. In a recent review by Subramani et al. [38] a comprehensive description of emerging desalination technologies is provided (both membrane and thermal-based) as well as of other alternative technologies that utilize means other than membranes or thermal technologies. A brief description of the above advances in membrane desalination follows, with emphasis given to SEC minimization through energy recovery devices and the development of emerging membrane technologies, alone or in hybrids with RO.

14.3.2 Technological Improvements to Minimize Losses and to Increase Energy Recovery

At present, the SEC in most new SWRO plants is three to four times greater than the SEC-*min* (Table 14.2). This is due to significant irreversibilities characterizing the main desalination process and the additional energy expenditure in the other parts of the plant. As shown in Figure 14.3, the main energy expenditure (>70%) occurs in the RO membrane desalination process, a significant amount (commonly >1 kWh/m^3) is also consumed in the intake, pre-treatment, post-treatment, and brine discharge stages of the seawater desalination plant. Out of that amount of energy, pre-treatment of raw seawater usually consumes the greatest part; it is noted that adequate pre-treatment is required to control membrane fouling and reduce the concomitant costly plant stoppage and membrane cleaning. Another energy consuming step is the post-treatment, for secondary disinfection of the water and/or removal of excessive boron and chlorides to meet standards for potable and irrigation applications. To reduce such species to acceptable levels, one strategy is to treat a part of or all of the product water by an additional second pass of reverse osmosis, thus adding to the total energy consumption and capital cost [39].

Table 14.2 Range of energy demand for various desalination technologies

Technology[a]	Feed Water Composition[b]	Specific Energy Demand (kWh/m^3)	Reference
MSF	SW	Electrical (SA or CG)[c]: 3.5–5.0	[13]
		SA: Thermal: 69.44–83.33	[14]
		CG: Thermal: 44.44–47.22	
		Electrical: 4.0–6.0	[15]
		Electrical equivalent for thermal energy[d]: 9.5–19.5	
		Total equivalent energy consumption: 13.5–25.5	
MED	SW	SA: Electrical: 0.5–1.5	[14]
		CG: Electrical: 1.5–2.5	[14, 16]
		SA: Thermal: 41.67–61.11	[17]
		CG: Thermal: 27.78	
		Electrical: 1.5–2.5	[15]
		Electrical equivalent for thermal energy[d]: 5.0–8.5	
		Total equivalent energy consumption: 6.5–11.0	
MED-TVC	SW	Electrical: 1.5–2.5	[15]
		Electrical equivalent for thermal energy[d]: 9.5–25.5	
		Total equivalent energy consumption: 11.0–28.0	
MVC	SW	Electrical: 7.0–12.0	[15]
		Electrical equivalent for thermal energy[d]: None	
		Total equivalent energy consumption: 7.0–12.0	
RO	BW	0.5–3.0	[18]
	SW	4.0–8.0 (3.0–4.0)[e]	[19]
		3.0–5.5	[15]
NF	BW	3.5	[20]
ED(R)	BW	0.5–7.0	[21]
		(15–17.0 for SW)	[14, 22]
		6.4	[23]
		1.9	[20]
		1.4–1.8	[24]
(M)CDI	BW	0.1–2.0	[25, 26]

[a]MSF: Multistage-Flash Distillation, MED: Multiple-Effect Distillation, MVC: Mechanical Vapor Compression, RO: Reverse Osmosis, NF: Nanofiltration, ED(R): Electrodialysis (Reversal), (M)CDI: (Membrane) Capacitive Deionization.
[b]BW: Brackish water, SW: Seawater.
[c]SA: stand alone; CG: co-generation.
[d]Electrical equivalent is the electrical energy which cannot be produced in a turbine because of the extraction of the heating steam.
[e]Including energy recovery system.

In relation to the minimization of the energy losses occurring during the RO process, the technological developments have focused on the optimization of different parts of the system, such as feed pumps, energy recovery devices, membrane and modules, as well as efficient system design.

Figure 14.3   Specific energy consumption (SEC) break down of a typical RO operating unit. Data retrieved from [40].

14.3.2.1 High efficiency pumps

In the early days of desalination, positive displacement (PD) pumps were common. PD pumps offered higher efficiencies than early centrifugal pumps when desalination plant sizes were relatively small. As desalination matured, however, users began to realize the limitations of positive displacement pumps, for scale-up to larger facilities, and began to adopt centrifugal pumps due to their inherently easier operation, maintenance and scalability to larger capacities. Centrifugal pump efficiencies have continued to increase due to technological innovations. One key innovation was the use of computational fluid dynamics (CFD) software to optimize the hydraulic component design. The use of CFD allowed pump manufacturers to precisely model the entire hydraulic space of the pump and reduce efficiency losses internally. Another key component was also the optimum design of pumps that can handle the high thrust loads imposed by the desalination process [41].

14.3.2.2 Energy recovery devices

In addition to using high efficiency pumps, energy recovery devices (ERDs) have gained ubiquitous popularity because they reduce the energy used in desalination systems and reduce costs. The first design introduced in the early eighties used systems based on a centrifugal high pressure pump (HPP), an engine and either, Francis or Pelton hydraulic turbines. In this system, the hydraulic energy in the reject stream is converted into rotational energy, in the form of mechanical shaft power. This rotational mechanical energy is transferred into another pumping device (or

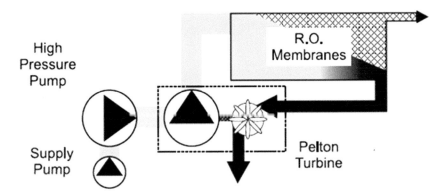

Figure 14.4 Simple schematic of energy recovery with energy recovery turbine and pump (based on Pelton wheel technology) [43].

to the same shaft of the initial HPP) that pressurizes the incoming stream (Figure 14.4). The efficiency of such a system allowed for SECs less than 5 kWh/m^3 [42].

Another type of centrifugal ERD, the hydraulic turbocharger, used for small and medium capacity SWRO plants at the beginning of the 1990s, is similar in operational concept to the Pelton turbine. The turbocharger and the HPP are not directly connected, providing a degree of flexibility in the operation of these devices. Also, turbochargers have a relatively small footprint and are easy to install but the overall efficiency of this ERD is typically 70–80% [44].

From the 1990s onward, several alternatives to centrifugal ERDs have been designed and tested. These kinds of devices use the principles of isobaric chambers for SWRO plants and are known as isobaric pressure exchangers [45]. The HPP size and its motor is substantially smaller than the Pelton wheel configuration. Similar to the centrifugal ERD, the ERDs based on isobaric chambers capture and recycle the energy from the brine (or waste) stream of the desalination process to boost the flow on the membrane feed side (Figure 14.5), thus, offering reduced RO energy consumption in comparison to other systems [28, 31]. For instance, in the case of an SWRO plant with a Pelton wheel energy recovery setup and an SEC of 4 kWh/m^3, the establishment of a new generation ERD could result in a total SEC of 2.35 kWh/m^3, with energy savings close to 41% [37]

There are several manufacturers of pressure exchanger devices nowadays. ERI$^®$, Calder (DWEER^{TM}) [47], RO Kinetic$^®$ ([44] and KSB

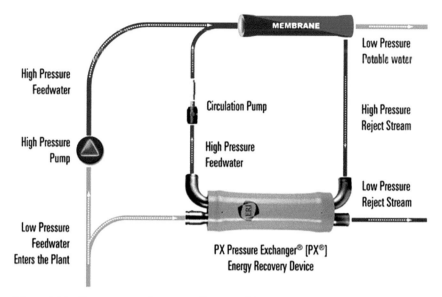

Figure 14.5   Reverse osmosis process diagram with energy recovery device (ERD) [46].

(SalTec DT) [48], Danfoss (iSave) [49] are the main manufacturers of this kind of equipment and are now competing for designing the best system. Other new ERDs are under development like Fluid Switcher (FS), Isobarix, Desalitech's CCD, etc. Most of the commercially available ERD achieve relatively similar net energy transfer efficiencies between 91 and 96%. These devices coupled with low-energy membranes can lead to an enormous advance in SEC reduction, which can drop down to 1.8–2.2 kWh/m^3 in new medium capacity SWRO plants [44, 50].

A general comparison between standard energy recovery turbines (ERTs) such as the Pelton wheel and isobaric ERD trains for large capacity SWRO plants is shown in Table 14.3. In this table, it is not intended to compare the performance of the aforementioned devices in existing plants, but rather to indicate their main differences at the design stage. The choice on the specific type of energy recovery device generally depends on the SWRO plant size, SWRO configuration, site specific conditions, project budget, fuel costs in the region, etc.

14.3.2.3 New generation membranes

It is well-known that the problem of fouling is inherent in RO membrane desalination, significantly impacting plant efficiency and unit product cost.

Table 14.3 Energy recovery turbines vs. isobaric energy recovery devices: a general comparison [37]

Comparison Criteria	Pelton Turbines	Isobaric ERDs
Typical train capacity	Up to 5000 m^3/day	Any capacity
High pressure pump capacity	No change	Can be reduced by 55–60%
Number of booster pumps required	None	1 for each device/train
Relevant civil work requirements	Yes	Little if any
Salinity increase of the feed water before membranes	No	Yes
Approximate efficiency	85%	95%
Relative price	Cheaper	More expensive
Typical SEC achieved	3–4 kWh/m^3	2–3 kWh/m^3
Total process energy saving	35–42%	55–60%

There are three general approaches, or combinations thereof, to deal with this problem; i.e., (i) development of fouling resistant membranes with tailored surface properties; (ii) employing an adequate pre-treatment operation (which entails somewhat increased energy consumption, and capital cost, but may reduce overall plant environmental impact); (iii) improved design and operation of membrane modules leading to reduced fouling and concentration polarization. The development of new, more efficient or fouling-resistant membranes with increased permeability is a challenging task because it requires the development of surface chemistries that resist the adhesion of a wide range of foulants while maintaining the high membrane permeability and selectivity necessary for seawater desalination.

In recent years, breakthroughs in nanotechnology (involving nanocomposites, nanotubes, graphene, biomimetic materials, etc.) has paved the way for the development of new generation membranes with superior properties for desalination (antifouling, increased permeability, higher salt rejection capacity, chlorine resistant, etc.). These new membranes often allow lower operating pressures, in effect reducing power requirements as well as capital costs since they may result in eliminating an additional stage. To date, however, most of these technologies are still under fundamental development. Although, the fabrication of advanced materials, such as carbon nanotubes [51, 52], graphene-based membranes [53–55] and aquaporins [53, 56] show promise due to their superior performance mainly in terms of water permeability while they maintain generally satisfactory salt rejection. However, their fabrication is still limited to small scales, and their usage might also require redesigning the membrane modules.

Feed spacers also play a very significant role in the optimization of membrane element design and overall performance. It is well known that properly designed feed spacers can mitigate membrane fouling (at least to some extent) and in particular the undesirable concentration polarization phenomena by inducing flow unsteadiness and increased shear stresses, which lead to an increased mass transfer coefficient at the membrane surface [57]. However, these rather thin spacers create a narrow feed-side channel gap which is largely responsible for increased pressure drop and energy losses of desalination plants. Therefore, the optimization of feed-spacer design is very important in the context of optimizing the design and performance of spiral wound membrane modules [57] in terms of satisfactory salt rejection and minimization of SEC.

14.3.2.4 Plant design and operation

The design and the configuration of a membrane-based unit have a significant influence on the specific energy cost of the desalination plant. In past decades, membrane unit configuration was usually based on a *two-stage* desalination mode, with six elements per pressure vessel. In recent years, with efforts to reduce energy consumption, in large and medium size plants of high salinity feed water, the preferred configuration is *single-stage*, with pressure vessels usually containing seven (or in some cases eight) elements [58]. For instance, in one study, a reduction of 2.5% in the energy requirements was reported in the case of single-stage compared to a two-stage unit [59].

Other designs of SWRO operations have also been considered. For example, cyclic desalination operations obviate the need for energy recovery devices by recirculating the pressurized feed until the desired percent recovery is achieved [60]. This semi-batch RO process has been commercialized as *closed-circuit desalination* (CCD) by Desalitech, LLC and reported to reduce desalination energy consumption by up to 20%. Although this design appears to reduce energy consumption, questions still remain regarding membrane stability (during the repeated loading and unloading of fluid from the recirculation loop) and the capital cost related to additional equipment requirements. Indeed, the CCD process requires additional pressure vessels that are used as side conduits to replenish the fresh feed-water and reject the brine, thus enabling the desalination cycle to be continuous. Although an ERD is not required in this case, additional control equipment (i.e., automatic valves) are needed for continuous operation.

There is significant activity to develop energy-efficient desalination technologies that are inherently less susceptible to fouling compared to high-pressure,
RO/NF membrane-based desalination methods. Along these lines, innovative
technologies such as membrane distillation, forward osmosis, adsorption
desalination, and (membrane) capaci- tive deionization have been investigated. These methods are characterized by various degrees of maturity and,
in most cases, lack sufficient long-term operating data. Thus, the conservative
water industry usually refrains from using such emerging technologies before
their viability has been sufficiently demonstrated.

Membrane distillation has lately gained popularity due to some unique
advantages. This process can be classified into four basic configurations,
depending on the methods to induce vapor pressure gradient across the
membrane and to collect the diffusing vapors from the permeate side. A
common feature of all the configurations is the direct contact of one side
of the hydrophobic membrane with the feed solution. Direct contact MD
(DCMD) has been the most widely studied option due to its inherent simplicity. However, vacuum MD (VMD) can be used for high output while
air gap MD (AGMD) and sweep gas MD (SGMD) are enjoy the benefit
of low energy losses and high performance ratio [61]. Some new configurations with improved energy efficiency, better permeation flux or smaller
foot print have also been proposed such as material gap MD (MGMD)
[62], multi-effect MD (MEMD) [63], vacuum-multi-effect membrane distillation (V-MEMD) [64] and permeate gap membrane distillation (PGMD)
[64]. Membrane distillation processes can be driven with low enthalpy heat,
including solar energy, geothermal energy, and waste grade energy associated
with low-temperature industrial streams. The hydrophobic membrane used in
the process allows the passage of vapors only and retains the non-volatile
species at the retentive side, thus the product obtained is theoretically 100%
pure from solid or non-volatile contaminants [65, 66]. Due to these attractive
features, membrane distillation has emerged as a potential contributor to the
third generation desalination processes to address some inherent drawbacks
of the established RO process. Recently, a lot or interest in membrane distillation commercialization has been realized; a list of the main suppliers
and developers active in its commercialization can be found in Drioli et al.
[66]. As an example, the Aquaver Company has recently commissioned one
of the world's first seawater MD-based desalination plants in the Maldives.
The plant uses the waste grade heat available from a local power plant and

has the capacity of 10,000 L/day [67]. Another interesting example of the fast-growing interest in MD system development can be found in an R&D program financially supported by the Korean Ministry of Land, Infrastructure and Transportation (MoLIT), and the Korean Agency for Infrastructure Technology Advancement (KAIA) in which the implementation of membrane distillation, valuable resource recovery, and pressure retarded osmosis (PRO) are the main goals (Figure 14.6). It is foreseen that an MD plant with a capacity of 400 m³/day will be realized together with a 200 m³/day PRO unit [68].

Electrodialysis and ED(R) are well proven electrochemical membrane technologies with a multitude of systems operating worldwide, especially for brackish water, tertiary wastewater treatment and industrial applications. However, their share in seawater desalination is very small in contrast to RO or thermal methods [3]. According to Fujifilm [69], there are currently approximately 1,000 ED and EDR installations worldwide. Roughly 10 to 20 new large EDR plants are being installed annually. An analysis of data concerning Japanese plants for seawater concentration by ED(R) (in the process of salt production) may be useful for estimating energy consumption in the process of fresh water obtained from seawater by means of electrodialysis. Reported energy consumption in these plants is 0.150–0.155 kWh/kg NaCl [70–72]. Considering desalination of seawater from a TDS level of 35 g/L, an estimate of 5.26 kWh/m³ of desalinated water may be obtained. In the

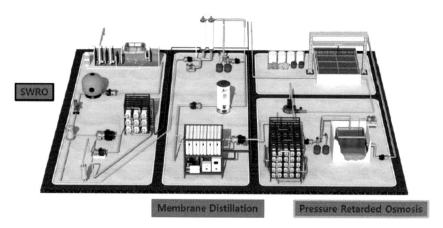

Figure 14.6   A conceptual design of third generation desalination scheme [68].

Abrera plant (close to Barcelona, Spain), which is the world's largest desalination plant using EDR technology (for production of drinking water from surface water, the Llobregat River), the reported average energy consumption (stacks and pumps) is smaller than 0.6 kWh/m^3. This plant started operating on a trial basis in June 2008, and came into the normal operation in April 2009. During the period of April 2009 to August 2010, more than 20 hm^3 were produced through the EDR line [22].

In addition to RO, FO has been proposed for removing salts from saline water since the 1970s [73]. In the early stages, however, most of the studies reported in the form of patents were based on the investigators' ideas, and few matured into operational systems. During the past decade, studies on FO for seawater/brackish water desalination have been revitalized and commercial FO membranes have become available. Notable progress has been made by researchers in the USA [74, 75], Singapore [76–78] and elsewhere [79, 80]. Although FO is unlikely to replace RO as the dominant desalination technology in the foreseeable future, it is expected to find a growing number of applications. A critical review on key aspects of the FO process has been recently published focusing on energy efficiency, membrane properties, draw solutions, fouling reversibility, and possible applications of this emerging technology [81]. This review analyzes the energy efficiency of the process (including evidence to disprove the common misguided notion that FO is a low-energy process) and highlights the potential use of low-cost energy sources.

Adsorption desalination (AD) is an emerging thermally driven process employing low temperature waste heat, that could be available from either renewable sources (geothermal or solar energy) or exhaust of industrial processes, to power the sorption cycle [82]. Broadly speaking, AD exploits the high affinity of water vapor (absorbate) towards silica gel (absorbent) due to extremely strong double bond surface forces developing between hydrophilic porous silica gel and water vapors [83]. In brief, AD involves water vaporization, followed by an adsorption/desorption in beds filled with silica gel and finally water condensation in the condenser. The adsorption/desorption cycle is performed in a pair of beds. During the first half cycle interval, water vapors produced in the evaporator are adsorbed by the silica gel; in the next half cycle, the saturated adsorbent is regenerated by the low temperature waste-heat stream. Finally, the desorbed vapors are condensed in a water-cooled condenser. Compared to other desalination technologies AD possesses some advantages such as exploitation of waste heat sources and low corrosion of the evaporator tubes due to low temperatures ($<35°C$)

prevailing during evaporation. In addition, AD has no moving parts which is translated to reduced maintenance costs, while it employs an environment-friendly adsorbent/adsorbate pair. Finally, it is noteworthy that this emerging technology produces high-grade potable water along with cooling power with only one source of heat and as a result it exhibits relatively low (less than 1.5 kWh/m^3) SEC [84]. Over the past ten years, Ng and co-workers [85–87] have dealt with the optimization of some key operating parameters, namely specific daily water production, cycle time, and performance ratio of this technology, using numerical simulation tools in parallel with laboratory tests. Recently, systematic efforts have also been made to integrate AD with other thermal-based desalination technologies (mainly Multi-Effect Distillation and Multistage Flush Distillation) for potable water production with low SEC [88, 89].

Capacitive de-ionization (CDI) is a technology similar to that of electrostatic double layer (EDL) super capacitors [90] but modified to operate in a "flow through" mode, while focus is on salt removal and not on charge storage. Details on the theoretical and technological background of CDI, the recent development of the so-called membrane capacitive deionization (MCDI) and the present attempts towards scaling up and commercialization are described in a number of reviews [26, 91–94]. CDI and MCDI are considered as energy-efficient processes because they are operated at low voltages (0.8–1.4V) without high-pressure pumps (as in the case of RO), thermal heaters (as in the case of thermal desalination) or high direct current voltage (as in the case of ED, EDR). In the original experiments of Farmer et al. [95, 96], an energy consumption value of 0.1 kWh/m^3 was obtained for brackish waters. This number serves as a figure of merit or preliminary benchmark for this technology. However, Welgemoed and Schutte [25], using a larger pilot plant unit based on similar carbon aerogel materials used before [95, 96], obtained a value of 0.59 kWh/m^3 to produce water with 500 mg/L of total dissolved salts from a 2,000 mg/L synthetic feed solution. In the case of MCDI, energy consumption is reported to be smaller than that of CDI [97], as it removes the salt ions, which are only a small percentage of the feed solution, as compared to most other desalination technologies that involve transferring water molecules, which account for the bulk of the feed solution. Moreover, in the ion desorption step, it is possible to recover energy due to the release of stored charge. Długołęcki and van der Wal [98] claim that up to 83% of the energy invested during the ion adsorption step can be recovered during the ion release step by using the constant current (CC) mode. Similarly, Demirer et al. [99] found a 63% energy recovery by using the constant potential (CV) mode. Recently, Zhao et al. [97] discerned the fraction of energy that can be recovered during the ion desorption step of MCDI, as a function of influent concentration, water flow rate and water recovery. Furthermore, in (M)CDI, membrane fouling appears to be of lesser significance, as compared to that occurring in reverse osmosis or electrodialysis, because no sustained concentrates tend to form. Finally, those CDI processes are considered environmentally friendly, although sufficient data are not available yet for a comprehensive environmental impact assessment.

14.3.2.6 Utilization of renewable energies

RO desalination process schemes exploiting renewable energies (solar, wind, geothermal, hydrostatic, wave and stream energy, as well as osmotic power) have attracted the attention of experts, designers, and researchers [23, 100, 101]. Current SWRO plants consume between 3 and 4 kWh/m^3 and emit between 1.4 and 1.8 kg CO_2 per cubic meter of produced water [27]. Renewable energy cannot reduce the energy footprint of desalination, but it can reduce the environmental footprint. It is, obviously, highly desirable to reduce greenhouse gas (GHG) emissions by reducing desalination dependence on fossil fuels that may also ultimately lead to a reduced cost of energy. Hence, to minimize GHG emissions, renewable energy sources could directly power SWRO desalination plants. Alternatively, indirect compensation or offset measures, such as the installation of renewable energy plants that feed energy into the grid, could also power desalination plants [102], which would resolve problems with intermittent and variable intensities of wind and solar sources. At the same time, desalination units could be operated in an intermittent way to match the variable availability of renewable resources. Due to the importance of this topic, the integration of renewable energies with desalination plants is described separately in another chapter.

14.4 SEC in Thermal Processes

Thermal processes include multi-stage flash (MSF), multiple-effect distillation (MED), and mechanical or thermal vapor compression (MVC, TVC); such units are often referred to as distillers, evaporators, or simply thermal desalination units. Thermal distillation is the oldest method used for desalination, with MSF considered as the backbone for water production, representing approximately 78% of global production capacity in 1999; at that time RO accounted for a modest 10%. However, improvements in RO efficiency have changed the trend. Specifically, in 2011, the market share had decreased to 27% for MSF, while RO accounted for 60% of worldwide capacity [3]. A significant percentage of global desalination capacity is covered by MED (approximately 8% in 2011), with an increasing trend. It is noteworthy that MED has greater compatibility with solar thermal energy [103]. MVC and TVC units represent a small fraction of the global desalination capacity and are generally applied in conjunction with other technologies, in particular with MED. These processes have a low fresh water production capacity (i.e., below 5,000 m^3/day, usually <100 m^3/day), and are used mainly at tourist resorts and industrial sites. Other thermal desalination processes, e.g., solar stills, humidification–dehumidification, freezing, etc., are only found on a pilot or experimental scale.

Thermal desalination processes are generally used in the following applications:

- In locations where energy costs are low or where waste heat sources are available.
- To treat highly saline waters.
- Where seawater has large seasonal variations in temperature and salinity together with high organic loads that are challenging for RO desalination.

The above reasons justify the popularity of distillation techniques in the arid areas of the Middle East and the countries of the GCC[1] in the Arabian Peninsula, where fossil fuels (oil and gas) are widely available [20, 104]. Countries that have significant domestic fossil fuel resources usually subsidize the provision of such fuels to power plants, thereby subsidizing the cost of electricity and steam used for thermal-based desalination technologies. Energy subsidies thus distort the choice of processes in favor of more energy-intensive technologies, such as MSF and MED. These systems are usually constructed in cogeneration plants where power and water are produced simultaneously. This is convenient because both systems require low pressure heating steam which can be easily extracted from the power plant at fairly low cost.

Another trend in GCC countries and elsewhere is the adoption of hybrid system plants that rely on both thermal and membrane-based processes for water and electricity production. Some of the advantages of this hybridization include flexibility in operation, less SEC, low intake/outfall construction cost,

---

[1]Gulf Cooperation Council; Bahrain, Kuwait, Oman, Qatar, Saudi Arabia and the United Arab Emirates.

high plant availability and better power and water matching [105]. The presence of RO compensates for the inflexibility of MSF/MED to follow variable demand and allows for blending of RO and MSF/MED product which reduces (or eliminates) the post treatment costs. A notable example is the plant located in Fujeirah in the United Arab Emirates which generates 650 MW of power, 295,100 m³/day of MSF desalinated water and 170,000 m³/day of RO desalinated water. Due to the importance of the dual-purpose power–desalination plants, a further description of this hybrid concept is given below.

14.4.1 Technological Improvements to Increase Energy Efficiency

Multi-stage flash plants require both electrical and thermal energy with relatively high SEC compared to other competing technologies such as MED and RO (Table 14.2). This is because water is produced by supplying energy for overcoming the latent heat of evaporation. Many authors in scientific and technical literature indicate that MSF has reached its maturity and there is no margin for further improvement; however, it appears that MSF performance and economics can still be improved if the top brine temperature (TBT) and the number of stages can be increased. The use of high TBT antiscalants, and NF for feed water pretreatment can increase TBT and improve system performance. In addition, the partial recycling of the low-pressure vapor can also contribute to improvements of MSF performance and reduction of driving thermal and electrical energies. Therefore, these could reduce specific CAPEX, OPEX, and water production costs [103].

The MED method, which is actually a modification of the MSF process, uses latent heat to produce secondary latent heat in each chamber. The use of double condensing-evaporation heat transfer mechanisms is highly efficient and saves energy (MED consumes only 0.8–1.25 kWh/m³ electrical energy). However, in the past, MED could not compete with MSF due to the scaling problem and the larger CAPEX and OPEX [12]. New designs with operation at lower TBT (~6°C), use of cheaper material and the use of thermal vapor compression seem to have mitigated this problem [104]. Moreover, by combining a source of waste heat with a thin-film distiller MED, the value of prime energy required to drive the process is closer to the minimum energy of separation than in any other desalting technique. Furthermore if electricity is generated 100% by renewable energy, using a distiller in this manner means that more renewable energy can be sent to the grid to reduce fossil fuel generation elsewhere.

In the vapor compression (VC) distillation process, the heat for evaporating the water comes from the compression of vapor rather than the direct exchange of heat from steam produced in a boiler. Two methods are used to compress vapor in order to produce enough heat to evaporate incoming seawater: a mechanical compressor, usually electrically driven, or a thermo-compressor. The mechanical vapor distillation (MVD) is inherently the most thermodynamically efficient process of single-purpose thermal desalination plants. Originally MVC, driven solely by electrical power, was pursued as a competitor to the newly introduced RO technology. However, operating experience has shown the dominance of the RO process. On average MVC consumes 7–12 kWh/m^3 of electrical power, to operate the system and other associated equipment including pumps, controls and auxiliaries. The hybridization (MED/MVC units) and partial use of waste heat for feed water preheating could improve the VC unit performance and reduce the specific fuel consumption. Moreover, the small size of the current MVC units, and the fact that only electrical power is required, makes it feasible to operate using various forms of renewable energy, i.e., wind, photovoltaics, etc.

14.4.2 Dual-Purpose Power–Desalting Water Plants

Boilers usually produce steam at a pressure greater than 20 bar, which is far higher than is required in the brine heater of a thermal desalination plant (i.e., usually not more than 3 bar). Hence, in a single-purpose plant, steam pressure has to be reduced, thereby losing energy efficiency. This loss is avoided in a dual-purpose plant, where water and electricity production are simultaneously pursued (i.e., cogeneration). Cogeneration may take various forms, and theoretically any form of energy production could be used. However, the majority of current and planned cogeneration desalination plants use either fossil fuels or nuclear power as their source of energy. Cogeneration plants normally work with MSF, MED, or hybrid MSF/MED-RO although using the preheated condenser cooling water as feed water to RO is also a possibility. In cases of increased water demand, the desalination units require feeding of direct low pressure steam from auxiliary boilers which tends to increase the specific energy requirements and the unit cost of water produced. In general, large scale hybrid MSF/MED–RO-Power plants allow for greater operating flexibility, demand matching and specific energy reduction. These plants have received significant attention recently with implementation of combined desalination and power plants at Fujairah I and Fujairah II in the

United Arab Emirates (UAE), Ras Al-Khair in Kingdom of Saudi Arabia and Az-Zour South in Kuwait [105].

The primary advantage of a dual-purpose plant is their potential for improved overall energy efficiency, due to significant savings in fuel, which is the major operating expense in both power and desalination operations [105]. Maximum advantage is obtained when both components of a dual-purpose plant (distillers and turbines) are operated as much as possible near their rated capacities. Hence, the size and characteristics of each plant-component should be carefully chosen.

The optimal design of a dual-purpose plant should take into account the individual site demands for electricity and water expressed in terms of the *power to water ratio* (PWR), the total network situation, the cost of energy, the capital costs of the plant etc. The design is eventually optimized for the lowest cost alternative, depending upon the ratios of water to base load power and peak to base load electricity demand. Due to the high cost of power–desalination plants, the countries in the Persian Gulf region are considering feasible alternatives with minimum investment and operating costs and optimal supply of water and power. However, the optimization problem is complicated by significant seasonal mismatch between water and power demands and the increasing future demand of water over power due to the fast rate of industrial and social development and continuous population growth. Furthermore, the choice of the optimal technology in cogeneration plants is made more difficult due to the large number of the combinations of desalination technologies and power facilities that could be coupled together. Hybridization of power systems with electrically driven desalination technologies presents promising design alternatives capable of minimizing PWR while satisfying the other constraints of an optimal selection of the power–desalting plant [106].

14.5 SEC in Hybrid Systems

14.5 SEC in Hybrid Systems

Weaknesses in certain processes can be reduced by other processes in an integrated system [107]. For instance, by coupling a coagulation process or a low pressure membrane filtration (UF or MF) with a high pressure membrane unit (NF or RO), the fouling problem can be reduced significantly (depending, of course, on the quality of the feed water).

Early suggestions for hybrid desalination were based upon elimination of the requirement for a second pass to the RO process so that the higher-salinity RO product could be combined with the better-quality product from a MSF or MED plant. This simple hybrid MSF/RO or MED/RO desalination power process results in several advantages, including extending RO membrane life, increased recovery rate, and decreased energy consumption. Energy consumption is reduced by feeding the cooling water from the distillation process to the RO unit; ultrafiltration (UF) and microfiltration (MF) can be also used to pretreat the feed stream as mentioned before. In another variant of this "classic hybrid scheme", the seawater RO plant's reject brine becomes the feed to the MSF or MED plant particularly if the seawater is softened by nanofiltration (NF) membranes, utilizing its high pressure, with a turbocharger, to boost the MSF plant's recirculation pump. By doing this, the conversion ratio of the hybrid NF/SWRO + MSF/MED plant is significantly increased [108].

Another attractive option is the application of Electrodialysis (ED) and Electodialysis Reversal (EDR) coupled with BWRO for seawater desalination. EDR could act as pre-desalination operation, for reduction of feed-water salt concentration to a desirable level. Subsequently, BWRO can be used as the final desalination process operating at relatively low pressure. The concentrate stream from BWRO, in an EDR/BWRO hybrid unit, has comparable (or lower) salinity than the feed water—when the feed water salinity is $\geq 3,000$ ppm and therefore ideal for recycling in the EDR and consequently increasing the overall recovery of drinking water from the feed water [109–111].

Membrane distillation offers the potential to use waste grade heat to produce pure water from a variety of sources ranging from brackish water, seawater to RO brine. From the efficiency point of view MD is particularly advantageous in the application to highly saline feed solutions (above 60 g/L TDS) where RO has osmotic pressure limitations. In an integrated system coupling MD to SWRO, the water recovery will be increased and waste heat from, combined-heat and power (CHP) can be utilized while the RO feed water can be used for cooling in the condenser side of the MD unit. An integrated SWRO/MD/CHP solution can thus provide energy and water simultaneously at very competitive costs and with increased overall water recovery leading to significantly reduced demand of electrical energy.

Forward osmosis hybrid systems provide energy cost savings, if alternative low-cost thermal energy can be used to power the post-FO separation process. For example, a thermolytic solution of ammonia–carbon dioxide has been proposed as a draw solution in FO, in which low-grade heat energy can be exploited to regenerate the draw solution by preferentially removing the solutes that are significantly more volatile than water [112]. Alternatively,

FO can also be coupled with MD to desalinate waters that are challenging for a standalone MD process. In a hybrid FO–MD system, FO can be employed to mitigate organic fouling and/or mineral scaling that are detrimental to the MD process, whereas the MD process separates the water and regenerates the draw solution using low-grade heat [113]. A detailed discussion on FO hybrid processes that utilize low-cost thermal energy for separation is made by Shaffer et al. [01].

14.6 Renewable Energy (RE) Utilization for Water Desalination

The integration of renewable resources in desalination and water purification is becoming increasingly attractive. This is particularly justified by the fact that areas with freshwater shortages usually have plenty of solar energy. There is a wide variety of technological combinations between desalination technologies and RE sources. The selection of the appropriate 'renewable source-desalination technology' combination has to be based on a variety of parameters such as the size, framework, conditions and purpose of the application.

Furthermore, Table 14.4 presents an overview of the most common or promising RE-desalination technologies, including their typical capacities, their energy demand, and the stage of development. Most technologies have already been tested rather extensively and the water generation costs are estimated based on operational experience and real data.

Finally, an overview of desalination processes driven by renewable energy resources worldwide as well as the breakdown of the renewable energy sources employed for sea and brackish water desalination are presented in Figure 14.7. It is quite clear that photovoltaic panels in conjunction with RO are the most common combination in desalination units operated with renewable sources. Furthermore, wind energy has been successfully combined with both RO and VC technologies. Although photovoltaic energy and wind energy are promising alternative energy sources to power desalination processes, the high cost of photovoltaic modules and the intermittent nature of wind energy are the main barriers for their applications. The most important challenge regarding the application of renewable energies is the intermittent power output generated compared to the steady energy demanded by desalination processes. This can be met with the use of hybrid renewable energy resource units and energy storage systems. An example is to combine the

Table 14.4   Possible combinations of renewable energy with desalination technologies [115]

	Typical Capacity	Energy Demand	Technical Development Stage Applications
Solar still	<0.1 m³/day	Solar passive	Applications
Solar Membrane Distillation	0.15–10 m³/day	Thermal: 150–200 KWh/m³	Advanced R&D
Solar/CSP Multiple Effect Distillation	>5000 m³/day	Thermal: 60–70 KWh/m³ Electrical: 1.5–2 KWh/m³	Advanced R&D
Photovoltaic – Reverse Osmosis	<100 m³/day	electrical BW: 0.5–1.5 KWh/m³ SW: 4–5 KWh/m³	Applications/ advanced R&D
Photovoltaic– Electrodialysis Reversed	<100 m³/d	Electrical: only BW: 3–4 KWh/m³	Advanced R&D
Wind–Reverse Osmosis	50–2000 m³/day	electrical: BW: 0.5–1.5 KWh/m³ SW: 4–5 KWh/m³	Applications/advanced R&D
Wind– Mechanical Vapor Compression	<100 m³/day	electrical: only SW: 11–14 KWh/m³	Basic research

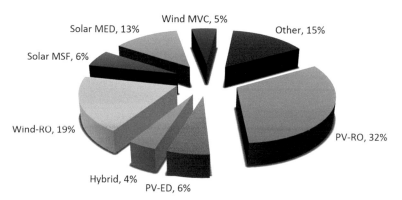

Figure 14.7   Distribution of renewable energy powered desalination technologies [114].

complementary features of wind and solar power as has been done in Oman, Mexico, Germany, and Italy. The application in large-scale projects still needs to be competitive [38, 114].

In general, desalination based on renewable energy sources is still expensive in comparison with conventional desalination, as both investment and power generation costs of renewable energy are higher. However, under certain circumstances – e.g., installations in remote areas where distributed energy generation (heat and power) is more appropriate than centralized energy generation, transmission and distribution – renewable desalination could compete with conventional systems. In addition, with the rapid decrease of renewable energy costs, technical advances and increasing number of installations, desalination by renewables is likely to achieve significantly reduced cost in the near future and become an important source of water supply for regions affected by water scarcity.

14.7 Conclusions and Future R&D

As desalination becomes widely adopted worldwide, energy consumption is and will remain a key issue. That is why today, developments in desalination technologies are specifically aimed at reducing energy consumption in addition to unit cost. Advancements include new and emerging technologies as low temperature distillation, membrane distillation, pressure retarded osmosis, and novel membranes (biomimetic, grapheme, CNT). Hybrid plants, especially those implementing MED and RO are gaining wider use in the Middle East, which has traditionally been home to facilities using more energy-intensive thermal technologies such as MSF. There is also a push to utilize renewable energy to power desalination plants; for example, Saudi Arabia has made a major investment in solar energy for desalination. Renewable energy, notably concentrated solar power (CSP) with thermal storage systems, can significantly contribute to reduce the currently used fossil fuels and associated CO_2 emissions. Other variable renewable energy sources, such as solar PV and wind power can also make significant contributions, possibly combined with energy storage systems. Desalination itself can be seen as a viable option to "store" available renewable energy, in situations when the latter exceeds the demand.

There is scope for future R&D on conventional desalination to focus on some of the following areas:

1. For RO plants: (i) increase the efficiency of pumps and power recovery devices; (ii) improve robustness of membranes and increase their tolerance for increased pressure, temperature and pollutants often found

in seawater; (iii) develop improved techniques for mitigating organic fouling and biofouling, thus leading to increased membrane lifetime; (iv) develop maintenance-free feed water pre-treatment systems that require a minimum amount of additives and chemicals;

2. For thermal plants: (i) improve the heat transfer coefficient through materials improvements; (ii) reduce the cost of equipment and materials, such as evaporators, heat transfer materials and intakes; (iii) develop alternative energy sources;

3. For all desalination technologies: (i) standardize plant sizes and design to reduce the need to design unique units for each site; and (ii) assess and reduce the environmental impacts of brine discharge;

4. More research and data collection is needed regarding the energy usage of desalination plants, especially cogeneration plants. Very few studies have included disaggregated energy usage in each "component" of the cogeneration plant, i.e., electricity and desalination;

5. Pursue applications of renewable energy desalination on a large scale. While a number of relatively small-scale studies have been conducted on renewable energy desalination (primarily solar), these studies are considered insufficient to give an overall picture of the potential for desalination.

The suggestions above are not meant to be exhaustive. Future avenues of research are sure to become available, which could be complementary to the above suggestions or could even push research and development into radically different directions.

References

[1] Joint Research Centre (JRC) (2012). *The Water and Energy Nexus, in Science for Water*. Thematic Report No. JRC71148. City of Brussels: Publication Office of the European Union.

[2] UN Water (2015). *Statistics*. Available at: http://www.unwater.org/statistics/statistics-detail/pt/c/211827/ [accessed November 2015].

[3] IRENA and IEA-ETSAP (2015). *Water Desalination Using Renewable Energy: Technology Brief*. Available at: http://www.irena.org/DocumentDownloads/Publications/IRENA-ETSAP%20Tech%20Brief%20I12%20Water-Desalination.pdf [last accessed August 2017].

[4] Klein G., K. M., Hall V., O'Brien T., Blevins B. (2005). *California's Water–Energy Relationship*. Sacramento, CA: California Energy Commission.

[5] United Nations World Water Assessment Programme (UN WWAP) (2014). *The United Nations World Water Development Report 2014.* Paris: UNESCO.

[6] International Energy Agency (IEA) (2012). *World Energy Outlook 2012.* Paris: International Energy Agency.

[7] Stockholm Environmental Institute (SEI) (2011). "Understanding the Nexus," In *Proceedings of the Background Paper for the Bonn 2011 Nexus Conference – The Water, Energy and Food Security Nexus: Solutions for the Green Economy* (Stockholm: Stockholm Environmental Institute).

[8] World Energy Council (2010). *Water for Energy.* London: World Energy Council.

[9] Spiegler, K. S., and El-Sayed, Y. M. (2001). The energetics of desalination processes. *Desalination* 134, 109–128.

[10] Shrivastava, A., Rosenberg, S., and Peery, M. (2015). Energy efficiency breakdown of reverse osmosis and its implications on future innovation roadmap for desalination. *Desalination* 368, 181–192.

[11] Elimelech, M., and Phillip, W. A. (2011). The future of seawater desalination: Energy, technology, and the environment. *Science* 333, 712–717.

[12] Cooley H., Gleick P., and G., W. (2006). *Desalination, with a Grain of Salt – A California Perspective.* Oakland, CA: Pacific Institute.

[13] Winter T., Pannell D. J., and L, M. (2008). *Economics of Desalination and its Potential Applications in Australia.* Crawley WA: University of Western Australia.

[14] Economic and Social Commission For Western Asia (ESCWA) (2009). *ESCWA Water Development Report 3 Role of Desalination in Addressing Water Scarcity.* New York, NY: United Nations.

[15] DESWARE (2015). *Encyclopedia of Desalination and Water Resources.* Available at: http://www.desware.net/ [last accessed August 2017]

[16] GWI Global Water Intelligence (GWI/IDA DesalData) (2015). *Market Profile and Desalination Markets, 2009–2014 Yearbooks and GWI.* Available at: http://www.desaldata.com/ [last accessed August 2017].

[17] Zhou, Y., and Tol, R. S. J. (2005). Evaluating the costs of desalination and water transport. *Water Resour. Res.* 41, 1–10.

[18] Ettouney, H. M., El-Dessouky, H. T., Faibish, R. S., and Gowin, P. J. (2002). Evaluating the economics of desalination. *Chem. Eng. Prog.* 98, 32–39.

[19] Sauvet-Goichon, B. (2007). Ashkelon desalination plant – A successful challenge. *Desalination* 203, 75–81.

[20] National Research Council (2008), *Committee on Advancing Desalination Technology, Desalination: A National Perspective.* Washington, DC: The National Academies Press.

[21] Xiujuan, C. H., Peigui, C. H., and Yongwen, N. T. (1995). Electrodialysis for the desalination of seawater and high strength brackish water. *Desalination Water Reuse* 4, 16–22.

[22] Valero, F., Barceló, A., and Arbós, R. (2011). "Electrodialysis technology," in *Theory and Applications, Desalination, Trends and Technologies*, ed. M. Schorr (Rijeka: InTech).

[23] Younos, T., and Tulou, K. E. (2005). Overview of desalination techniques. *J. Contemp. Water Res. Educ.* 132, 3–10.

[24] Bennett, B., Park, L., and Wilkinson, R. (2010). *Embedded energy in Water Studies – Study 2: Water Agency and Function Component Study and Embedded Energy – Water Load Profiles.* San Francisco, CA: California Public Utilities Commission.

[25] Welgemoed, T. J., and Schutte, C. F. (2005). Capacitive Deionization Technology^{TM}: an alternative desalination solution. *Desalination* 183, 327–340.

[26] Anderson, M. A., Cudero, A. L., and Palma, J. (2010). Capacitive deionization as an electrochemical means of saving energy and delivering clean water. Comparison to present desalination practices: Will it compete? *Electrochim. Acta* 55, 3845–3856.

[27] Fritzmann, C., Löwenberg, J., Wintgens, T., and Melin, T. (2007). State-of-the-art of reverse osmosis desalination. *Desalination* 216, 1–76.

[28] Shakib, S. E., Amidpour, M., and Aghanajafi, C. (2012). A new approach for process optimization of a METVC desalination system. *Desalination Water Treat.* 37, 84–96.

[29] Sayyaadi, H., and Saffari, A. (2010). Thermoeconomic optimization of multi effect distillation desalination systems. *Appl. Energy* 87, 1122–1133.

[30] El-Sayed, Y. M. (2001). Designing desalination systems for higher productivity. *Desalination* 134, 129–158.

[31] Mussati, S. F., Aguirre, P. A., and Scenna, N. J. (2003). Novel configuration for a multistage flash-mixer desalination system. *Ind. Eng. Chem. Res.* 42, 4828–4839.

[32] Cardona, E., Culotta, S., and Piacentino, A. (2003). Energy saving with MSF-RO series desalination plants. *Desalination* 153, 167–171.

[33] Ludwig, H. (2004). Hybrid systems in seawater desalination – Practical design aspects, present status and development perspectives. *Desalination* 164, 1–18.

[34] Seacord, T., MacHarg, J., and Coker, S. "Affordable desalination Collaboration 2005 results," in *Proceedings of the American Membrane Technology Association Conference*, Stuart, FL.

[35] Semiat, R. (2008). Energy issues in desalination processes. *Environ. Sci. Technol.* 42, 8193–8201.

[36] Subramani, A., Badruzzaman, M., Oppenheimer, J., and Jacangelo, J. G. (2011). Energy minimization strategies and renewable energy utilization for desalination: a review. *Water Res.* 45, 1907–1920.

[37] Peñate, B., and García-Rodríguez, L. (2012). Energy optimisation of existing SWRO (seawater reverse osmosis) plants with ERT (energy recovery turbines): Technical and thermoeconomic assessment. *Energy* 36, 613–626.

[38] Subramani, A., and Jacangelo, J. G. (2015) Emerging desalination technologies for water treatment: a critical review. *Water Res.* 75, 164–187.

[39] Dreizin, Y., Tenne, A., and Hoffman, D. (2008). Integrating large scale seawater desalination plants within Israel's water supply system. *Desalination* 220, 132–149.

[40] Voutchkov, N. (2012). "Energy use for seawater desalination – current status and future trends," in *Water-Energy Interactions in Water Reuse*, eds V. Lazarova, K.-H. Choo, and P. Cornel (London: IWA Publishing), 227–241.

[41] Kadaj, A. (2011). *High Pressure Pumping*. Avalaible at: http://www.pumpsandsystems.com/topics/pumps/pumps/high-pressure-pumping [last accessed August 2017].

[42] Woodcock, D. J., and Morgan White, I. (1981). The application of pelton type impulse turbines for energy recovery on sea water reverse osmosis systems. *Desalination* 39, 447–458.

[43] AES (2015). *Environmental and Process Engineering*. Available at: http://www.aesarabia.com/energy-recovery-systems [last accessed August 2017].

[44] Peñate, B., de la Fuente, J. A., and Barreto, M. (2010). Operation of the RO Kinetic® energy recovery system: description and real experiences. *Desalination* 252, 179–185.

[45] Stover, R. L. (2007). Seawater reverse osmosis with isobaric energy recovery devices. *Desalination* 203, 168–175.

[46] Anderson, W., Stover, R., and Martin, J. (2009). Keys to high efficiency, reliable performance and successful operation of SWRO processes. IDA World Congress – Atlantis, The Palm – Dubai, UAE November 7–12, 2009, REF: IDAWC/DB09-084. Available at: http://www.energyre covery.com/resources/?filter_resource-sector=&filter_resource-category =white-paper&filter_resource-product=px-pressure-exchanger&filter_ resource-location=&search_value= [last accessed August 2017].

[47] Schneider, B. (2005). Selection, operation and control of a work exchanger energy recovery system based on the Singapore project. *Desalination 184*, 197–210.

[48] Bross, S., and Kochanowski, W. (2007). SWRO core hydraulic system: extension of the SalTec DT to higher flows and lower energy consumption. *Desalination* 203, 160–167.

[49] DANFOSS company. Available at: http://www.isave.danfoss.com [last accessed August 2017].

[50] Dundorf, S., Macharg, J., and Seacord, T. F. (2007). "Optimizing lower energy seawater desalination, the affordable desalination collaboration," in *IDA World Congress*, Maspalomas.

[51] Humplik, T., Lee, J., O'Hern, S. C., Fellman, B. A., Baig, M. A., Hassan, S. F., et al. (2011). Nanostructured materials for water desalination. *Nanotechnology* 22.

[52] Ahn, C. H., Baek, Y., Lee, C., Kim, S. O., Kim, S., Lee, S., et al. (2012). Carbon nanotube-based membranes: fabrication and application to desalination. *J. Ind. Eng. Chem.* 18, 1551–1559.

[53] Pendergast, M. M., and Hoek, E. M. V. (2011). A review of water treatment membrane nanotechnologies. *Energy Environ. Sci.* 4, 1946–1971.

[54] Yang, H. Y., Han, Z. J., Yu, S. F., Pey, K. L., Ostrikov, K., and Karnik, R. (2013). Carbon nanotube membranes with ultrahigh specific adsorption capacity for water desalination and purification. *Nat. Commun.* 4.

[55] Mi, B. (2014). Graphene oxide membranes for ionic and molecular sieving. *Science* 343, 740–742.

[56] Kaufman, Y., Berman, A., and Freger, V. (2010). Supported lipid bilayer membranes for water purification by reverse osmosis. *Langmuir* 26, 7388–7395.

[57] Karabelas, A. J., Kostoglou, M., and Koutsou, C. P. (2015). Modeling of spiral wound membrane desalination modules and plants – review and research priorities. *Desalination* 356, 165–186.

[58] Petry, M., Sanz, M. A., Langlais, C., Bonnelye, V., Durand, J. P., Guevara, D., et al. (2007). The El Coloso (Chile) reverse osmosis plant. *Desalination* 203, 141–152.

[59] Wilf, M., and Bartels, C. (2005). Optimization of seawater RO systems design. *Desalination* 173, 1–12.

[60] Efraty, A. (2010). U.S. Patent 7,695,614 B2.

[61] Summers, E. K., Arafat, H. A., and Lienhard, V. J. H. (2012). Energy efficiency comparison of single-stage membrane distillation (MD) desalination cycles in different configurations. *Desalination* 290, 54–66.

[62] Francis, L., Ghaffour, N., Alsaadi, A. A., and Amy, G. L. (2013). Material gap membrane distillation: a new design for water vapor flux enhancement. *J. Membrane Sci.* 448, 240–247.

[63] Liu, R., Qin, Y., Li, X., and Liu, L. (2012). Concentrating aqueous hydrochloric acid by multiple-effect membrane distillation. *Front. Chem. Sci. Eng.* 6, 311–321.

[64] Zhao, K., Heinzl, W., Wenzel, M., Büttner, S., Bollen, F., Lange, G., Heinzl, S., and Sarda, N. (2013). Experimental study of the memsys vacuum-multi-effect-membrane-distillation (V-MEMD) module. *Desalination* 323, 150–160.

[65] Camacho, L. M., Dumée, L., Zhang, J., Li, J. D., Duke, M., Gomez, J., and Gray, S. (2013). Advances in membrane distillation for water desalination and purification applications. *Water (Switzerland)* 5, 94–196.

[66] Drioli, E., Ali, A., and Macedonio, F. (2015). Membrane distillation: Recent developments and perspectives. *Desalination* 356, 56–84.

[67] Available at: http://www.aquaver.com [last accessed August 2017].

[68] Available at: http://www.globalmvp.org [last accessed August 2017].

[69] FUJIFILM (2015). *Membrane Development*. Available at: http://www.fujifilmmembranes.com/water-membranes/technology [last accessed August 2017].

[70] Tanaka, Y. (1999). Regularity in ion-exchange membrane characteristics and concentration of sea water. *J. Membrane Sci.* 163, 277–287.

[71] Yamamoto, M., Hanada, F., Funaki, S., and Takashima, K. (2000). "A new electrodialyzer technique for the salt production by ion-exchange membrane," in *Proceedings of the 8th World Salt Symposium*, Vol. 1, ed. R. M. Geertman (Amsterdam: Elsevier), 647–652.

[72] Asahi Glass Co., Ltd. (2002). Brochure.

[73] Zhao, R., Biesheuvel, P. M., and Van Der Wal, A. (2012). Energy consumption and constant current operation in membrane capacitive deionization. *Energy Environ. Sci.* 5, 9520–9527.

[74] Cath, T. Y., Childress, A. E., and Elimelech, M. (2006). Forward osmosis: principles, applications, and recent developments. *J. Membrane Sci.* 281, 70–87.

[75] Hoover, L. A., Phillip, W. A., Tiraferri, A., Yip, N. Y., and Elimelech, M. (2011). Forward with osmosis: emerging applications for greater sustainability. *Environ. Sci. Technol.* 45, 9824–9830.

[76] Ling, M. M., and Chung, T. S. (2011). Desalination process using super hydrophilic nanoparticles via forward osmosis integrated with ultrafiltration regeneration. *Desalination* 278, 194–202.

[77] Yang, Q., Wang, K. Y., and Chung, T. S. (2009). Dual-layer hollow fibers with enhanced flux as novel forward osmosis membranes for water production. *Environ. Sci. Technol.* 43, 2800–2805.

[78] Su, J., Yang, Q., Teo, J. F., and Chung, T. S. (2010). Cellulose acetate nanofiltration hollow fiber membranes for forward osmosis processes. *J. Membrane Sci.* 355, 36–44.

[79] Valladares Linares, R., Li, Z., Yangali-Quintanilla, V., Ghaffour, N., Amy, G., Leiknes, T., and Vrouwenvelder, J. S. (2016). Life cycle cost of a hybrid forward osmosis – low pressure reverse osmosis system for seawater desalination and wastewater recovery. *Water Res.* 88, 225–234.

[80] Valladares Linares, R., Li, Z., Sarp, S., Bucs, S., Amy, G., and Vrouwenvelder, J. S. (2014). Forward osmosis niches in seawater desalination and wastewater reuse. *Water Res.* 66, 122–139.

[81] Shaffer, D. L., Werber, J. R., Jaramillo, H., Lin, S., and Elimelech, M. (2015). Forward osmosis: Where are we now? *Desalination* 356, 271–284.

[82] Ng, K. C., Thu, K., Kim, Y., Chakraborty, A., and Amy, G. (2013). Adsorption desalination: an emerging low-cost thermal desalination method. *Desalination* 308, 161–179.

[83] Ghaffour, N., Bundschuh, J., Mahmoudi, H., and Goosen, M. F. A. (2015). Renewable energy-driven desalination technologies: a comprehensive review on challenges and potential applications of integrated systems. *Desalination* 356, 94–114.

[84] Thu, K., Ng, K. C., Saha, B. B., Chakraborty, A., and Koyama, S. (2009). Operational strategy of adsorption desalination systems. *Int. J. Heat Mass Trans.* 52, 1811–1816.

[85] Kim, Y.-D., Thu, K., Ng, K. C., Amy, G. L., and Ghaffour, N. (2016). A novel integrated thermal-/membrane-based solar energy-driven hybrid desalination system: Concept description and simulation results. *Water Res.* 100, 7–19.

[86] Thu, K., Chakraborty, A., Kim, Y. D., Myat, A., Saha, B. B., and Ng, K. C. (2013). Numerical simulation and performance investigation of an advanced adsorption desalination cycle. *Desalination* 308, 209–218.

[87] Saha, B. B., Koyama, S., Kashiwagi, T., Akisawa, A., Ng, K. C., and Chua, H. T. (2003). Waste heat driven dual-mode, multi-stage, multi-bed regenerative adsorption system. *Int. J. Refrig.* 26, 749–757.

[88] Thu, K., Kim, Y. D., Amy, G., Chun, W. G., and Ng, K. C. (2013). A hybrid multi-effect distillation and adsorption cycle. *Appl. Energy* 104, 810–821.

[89] Ng, K. C., Thu, K., Oh, S. J., Ang, L., Shahzad, M. W., and Ismail, A. B. (2015). Recent developments in thermally-driven seawater desalination: energy efficiency improvement by hybridization of the MED and AD cycles. *Desalination* 356, 255–270.

[90] Simon, P., and Gogotsi, Y. (2008). Materials for electrochemical capacitors. *Nat. Mater.* 7, 845–854.

[91] AlMarzooqi, F. A., Al Ghaferi, A. A., Saadat, I., and Hilal, N. (2014). Application of capacitive deionisation in water desalination: a review. *Desalination* 342, 3–15.

[92] Biesheuvel, P. M., and van der Wal, A. (2009). Membrane capacitive deionization. *J. Membrane Sci.* 346(2), pp. 256–262.

[93] Porada, S., Zhao, R., Van Der Wal, A., Presser, V., and Biesheuvel, P. M. (2013). Review on the science and technology of water desalination by capacitive deionization. *Progr. Mat. Sci.* 58, 1388–1442.

[94] Oren, Y. (2008). Capacitive deionization (CDI) for desalination and water treatment – past, present and future (a review). *Desalination* 228, 10–29.

[95] Farmer, J. C., Fix, D. V., Mack, G. V., Pekala, R. W., and Poco, J. F. (1996). Capacitive deionization of NaCl and NaNO3 solutions with carbon aerogel electrodes. *J. Electrochem. Soc.* 143, 159–169.

[96] Farmer, J. C., Fix, D. V., Mack, G. V., Pekala, R. W., and Poco, J. F. (1996). Capacitive deionization of NH4ClO4 solutions with carbon aerogel electrodes. *J. Appl. Electrochem* 26, 1007–1018.

[97] Zhao, R., Porada, S., Biesheuvel, P. M., and Van der Wal, A. (2013). Energy consumption in membrane capacitive deionization for different water recoveries and flow rates, and comparison with reverse osmosis. *Desalination* 330, 35–41.

[98] Długołęcki, P., and Van Der Wal, A. (2013). Energy recovery in membrane capacitive deionization. *Environ. Sci. Technol.* 47, 4904–4910.

[99] Demirer, O. N., Naylor, R. M., Rios Perez, C. A., Wilkes, E., and Hidrovo, C. (2013). Energetic performance optimization of a capacitive deionization system operating with transient cycles and brackish water. *Desalination* 314, 130–138.

[100] Mathioulakis, E., Belessiotis, V., and Delyannis, E. (2007). Desalination by using alternative energy: Review and state-of-the-art. *Desalination* 203, 346–365.

[101] Charcosset, C. (2009). A review of membrane processes and renewable energies for desalination. *Desalination* 245, 214–231.

[102] Lattemann, S., and Höpner, T. (2008). Environmental impact and impact assessment of seawater desalination. *Desalination* 220, 1–15.

[103] Mezher, T., Fath, H., Abbas, Z., and Khaled, A. (2011). Techno-economic assessment and environmental impacts of desalination technologies. *Desalination* 266, 263–273.

[104] Buros, O. K. (2000). *The ABCs of Desalting*. Topsfield, MA: International Desalination Association.

[105] Hamed, O. A. (2005). Overview of hybrid desalination systems – Current status and future prospects. *Desalination* 186, 207–214.

[106] Helal, A. M. (2009). Hybridization – A new trend in desalination. *Desalination Water Treat.* 3, 120–135.

[107] Ang, W. L., Mohammad, A. W., Hilal, N., and Leo, C. P. (2015). A review on the applicability of integrated/hybrid membrane processes in water treatment and desalination plants. *Desalination* 363, 2–18.

[108] Awerbuch, L. (2014). Hybrid desalination: the best of both worlds? *Water Wastew. Int.* 20–24.

[109] Pilat, B. (2001). Practice of water desalination by electrodialysis. *Desalination* 139, 385–392.

[110] Turek, M. (2003). Cost effective electrodialytic seawater desalination. *Desalination* 153, 371–376.

[111] Post, J. W., Huiting, H., Cornelissen, E. R., and Hamelers, H. V. M. (2011). Pre-desalination with electro-membranes for SWRO. *Desalination Water Treat.* 31, 296–304.

[112] McCutcheon, J. R., McGinnis, R. L., and Elimelech, M. (2005). A novel ammonia-carbon dioxide forward (direct) osmosis desalination process. *Desalination* 174, 1–11.

[113] Xie, M., Nghiem, L. D., Price, W. E., and Elimelech, M. (2013). A forward osmosis-membrane distillation hybrid process for direct sewer mining: System performance and limitations. *Environ. Sci. Technol.* 47, 13486–13493.

[114] Eltawil, M. A., Zhengming, Z., and Yuan, L. (2009). A review of renewable energy technologies integrated with desalination systems. *Renew. Sustain. Energy Rev.* 13, 2245–2262.

[115] Papapetrou, M., Wieghaus, M., and Biercamp, C. (2010). *Roadmap for the Development of Desalination Powered by Renewable Energy, Promotion of Renewable Energy for Water Desalination.* Available at: http://wrri.nmsu.edu/conf/conf11/prodes_roadmap_online.pdf

CHAPTER 15

Brine Management

Christopher Bellona[1]

[1]Colorado School of Mines, Golden, CO, USA
E-mail: cbellona@mines.edu

15.1 Introduction

Desalination of unconventional water supplies such as seawater is becoming an increasingly important option for water supply. One of the most divisive issues related to desalination is the management and disposal of the reject stream produced during treatment. Desalination facilities primarily rely on discharge to wastewater treatment plants (potable reuse applications and inland brackish water desalination), deep-well injection (inland brackish water desalination), and ocean discharge (seawater desalination, inland brackish water desalination, and indirect potable reuse applications). These disposal options are often scrutinized because of concerns regarding human and environmental health, impacts on receiving water quality, and inability of wastewater treatment plants to accommodate increased salinity. In addition, the costs associated with reject disposal can be a significant portion of the overall cost depending on the volume of reject stream and type of discharge required. In order to alleviate the aforementioned issues, and recover more water at a desalination facility, various strategies and technologies have been developed to minimize reject disposal with the ultimate goal of zero liquid discharge (ZLD).

Recent estimates indicate that the capacity of desalination plants operating worldwide will soon reach 100 million cubic meters per day of desalinated water. If it is assumed that the average recovery of these facilities is 50%, an equivalent flow rate of 100 million cubic meters of reject will require disposal per day. As such, there has been significant interest in reject management and the development of approaches to reduce the volume produced and more sustainable disposal practices. The purpose of this chapter is to highlight and summarize this topic and inform practitioners on the work being performed to minimize desalination reject impacts on the environment.

15.2 Desalination Brine/Concentrate Characteristics and Disposal

The characteristics of desalination reject are dependent upon the raw water quality, the type and recovery of the desalination process, and whether or not pretreatment chemicals are added during the process. Pretreatment chemicals include disinfectants, acids, and antiscalants. Additionally, dilution of reject can be accomplished through several approaches including co-locating desalination facilities with power generating facilities and blending reject with process water. For thermal processes, brine composition is mainly a function of recovery as the product water is extremely pure. The composition of membrane system brine/concentrate is a function of the membrane system recovery and solute rejection. The solute concentration factor (CF) can be calculated as:

$$CF = \left(\frac{1}{1 - R_w} \right) [1 - R_w(1 - R_s)] \qquad (15.1)$$

where, R_w is the product water recovery and R_s is the rejection or removal of the constituent of interest. Seawater desalination facilities typically operate at recovery values between 30 and 50% while brackish water and wastewater reclamation facilities can operate up to 85% recovery. The impact of recovery and rejection on solute concentrations is presented in Figure 15.1. As brine/concentrate becomes more concentrated through a desalination system, sparingly soluble salts reach their solubility limit and can precipitate

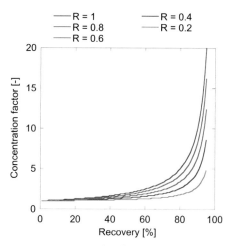

Figure 15.1   Brine/concentrate concentration factor as a function of recovery and rejection.

within the system, in pipes and associated equipment and in downstream treatment processes. Precipitation can have a detrimental effect on thermal and membrane-based desalination processes and is one of the main reasons why product water recoveries are limited.

15.2.1 Seawater Desalination

Probably, the greatest challenge toward managing and minimizing seawater desalination brine is elevated concentrations of ions, particularly monovalent salts which are difficult to remove, and sparingly soluble salts which cause scaling in downstream processes. As mentioned, seawater desalination facilities are typically operated at recoveries between 30 and 50% and, therefore, produce a brine stream that is up to two times the concentration of seawater (Table 15.1). This waste stream is corrosive, requiring compatible materials for conveyance systems, and contains chemicals added upstream of the desalination system including pretreatment chemicals such as acids,

Table 15.1   Example seawater reverse osmosis brine measured water quality ($n = 168$ with exception of bromide and fluoride)

Parameter or Ion	Unit	Average Value	Standard Deviation
pH	–	7.976	0.19
Turbidity	NTU	0.361	0.43
Conductivity	mS/cm	62.321	11.61
Total dissolved solids (TDS)	ppt	45.272	9.63
Alkalinity	ppm	140.612	25.14
Ammonia	ppm	0.027	0.03
Nitrate	ppm	0.040	0.02
Nitrogen oxides	ppm	0.041	0.03
Total nitrogen (as N)	ppm	0.299	0.68
Total phosphorous (as P)	ppm	0.129	0.26
Calcium	ppt	0.550	0.10
Magnesium	ppt	1.744	0.35
Sodium	ppt	14.070	3.12
Potassium	ppt	0.650	0.16
Bromide	ppm	70.800	NA
Bicarbonate	ppt	0.132	0.03
Carbonate	ppt	1.920	6.83
Sulfate	ppt	3.874	0.80
Chloride	ppt	25.264	5.22
Fluoride	ppm	2.200	2.53

ppt, parts per thousand; ppm, parts per million; NTU, nephelometric turbidity units; mS/cm, millisiemens per centimeter.

disinfectants, coagulants, antifoaming agents and antiscalant chemicals. While technologies exist for further treatment of seawater desalination brine, it is significantly cheaper to return it to the ocean.

One main issue cited by critics of seawater desalination facilities is the potential negative impacts of brine disposal on marine organisms. Because salts are concentrated during desalination, the resulting brine has a density greater than seawater and the potential to create plumes at the bottom of the ocean, which could impact bottom-dwelling organisms. Brines diluted by other waste streams can also form neutral or buoyant plumes that can negatively impact marine life. In particular, high salinity, increased alkalinity, elevated temperature, and presence of trace metals and pretreatment chemicals have been cited as potentially causing adverse environmental impacts. Although adverse health impacts to marine life have been demonstrated in several studies, ocean discharge remains the industry standard for seawater desalination facilities. Strategies to mitigate adverse impacts caused by brine discharge include blending brine with wastewater or power plant cooling water, multiple-port diffusers to prevent plume formation, and flow augmentation using untreated wastewater. The drivers for seawater desalination brine treatment and/or minimization include the recovery of additional drinking water, reduction of brine impacts on the environment, and the extraction of salable or usable commodities. However, besides a few installations, seawater desalination brine treatment is rarely performed.

15.2.2 Brackish Water Desalination

Brackish water is loosely defined as having intermediate salinity compared to seawater and conventional drinking water sources. The chemical composition of brackish water depends on the source (e.g., estuarine water or groundwater) as well as the fact that many established brackish water desalination facilities are located in inland areas making brine/concentrate disposal difficult. Brackish water quality can vary significantly and depends on the surrounding geologic formation, origin, hydrogeological recharge mechanisms, anthropogenic impacts, and influence of seawater on the aquifer (Table 15.2).

Brackish water desalination reject management strategies vary widely depending on the salinity of the feed water composition, recovery of the desalting system and resulting brine composition, proximity to seawater and/or subsurface disposal wells, and climactic conditions. Mickley [1]

Table 15.2 Major ion composition of brackish groundwater

Ion	Groundwater Tularosa Basin, NM	Groundwater Las Vegas, NV	Groundwater Hueco Bolson, TX	Oil and Gas Prod. Water Eddy Co., NM
Sodium	114	755	116	3430
Potassium	2	72	7	NA
Calcium	420	576	136	600
Magnesium	163	296	33	171
Chloride	170	954	202	4460
Nitrate	10	31	NA	NA
Phosphate	0	NA	NA	NA
Sulfate	1370	2290	294	2660
Bicarbonate	270	210	190	488
Silicon dioxide	22	77	31	NA
TDS	2630	5270	1200	11900

Source: Adapted from [2].

All values in units of mg/L; TDS, total dissolved solids.

conducted a survey of USA desalination facilities[1] and reported that the predominant method was surface water disposal (mainly ocean discharge) followed by wastewater treatment plants, injection wells, evaporation, and recycle and reuse. In arid regions of the world, such as the Middle East, inland desalination plants often rely on evaporation ponds due to high evaporation rates. Increased reliance on brackish water desalination, more stringent regulations, environmental concerns, and limited disposal options has resulted in significant interest in brine/concentrate minimization strategies.

15.2.3 Desalination for Potable Water Reuse

With dwindling water supplies in the United States and many regions worldwide, potable reuse is becoming an increasingly important component of water resource management. Potable reuse treatment methods have evolved over the past 30 years and the use of an integrated membrane system (IMS, i.e., microfiltration (MF) or ultrafiltration (UF) followed by reverse osmosis (RO)) coupled with advanced oxidation processes (AOPs) has emerged as the industry standard [3]. Due to relatively low wastewater effluent salinity, potable reuse RO systems can operate at recoveries between 70 and 85%,

---

[1]Desalination facilities included seawater, softening, and potable reuse plants however, the number of seawater facilities was low.

As a result, concentrate from reuse facilities is approximately six times more concentrated as the feed water and must be disposed of in an appropriate manner (Table 15.3). In addition to salts, potable reuse concentrate also contains wastewater derived organic contaminants and other chemicals discharged to sanitary sewer systems. There is increased interest in the impact that these chemicals have on the environment. The majority of reuse facilities are

Table 15.3   Wastewater reverse osmosis concentrate measured water quality at two wastewater reclamation facilities

| | | Facility 1 | | Facility 2 |
| | | | | |
Analyte	Detection Limits	RO Concentrate ppm ($n = 2$)	Average Deviation	RO Concentrate ppm ($n = 1$)
Aluminum	0.018	0.16	0.05	0.10
Antimony	0.016	BDL	NA	BDL
Arsenic	0.039	BDL	NA	BDL
Barium	0.0003	0.36	0.09	0.19
Beryllium	0.0002	BDL	NA	BDL
Boron	0.053	0.66	0.04	BDL
Cadmium	0.001	BDL	NA	0.02
Calcium	0.01	439.44	109.87	482.00
Chromium	0.003	BDL	NA	BDL
Cobalt	0.004	BDL	NA	0.01
Copper	0.003	0.03	0.00	0.02
Iron	0.002	0.33	0.08	0.77
Lead	0.013	BDL	NA	BDL
Lithium	0.003	0.06	0.03	0.12
Magnesium	0.0004	104.31	26.08	156.62
Manganese	0.0003	0.33	0.10	0.34
Molybdenum	0.002	0.09	0.03	0.12
Nickel	0.002	BDL	NA	0.04
Phosphorus	0.1	1.85	0.45	2.78
Potassium	0.19	118.26	29.50	111.91
Selenium	0.04	BDL	NA	BDL
Silica	0.11	71.44	13.53	61.49
Silver	0.004	BDL	NA	BDL
Sodium	0.005	1148.88	224.89	1398.53
Strontium	0.023	4.73	1.23	4.23
Tin	0.01	BDL	NA	BDL
Titanium	0.001	BDL	NA	BDL
Vanadium	0.001	BDL	NA	0.01
Zinc	0.001	0.21	0.07	NA

The RO system was operated at approximately 85% recovery (n = number of samples); BDL, below detection limit; NA, not available; RO, reverse osmosis.

located in coastal areas where ocean discharge is available. However, recent droughts in the US have pushed the development of inland reuse treatment facilities at which, concentrate management strategies are becoming more important.

15.3 Technologies for Brine/Concentrate Treatment

15.3.1 Reverse Osmosis and Nanofiltration

The fundamentals of high-pressure membranes (RO and NF) are covered in previous chapters. An important aspect of desalination brine/concentrate management is the product water recovery of the membrane system. Membrane system recovery is governed by a number of factors, including feed water ionic composition and salinity, level of pretreatment, and optimum conditions for minimizing energy consumption. Because of the high osmotic pressure of seawater, the recovery of seawater RO (SWRO) systems is far less than less saline sources. Brine/concentrate management strategies (with the exception of seawater systems) often involve increasing membrane system recovery through various methods, the most common being the use of secondary RO (i.e., the second RO treating the primary RO brine/concentrate stream, also known as two-stage desalination). For certain brackish groundwater applications, the implementation of secondary RO requires removal of sparingly soluble salts (e.g., $CaCO_3$) through a precipitation process (lime softening, Figure 15.2). Recently, a water reuse facility in Southern California added a secondary RO system to allow for plant expansion while maintaining the same concentrate flow rate that is subsequently sent to a wastewater treatment plant. Beyond RO, significant research has recently been performed to identify and develop treatment technologies that can treat brines with high salt concentrations. More discussion is provided in following sections.

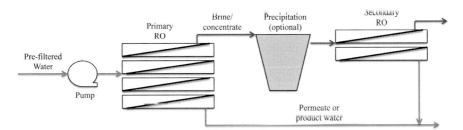

Figure 15.2 Secondary (two-stage) RO system with intermediate softening.

Compared to RO, NF has been proposed only for certain niche applications in the desalination industry. These applications include a dual NF system that was extensively piloted by the Long Beach Water Department in Southern California and NF as a pretreatment to RO or MSF for seawater desalination. Hassan et al. [4] proposed a desalination process using NF as a pretreatment step to SWRO or MSF, which has subsequently been pilot- and demonstration-tested at the Umm Lujj facility in Saudi Arabia. The advantage of NF pretreatment is reported to be the reduction in sparingly soluble salts, microorganisms, organic foulants, turbidity, and TDS, which allows the SWRO to operate at lower energy consumption and higher recovery (70%) and the MSF at a higher brine temperature and recovery (80%). Subsequent pilot testing indicated that hybridization of RO and MSF desalination processes by introducing NF could reduce typical SWRO desalination costs by approximately 30% [5]. Based on these results, the Umm Lujj desalination facility was retrofitted with NF pretreatment for long-term demonstration testing of the NF/SWRO process [6, 7]. Macedonio et al. [8] conducted a cost and energy analysis on several integrated desalination processes (RO, MSF, and membrane crystallization [MCr]), including NF pretreatment and reported that although NF can significantly increase the recovery of desalination systems, the cost savings are only marginal when energy recovery devices are used. However, the application of MCr to treat NF brine has been proposed as a method to yield salable products such as calcium carbonate, sodium chloride, and magnesium sulfate [9, 10].

15.3.2 Electrodialysis and Electrodialysis Reversal

A number of treatment schemes using ED and EDR have been proposed for desalination, RO brine/concentrate minimization, ZLD applications, and salt production from brine and concentrates [11–16]. Typically, in these processes, ED or EDR is proposed to treat desalination system brine or concentrate to increase system recovery and for further concentration prior to evaporation or crystallization for salable salt recovery [12, 13, 17]. Most of the studies evaluating the latter approach are limited to laboratory-scale investigations; however, such an approach is used to produce food-grade salt from seawater in Japan, Korea, and Taiwan. Tanaka et al. [12] analyzed costs and energy requirements associated with sodium chloride production from RO brine using ED and evaporation and reported that it required 80% of the energy associated with sodium chloride production from seawater.

Several studies have investigated EDR as a means of increasing the product water recovery for brackish water desalination applications [15, 18–20].

This goal can be achieved either by operating the EDR as the primary desalination technology at higher recovery than an RO counterpart, or by treating RO brine with EDR. EDR has several benefits over RO including the potential to be operated at higher recovery ($\sim$90–94%), less pretreatment requirements, lower fouling propensity towards certain constituents (e.g., silica and particles), and greater chlorine resistance. However, EDR generally only compares favorably to RO on an economic basis for brackish water with 1,000–3,000 mg/L TDS. As an example, Juby et al. [18] conducted a comprehensive evaluation of multiple approaches to ZLD including lab- and pilot-scale studies and an economic evaluation of various options. The researchers reported that RO brine treatment with EDR was significantly more expensive (total annual costs) compared to a secondary RO system.

An additional ED technology receiving considerable recent attention for brine/concentrate minimization is termed electrodialysis metathesis (EDM). EDM differs from ED in that it has two diluting streams (EDM feed and NaCl) and two concentrating streams that are normally called mixed Cl and mixed Na (Figure 15.3). The EDM feed can be brackish water or RO brine/concentrate. Anions from the EDM feed mix with Na from NaCl to

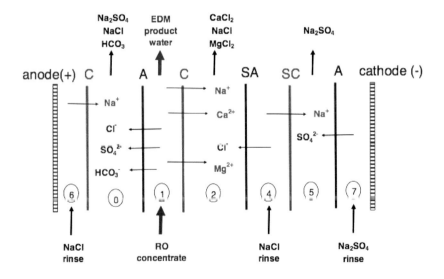

C = ordinary cation exchange membrane
A = ordinary anion exchange membrane
SC = monovalent selective cation exchange membrane
SA = monovalent selective anion exchange membrane

Figure 15.3   Schematic of the EDM process (from [23]).

produce various sodium salts (Na_2SO_4, NaCl, $NaHCO_3$, etc.). Cations from the EDM feed mix with Cl from NaCl to produce various chloride salts ($CaCl_2$, $MgCl_2$, and NaCl). This configuration is designed to separate ions from the raw water into two streams of highly soluble salts (sodium salts and chloride salts) and, therefore, achieve high recoveries when treating RO concentrate [21]. EDM system recoveries have been reported at 99% when raw water TDS is less than 2000 mg/L. Veolia has purchased the rights to the technology and is apparently developing a commercial EDM system called Zero Discharge Desalination [22]. Several sources of information have alluded that sequential precipitation processes could be performed on EDM concentrate to obtain salable products (e.g., calcium carbonate, calcium sulfate, and sodium sulfate); however, limited information is available regarding feasibility [22, 23]. Recently, a Denver Colorado-based membrane research group piloted the EDM system and assessed the feasibility of ZLD with the current system in several process configurations. When operating, the researchers were able to achieve overall desalination system recovery values greater than 96%. However, EDM failure caused by membrane scaling, current leakage through the membrane, and other issues limited the amount of time the system could be operated.

15.3.3 Evaporation Ponds and Solar Evaporation

Although not very common in the United States, evaporation ponds are a popular brine/concentrate disposal option in arid regions with high evaporation rates and inexpensive land. According to Mickley [24], the cost of evaporation ponds becomes excessive when brine/concentrate flow rates exceed 0.3 mgd. Good reviews and analysis of the use of evaporation ponds for desalination brine/concentrate disposal can be found in several references [1, 24–26]. Although solar salt is commonly produced from evaporation ponds, the literature does not suggest that this method has been significantly explored for salt production from desalination brine/concentrate. One exception is a reported case in Israel where seawater desalination brine is sent to a solar salt works which will be further discussed in Section 15.5.8.

An additional evaporation technology called wind-aided intensified evaporation technology (WAIV) has been developed to overcome the issues associated with evaporation ponds. The process improves evaporation rates by employing wind to reduce humidity at the evaporation surface. Studies have demonstrated WAIV evaporation rates can be improved by up to 90% compared to conventional evaporation ponds [27]. Advantages to the process

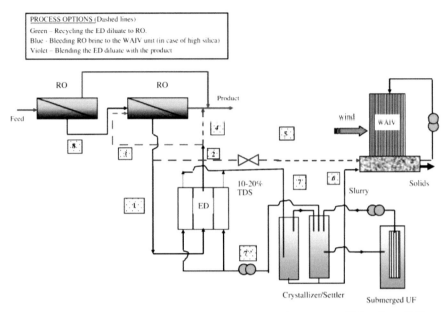

PROCESS OPTIONS (Dashed lines)

Green – Recycling the ED diluate to RO.
Blue – Bleeding RO brine to the WAIV unit (in case of high silica)
Violet – Blending the ED diluate with the product

Figure 15.4   ZLD approach combining RO, ED, crystallization and WAIV (from [15]).

include reduced requirements for pond siting and engineering and a smaller footprint. WAIV has been evaluated for brackish water ZLD systems in conjunction with secondary RO, ED and crystallization (Figure 15.4) [15].

15.3.4 Distillation and Evaporation Systems

There are a number of distillation/evaporation processes that can be used for the goal of brine/concentrate minimization, ZLD, or extraction of constituents from waste streams. For the extraction of materials, evaporators can be used to precipitate salts and, when used upstream of other processes (e.g., crystallizers) can increase solute concentrations and reduce the volume of brine/concentrate being treated. The selection of one evaporator design over another is based on a number of factors including cost, energy requirements, access to waste heat, flow rate, and feed water dissolved solids concentration. Several distillation and evaporation processes are discussed in preceding chapters. Both multiple-effect distillation (MED) and multiple-stage flash (MSF) have been proposed in treatment schemes for increasing system reco very and the extraction of constituents from desalination brine/concentrate. For example, MED and MSF have been used or proposed as evaporation steps after ED for production of food-grade salt [12], and sodium hydroxide, and chlorine [28, 29]. However, these processes have high capital, and operational and maintenance costs and are not commonly used for flow rates less than at least 1 mgd. As opposed to MED and MSF, mechanical vapor compression (MVC) and thermal vapor compression (TVC) evaporators can be used to treat small flows of highly concentrated brines and achieve high recoveries [11]. Often termed brine concentrators, similar to MED and MSF, energy requirements increase with increasing feed stream concentration, and energy requirements for MVC and TVC are reportedly between 8.5 and 23 kWh/m^3. Brine concentrators have been considered for ZLD systems [11] and the production of salt from RO brine/concentrate [29]. Juby et al. [18] and Mickley [30] showed that the use of a brine concentrator to achieve near ZLD increases total annual costs of a desalination system significantly.

15.3.5 Membrane Distillation/Crystallization

Membrane distillation (MD) is a combined thermal evaporation–membrane separation process that relies on a vapor pressure gradient across a membrane to evaporate and transport the feed solution to the distillate side of the membrane where it condenses (Figure 15.5). Because nonvolatile solutes cannot penetrate the membrane, as the feed solution becomes concentrated, substances preferentially crystallize on the membrane surface (MCr). Once crystallized, solids can be removed from the membrane surface through solvent rinsing. Laboratory-scale research has demonstrated that the MD and MCr process can operate at very high TDS concentrations and requires lower temperatures than other thermal processes.

Permeate flux is only moderately impacted by crystal growth on the membrane surface, and flux can be restored through membrane rinsing [9, 10, 31, 32]. Several researchers have claimed that integrating SWRO with

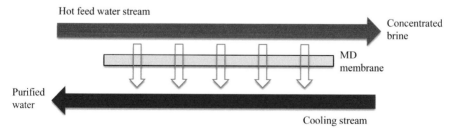

Figure 15.5 Schematic of membrane distillation.

MCr could lead to system recoveries of 90% or more and the recovery of precipitated salts [33]. Ji et al. [32] evaluated MCr for the treatment of RO brine and production of sodium chloride and reported that approximately 21 kg of sodium chloride could be produced from 1 m^3 of RO concentrate. Although MCr operation was stable over a period of 100 hours, the researchers noted that the presence of organic matter had a negative impact on sodium chloride production (crystal growth was reduced by 15–25%).

15.3.6 Precipitation and Crystallization

Crystallizers are used for the production of bulk and high purity compounds and materials in a variety of industries including chemical, pharmaceutical, and food applications. Industrial crystallization is well developed, and a number of different types of crystallizer designs exist [34]. The selection of one crystallizer design over another is dependent upon a number of factors, including desired product size, quality, process economics, and scale of operation [88]. Based on the available literature, fluidized bed crystallizers (FBCs) and forced-circulation crystallizers have been evaluated or proposed in ZLD systems and intermediate softening or chemical extraction applications for desalination [35]. Other crystallization processes proposed for brine/concentrate minimization or chemical extraction from desalination brine/concentrate include eutectic freezing crystallization [29] and MCr [32, 36, 37]. The use of crystallizers for ZLD systems would likely require pre-concentration through the use of a brine concentrator.

Fluidized bed crystallizers use seeding material (e.g., sand and crystals) to promote secondary nucleation and crystal growth [38]. Depending on the ionic solid being produced, crystallization can be induced through evaporation or chemical reagents. Large seed crystal surface areas in FBCs allow for crystal growth at lower supersaturations than would be feasible in the absence of seed material [35, 38]. A typical FBC operates in an upflow configuration to fluidize the seeding materials and crystals without washing them out in the process effluent (Figure 15.2). Within the FBC there are two zones, the lower zone containing the fluidized bed (approximately 25% solids) and the upper zone containing fine crystals with a lower solids content [39]. For softening applications, fine silica is typically used as seeding material, and crystals produced in FBCs are reported to be between 0.3 and 10 mm in diameter [34, 38]. As crystals grow, they sink toward the bottom of the reactor, where they can be continuously or periodically removed. The drained crystals typically have high solids content (~90%) and are easily dewatered [38].

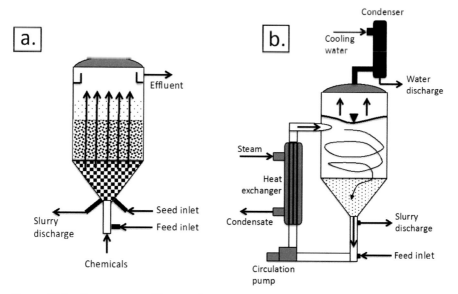

Figure 15.6 Common crystallizer configurations: (a) fluidized-bed crystallizer, (b) forced-circulation crystallizer.

Fluidized bed crystallizers are commonly used in industry to produce potassium chloride, ammonium sulfate, and sodium borate [34]. Bond and Veerapaneni [35] reported that calcite crystals produced in FBC softening applications in the Netherlands are reused for various beneficial purposes. Several recent studies have evaluated FBCs for intermediate softening prior to secondary RO, and purportedly a utility in Southern California (Western Municipal Water District) is upgrading its desalination process to include intermediate softening with an FBC [40]. The municipality plans to produce calcite crystals to be sold to a construction material producer. For this application, clarification (with coagulant addition) post-FBC was required to meet the calcium, hardness, silica, and turbidity removal goals prior to secondary RO.

Forced-circulation (FC) crystallizers use evaporation to induce crystallization. FC crystallizers consist of a main body with a recirculation loop through a heat exchanger to vaporize the liquid within the tank (Figure 15.2). According to Bennett [34], FC crystallizers are the most widely used and least expensive type of equipment available for the production of sodium chloride, sodium sulfate, and sodium carbonate.

Chemical precipitation can be an additional and important component of desalination systems designed to achieve high recoveries. Different schemes have been developed but typically involve the addition of sodium hydroxide, lime, or soda ash and coagulants/polymers followed by clarification. Proprietary high recovery RO systems employing intermediate softening include OPUS (Veolia Water Solutions and Technology, France), and Advanced Reject Recovery of Water (ARROW, O'Brien and Gere, Australia). The objective of chemical precipitation is to remove sparingly soluble salts so that they do not precipitate in the secondary RO system. An additional proprietary high recovery system called Slurry Precipitation and Reverse Osmosis (SPARRO) utilizes tubular RO membranes in which a seed crystal slurry is circulated through. During high recovery operation, scalants precipitate on the seed crystals instead of the membrane. The aforementioned systems have been demonstrated to operate at greater than 90% recovery depending on the brine TDS and water chemistry.

15.3.7 Ion Exchange

Ion-exchange processes use the interchange or exchange of ions through electrostatic interactions between two phases (i.e., water and ion-exchange material) to separate ions. The exchange of ions in solution is dependent upon numerous factors but largely depends on the concentration of ions in solution and the affinity of the ions for the ion-exchange material relative to the solution phase. Various natural materials (e.g., inorganic solids) act as ion-exchange resins; however, ion-exchange resins have also been engineered from various polymers to provide high affinity and selectivity for specific ions. Hundreds of ion-exchange resins have been developed over the past 100 years using various polymers and ligands, and a good historical review of ion-exchange resin development can be found in Alexandratos [41].

Several proprietary high-recovery RO systems use ion-exchange to treat brine from a primary RO prior to secondary RO treatment (e.g., OPUS, ARROW, and high-efficiency reverse osmosis (HERO)). For these systems, ion-exchange is used to remove sparingly soluble salts such as barium sulfate, calcium carbonate, calcium sulfate and calcium phosphate so scaling during primary or secondary RO treatment is minimized. In many cases ion-exchange treatment is used in conjunction with chemical precipitation to target a wide variety of scalants.

A large amount of research has been conducted to develop and optimize ion-exchange materials and resins for the separation of valuable constituents from aqueous solutions. A comprehensive review of ion-exchange processes developed for metal extraction from seawater (up until 1984) can be found in Schwochau [42]. Schwochau's review demonstrates that significant effort has been put towards the development of ion-exchange resins for metal extraction from seawater, particularly lithium, magnesium, and uranium. An additional and more recent review was also given by Khamizov et al. [43]. Ion-exchange resins have been proposed or evaluated for the extraction of ions from seawater and desalination brine, including ammonia, boron, cesium, lithium, magnesium, molybdenum, phosphorus, potassium, rubidium, and uranium.

15.3.8 Other Notable Processes

A number of additional processes that have been or are being developed to minimize operational difficulties and reduce costs and energy requirements when treating high salinity water exist. Variants of NF and RO processes use novel methods to reduce fouling and scaling including the vibratory shear enhanced processing (VSEP) system and vortex enhanced separation systems. The VSEP system uses vibration to minimize foulant deposition on the membrane while the vortexing system uses mechanical mixing on the membrane surface. Forward osmosis (i.e., pulling water from brine using osmotic pressure) has gained recent attention and has been tested for treating brackish water desalination brine with mixed results [11]. Capacitive deionization uses porous electrodes to pull ions out of solution and hold them within an electrode material until regeneration occurs.

15.4 Implementation of Brine/Concentrate Minimization

While there has been a significant amount of research on brine/concentrate management strategies and technologies, to date there have been few full-scale applications. This can most likely be explained by the fact that very high recovery or ZLD systems have high capital, operational and maintenance costs, are operationally complex, and often require significant chemical inputs. Therefore, it appears that when brine/concentrate management is a major issue and utilities are faced with the choice between desalination and alternative water supplies or approaches, the latter is often selected. Alternative approaches often include conservation measures, and when possible, alternative water resource development or treatment technology selection, and/or wastewater recycling.

There are several good references regarding the costs and constraints related to brine/concentrate management schemes. Juby et al. [18] performed a study for the Eastern Municipal Water District (EMWD) in California to determine the best brackish water desalination brine management strategy to increase potable water recovery and decrease the volume of brine requiring disposal. A number of treatment train scenarios were developed (Figure 15.7) and an economic and energy analysis was performed for treating the primary RO brine (flow rate of 1 mgd, TDS of 6,000 mg/L) towards the goal of ZLD. Brine treatment processes considered included lime softening, secondary RO (2 RO), EDR, thermal brine concentration (BC), the SPARRO process, crystallization, evaporation ponds, commercial extraction of valuable materials, and disposal to an ocean discharge brine pipeline.

Results of the economic assessment for the various treatment schemes illustrated in Figure 15.7 are provided in Table 15.4. Besides commercial extraction options, ZLD options analyzed had similar total annual costs to one another at nearly $0.016/gal ($4.23/m^3). If extracted salts could be sold, the costs of commercial extraction approaches are reportedly competitive with alternate ZLD scenarios. It is worth noting that options which use brine concentrators result in high-energy requirements and operational and

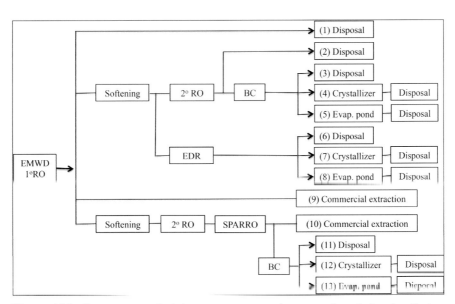

Figure 15.7 Reverse osmosis brine management schemes evaluated for the Eastern Municipal Water District (EMWD) in California. Figure adapted from Juby et al. [18].

Table 15.4   Economic and energy analysis results for alternative brine volume reduction options

Alternative	Process	Total Capital Cost (MD)	Annualized Capital Cost (MD)	Annual O&M Cost (MD)	Total Annual Cost (MD)	Power Consumption (MWh/yr)
1	Baseline	8.1	–	–	1.1	–
2	Softening+2°RO	14.8	1.3	1.0	2.3	1.1
3	Softening+2°RO+BC+Disposal	32.1	2.8	2.7	5.5	9.4
4	Softening+2°RO+BC+Crystallization(ZLD)	29.2	2.5	3.4	6.0	10.8
5	Softening+2°RO+BC+Evap.Pond(ZLD)	25.6	2.2	3.6	5.9	9.4
6	Softening+EDR+BC+Disposal	32.6	2.8	2.8	5.7	10.9
7	Softening+EDR+BC+Crystallization(ZLD)	29.1	2.5	3.4	6.0	12.1
8	Softening+EDR+BC+Evap.Pond(ZLD)	25.8	2.3	3.6	5.9	10.9
9	Commercial extraction	25.5	2.2	6.5	8.8	2.3
10	Softening+2°RO+SPARRO+Commercial extraction	27.8	2.4	4.3	6.8	2.0
11	Softening+2°RO+SPARRO+BC+Disposal	35.3	3.1	2.3	5.4	6.6
12	Softening+2°RO+SPARRO+BC+Crystallizer(ZLD)	30.6	2.7	2.7	5.3	7.5
13	Softening+2°RO+SPARRO+BC+Evap.Pond(ZLD)	30.0	2.4	2.7	5.2	5.6

Source: Table adapted from Juby et al. [18]

maintenance costs. Alternatives 12 and 13 have lower energy requirements and costs due to the reduction in the flow rate sent to the brine concentrator, although the SPARRO system is still in the research and development stages. The researchers also found that the use of crystallization and landfilling was cost competitive to evaporation ponds. By far, the most economical brine minimization approach was found to be softening followed by secondary RO, which would still require final brine disposal. It is also worth noting that the researchers evaluated MD and forward osmosis for brine minimization both of which were either cost prohibitive and/or not mature enough for actual application.

Mickley [30] performed a similar study looking at the costs, energy requirements, and factors affecting various ZLD approaches including:

1. Primary RO followed by thermal brine concentration and evaporation ponds.
2. Primary RO followed by thermal brine concentration, crystallization, evaporation ponds and landfilling.
3. Primary RO followed by lime softening, secondary RO, thermal brine concentration, evaporation ponds and landfilling.
4. Primary RO followed by lime softening, secondary RO, thermal brine concentration, crystallization, evaporation ponds and landfilling.
5. Primary RO followed by lime softening, secondary RO, evaporation ponds and landfilling.

Similar to Juby et al. [18], the authors demonstrated that pre-concentration through secondary RO prior to thermal brine concentrations could significantly reduce capital and operating costs of ZLD systems. As such, options 3 through 5 above were more cost effective than options 1 and 2 with option 3 generally being the most cost effective (total annualized costs) across different flow rates and salinities. However, the economics of the secondary RO and lime softening system is highly dependent upon feedwater salinity and becomes less cost effective at higher salinities. In addition, higher calcium concentrations increased estimated costs due to increased chemical use and sludge disposal. Treatment trains that did not use crystallization tended to be more cost-effective than those that did. From almost all of the evaluated options, evaporation ponds and landfilling comprised the greatest portion of treatment train capital costs. It is worth noting that both Mickely and Juby et al. caution that their reported cost assessments can have significant associated error and that there are a number of additional factors that affect the viability of high-recovery and ZLD systems. Factors include, but are not limited to, salinity changes throughout the treatment train, actual chemical inputs required, corrosion, and operation and maintenance issues.

15.5 Extraction of Constituents from Brine/Concentrate

The extraction of important commodities from seawater has been practiced for thousands of years (i.e., sodium chloride), and was once the major source of magnesium and bromine in the US. Given its massive volume, the ocean has long been seen as a source of additional valuable commodities and more recently, researchers have postulated that commodity extraction from desalination brine could be viable due to elevated concentrations of various compounds and ions. If viable, utilities could potentially reduce treatment costs while improving the potential for brine volume minimization and increased recovery of fresh water. Even without economic gains, a utility could potentially reduce costs of brine treatment and possibly recover chemicals used at the treatment plant. The following sections summarize work performed on the extraction of various constituents from brackish water, seawater and desalination brine/concentrate. Focus will be placed on constituents that have been recovered from seawater and those that have garnered significant research attention.

15.5.1 Bromine

Bromine was once produced from seawater in large quantities to meet the demand created by the advent of antiknock gasoline [44]. The main process produced hydrobromic acid (HBr) using a method by which bromine was stripped from solution using gaseous chlorine (termed the chlorine blowout method), subsequently converted to HBr using sulfur dioxide, and captured using a mist elimination system. This process was ultimately halted due to increased competition in the bromine market, reduced bromine demand, and the discovery of more concentrated bromine sources. More recently, several researchers have proposed or evaluated processes to extract bromine from desalination brine [17, 43, 45, 46]. Davis [17] proposed using EDR to separate sodium, chloride, potassium, and bromide followed by a crystallization process to precipitate sodium chloride. The remaining concentrated solution would then be treated using the chlorine blowout process.

15.5.2 Calcium

There are a number of salable calcium commodities including calcium carbonate, calcium hydroxide, calcium sulfate, and calcium chloride. While calcium sulfate and chloride prices are relatively low, high purity calcium carbonate and hydroxide are relatively valuable. There is increased interest in calcium removal from RO brine to increase RO system recovery through secondary RO, as well as its extraction and sale to recoup desalination costs [18, 35, 40]. This is typically performed by precipitating calcium carbonate in fluid bed crystallizers but also can be accomplished through thermal/mechanical evaporation. Currently, the Western Municipal Water District in California is expanding one of its desalination facilities (Arlington Desalter) to include a fluidized bed crystallizer for calcium carbonate removal and subsequent recovery [40]. The primary purpose of the crystallizer is to remove calcium carbonate prior to a secondary RO system, and allow for plant expansion without increasing the brine flow rate to the ocean discharge pipeline.

15.5.3 Chlorine and Sodium Hydroxide

Chlorine and sodium hydroxide are primarily produced from concentrated and relatively pure sodium chloride solutions. The three main technologies for producing chlorine and sodium hydroxide use electrolysis, and are either membrane, diaphragm or mercury electrolytic cells. For seawater and impure brines however, relatively simple electrolytic cells (e.g., plate, tubular, or disc electrolyzers) are primarily employed for chlorine and sodium hydroxide production. Various companies produce these on-site chlorine generation systems to produce free chlorine from seawater for coastal industries, oil platforms, ships, and their use has been proposed for desalination facilities. These systems however, would likely not allow for sodium hydroxide recovery, have relatively high energy demand (2.5–5 kWh/kg of equivalent free chlorine), and produce low chlorine concentrations (up to approximately 2,000 mg/L). The generation of chlorine from seawater or desalination brine for disinfection of drinking water at desalination facilities has been employed at various plants. Spagnoletto [47] raised the point that produced chlorine solutions may require sterilization and proposed an ultrafiltration system upstream of the chlorine generation system.

Approximately 14 million tons of chlorine and sodium hydroxide are produced by the US chlor-alkali industry annually and their combined value is estimated between $500–$600/ton [48]. While not demonstrated, several researchers have analyzed systems for producing chlorine gas, hydrogen, and high-purity sodium hydroxide from desalination brine. Kim [29] analyzed a production system using MSF for brine concentration followed by electrolysis and reported a sodium hydroxide and chlorine production cost of \$166/ton. This value is likely lower than reality, due to the capital and operational costs of MSF and the electrolytic cell, and the likely need for brine purification prior to electrolysis. Realizing the need for brine purification, Melian-Martel et al. [49] developed a scheme to concentrate and purify brine using precipitation, clarification, filtration, thermal evaporation and ion-exchange prior to electrolysis. The authors estimated energy requirements of the electrolysis system to be between 2,100 and 2,600 kWh/ton-Cl_2, which is close to reported values for membrane electrolytic cells.

15.5.4 Lithium

A large amount of research has been conducted to extract lithium from seawater owing to the increase in lithium demand, uncertainty regarding long-term lithium availability, and uneven global lithium distribution [50]. The main technology evaluated for lithium extraction has been adsorption, with researchers attempting to develop highly selective materials with high adsorption capacity. In addition, membrane based technologies, specifically liquid membrane systems, have also been evaluated. A summary of the effectiveness and selectivity of recently evaluated lithium extraction technologies and associated research findings can be found in Bellona et al. [51].

There are several major challenges associated with lithium extraction from seawater and desalination brine. First, the concentration of lithium in desalination brine is expected to be low as the average seawater concentration is 174 micrograms per liter [52]. Second, while sorption capacities of engineered materials have improved significantly, the volume of water needed to extract an appreciable amount of lithium is immense and requires an equally immense amount of adsorbent material. Nevertheless, given the anticipated boom in electric vehicles and lithium-ion batteries, there is a drive for the development of processes to use the ocean as a source of lithium.

15.5.5 Magnesium

Magnesium was once almost exclusively derived from seawater and numerous seawater extraction facilities exist worldwide due to its relatively high concentration in seawater [44]. Economically valuable magnesium compounds include various grades of magnesium oxide (i.e., magnesia, MgO), magnesium carbonate ($MgCO_3$), magnesium chloride ($MgCl_2$), magnesium hydroxide ($Mg(OH)_2$), and magnesium sulfate ($MgSO_4$). Typical seawater magnesium facilities recover magnesium hydroxide through successive precipitation steps using lime or dolime for precipitation of magnesium hydroxide followed by washing and a filtration step to concentrate the magnesium hydroxide slurry [44, 53, 54]. High-purity magnesium products (i.e., magnesia oxide, magnesium hydroxide, magnesium chloride) are relatively valuable commodities and although commodity information is difficult to obtain, are worth in the range of $200–$500 per ton. Although desalination facility brine is concentrated relative to seawater (potentially greater than a concentration factor of two), there is little evidence indicating that this would improve the economics of extraction considerably. Several researchers have explored the concept of extracting magnesium from desalination brine to both improve the recovery of desalination facilities as well as for economic gain [17, 28, 45, 55, 56]. In general, most of these studies were limited to desk-top feasibility analyses and small-scale laboratory experiments. There is some evidence that desalination facilities in China are implementing processes to recover potassium and magnesium from RO brine for fertilizer [57].

Several alternate approaches have been proposed in recent years for producing magnesium compounds. Abdel-Aal et al. [58] proposed an MSF and separation process for recovery of water, sodium chloride and magnesium chloride-rich brines from desalination facilities in Saudi Arabia. Ohya et al. [56] proposed a series of processes to recover salts from an integrated desalination system comprised of UF and NF pretreatment followed by RO. Production of magnesium in the system would require a precipitation step to remove calcium carbonate followed by an evaporation or crystallization step for recovery of magnesium sulfate. Drioli et al. [10, 59] investigated several integrated processes (similar to the process proposed by Ohya et al. [56]) to extract calcium carbonate (calcite), sodium chloride and magnesium sulfate (epsomite) from NF and RO retentate during desalination using NF pretreatment followed by RO. The integrated processes evaluated consisted of variants of NF, RO, MCr, and chemical precipitation. Economic analyses conducted both by Drioli et al. [10] and Kim [29] indicated that several of these configurations could produce a profit after the sale of the extracted salts although final brine disposal costs would need to be evaluated.

Other proposed processes include integrated systems for the recovery of additional commodities along with magnesium. Davis [17] recently evaluated the extraction of magnesium (as $Mg(OH)_2$), salt and bromine from RO brine during a laboratory-scale study. The proposed process consisted of ED treatment of RO brine to separate monovalent and divalent ions followed by a precipitation step using sodium hydroxide to yield magnesium from

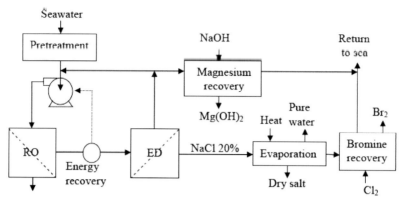

Figure 15.8 Proposed process for extraction of bromine, salt and magnesium from RO brine (from [17]).

the ED diluate (Figure 15.8). The SAL-PROC process (Geoprocessors) has been discussed [60, 61] as a technology to reclaim sodium chloride, calcium chloride, calcium sulfate and magnesium hydroxide from desalination brine and other high salinity waste streams. SAL-PROC is an integrated process consisting of sequential chemical precipitation steps that purportedly requires no hazardous chemicals and has been demonstrated at pilot-scale for treating saline waters to produce salable by-products and significantly reduce the volume of brine to be disposed [61].

Magnesium extraction may only be viable in the case that a desalination facility recoups some costs through the sale of magnesium, and simultaneously increases recovery while reducing brine disposal. The re-mineralization of desalinated water has also received attention due to the issues regarding the stabilization of RO permeate at desalination facilities [62, 63]. One proposed scheme is to separate magnesium and calcium from a portion of desalination feedwater using nanofiltration and to blend the NF concentrate into the finished water. The proposed NF process was calculated to be significantly cheaper (at least two orders of magnitude) than adding chemicals.

15.5.6 Nitrogen and Phosphorous

Recovery of nitrogen and phosphorous from seawater and brackish water desalination brine is unlikely to be feasible. However, recovery from potable reuse RO concentrate has garnered some attention. Precipitation of nitrogen and phosphorous from wastewater in the form of struvite ($MgNH_4PO_4$) has been researched extensively as it serves as a slow release fertilizer. For wastewater streams with low phosphate concentrations, several researchers have utilized fixed-bed sorption processes to remove the phosphate followed by regeneration, and subsequent precipitation of struvite from the regenerent solution [64–66]. Precipitation of struvite from the regenerent solution typically requires the addition of magnesium and ammonia. However, the volume is much smaller than the volume of wastewater treated, and the regenerent solution can be reused once struvite is precipitated.

Additional processes for recovering nitrogen include air-stripping, electrodialysis, and ion-exchange. The capital and operating costs of these technologies may preclude the extraction of nitrogen due to relatively low concentrations. However, the demand for traditional and unconventional sources of phosphorous is expected to increase in the near future due to population growth, and increased fertilizer use in food and biofuel production [67, 68]. Kumar et al. [69] investigated phosphate selective resins for the extraction of phosphate from the concentrate of an RO system operating on microfiltered wastewater effluent. Once the selective ligand exchange resins were exhausted with respect to phosphate, they were regenerated with a sodium chloride solution from which, struvite was precipitated by adding ammonium and magnesium chloride (molar ratio of $P:Mg:NH_4$ was 1:1.5:1). The process was evaluated using concentrate samples with phosphate concentrations of 10–20 mg-P/L and found feasible for both removing phosphate from RO concentrate with relatively high total dissolved solids, and producing struvite.

15.5.7 Potassium

Potassium is the sixth most abundant mineral in seawater and is found at a concentration of approximately 0.4 ppt [52]. Potassium compounds are currently produced from seawater; however, they are generally produced as a byproduct from solar salt operations which makes up a very small percentage of worldwide potassium compound production. As an alternative to solar evaporation/precipitation, several researchers have suggested that potassium could be produced from RO concentrate using an evaporation/crystallization process [45, 70, 71]. Although no physical evaluations were conducted, Mohammadesmaeili et al. [71] performed thermodynamic phase equilibrium calculations to determine the evaporation requirements to produce potassium compounds from secondary RO concentrate. After an evaporation/crystallization step to produce sodium chloride (concentration factor of approximately 10), glaserite ($K_3Na(SO_4)_2$) could theoretically be produced through an additional evaporation/crystallization step by reducing the volume by 93%. The researchers indicate that a 25% glaserite/75% halite product would be produced, but cautioned that evaporation experiments should be performed to verify the evaporation path used in the calculations. An alternative process uses potassium selective ion-exchange resin and according to Khamizov et al. [72], the US, China and Japan have attempted to develop technologies to recover potassium from seawater for fertilizers. Based on Khamizov et al. [72], a variant of this process is being pilot tested through a collaboration of the Vernadsky Institute (Moscow, Russia) and the King Abdulaziz University (Jeddah, Saudi Arabia).

15.5.8 Sodium

Commercially available sodium chloride is primarily produced from solar salt operations (evaporation ponds), underground mining operations, or by evaporation of concentrated brine obtained through solution mining [73]. A number of researchers have proposed producing sodium chloride from desalination brine although in most cases, salt production was never evaluated on an appreciable scale. Compared to conventional methods, several alternate treatment schemes and technologies have been proposed or evaluated including evaporation technologies [58], evaporation followed by crystallization [19, 74], membrane crystallization [33, 59], ED/EDR followed by evaporation [12, 19], and evaporation ponds [61, 75].

In the absence of appreciable halite deposits and available land, and unfavorable conditions for evaporation ponds, Japan has commercially produced salt from seawater by employing ED for ion separation followed by evaporation using MED or vacuum pan evaporators. Reportedly, ED or EDR is used to concentrate sodium chloride in seawater up to 200 g/L prior to evaporation and crystallization using MED [73]. Tanaka et al. [12] conducted a simulation on the ED/MED process for salt production, and concluded that using SWRO desalination brine would reduce energy requirements 80% compared with using seawater. Davis [17] recently proposed a similar process for ZLD systems, and has developed the EDM system which the author claims could be integrated with evaporation to yield sodium chloride. Unfortunately, no information could be acquired that demonstrated that EDM could be an effective method for the precipitation of sodium chloride.

In addition to Japan, a publication by Ravizky and Nadav [75], describes a partnership between Mekorot Water Company (Israel) and Israel Salt Company for SWRO disposal in evaporation ponds and subsequent production of food-grade salt. In the process, the entire brine stream from the SWRO desalination facility is blended with seawater and sent to a series of evaporation ponds. According to the authors, this arrangement has led to a 30% increase in salt production compared to seawater alone. MCr has also been investigated for sodium chloride production, and Drioli et al. [59] reportedly produced relatively pure salt crystals from a synthetic NF concentrate solution containing calcium and magnesium. MCr is a technology that is currently in the developmental stages; however, the advantages reportedly include operation at low temperatures and the capability of manipulating system conditions to selectively crystallize salts from multicomponent solutions [23].

Literature on the production of sodium carbonate and sodium sulfate from desalination brines is not readily available. Both compounds can be produced from concentrated sodium chloride solutions; however, these methods are not commonly used and production from highly concentrated brines and mined minerals are more common. Generally, discussions on the recovery of carbonate and sulfate solids are discussed in reference to calcium carbonate, and magnesium or calcium sulfate, respectively.

15.5.9 Other Notable Commodities

Beyond the commodities discussed in the previous sections, there is available literature on the extraction of various other components from either seawater or desalination brine. Because boron removal is problematic in RO desalination systems, boron selective ion-exchange resins have been developed which could allow for boron recovery if a significant demand was identified. Several researchers [70, 76, 77] have proposed that extracting rubidium, which is very valuable, from desalination brine would significantly improve the economics of desalination. A considerable amount of research has also gone into uranium recovery from seawater with research supposedly being conducted in India, Japan, and Korea [78, 79]. These extraction schemes are likely unfeasible for the foreseeable future due to a very low demand for rubidium and the cost of uranium being significantly less than the cost of extraction. Other constituents studied include cobalt, germanium, gold, molybdenum, rare earth metals, strontium and silica although the low concentrations in seawater is likely to inhibit their commercial extraction.

15.5.10 Economic Considerations

Without economic incentive, it is unlikely that utilities would take on the burden of producing salable commodities at a desalination facility. A number of researchers have conducted studies to identify the technical feasibility of commodity extraction with only a limited focus on economic viability. Recently Shahmansouri et al. [80] performed a comprehensive literature review on the technical and economic viability of extracting most of the ions found in seawater from desalination brine or concentrate. A simple economic assumption that a facility would have to recoup $33,000 per year to pay a single worker was used to weed out potentially viable commodities. Once paired down, an evaluation of the commodity market and technical recovery approaches was performed to better understand feasibility. It was found that under certain advantageous conditions that the recovery of sodium chloride, chlorine and sodium hydroxide, potassium and magnesium could potentially lead to a return on investment in less than ten years. However, the economics of mineral extraction are strongly tied to product purity and pricing and it is currently unclear whether or not high purity products could be produced from desalination brine. Further purification would result in further costs that would render production uneconomical.

15.6 Conclusion

Desalination of unconventional drinking water supplies (i.e., seawater, brackish water, wastewater) has emerged as an important component of water supply management and ensuring an adequate amount of drinking water in many regions of the world. The disposal of desalination brine or concentrate is one of the most divisive issues related to desalination and mainly limits adoption to coastal communities where ocean discharge can be practiced. The need for desalination in regions with few disposal options has led to the development of approaches to minimize brine or concentrate volumes mainly through treatment and optimizing product water recovery. Due to issues related to desalination waste disposal (e.g., high salinity, presence of chemicals, large volumes) there has been a recent push to move towards ZLD desalination systems.

Although ZLD and near-ZLD systems and approaches have received significant attention in the past decade, there are few actual real-world adoptions. Volume reduction of relatively high salinity brine often precludes the use of RO membranes and requires alternate approaches such as thermal evaporation processes and crystallizers which have high capital, and operational and maintenance costs. While several promising technologies have emerged for solving the difficulties related to brine treatment, widespread adoption has not yet occurred and cost analyses have demonstrated that no magic bullet currently exists. When possible, practitioners have favored intermediate softening of the brine stream followed by secondary RO. Landlocked states in the United States have generally favored alternatives to desalination including conservation, development of alternative water supplies, and wastewater recycling for irrigation to name a few.

A possible approach to promoting further brine treatment and volume reduction is the recovery of valuable materials that could be sold to recoup costs. Because the ocean was once (and still is to a lesser extent) a source for a variety of commodities, a significant amount of research has been performed to develop systems to extract materials from desalination brine and improve the prospects for further treatment. Although many of the proposed schemes appear unfeasible from both a technical and economic standpoint, there may be some benefits of recovering calcium, chlorine and sodium hydroxide, and magnesium compounds at a desalination facility. There have been limited instances of the simultaneous production of commodities and reduction of brine volumes to date; however, the concept is likely to become more important as desalination becomes more common.

References

[1] Mickley, M. C. (2006). *Membrane Concentrate Disposal: Practices and Regulation*. Salt Lake City, UT: U.S. Department of the Interior, Bureau of Reclamation, 298.

[2] Brady, P. V., et al. (2005). Inland desalination: challenges and research needs. *J. Contemp. Water Res. Educ.* 132, 46–51.

[3] Christopher, B., et al. (2008). Comparing nanofiltration and reverse osmosis for drinking water augmentation. *JAWWA* 100, 102–116.

[4] Hasson, D., Sidorenko, G., and Semiat, R. (2010). Calcium carbonate hardness removal by a novel electrochemical seeds system. *Desalination* 263, 285–289.

[5] Al-Sofi, M. A. K., et al. (2000). Optimization of hybridized seawater desalination process. *Desalination* 131, 147–156.

[6] Al-Amoudi, A. S. and Farooque, A. M. (2005). Performance restoration and autopsy of NF membranes used in seawater pretreatment. *Desalination* 178, 261–271.

[7] Hassan, A. M., et al. (1998). A new approach to membrane and thermal seawater desalination processes using nanofiltration membranes (Part 1). *Desalination* 118, 35–51.

[8] Macedonio, F., Curcio, E., and Drioli, E. (2006). Integrated membrane systems for seawater desalination: energetic and exergetic analysis, economic evaluation, experimimental study. *Desalination*, 203, 260–276.

[9] Creusen, R., et al. (2013). Integrated membrane distillation–crystalliza tion: process design and cost estimations for seawater treatment and fluxes of single salt solutions. *Desalination* 323, 8–16.

[10] Drioli, E., et al. (2006). Integrating membrane contactors techology and pressure-driven membrane operations for seawater desalination: energy, exergy and costs analysis. *Chem. Eng. Res. Des.* 82, 209–220.

[11] Juby, G. J. G., et al. (2008). "*Evaluation and selection of available processes for a zero-liquid discharge system for the Perris, California, Ground Water Basin*," in *Desalination and Water Purification Research and Development Program Denver*, ed. U.S.D.o.t.I.B.o. Reclamation.

[12] Tanaka, Y., et al. (2003). Ion-exchange membrane electrodialytic salt production using brine discharged from a reverse osmosis seawater desalination plant. *J. Membr. Sci.* 222, 71–86.

[13] Turek, M. (2003). Dual-purpose desalination-salt production electrodialysis. *Desalination* 153, 377–381.

[14] Seigworth, A., Ludlum, R., and Reahl, E. (1995). Case study: integrating membrane processes with evaporation to achieve economical zero liquid discharge at the Doswell Combined Cycle Facility. *Desalination* 102, 81–86.

[15] Oren, Y., et al. (2010). Pilot studies on high recovery BWRO-EDR for near zero liquid discharge approach. *Desalination* 261, 321–330.

[16] Jordahl, J. (2006). *Beneficial and Nontraditional Uses of Concentrate*. Englewood, CO: CH2M Hill, 217.

[17] Davis, T. A. (2006). *Zero Discharge Seawater Desalination: Integrating the Production of Freshwater, Salt, Magnesium, and Bromine*. U.S. Bureau of Reclamation Research Report No. 111.

[18] Juby, G. J. G., et al. (2008). Optimize brackish groundwater: treating RO Brine to increase recovery and reduce disposal volume. *IDA J.* 1, 64–73.

[19] Turek, M., Dydo, P., and Klimek, R. (2005). Salt production from coal-mine brine in ED – evaporation – crystallization system. *Desalination* 184, 439–446.

[20] Zhang, Y., et al. RO concentrate minimization by electrodialysis: Techno-economic analysis and environmental concerns. *J. Environ. Manag.* 107, 28–36.

[21] Bond, R., et al. (2011). Zero liquid discharge desalination of brackish water with an innovative form of electrodialysis: electrodialysis metathesis. *Fla. Water Res. J.* 36–44.

[22] Veolia Water. 2010: Zero discharge desalination.

[23] Veerapaneni, V., et al. (2011). "Emerging desalination technologies – an overview," in *Proceedings of the WateReuse Symposium*, Phoenix, AZ, 34.

[24] Mickley, M. (2008). *Treatment of Concentrate*. Washington, DC: U.S. Bureau of Reclamation.

[25] U.S. Bureau of Reclamation (2000). *Evaluation of Two Concentrate Disposal Alternatives for the Phoenix Metropolitan Area: Evaporation Ponds and Discharge to the Gulf of California*. Washington, DC: U. S. Bureau of Reclamation.

[26] Ahmed, M., et al. (2000). Use of evaporation ponds for brine disposal in desalination plants. *Desalination* 130, 155–168.

[27] Gilron, J., et al. (2003). WAIV—wind aided intensified evaporation for reduction of desalination brine volume. *Desalination* 158, 205–214.

[28] Al-mutaz, I. S., and Wagialla, K. M. (1988). Techno-Economic feasibility of extracting minerals from desalination brines. *Desalination* 69, 297–307.

[29] Kim, D. H. (2011). A review of desalting process techniques and economic analysis of the recovery of salts from retentates. *Desalination* 270, 1–8.

[30] Mickley, M. (2008). *Survey of High-Recovery and Zero Liquid Discharge Technologies for Water Utilities*. Alexandria, VA: WateReuse Research Foundation, 180.

[31] Jean-Pierre Mericq, S. L. (2010). Corinne Cabassud Vacuum membrane distillation of seawater reverse osmosis brines. *Water Res.* 44, 5260–5273.

[32] Ji, X., et al. (2010). Membrane distillation-crystallization of seawater reverse osmosis brines. *Sep. Purif. Technol.* 71, 76–82.

[33] Drioli, E., Di Profio, G., and Curcio, E. (2012). Progress in membrane crystallization. *Curr. Opin. Chem. Eng.* 1, 178–182.

[34] Bennett, R. C. (2002). "Crystallizer selection and design," in *Proceedings of the Handbook of Industrial Crystallization*, ed. A. S. Myerson (Oxford: Butterworth-Heinemann).

[35] Bond, R., and Veerapaneni, S. (2007). *Zero Liquid Discharge for Inland Desalination*. Overland Park, KS: Black and Veatch, 1–233.

[36] Gryta, M. (2002). Concentration of NaCl Solution by membrane distillation integrated with crystallization. *Sep. Sci. Technol.* 37, 3535–3558.

[37] Macedonio, F., et al. (2010). Wind-Aided Intensified evaporation (WAIV) and membrane Crystallizer (MCr) integrated brackish water desalination process: advantages and drawbacks. *Desalination*.

[38] Harms, W., and Bruce Robinson, R. (1992). Softening by fluidized bed crystallizers. *J. Environ. Eng.* 118, 513–529.

[39] Belcu, M., and Turtoi, D. (1996). Simulation of the fluidized-bed crystallizers (II) productivity related aspects. *Crystal Res. Technol.* 31, 1025–1031.

[40] Shih, W., Yallaly, B., Marshall, M., DeMichele, D. (2013). *Chino II Desalter Concentrate Management Via Innovative Byproduct*. San Anotonio, TX: *American Membrane Technology Association*.

[41] Alexandratos, S. D. (2008). Ion-exchange resins: a retrospective from industrial and engineering chemistry research. *Ind. Eng. Chem. Res.* 48, 388–398.

[42] Schwochau, K. (1984). "Extraction of metals from sea water," in *Topics in Current Chemistry*. Berlin: Springer, 91–133.

[43] Khamizov, R. K., Muraviev, D., and Warshawsky, A. (1995). "Recovery of valuable mineral components from seawater by ion exchange and sorption methods," in *Ion Exchange and Solvent Extraction. A Series of Advances*, eds J. Marinsky and Y. Marcus (New York, NY: Marcel Dekker).

[44] Mero, J. L. (1965). *The Mineral Resources of the Sea*. Vol. 1. Amsterdam: Elsevier.

[45] Al-Mutaz, I. S. (1987). By-product recovery from Saudi desalination plants. *Desalination* 64, 97–110.

[46] El-Hamouz, A. M. and Mann, R. (2006). Chemical reaction engineering analysis of the blowout process for bromine manufacture from seawater. *Ind. Eng. Chem. Res.* 46, 3008–3015.

[47] Spagnoletto, G. (2008). Innovation in food grade hypochlorination generation and injection plant at Al Taweelah site. *Desalination* 182, 259–265.

[48] Mansfield, C. A., Depro, B. M., and Perry, V. A. (2000). *The Chlorine Industry: A Profile*. Washington, DC: U.S. EPA, 47.

[49] Melian-Martel, N., Sadhwani, J. J., and Ovidio Perez Baez, S. (2011). Saline waste disposal reuse for desalination plants for the chlor-alkali

industry: the particular case of pozo izquierdo SWRO desalination plant. *Desalination* 281, 35–41.

[50] Grosjean, C., et al. (2012). Assessment of world lithium resources and consequences of their geographic distribution on the expected development of the electric vehicle industry. *Renew. Sustain. Energy Rev.* 16, 1735–1744.

[51] Bellona, C., et al. (2015). *Selection of Salt, Metal, Radionuclide, and Other Valuable Material Recovery Approaches*. Alexandria, VA: WateReuse Research Foundation.

[52] Pilson, M. E. Q. (1998). *An Introduction to the Chemistry of the Sea*, in *An Introduction to the Chemistry of the Sea*, ed. R. A. McConnin (Upper Saddle River, NJ: Prentice-Hall Inc.), 207–232, 341–350.

[53] Austin, G. T., and Basta, N. (1998). *Sre Shreves Chemical Process Industries Handbook*. Fifth Edn. New York City, NY: McGraw-Hill Professional.

[54] Kramer, D. A. (2001). *Magnesium, Its Alloys and Compounds*.

[55] Le Dirach, J., Nisan, S., and Poletiko, C. (2005). Extraction of strategic materials from the concentrated brine rejected by integrated nuclear desalination systems. *Desalination* 182, 449–460.

[56] Ohya, H., Suzuki, T., and Nakao, S. (2001). Integrated system from complete usage of components in seawater: a proposal of inorganic chemical combinat on seawater. *Desalination* 134, 29–36.

[57] Unknown *Chinese desal turns green*. Global Water Intelligence, 2007.

[58] Abdel-aal, H. K., et al. (1990). Recovery of mineral salts and potable water from desalting plant effluents by evaporation. Part II. Proposed simulation system for salt recovery. *Sep. Sci. Technol.* 25, 437–461.

[59] Drioli, E. E., et al. (2004). Integrated system for recovery of $CaCO_3$, NaCl and $MgSO_4 \cdot 7H_2O$ from nanofiltration retentate. *J. Membr. Sci.* 239, 27–38.

[60] Ahmed, M., et al. (2001). Integrated power, water and salt generation: a discussion paper. *Desalination* 134, 37–45.

[61] Ahmed, M., et al. (2003). Feasibility of salt production from inland RO desalination plant reject brine: a case study. *Desalination* 158, 109–117.

[62] Delion, N., Mauguin, G., and Corsin, P. (2004). Importance and impact of post treatments on design and operation of SWRO plants. *Desalination* 165, 323–334.

[63] Birnhack, L., and Lahav, O. (2007). A new post-treatment process for attaining Ca^{2+}, Mg^{2+}, SO_4^{2-}, and alkalinity criteria in desalinated water. *Water Res.* 41, 3989–3997.

[64] Liberti, L., Boari, G., and Passino, R. (1978). Phosphates and ammonia recovery from secondary effulents by selective ion exchange with production of a slow-release fertilizer. *Water Res.* 13, 65–73

[65] Sengupta, S., and Pandit, A. (2011). Selective removal of phosphorus from wastewater combined with its recovery as a solid-phase fertilizer. *Water Res.* 45, 3318–3330.

[66] Zhao, D., and Sengupta, A. K. (1998). Ultimate removal of phosphate from wastewater using a new class of polymeric ion exchangers. *Water Res.* 32, 1613–1625.

[67] Sverdrup, H. U., and Ragnarsdottir, K. V. (2011). Challenging the planetary boundaries II: assessing the sustainable global population and phosphate supply, using a systems dynamics assessment model. *Appl. Geochem.* 26(Suppl. 0), S307–S310.

[68] Vaccari, D. A. (2009). *Phosphorous Famine: The Threat to Our Food Supply*. Scientific American.

[69] Kumar, M., et al. (2007). Beneficial phosphate recovery from reverse osmosis (RO) concentrate of an integrated membrane system using polymeric ligand exchanger (PLE). *Water Res.* 41, 2211–2219.

[70] Jeppesen, T., et al. (2009). Metal recovery from reverse osmosis concentrate. *J. Cleaner Prod.* 17, 703–707.

[71] Mohammadesmaeili, F., et al. (2010). Mineral recovery from inland reverse osmosis concentrate using isothermal evaporation. *Water Res.* 44, 6021–6030.

[72] Khamizov, R. K., Ivanovand, N. A., and Tikhonov, N. A. (2011). "Dual temperature methods of separation and concentration of elements in ion exchange columns," in *Ion Exchange and Solvent Extraction*, ed. A. K. SenGupta (Boca Raton, FL: CRC Press), 171–231.

[73] Kasedde, H., et al. *A State of the Art Paper on Improving Salt Extraction from Lake Katwe Raw Min Uganda*.

[74] Turek, M., Dydo, P., and Klimek, R. (2008). Salt production from coal-mine brine in NF – evaporation – crystallization system. *Desalination* 221, 238–243.

[75] Ravizky, A., and Nadav, N. (2007). Salt production by the evaporation of SWRO brine in Eilat: a success story. *Desalination* 205, 374–379.

[76] Gilbert, O., et al. (2010). Evaluation of selective sorbents for the extraction of valuable metal ions (Cs, Rb, Li, U) from reverse osmosis rejected brine. *Solvent Extract. Ion Exchange* 28, 543–562.

[77] Le Dirach, J., Nisan, S., and Poletiko, C. (2005). Extraction of strategic materials from the concentrated brine rejected by integrated nuclear desalination systems. *Desalination* 182, 449–460.

[78] Rao, L. (2011). *Recent International R&D Activities in the Extraction of Uranium from Seawater*. Berkeley, CA: Lawrence Berkeley National Laboratory, 1–20.

[79] Tamada, M. (2009). "Current status of technology for collection of uranium from seawater," in *NA*, NA: Environment and Industrial Materials Research Division, 1–9.

[80] Shahmansouri, A., et al. (2015). Feasibility of extracting valuable minerals from desalination concentrate: a comprehensive literature review. *J. Cleaner Prod.* 100, 4–16.

CHAPTER 16

Environmental Impacts of Desalination Plants

Karim Bourouni[1]

[1]Faculty of Engineering and Applied Sciences, Mechanical and Industrial
Engineering Department, ALHOSN University, Abu Dhabi,
United Arab Emirates
E-mail: k.bourouni@alhosnu.ae

Abstract

D espite the many benefits desalination technologies have to offer, there are concerns over potential negative impacts on the environment. Key issues are the brine and chemical discharges to the marine environment, the emissions of air pollutants and the energy demand of the processes.

To ensure sustainable utilization of desalination, the impacts of each major desalination project should be investigated and mitigated. This can be achieved by optimizing the project location and undertaking specific environmental impact assessment (EIA) studies. On the other hand, the benefits and impacts of different water supply options should be balanced on the scale of regional management plans. In this context, the present chapter intends to present a summary of the key environmental concerns of desalination.

This chapter presents an overview on the Health, Safety and Environmental (HSE) issues associated with the intakes, outfalls, and environmental impacts of desalination plants. Moreover, a segment on the comparative LCA (Life Cycle Assessment) of the various desalination options is included.

16.1 Introduction

While the social and economic benefits of desalination are recognized by providing a seemingly unlimited, constant supply of high-quality drinking water without impairing natural freshwater ecosystems, scientific, and public concerns have been raised over potential adverse impacts of desalination on the environment. The list of potential impacts of desalination plants on the environment is long, and in some aspects, similar to any other development project.

Key concerns are the brine and chemical discharges, which may impair coastal water quality and affect marine life, air pollutant emissions attributed to the energy demand to drive the process, and the emission of greenhouse gases. The major concerns for the marine environment are the construction of intake and outfall structures, which may provoke the temporary or permanent destruction of coastal habitats, and the impingement and entrainment of marine organisms with the intake water. This, in turn, may have a negative effect on ecosystem population dynamics. Other major concerns for the marine environment are the brine and chemical discharges into the sea, which may affect water, sediments, and marine life if intakes and outfall structures are not well designed and managed [1]. Moreover, several other issues of possible concern should be taken into account, such as hazards related to the use and storage of different chemicals. A large majority of the potential environmental impacts during construction and operation of desalination plant facilities are of a local nature. Also, although the potential environmental impacts associated with construction activities are generally temporary and reversible, the impacts during operation are mostly continuous.

The problem of land utilization is general to every major industrial project. Seawater desalination plants are located at coastal sites which are a particular sensitive environmental habitat with many social and economic aspects. Thus, finding an appropriate plant location has to be done with great care in order to minimize different impacts.

Despite great effort in order to reduce the overall energy consumption, in particular for reverse osmosis (RO) plants (by using energy recovery devices), desalination remains an energy-intensive technology. Global warming is becoming a major problem; hence the carbon dioxide production caused by desalination plants, due to their consumption of fossil energy, is another major environmental issue.

The desalination process byproduct is concentrated salt water containing a mixture of chemicals used during plant operation. The composition of this brine depends on the desalination technology, the operational parameters, the component materials and the pretreatment method. This chapter will focus on the impacts that discharging this effluent to the natural environment could have. The concentrations of different pretreatment chemicals in multi-stage flash (MSF) and RO brine streams are important for the marine environment.

The high salinity of RO brine as well as the high discharge temperatures and copper concentrations of MSF effluents present potential negative
impacts.

In order to avoid uncontrollable and unsustainable development of coastal
regions, desalination activity should be more integrated into management
plans that regulate water resource utilization and desalination activities on
a regional scale [2].

16.2 Health Issues

Drinking water systems should be able to produce and provide safe drinking
water that satisfies all quality specifications. Currently, several technologies
can produce high quality water from non-freshwater sources. The WHO
guidelines for drinking-water quality provide comprehensive information in
this respect [3]. These guidelines can be applied equally to conventional
or desalinated water. However, desalinated water provides some additional
issues to be considered, in respect to both chemical and microbial components. These issues are due to the nature, origin and typical locations of
desalination facilities.

Seawater is enriched with several beneficial chemicals found in drinking
water such as calcium, magnesium and sodium; however, it is deficient of
other essential ions such as zinc, copper, chromium, and manganese. All
desalination facilities, regardless of the technology used, reduce the quantities
of all mineral ions in drinking water. Hence, people who are used to consume desalinated water may be receiving smaller amounts of some nutrients
compared to people who consume water from more traditional sources.
Nonetheless, the deficiency of ions from the drinking water is not a serious
problem. Since desalinated water can be stabilized by the addition of lime
or blending, some of these ions will be re-injected during the desalination
post-treatment process [3].

16.2.1 Aesthetics and Water Stability

Although these parameters are not directly related to health issues, aesthetic
factors such as taste, odor, and turbidity affect palatability and thus consumer
acceptance, and indirectly health.

The control of alkalinity, hardness, and pH has economic consequences
and also determines the corrosion of metals and other pipe components

during distribution. Chemical additives and blending are used to adjust these parameters to the standard values.

16.2.2 Blending Waters

Blending is used to increase the TDS and improve the desalinated water stability. Components of the blending water can also affect the quality and safety of the produced water because no further treatment beyond residual disinfection will be applied. Some contaminants may be best controlled by the selection of blending water pretreatment. It is possible that some of the microorganisms in the blending water could be resistant to the residual disinfectant and could contribute to biofilms, or could be inadequately represented by surrogate microbial quality measurements such as *Escherichia coli* or heterotrophic plate counts [4].

16.2.3 Nutritionally Desirable Components

Desalinated water can be stabilized by adding lime and other chemicals. Although drinking water cannot be considered as a significant source of minerals in the daily diet, there is a legitimate question as to the optimal mineral balance of drinking water to assure quality and health benefits. Dietary calcium and magnesium are considered as important health factors, as well as certain trace metals, and fluoride which is considered to be beneficial for dental and skeletal health by most authorities [4, 5].

16.2.4 Chemicals and Materials Used in Water Production

Chemicals used in desalination and standard water production processes are similar; however, in the desalination process they may be used in greater amounts and under different conditions. Due to the operational conditions in desalination plants (heating, adding chemicals, concentrated salts, etc.), metals and other components could be subjected to excessive thermal and corrosion stresses compared to conventional water production facilities. The guidelines (GDWQ) developed by the WHO in 2011 [6] recommend some chemicals and materials used in drinking water treatment. GDWQ provides quality and dosing specifications, and encourages the establishment of processes to develop guidelines for quality and safety of direct and indirect additives by national and international institutions. These guidelines encourage governments to establish systems for specifying the appropriateness and quality of additives used in desalination, or to adopt existing recognized standards for the products that would be adapted to desalination conditions.

16.3 Safety Issues

Desalination facilities do not have specifically outstanding safety issues. In fact, the existing risks are more or less common with other industrial plants (power plants, oil plants, etc.).

16.4 Environmental Impacts

Among the impacts of a desalination plant, there are those that are limited to the construction phase and those that are related to the operation phase. The impacts start with the transformation of land use, and then continue with visual impact and noise pollution to be extended to emissions to the atmosphere and releases to water as well as other potential damages to the environment. The construction and operating activities can result in a series of impacts on coastal areas, in particular affecting the quality of air and water, the marine flora and fauna, disruption of important ecosystems (sand dunes, sea grasses and other vulnerable habitats as a result of the location chosen for the journey pipes), dredging and removal of cuttings that result in noise and barriers to public access and recreation. The most important of these impacts regard the quality of surrounding water, which can influence flora, fauna, and marine ecosystems.

Despite the fact that different technologies have been developed for desalination (RO, distillation, electrodialysis, etc.) they have all been designed to remove almost all of the constituents in seawater (salts in particular). This results in the generation of a waste stream (brine) for all technologies, that has a similar chemical composition as the feed seawater but with a concentration 1.2–3 times higher [7]; in addition to the chemicals used in the pre-processing steps. A range of chemicals and additives are used during desalination, to prevent or combat scaling or biofouling that otherwise would disturb desalination system operation. The constituents present in the waste stream discharged by desalination plants depends to a large extent on the quality of the feedwater, quality of the freshwater produced, and the adopted desalination technique. Discharges from desalination plants include concentrated disinfectants and antifouling agents, in addition to warm water and aqueous effluents such as distillates' ejector condensates [8].

Based on the recommendations of the California Coastal Commission in 2004 it was concluded that *"the most significant potential direct adverse environmental impact of seawater desalination is likely to be on marine organisms: This impact is due primarily to the effects of the seawater intake and discharge on nearby marine life; however, these effects can be avoided or minimized through proper facility design, siting, and operation"* [9].

On the other hand, the US Clean Water Act underlined that *"the location, design, construction and capacity of cooling water intake structures reflect the best technology available for minimizing adverse environmental impact"* [9].

16.4.1 Seawater Intakes and Pretreatment

Seawater contains particles and substances which are potentially harmful for the components of desalination facilities and processes. Solid particles can create coagulation and deposition, biological substances can provoke fouling, dissolved solids can generate scaling, and material corrosion can be accelerated. To reduce these harmful effects, the intake system should be chosen carefully. Moreover, the intake position at the site should ensure the best quality of water, and the most robust materials have to be considered. In most cases, cleaning systems need to be installed because the quality of raw water is not adequate for plant operation. Filters are integrated to purify the water as far as possible and chemicals are dosed to adjust water parameters in order to meet operational requirements.

16.4.1.1 Intakes

In most cases water collectors are located on the seashore, in a close vicinity to the ocean. To feed desalination plants two main types of intake facilities can be used: subsurface intakes (wells, infiltration galleries, etc.) and open intakes. Wells can be configured as either horizontal or vertical (Figure 16.1). A comparison between seawater collected using wells and seawater collected using open intakes shows that the former is of better quality in terms of solids, oil and grease, natural organic matter, and aquatic microorganisms. For beach wells the sandy soil is used as natural prefiltration and thus delivers a better feedwater quality. In addition, the danger of impingement

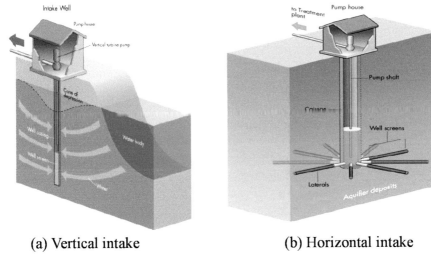

(a) Vertical intake   (b) Horizontal intake

Figure 16.1 Different kinds of beach wells [10].

and entrainment is avoided. Sometimes, beach intakes may also yield lower salinity for feedwater. Horizontal wells are usually costlier than the vertical beach wells; on the other hand, the productivity of vertical wells is relatively small (typically, 400–4000 m^3/day). For this reason, the use of vertical wells for large plants is less favorable. Higher intake volumes can be delivered by horizontal directional drilling (HDD). For this technique, pipelines under the seabed are installed allowing the water to be pre-filtered by the geological layers, to be collected in sufficient quantities, independent of currents, waves and tides.

However, HDD cannot be used in all geologic conditions, and beach wells installed in this way can be difficult to construct and maintain. Furthermore, there is a risk of saltwater intrusion into the ground water when beach wells and HDD are used [9].

When feedwater is taken directly from the sea, by open water intakes, via pipes, this approach allows a theoretically unlimited raw water stream. When open water intakes are used, there is an important risk of impingement and entrainment of fish and other animals caused by the strong water suction. Another risk is that very small organisms that can pass through the screens are sucked into the plant causing a deterioration of feed water quality [11]. Concerning the environmental impacts of intake systems, the following effects can be caused by open seawater intakes:

- loss of eggs and larvae of fish and invertebrate species,
- loss of spores from algae and seagrasses, and
- loss of phytoplankton and zooplankton.

Two major phenomena can take place at the intakes: (i) entrainment corresponding to drawing smaller marine organisms into the plant with the seawater, and (ii) impingement that happens usually in open intakes and results in the loss of larger marine organisms when these collide with screens at the intake. Suffocation, starvation, or exhaustion due to being pinned up against the intake screens or from the physical force of jets of water used to clear debris off screens are the main causes of mortality in impingement [12]. Impingement may also be a significant cause of mortality for protected marine species, such as sea turtles or sea snakes. Gulamhusein et al. [13] have reported the intrusion of a three-meter-long whale shark into the intake of a SWRO plant in the Red Sea after a storm which damaged the intake structure. In the case of other facilities co-located near the desalination plants such as power plants, entrainment and impingement effects are difficult to estimate and cumulative effects analysis should be performed.

16.4.1.2 Pretreatment

When the quality of raw water does not meet the criteria for the plant, pretreatment has to be applied in order to avoid operation problems during desalination. Due to the pretreatment, which involves chlorination at the intakes to control marine growth, and the removal of suspended solids, it must be assumed that the survival rate of organisms which are drawn into the plant is minimal. Chemical pretreatment is the most commonly used technique in seawater desalination plants [2]. It allows to solve and avoid several problems (suspended particles, fouling, scaling, corrosion and foaming). Nevertheless, the following should be noted:

1. Feedwater contains suspended particles that can foul RO membranes. Adding coagulants such as ferric chloride or polyelectrolytes to the water can force the particles to form bigger agglomerations so that they can be filtered with dual media and cartridge filters. In addition, mechanical flocculation can be achieved through slow mixing using turbines or propellers [14].
2. Organic material in the feedwater (fine unfiltered particles and bacteria) which settle on surfaces may cause fouling. This can cause blockage and destruction of RO membranes and reduce the heat transfer coefficient and the overall operating efficiency GOR (Gain Output Ratio) in MSF

and MED plants. Biocides, most commonly chlorine, are usually added continuously to the feedwater, which restricts biological growth, and thereby reduces bio-fouling. In order to stop all biological activity, shock chlorination with higher dosages is carried out in regular intervals.

3. In RO plants scale formation decreases RO membrane performance. In thermal plants (MSF and MED) scale formation promotes corrosion and reduces the heat transfer and thus the GOR. Usually, acids and antiscalant chemicals are dosed in order to control scale formation. Regular pretreatment cannot ensure a complete elimination of fouling and scaling and fine films will form eventually. Therefore, additional regular chemical cleaning with acids and a mix of other chemicals has to be carried out.

4. Corrosion is a major problem in thermal plants. It is enhanced by high salinity and temperatures, oxygen and chlorine. In particular, copper-nickel alloys which are the predominant materials in evaporators due to their high heat transfer coefficient, are vulnerable to corrosion. These sensitive metals can be protected by dosing anti-corrosive chemicals and/or by depleting the feedwater of oxygen by using oxygen scavengers.

5. Foaming is an exclusive problem of thermal plants. It occurs when dissolved organics concentrate on the water surface due to the water movement, which increases the risk of salt intrusion into the distillate. Foaming can be avoided by using antifoaming agents that reduce the tension in the surface and destroy the surface films.

16.4.2 Reject Streams and Outfalls (Impact of Brine Discharge)

The main environmental impact of desalination is the possible negative effects on the marine environment. Enclosed and shallow sites with abundant marine life are considered to be more sensitive to desalination reject streams than exposed and open sea locations, where dilution and dispersion of discharges are more effective [15].

In order to carry out a comprehensive impact analysis of reject streams several aspects should be considered. For example, physical properties such as salinity and temperature play an important role. Both of them change the effluent density and thus, have an effect on its flow characteristics and the impact area. The desalination technology and the pretreatment applied have a significant influence on the physio-chemical properties of the discharged brine (Table 16.1).

Table 16.1   Typical effluent properties of reverse osmosis (RO) and multi-stage flash (MSF) seawater desalination plants [2, 16]

	RO	MSF
Physical properties		
Salinity	Up to 65,000–85,000 mg/L	About 50,000 mg/L
Temperature	Ambient seawater temperature	+5 to 15°C above ambient
Plume density	Negatively buoyant	Positively, neutrally or negatively buoyant depending on the process, mixing with cooling water from co-located power plants and ambient density stratification
Dissolved oxygen (DO)	If well intakes used: typically, below ambient seawater DO because of the low DO content of the source water If open intakes used: approximately the same as the ambient seawater DO concentration	Could be below ambient seawater salinity because of physical deaeration and use of oxygen scavengers
Biofouling control additives and by-products		
Chlorine	If chlorine or other oxidants are used to control biofouling, these are typically neutralized before the water enters the membranes to prevent membrane damage	Approx. 10–25% of source water feed dosage, if not neutralized
Halogenated organics	Typically, low content below harmful levels	Varying composition and concentrations, typically trihalomethanes
Removal of suspended solids		
Coagulants (e.g., iron-III-chloride)	May be present if source water is conditioned and the filter backwash water is not treated. May cause effluent coloration if not equalized prior to discharge.	Not present (treatment not required)

Table 16.1 Continued

Coagulant aids (e.g., polyacrylamide)	May be present if source water is conditioned and the filter backwash water is not treated	Not present (treatment not required)
Scale control additives		
Antiscalants	Typically, low content below harmful levels	Typically, low content below harmful levels
Acid (H_2SO_4)	Not present (reacts with seawater to cause harmless compounds, i.e. water and sulfates; the acidity is consumed by the naturally alkaline seawater, so that the discharge pH is typically similar to or slightly lower than that of ambient seawater).	Not present (reacts with seawater to cause harmless compounds, i.e. water and sulfates; the acidity is consumed by the naturally alkaline seawater, so that the discharge pH is typically similar to or slightly lower than that of ambient seawater)
Foam control and additives		
Antifoaming agents (e.g., polyglycol)	Not present (treatment not required)	Typically, low content below harmful levels
Contaminants due to corrosion		
Heavy metals	May contain elevated levels of iron, chromium, nickel, molybdenum if low-quality stainless steel is used.	May contain elevated copper and nickel concentrations if inappropriate materials are used for the heat exchangers
Cleaning chemicals		
Cleaning chemicals	Alkaline (pH 11–12) or acidic (pH 2–3) solutions with additives such as: detergents (e.g., dodecylsulfate), complexing agents (e.g., EDTA), oxidants (e.g., sodium perborate), biocides (e.g., formaldehyde)	Acidic (pH 2) solution containing corrosion inhibitors such a benzoatriazole derivates

Desalination brine has higher salinity than feedwater in both RO and thermal plants; however, there is a temperature increase only in thermal plant discharges. Furthermore, the different types of chemicals used during pretreatment and cleaning as well as corroded metals must be considered. In this section, the different impact categories are analyzed considering the differences between the two main desalination technologies MSF and RO.

While biocides and antifoaming additives are generally only found in the brine discharges of distillation plants, chemical residues of antiscalants are present in both waste streams. Both kinds of brine streams contain varying but relatively low concentrations of metals from corrosion. In distillation plants, copper–nickel heat exchangers are widely used. This material is very sensitive to corrosion and copper contamination may be a concern in the brine streams of distillation plants.

At RO plants, backwash water from coagulation and media filtration is sometimes combined with the brine, resulting in reject streams containing coagulants. Moreover, the brine discharges, especially of RO plants, may contain spent cleaning solutions if these are diluted by the brine and discharged to the sea.

The environmental impacts of the single brine stream are discussed here, however, it is worth noting that the whole brine effluent is a mix of these pollutants, and that their combination may have synergetic effects on marine life. Tables 16.2–16.4 provide a summary of the single physical and chemical effluent parameters discussed in this chapter.

Table 16.2   Effluent properties of SWRO, MSF and MED plants with conventional process design: physical parameters [10]

Physical Parameters	Seawater Reverse Osmosis (SWRO)	Multi-Stage Flash (MSF)	Multi-Effect Distillation (MED)
Cooling water	No use of cooling water in the process, but RO plants may receive their intake water from cooling water discharges of power plants	• MSF/MED: it is assumed that the waste streams from the desalination process and cooling water are typically combined, i.e., the brine is diluted with major amounts of cooling water from the desalination process. Further dilution with cooling water from power plants may occur but is not considered here.	

Table 16.2   Continued

Salinity (S) mg/L	• SWRO: 65–85 • BWRO: 1–25	• Cooling water S: ambient • Brine S: typically 60–70 • Combined S: typically 45–50	• Cooling water S: ambient • Brine S: typically 60–70 • Combined S: typically 50–60
Temperature (T)	• If subsurface intakes are used: may be below ambient T due to a lower T of the source water • If open intakes are used: close to ambient • If power plant cooling waters are used as source: above ambient	• Brine T: 3–5°C above ambient • Cooling water T: 8–12°C above ambient • Combined T: 5–10°C above ambient	• Brine T: 5 25°C above ambient • Cooling water T: 8–12°C above ambient, up to 20°C possible • Combined T: 10–20°C above ambient
Plume density (ρ)	• Higher than ambient and hence is negatively buoyant, sinking to the bottom	• MSF/MED: plume can be positively, neutrally or negatively buoyant depending on the process design and mixing with cooling water before discharge, typically positively buoyant due to large cooling water flows	
Dissolved Oxygen (DO)	If subsurface intakes are used: may be below ambient DO due to a lower DO of the source water If open intakes are used and if chlorine reducing agent is not overdosed, close to ambient	• Brine: below ambient due to deaeration and oxygen scavengers Cooling water: close to ambient (minor decreases of DO possible because of increases in temperature, as oxygen is less soluble in warmer water) combined: mixing of brine with cooling water increases the DO content of the combined effluent close to ambient, as turbulent mixing allows oxygen take-up from air	

Table 16.3   Effluent properties of SWRO, MSF and MED plants with conventional process design: chemical parameters [10]

Biofouling	Seawater Reverse Osmosis (SWRO)	Multi-Stage Flash (MSF)	Multi-Effect Distillation (MED)
Oxidants • Mainly chlorine • Chlorine dioxide used in some plants	• usually low-level chlorination with a dosage of 1–2 mg/L to feedwater, added intermittently or continuously		
	• SWRO: oxidants typically removed with sodium bisulfite (2–4 times the dosage of the oxidizing agent)	• MSF/MED: chlorine typically not removed by a dechlorination step • Discharge concentration of residual chlorine often reduced to 10–25% of the dosage level due to the oxidant's demand in the seawater • Both the brine and the cooling water contain residual chlorine	
Halogenated organic by-products, typically trihalomethanes (THMs)	• SWRO: by-products may form during chlorination but levels are probably low due to dechlorination	• MSF/MED: chlorination of seawater results in varying composition and concentrations of halogenated (chlorinated and brominated) organic by-products, mainly THMs such as bromoform	
	• all processes: use of chlorine dioxide has been reported to reduce the risk of by-product formation		
FeCl$_3$ or FeSO$_4$ coagulants (1–30 mg/l) • Coagulant aids (<1–5 mg/l) such as poly-acrylamide	• if filter backwash is discharged to surface waters: may cause turbidity; iron salts may cause effluent coloration ("red brines"), and increased sedimentation rates	• MSF/MED: treatment not applied/coagulation not necessary	

Scale control additives (used in all desalination processes, can be a blend of several different antiscalants in combination with acid)

| Polymers (1–2 mg/L) polymeric or polyacrylic acids, phosphonates | • dosage/discharge concentration below toxic levels for invertebrate and fish species
• some products are classified as being "harmful" to algae, presumably due to a nutrient inhibition effect, but dosing levels (1–2 mg/l) are still an order of magnitude below harmful levels (20 mg/L)
• some products classified as "inherently" biodegradable: increased residence times in surface waters | | |
| | | • MSF/MED: antiscalants only present in the brine, not in the cooling water | |

Table 16.3 Continued

Phosphates (2 mg/l)	• SWRO: still used on a limited scale	• MSF/MED: as phosphates are not stable at high T, polymers are preferred
	• may cause eutrophication near outlets, as phosphates are easily hydrolyzed to orthophosphates, a major nutrient	
Acid (H_2SO_4) • dosage: 20–100 mg/L	• still applied in SWRO plants • lowers pH from 8–8.3 (ambient seawater) to 6.7 • acidity quickly neutralized by seawater alkalinity	
Foam control additives		
Antifoaming agents (e.g. polyglycol)	• Treatment not applied	• Typically low dosage (0.1 mg/l), which is below harmful levels • Used in all distillation processes, but primarily in MSF • Antifoam only present in the brine, but not in the cooling water
Corrosion		
Heavy metals	• usually corrosion-resistant stainless steels and plastics used • concentrate may contain low levels of iron, nickel, chromium and/or molybdenum if low-quality steel is used	• metallic equipment usually made from carbon or stainless steel and copper nickel alloys • concentrate may contain iron and copper; increased copper levels may be of concern but the available data is limited	• metallic equipment usually made from carbon and stainless steel, aluminum and aluminum brass, titanium, or copper nickel • lower corrosion rates than in MSF reported, but no extensive data on brine contamination levels available
Corrosion prevention	• not necessary if proper materials are used	• as the feed water is deaerated, the brine is also deaerated before mixing with cooling water, which is not deaerated • in MSF plants, the feed water may be treated with oxygen scavengers (e.g. sodium bisulfite), which may also remove residual chlorine as a side effect	

(*Continued*)

Table 16.3 Continued

Biofouling	Seawater Reverse Osmosis (SWRO)	Multi-Stage Flash (MSF)	Multi-Effect Distillation (MED)
Cleaning solutions (if discharged into surface waters)			
Cleaning chemicals (used intermittently)	Alkaline (pH 11–12) or acidic (pH 2–3) solutions containing cleaning additives, e.g.: • Detergents (e.g., dodecylsulfate) • Complexing agents (e.g. EDTA) • Oxidants (e.g., sodium perborate) • Biocides (e.g., formaldehyde)	• Acidic (low pH) washing solution which may contain corrosion inhibitors such as benzotriazole derivates	

16.4.2.1 Salinity

The salinity of desalination brine depends on the salinity of the feedwater and the recovery rate and can highly exceed the natural salinity level in the ocean. The recovery rate of a desalination plant is defined as the ratio of produced freshwater volume to the feedwater volume (Equation (16.1)). Increasing the recovery rate decreases brine volumes, but results in higher brine salinities.

$$\text{Recovery rate} = \frac{\text{Fresh water volume}}{\text{Feed water volume}} \tag{16.1}$$

The recovery rates of SWRO plants are higher than distillation plants and therefore they result in higher brine salinities, which typically vary between 65 and 85 g/L. Although in MSF distillation plants the salinity of brine blow-down from the last stage may reach 70 g/L, the brine is effectively diluted with a threefold amount of cooling water. By applying efficient dilution in distillation plants the salt concentration of the reject stream is rarely more than 15% higher than the salinity of the receiving water, while the SWRO brine may contain twice the seawater concentration [15]. Thus, RO effluents have a higher density and can affect the benthic species more, whereas MSF effluents affect the open water organisms.

Salinity is a vital environmental parameter for marine life. In general, toxicity will depend on the sensitivity of a species to high salinity levels, its life cycle stage, the exposure time and the natural variations of habitat salinity to which the species is adapted. Different investigations highlight that constant salinity levels above 45 g/L alter the benthic community and reduce the diversity of organisms [17]. Most organisms are able to cope with short salinity peaks of up to 50 g/L and can adapt to long-term variations of 1–2 g/L. On the other hand, some organisms have very low levels of tolerance to salinity variation. For example, a salinity of 43 g/L can already be lethal for corals [10]. In typical RO plants, the reject stream significantly exceeds the indicated tolerance levels for corals and must be classified as dangerous until it is sufficiently diluted. Turbidity can also be increased by high salinity which can disrupt the photosynthesis process. Hence, higher salt concentration levels and less sunlight increase the mortality rate of plankton species and reduce the variety of other immobile organisms. The tolerance varies greatly between the different species (which is true also for fish). Less tolerant species will be deprived of their natural habitat and will disappear from the place of impact [17]. Höpner and Lattemann [17] found that a salinity of 45 g/L can lead to a 50% mortality rate of some species after two weeks. These levels of salinity can easily be reached around the immediate vicinity of the discharge point.

16.4.2.2 Temperature

In RO plants the temperature of reject stream might be marginally higher than ambient values and can be neglected. On the other hand, discharges from thermal desalination plants have higher temperatures and may have an effect on species' distribution by changing the annual temperature profiles at the discharge site. In extreme cases, a die-off of sessile marine species can be caused by thermal discharge. MSF plants generate high thermal emissions and discharge the brine at a maximum of 10–15°C above ambient seawater temperature, after dilution by cooling water. Increased temperatures reduce the solubility of oxygen in water, which can be harmful for species in case of significant gaps. In winter, the temperature increase can boost biological activity; however, it induces thermal stress when critical values are exceeded in the summer. Warm water may attract or repel marine organisms, promoting the dominance of species that are more adapted to higher temperatures [10]. Thermal impact is generally a minor problem in hot regions, such as the Middle East and North Africa region (MENA), where large annual temperature changes are a natural phenomenon. The highest stress will be in temperate environments not adapted to rapid variations in temperature [15].

16.4.2.3 Antifouling additives

Chlorine is the most commonly antifouling additive used in desalination. Because it is a broad-effect agent, chlorine can have different impacts on marine organisms. Moreover, chlorine is highly reactive and can participate in chemical reactions such as the halogenations of organic compounds. In order to prevent fouling, a typical chlorine or hypochlorite dose less than 1 mg/L is applied in both MSF and RO plants [15]. Higher doses may be added over shorter periods for shock chlorination. In RO plants the polymeric membranes must be protected by dechlorination of the feedwater prior to contact with the membranes. However, trace residual chlorine levels can persist in the brine and the problem of forming toxic halogenated organic compounds remains [15]. Nevertheless, for MSF plants, the impacts of chlorine are more significant since usually no dechlorination process is applied. In addition, larger feedwater volumes are required by MSF plants, which increase the loads of chlorine and its by-products. Concentrations of chlorine in the mixing zone of MSF plants have been reported to be around 100 μg/L [15]. The mixing zone is the area around the discharge point in which the brine and its constituents are diluted by the surrounding water. It has been shown that chlorine is toxic at concentrations of a few micrograms per liter only [10]. For context, the plankton photosynthesis process can be seriously reduced at concentrations of only 20 μg/L. The composition of marine organisms can change and their variety can be reduced at levels of 50 μg/L. The known lethal values for fish species range between 20 and several hundred μg/L [10]. Toxic chlorine concentration levels for a range of species are presented in Figure 16.2. It can be seen that the reported chlorine concentrations in MSF effluents and in the mixing zone are acutely toxic for many of the examined marine organisms.

16.4.2.4 Residual biocides

In order to control biogrowth on the screens, in the intake pipe and in the pretreatment line chlorine is added at the intake in most desalination plants.

Inside the desalination plant, several reactions take place between organic seawater constituents. These reactions combined with abiotic degradation (decomposition) increase the oxidant demand of seawater inducing a decline in the chlorine concentration.

The World Bank Pollution Prevention recommends a maximum total residual chlorine/bromine concentration of 0.2 mg/L for effluents from thermal power plants [18]. In the case of shock chlorination application, a maximum value of 2 mg/L for up to 2 h should be applied. Moreover,

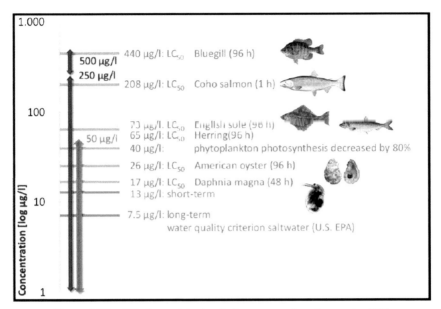

Figure 16.2   Chlorine toxicity levels for a range of marine species [17].

shock chlorination should not be repeated more frequently than once every 24 h. The BAT and World Bank reference values are not compulsory, and discharge regulations are normally established at the national or site-specific level. While some countries may adopt these recommendations, others may follow more lenient or strict regulations. Worldwide, regulatory authorities exert pressure on industries to decrease their utilization of chlorine, driven by the acute toxicity of chlorine and the formation of chlorination by-products [19] In order to decrease the discharge of free residual oxidants, the required chlorine dosage should be established based on target species' behavior and the seawater quality parameters. The main environmental benefit of pulse chlorination is that utilization of chlorine can be significantly reduced (by 30–50%) while ensuring an effective control of macrofouling.

In conclusion, no alternative biocide has gained wide acceptance over chlorination in desalination applications so far. The environmental assessment of seawater chlorination includes two different groups of chemicals, the oxidants and the chlorination byproducts. These two groups differ in terms of ecotoxicity, bioaccumulation, and biodegradation. In fact, the free oxidants' toxicity is high, even at low concentrations. However, they decompose quickly and do not bioaccumulate. Therefore, toxicity to non-target organisms will be limited to the mixing zone. Meanwhile, some of the chlorination by-products can be expected to be more persistent, to bioaccumulate, and/or to show a chronic mutagenic and carcinogenic toxicity, possibly also beyond the mixing zone.

16.4.2.5 Coagulants

In RO plants, there is a real need for the removal of suspended particles. In order to enhance the removal, coagulants such as ferric chloride (at dosages of 1–30 mg/L) or polyelectrolytes like polyacrylamide (at about 1–4 mg/L) are usually added to the intake water. The dosage depends on the amount of suspended particles in the water.

In most desalination plants, the agglomerated particles are filtered by media filters and periodically backwashed into the sea. Coagulants are non-toxic in the concentrations applied in RO plants. Their environmental impact concerns only the possible disturbance of photosynthesis processes due to an increase in turbidity during backwash of the coagulated sludge and by coagulant enrichment in sediments. In the Ashkelon RO plant in Israel (330,000 m^3/day) a dosing rate of 3 mg/L of ferrous coagulant is applied. The plant produces a highly turbid, red colored effluent during backwash which is applied every hour for 10–15 min. It was shown that the effect of coagulants can be reduced by treating or diluting the backwash with feedwater prior to discharge.

In new RO plants in Australia and the United States, sludge treatment is applied. Land deposition of the filtered sludge is an alternative but a cost estimated at 1–5 US-cents/m^3 should be added to the water price [17].

16.4.2.6 Antiscalants

Scale formation can be controlled either by the addition of sulfuric or hydrochloric acid (against carbonate scales), the dosing of special scale inhibitors (known as antiscalants), or a combination of both. Acids react stoichiometrically with calcium carbonate and must therefore be added in relatively high concentrations (20–50 mg/L) to the feed water. The pH of the acidified feed is usually between 6.8–7.0, while the natural seawater pH is about 7.8–8.1 depending on its alkalinity and salinity. The effect of this reduced pH can be minimized if the outfall is located in an area that provides good mixing conditions. The good buffering capacity of seawater allows for a fast neutralization of any residual acidity by mixing the brine with the surrounding seawater [10]. Antiscalants, with low and non-stoichiometric doses of 1–2 mg/L, delay the nucleation process and impair crystal growth of scales. The main generic groups are polyphosphates, phosphonates, and organic polymers with multiple carboxylic groups. Polyphosphate antiscalants are easily hydrolyzed to orthophosphate, which is an essential nutrient for primary producers. The use of polyphosphates as an antiscalant may cause a nutrient surplus inducing an increase in primary production at the discharge point. This may lead to oxygen depletion when the plants die and their biomass decays. These effects of eutrophication were observed at the outlets of some larger distillation plants that used polyphosphate antiscalants in the 1990s [20, 21]. Nowadays, the utilization of polyphosphates in distillation plants is limited due to their instability at higher temperatures, but they might still be in use at some SWRO plants [17]. On the other hand, the phosphonates and organic polymers are rather resistant to biological, chemical, and physical degradation and generally have a slow to moderate rate of elimination from the environment through abiotic and biotic degradation processes. A large portion of antiscalants are classified as "inherently biodegradable", with half-lives of about one month or longer. They do not present any danger to invertebrate and fish species as the dosing levels are considerably lower than the concentrations at which acutely toxic effects can be observed.

16.4.2.7 Metals

Heavy metal discharge as a consequence of corrosion is a main concern in thermal desalination plants due to the high temperatures involved. Depending on the materials used for the tubes of heat exchangers and vessels (copper, nickel, iron, zinc, and other heavy metals) sizable quantities of metals are corroded and discharged [15]. For economic reasons and heat transfer considerations, the prevailing alloy in heat exchanger tubes used in desalination is copper-nickel. On the other hand, this material has poor corrosion resistance and is considered as the highest heavy metal pollutant in MSF plants. In RO plants, non-metal materials are used and highly resistant stainless steel materials are ubiquitous. In the RO effluent traces of iron, nickel, chromium, and molybdenum can be detected; however, the concentrations remain non-critical. The average copper concentration of the oceans lies at a minimum of 0.1 μg/L. Copper concentrations in MSF effluents were found to be in the range of 15–100 μg/L. The tolerance toward copper pollution is not yet entirely known for all species. Copper can be toxic at higher concentrations, causing enzyme inhibition in organisms and reducing growth and reproduction [10]. Figure 16.3 illustrates the toxicity levels for a range of marine organisms. Despite the discharged concentrations, that can be higher than natural levels in the mixing zone, the risk of acute toxicity is generally low. However, there is a higher risk of accumulation and long term effects.

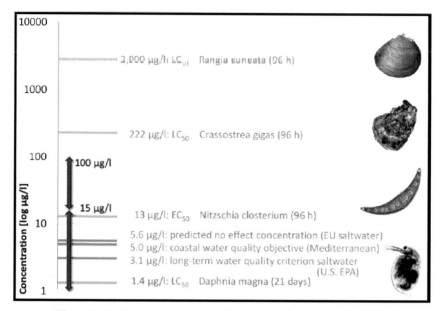

Figure 16.3   Copper toxicity levels for a range of marine species [17].

Copper compounds tend to settle down and accumulate in the sediments, can be absorbed by benthic organisms and even eventually be transferred into the food chain. With respect to bioaccumulation, the main point of concern becomes the discharged loads instead of the concentrations. To reduce impacts of brine discharges the US Environmental Protection Agency (EPA) recommends a maximum concentration of 8.2 μg/L for long term exposure. With adequate dilution at the discharge point most effluents reach this level very close to the outfall.

Despite the high mobility of Nickel in water, the majority of the load will accumulate in the sediments around the outfall. Adverse effects of accumulation cannot be excluded. Acid cleaning can also have an effect on metal deposits; in fact, corrosion rates will most likely increase during the process of acid cleaning although no specific data is available. Moreover, low pH values increase the mobility of discharged metals and thus, they become more harmful for the environment. Stainless steel materials contain mainly iron and lower amounts of chromium, nickel and molybdenum. The toxicity and overall discharge concentrations are considered to be harmless. Concentrations might increase through pitting and poor process control.

16.4.2.8 Antifoaming agents (Thermal plants only)

To reduce foaming in thermal plants, antifoaming agents can be added to the feedwater. Commonly, polyglycols and fatty acids are used as antifoaming agents, with typical dosages of about 0.1 mg/L. The dosage rate depends mainly on the raw water quality and its organic composition. Polyglycols are not toxic, but may transform into a polymerized state which is more persistent in the environment. However, due to the low concentrations used in desalination plants, polyglycols are of minor concern for the marine environment [2].

16.4.2.9 Cleaning chemicals

Although much effort and care is invested into the design of pretreatment systems, RO membranes and MSF-tubing systems and boilers are cleaned periodically in order to remove residual deposits. In order to avoid irreversible membrane damage, cleaning is required. Fouling inside distillation plants can reduce heat transfer, enhance corrosion and cause material failures. Cleaning intervals are established on a case by case basis depending on the ambient seawater conditions and the efficiency of the pretreatment process applied in the plant. Cleaning is typically applied in intervals varying from one to two years in SWRO plants operating on beach-well water and more frequently, up to several times per year, in SWRO plants operating on surface seawater [2]. Acidic solutions (pH 2–3) are used for removal of scales, metal oxides and inorganic colloids. Alkaline solutions (pH 11–12) are applied to remove biofilms as well as organic and inorganic colloids. The impacts of brine discharge volumes of cleaning solutions are higher for thermal plants. The discharge of cleaning solutions should, at least, be done by gradually mixing with the brine. In most cases the cleaning solutions are discharged into the sea without any treatment. The high pH values of cleaning solutions can be a hazard to the marine ecosystem depending on the discharged volumes and the degree of degradation at the discharge point. It was shown that LC50 mortality for certain fish species in an HCl solution of pH 2–2.5 is reached after 48 h [17]. Thankfully, residual acidity and alkalinity are often quickly neutralized by seawater.

Other issues are encountered because of the additives which are dosed to the cleaning solutions. These differ according to the desalination process. In the case of thermal plants, the chemical impacts are relatively low as only corrosion inhibitors like benzotriazole are dosed. Benzotriazole has low toxicity but is quite persistent and its degradation in seawater is very slow. It can accumulate in the sediments. However, its tendency for accumulation in organisms is low. Comparatively, a much more diverse and more harmful mix of chemicals is used for the chemical cleaning process of RO membranes. These agents commonly recommended by most membrane manufacturers are:

- Disinfectants like formaldehyde and isothiazole

Disinfectants are biocides used for removing biological films from membranes and are acutely toxic in the marine environment. In the case of formaldehyde, LC50 levels of only 0.1 mg/L were found for certain species. A seawater volume of more than 58,000 m^3 would be exposed to lethal concentrations if a common disinfection solution with 1% formaldehyde is discharged.

- Sulfonate detergents like sodium dodecylsulfate (NA-DDS)

Toxicity is in the middle range, with LC50 levels of NA-DDS ranging between 1 and 10 mg/L for many marine species. Pretty good degradability at 80% in a couple of days is documented. Complexing agents reduce the water hardness and remove scale deposits. Also, the detergents which are used for the removal of colloids can provoke a disruption of the intercellular membrane system in organisms.

- Complexing agents like Ethylene Diamine Tetraacetic Acid (EDTA).

The toxicity of EDTA is low but it is poorly degradable (only 5% in 3 weeks). Although the cleaning volumes are much lower in RO than in MSF, the toxicity of its constituents makes RO cleaning solutions more dangerous for the marine ecosystem [10, 19].

16.4.3 Air Quality Impacts

Air quality impacts of desalination projects concern the use of energy during the manufacturing and transportation of materials, the construction of the plant facilities and associated infrastructure, and most importantly, plant operation. During operation, the energy consumption includes the electrical or/and thermal energy produced on site or taken from external sources, such as the electricity grid. The total energy demand of the desalination facility includes the energy required to drive the desalination process, energy for pumping and pretreatment, for heating and air conditioning, and for lighting and office supplies. The specific energy demand refers to the energy demand of the desalination process which depends on the choice of the process and pretreatment: for one cubic meter of water produced, up to

12 kWh of thermal energy and 3.5 kWh of electrical energy is required in MSF plants, which have a maximum operation temperature of 120°C. These consumptions are lower for MED plants, which operate at lower temperatures (<70°C) and require up to 6 kWh of thermal and 1.5 kWh of electrical energy per cubic meter of produced water. The RO process is the most energy efficient desalination technology, in fact it requires at most between 4 and 7 kWh/m^3 depending on the size of the plant and energy recovery systems installed (in newer systems it can even be lower, as discussed in Chapter 14) [15]. Environmental concerns associated with the energy demand and thus indirectly associated with the desalination process are the fuel source and its transportation, and the emission of air pollutants and utilization of cooling water for electrical power generation.

16.5 Mitigating the Impact of Desalination on the Environment

The evaluation of the impacts of any desalination plant should be carried out and the adverse effects should be mitigated as much as possible. The most adequate tool for this purpose is the environmental impact assessment (EIA), which is a systematic procedure for identifying and evaluating all potential impacts of a proposed project. The EIA could be used also to develop appropriate mitigation measures and alternatives, such as modifications to the process or alternative project sites. As an EIA is project-and-location specific, it is out of the scope of this chapter to present all the environmental impact issues. Only the key impact categories will be presented in the following sections.

16.5.1 Source Water Intake

In order to mitigate the impacts of open intakes, a combination of differently meshed screens and a low intake velocity should be considered. This can minimize the impingement and entrainment of larger organisms, such as fish or turtles. The entrainment of smaller organisms, plankton, eggs and, larvae can be minimized by locating intakes away from productive areas (deeper waters, offshore, or underground). In the case of underground intake, beach wells can be used. Beach well intakes are adaptable to small or medium-sized desalination plants only. Since the quality of intake water is often better in these locations than in near shore and surface waters, only minimal or no chemical pretreatment may be required. On the other hand, the initial soil disturbance during construction of below ground intakes or long pipelines may be higher, especially when this involves drilling or excavation activities.

In case of large plants co-location of desalination and power plants should be considered where possible. The total intake water volume can be reduced since the cooling water from the power plant can be used as feedwater to the desalination plant. This allows minimizing the impacts from entrainment and impingement, the usage of chemicals, construction, and land use.

16.5.2 Reject Streams

Different approaches can be applied to mitigate the environmental effects of the waste discharges. To minimize impacts from high salinity, brine from desalination plants can be pre-diluted with other waste streams where applicable (e.g., power plant cooling water). Impacts from high temperature at the outfall can be avoided, in one way by dissipating maximum heat from the waste stream to the atmosphere before entering the receiving body of water. This can be achieved for example by using cooling towers. Another way is to achieve maximum dilution following discharge. Mixing and dispersal of the discharge plume can be improved by installing a diffuser system, and by locating the discharge in a favorable oceanographic site dissipating the heat and salinity load quickly. To analyze the plume dissipation at a specific project site, the environmental and operational conditions should be investigated by hydrodynamic modeling, combined with salinity and temperature measurements. This allows density calculations before and during operation of the desalination plant.

Negative impacts from chemicals can be reduced by different ways such as: (i) upstream discharge treatment, (ii) substitution of dangerous substances, and (iii) by using alternative treatment options. In particular, biocides such as chlorine, which may acutely affect non-target organisms in the discharge site, could be replaced or treated prior to discharge. Different chemicals can be used to effectively remove chlorine from reject streams, such as sodium bisulfite in RO plants and sulfur dioxide or hydrogen peroxide in thermal plants [21]. Filter backwash waters can be treated by sedimentation, dewatering and land-deposition, while cleaning solutions can be treated on-site in special treatment facilities or discharged to a sanitary sewer system.

Several efficient and economical technologies have been developed to reduce the impact of brine discharges on the marine environment. The need for most chemicals can be drastically reduced by using modern physical water pretreatment with ultrafiltration membranes or sponge ball systems. Residual antiscalant chemicals can be replaced by more biocompatible chemicals. Copper pollution can be eliminated by using duplex and super duplex steel components resistant to corrosion. Corrosion can be fought by reducing the oxygen levels of water during the desalination process. The chemical sodium bisulfite, also used for dechlorination in RO processes, can be applied as an oxygen scavenger in MSF plants. Other corrosion inhibitors like benzo-triazole are particularly used during chemical cleaning [2]. The impact of high salinity and temperature can be mitigated by discharging the brine via multiport diffuser outfalls. Overall, the combination of all these measures contributes to significantly reducing the environmental impacts of brine dis-charges. The costs of the investments required for these techniques are in the same range as conventional plants, even without monetizing the benefits of reduced marine pollution.

Non-chemical treatment options can be adopted for disinfection. One option consists of irradiating the intake water with UV light at 200–300 nm wavelength. Using UV-light allows avoiding storage, handling, and disposal of toxic chemicals; however, UV irradiation has not been found to be an effective pretreatment for larger desalination plants.

To conclude, several technical options can be applied to mitigate environ-mental impacts, including advanced systems for the intake of the seawater and the diffusion of the waste products, nonchemical pretreatment options such as UF and MF, and wastewater treatment technologies. Equally or even more important than the technical options, however, is the selection of a proper site for a desalination project.

16.5.3 Energy Use

Energy consumption is a main cost factor in water desalination. Great effort has been done to reduce this consumption by introducing several innovations such as development of energy recovery devices (ERD), or variable frequency pumps and pre-treatment techniques in RO plants. For example, a very low specific energy consumption of under 2.3 kWh/m^3 has been reported for a seawater desalination plant using an energy recovery system consisting of a piston type accumulator and a low-pressure pump [22]. Furthermore, the potential for renewable energy use (solar, wind, geothermal, and biomass) has been, and is being, investigated to minimize impacts on air quality and climate change. As seen in Chapter 13, using renewable energy sources for powering desalination processes is a promising option especially in remote and arid regions where the use of conventional energy is costly or unavailable. RO is one of the most suitable desalination processes to be coupled with different sources such as solar and wind.

16.5.4 Site Selection for Impact Mitigation

When a site for a desalination project is to be selected, a large number of characteristics related to the site must be considered depending on the specific operational aspects of the considered plant. In order to minimize the impacts of the desalination plant on the environment, the biological and oceanographic site features should be, at least, taken into account [23]. It is recommended to avoid ecosystems or habitats, if they are unique within a region or worth protecting on a global scale, inhabited by protected, endangered or rare species, important in terms of their productivity or biodiversity, or if they play an important role as feeding or reproductive areas in the region. Moreover, the site should provide sufficient capacity for the dilution and dispersion of the brine and for dilution, dispersion, and degradation of any residual chemicals. Water circulation and exchange rate (function of currents, tides, surf, water depth, and shoreline morphology) will play an important role in determining the load and transport capacity of a site. In general, exposed rocky or sandy shorelines with strong currents and surf are preferred over shallow, sheltered sites with little water exchange.

The oceanographic conditions are important criteria for site selection because they determine the residence time of residual pollutants and the time of exposure of marine life to these pollutants. Furthermore, the site should be close to the sea, to water distribution networks and to consumers to avoid construction and land-use of pipelines and pumping efforts for water distribution. Connection with other infrastructure, such as the power grid, road and communication networks should be conducted with minimal impact. To the extent possible, conflicts with other uses and activities, especially recreational and commercial uses, shipping, or nature conservation, should be avoided.

16.6 Avoiding Possible Disturbances

Site selection can minimize the impacts of the desalination plant on the environment, but can also reduce the impacts of the environment on the desalination plant. In order to keep the impacts on the desalination process at their minimum, the site should provide a good and reliable water quality, taking seasonal variations and periodic events into account. Generally raw waters that are subject to anthropogenic pollution caused by municipal, industrial, shipping, or other wastewater discharges should be avoided.

Raw water with naturally poor quality should also be avoided. This is especially the case for locations with high concentrations of particulate and dissolved organic matter, a high biological activity and thus fouling potential, or the potential for contamination of the intake water quality due to periodically recurring algal blooms.

Intakes that are located further offshore and in deep water layers and thus away from land-based sources of pollution and areas of high biological productivity often provide a more stable and reliable water quality than near shore surface waters. This is also the case for underground intakes, such as beach wells, where the surrounding sediment layers naturally pre-filter the incoming seawater. Moreover, when selecting a site, the risk for oil pollution should be taken into account as oil films can cause serious damage inside a desalination plant and oil contaminants may affect the product water quality [10]. Finally, high-risk areas (e.g., near major shipping routes) should be avoided if possible.

16.7 Life Cycle Assessment of Desalination Technologies

Life Cycle Assessment (LCA) is a useful tool to evaluate the environmental impacts while taking into consideration all the aspects of the entire process, starting from raw material mining and processing, to transport, to manufacturing, to operations and maintenance, and all the way through to the final waste disposal and recycling. In other words, according to the ISO 14040 [24], "LCA is a cradle-to-grave approach". Figure 16.4 illustrates the concepts of LCA applied to desalination [25].

The main environmental impacts of desalination are attributed to the consumption of natural resources and discharge of pollutant emissions through infrastructure construction, energy generation, chemical production, membrane fabrication, and waste management.

16.7.1 LCA Methodology

As defined in the ISO 14040 and ISO 14044 standards [24], an LCA has four phases (Figure 16.4).

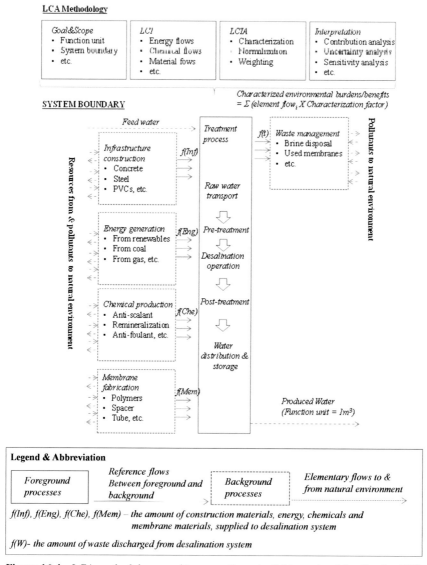

Figure 16.4   LCA methodology used to assess the potential impacts of desalination [25].

16.7.1.1 Phase 1: Goal and scope definition

The main objective of the first phase "goal and scope definition" is to define the functional unit and system boundary. The functional unit describes the primary purpose of a system and enables different systems to be treated as functionally equivalent [26]. In desalination LCA studies, the functional unit is often defined as 1 m^3 of desalinated water. Boundary selection determines the processes and activities included in the LCA study. The determination of system boundary is often affected by factors such as the purpose of the study, geographic area affected, relevant time horizon, etc. [27].

16.7.1.2 Phase 2: Life cycle inventory

"Life cycle inventory" (LCI) analysis is a methodology for estimating the consumption of resources, the quantities of waste flows and emissions caused by, or otherwise attributable to, a product's life cycle [28].

16.7.1.3 Phase 3: Life cycle impact assessment

Life cycle impact assessment (LCIA) consists of characterization, normalization, and weighting of impacts. The characterization is a compulsory step; however, the latter two are considered optional. The characterization step evaluates impact in terms of several impact categories (such as water use, climate change, toxicological stress, land use, etc.) and, in some cases, in an aggregated way (such as years of human life lost due to carcinogenic effects, etc.) [28].

16.7.1.4 Phase 4: Interpretation

After the impact assessment, the "interpretation" allows the LCA to guide decision-makers by providing a better understanding of uncertainties and assumptions. ISO 14044 [29] recommends several checks to determine the level of confidence in the final results, including contribution analysis, uncertainty analysis, sensitivity analysis, etc.

16.7.2 Main Results from Desalination LCA Studies

In previous investigations, the main results have shown that:

1. RO is the lowest polluting desalination technology, and its environmental load is one order of magnitude lower than thermal desalination (MSF and MED) with conventional boilers (90% efficiency).
2. All desalination technologies are energy intensive, therefore, during the operation stage the environmental load is much higher than the plant construction stage. In fact, more than 90% of the energy is used during the operation stage of the life cycle.
3. An appropriate integration of thermal desalination with energy production systems (conventional power plant, combined cycle, hybrid plant,

residual heats, etc.) allows a significant reduction in the environmental loads (about 75%) [30].

4. In the case of RO integration the reduction is more than 30% on average, depending on the electricity production technology (cogeneration, internal combustion engine, or combined cycle). Principal airborne emissions and overall scores related to the different desalination technologies are illustrated respectively in Tables 16.4 and 16.5.

5. Applying an electricity production model mainly based on renewable energies instead of fossil fuels for all desalination technologies, allows a reduction of about 80–85% of fossil fuel usage.

16.8 Conclusions

In this chapter, the different environmental impacts of desalination technologies were investigated. Desalination has been described as a water treatment process that is "chemically, energetically and operationally intensive," so costs as well as environmental concerns are still an obstacle to the widespread use of desalination technologies today. The main marine impacts of RO and MSF plants are summarized in Table 16.6.

Table 16.4   Airborne emissions for MSF and MED and different integration arrangement of RO [25]

	kg CO_2/m^3 Produced Water	g NO_x/m^3	g NMVOC/m^3	g SO_x/m^3
MSF (TC-CB-EM)	23.41	28.30	8.20	28.01
MSF (CCC-EM)	9.41	10.88	3.13	11.34
MSF (TC-CB-EM)	1.98	4.46	1.27	14.96
MED (TC-EM)	18.05	21.43	6.10	26.31
MED (CCC-EM)	7.01	8.16	2.25	15.74
MED (DWH-EM)	1.19	2.53	0.62	19.59
RO (EM, 4 kWh/m^3)	1.78	4.05	1.15	11.13
RO (SC)	2.79	3.38	0.93	3.25
RO (ICE)	2.13	2.61	0.65	2.86
RO (CC)	1.75	2.05	0.59	2.79

TC, thermal consumption; CB, conventional boiler; CCC, cogeneration combined cycle; DWH: driven waste heat; SC: steam cycle; ICE: internal combustion engine; CC: combined cycle; EM: European model.

Table 16.5 Overall scores for MSF and MED and different integration arrangement of RO, evaluated with different indicators: Eco-Indicator 99, Eco-Points 97, and the CML2 baseline 2000 [25]

	EI 99 (MPoints/hm^3)	Eco 97 (GPoints/hm^3)	CML 2 Baseline 10^{-5}/hm^3
MSF (TC-CB-EM)	1.622	9.20	4.43
MSF (CCC-EM)	0.670	3.70	1.45
MSF (TC-CB-EM)	0.104	1.66	1.55
MED (TC-EM)	1.269	7.44	3.08
MED (CCC-EM)	0.507	3.27	1.04
MED (DWH-EM)	0.069	1.48	0.80
RO (EM, 4 kWh/m^3)	0.088	1.30	1.38
RO (SC)	0.199	1.08	0.42
RO (ICE)	0.152	0.85	0.31
RO (CC)	0.126	0.72	0.26

Based on our analysis, a classification can be established for pollutants in order to display their potential harmfulness in different desalination processes (Table 16.7). The classification takes into account the outlined results about: toxicity, applied dosages, degradability, and process relevance.

In conclusion, it is important to emphasize that greater efforts should be made to reduce or eliminate the discharge of pollutants classified as "critical" and "very critical".

This can be achieved by the application of environmental science to preserve the natural environment and resources and to limit the negative impacts of human activity [10]. A project has to be designed and operated according to environmental criteria in order not to knock-on the problem from freshwater scarcity to damaging marine ecosystems. Sustainable desalination is not a distant utopian dream and can be achieved with existing technologies. For example, desalination projects in Australia, including Sydney, Perth, or the Gold Coast project, have incorporated environmental protection measures that will encourage others to follow in their footsteps.

On the other hand, communities should accept that in the future, in addition to the usual construction and operating costs, water costs will include costs that are necessary to reduce the environmental footprint through environmental studies, advanced technology, or compensation measures.

Table 16.6   Environmental impacts of RO and MSF effluents [31]

Effluent Characteristic	Process Relevance	Environmental Impact
Salinity	RO ($\approx$70 mg/l) MSF ($<$50 mg/l)	Can be harmful: reduces vitality and biodiversity at higher values; harmless after initial dilution
Temperature	MSF (+10 to 15°C)	Can be harmful; can have local impact on biodiversity; minor concern in arid regions
Density	RO (higher) MSF (lower)	RO effluents denser than seawater $\rightarrow$ affect benthic species; MSF density depends on process parameters
Chlorine	MSF (up to 2 mg/L)	Toxic in vicinity of discharge point; possible chronic effects
Antiscalants	RO ($\approx$2 mg/L) MSF ($\approx$2 mg/L)	Poor or moderate degradability + high total loads $\rightarrow$ accumulation, chronic effects, unknown side-effects
Coagulants	RO (1–30 mg/L)	Non-toxic; increased local turbidity $\rightarrow$ may disturb photosynthesis: possible accumulation in sediments
Antifoaming Copper	MSF (0.1 mg/L) MSF (15–100 mg/L)	Non-toxic; good degradability Low-acute toxicity for most species; danger of accumulation and long-term effects; bioaccumulation
Other metals (Fe, Cr, Ni, and Mb)	RO MSF	Only traces; partly natural seawater components; no toxic or long term effects (except for Ni in MSF)
RO cleaning solutions	Disinfectants, detergents, and complexing agents	Disinfectants highly toxic at very low concentrations; detergents moderate toxicity; complexing agents very poorly degradable
MSF cleaning solutions	Corrosion inhibitor	Low toxicity; poor degradability

Table 16.7   Potential harmfulness of major pollutants in MSF and RO effluents [31]

	MSF	RO
Very Critical	Chlorine	
Critical	Temperature, antiscalants, copper, and THMs	Salinity, antiscalants, and THMs
Less Critical	Salinity, cleaning solution, and Nickel	Coagulants
Non-Critical	Antifoaming and other metals	Temperature and metals

The cost of water production will certainly be increased by environmental protection measures; however, as shown by Australian projects, sustainable solutions are economically viable.

In the end, decisions about desalination developments stem from local circumstances including the need for water and available alternatives to desalination, the costs of the project and financing options, the significance of environmental impacts, and the definition of significance.

It is obvious that all the world's water problems will not be solved by desalination only. Nonetheless, modern SWRO projects seem to be a more sustainable alternative to many existing water supply schemes and can relieve pressure from freshwater ecosystems.

To encourage developers to implement new, environmentally friendly technologies more ongoing research and demonstration projects are required to gain experience, knowledge and trust. At the same time, political incentives through policies or financial support should be applied in order to implement the best available technologies.

The Fresh Keeper Concept

Arjen van Nieuwenhuijzen, Jaap Klein, Cor Merks
and Ebbing van Tuinen

Witteveen+Bos Consulting Engineers, Deventer, The Netherlands

All around the world and especially in densely populated coastal areas, groundwater resource management is considered of high relevance and has become crucial in some coastal areas to overcome problems including:

- Over-exploitation of water resources, especially in arid and semi-arid areas;
- Increased demand due to economic development and population growth;
- Quality deterioration of the available surface water resources;
- Insufficient knowledge of the aquifer architecture and processes to design effective management programs;
- Climate change and sea level rise.

When freshwater is pumped from a well, the freshwater–saltwater interface from under the bottom of the well moves vertically upwards. If the pumping rate is below a certain critical value, a stable cone will develop without intrusion into the well. But when the pumping rate reaches and surpasses the critical value, the cone will be unstable, and the interface will rise abruptly causing the discharge to become salty. This phenomenon is referred to as upconing. In many coastal zones the balance between water availability and demand has reached a critical level, due to over-abstraction and prolonged periods of low rainfall or drought. As a result, salinization of fresh coastal aquifers due to the upconing of brackish water is increasingly observed in delta areas around the world. Without changes in present policies, management and modes of operation for water supply, climate change will almost certainly exacerbate the upconing of brackish water and/or saltwater intrusion in the future, due to sea level rise and an increase in the frequency and intensity of hydrological droughts.

Brackish and saline groundwater can only be used for potable water, irrigation and process water by relatively expensive and energy-intensive treatment (i.e. desalination) processes. Disposal of concentrate streams resulting from desalination processes pose environmental and cost concerns, and often determine whether a desalination project is viable, especially for inland communities.

Battling problems related to salinization of aquifers is not a new phenomenon. In the Netherlands, a country with abundant freshwater, various schemes with managed aquifer recharge have been implemented along the coastline for maintaining fresh coastal aquifers since decades ago (Figure 16.5). Furthermore, the capacity of some public supply well fields had to be reduced dramatically to limit the upconing of brackish water, which, ultimately, would result in 'full capacity' implementation of brackish water reverse osmosis as an *effect-oriented remedy*.

In the last decade, the Dutch water utilities Vitens and Brabant Water, together with KWR Watercycle Research Institute have developed a highly effective *source-oriented remedy* against salinization [23, 24] of fresh groundwater wells together with a sustainable solution for concentrate disposal. This Fresh Keeper concept consists of three components (numbers correspond to the schematic shown in Figure 16.6 on the right):

1. Fresh and brackish groundwater are separately abstracted from the same borehole by means of two filter screens.
2. The abstracted brackish groundwater is used as an additional source for water supply, after desalination: Brackish Water Reverse Osmosis (BWRO).

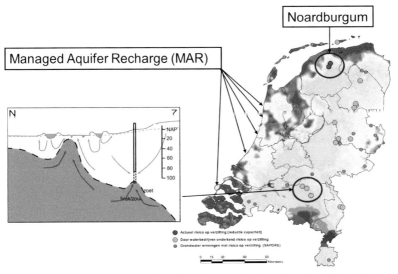

Figure 16.5 The Netherlands as a city of the future situation; the salty delta with abundant fresh surface and groundwater resources, and measures taken in response to salinization, climate change and sea level rise [22, 23].

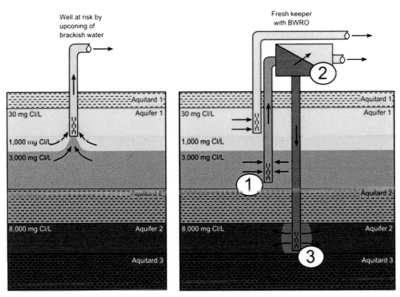

Figure 16.6 The schematic on the left presents a well field at serious risk of upconing of brackish groundwater (salinization) due to over-exploitation. The schematic on the right presents the solution: the Dutch Fresh Keeper concept [23, 24].

3. The BWRO concentrate is disposed through deep-well injection into a deep geological formation in a confined aquifer having similar salinity levels as the BWRO concentrate. For treatment facilities near the coast, discharge back into the ocean could also be a sustainable solution.

This Fresh Keeper concept has been successfully applied in small-scale (single well) pilots by the Dutch water utilities Vitens and Brabant Water. At both inland locations with a specific geo(hydro)logical setting, salinization of fresh groundwater resources was effectively stopped and reversed by abstraction of the upconing brackish groundwater. The brackish anoxic groundwater proved to be excellent RO feed water which could be operated at acceptable energy levels. The additional capacity for water supply from the BWRO also adds value to the concept. Deep-well concentrate injection proved to be a sustainable solution at these two locations. Based on these pilots, Vitens is currently investigating the potential of this technology for a full-scale (well field) application near Noardburgum.

References

[1] IDA (2006). *IDA Worldwide Desalting Plant Inventory*, No. 19 in MS Excel format. Oxford: Media Analytics Ltd.

[2] Lattemann, S., and Hoepner, T. (2003). *Seawater Desalination, Impacts of Brine and Chemical Discharges on the Marine Environment*. L'Aquila: Desalination Publications, 142.

[3] WHO (2015). *Guidelines for Drinking-Water Quality*, 3rd Edn. Geneva: World Health Organization.

[4] Cotruvo, J. A. (2006). Health aspects of calcium and magnesium in drinking water. *Water Cond. Purif.* 48, 40–44.

[5] Cotruvo, J. A. (2005). Water desalination processes and associated health issues. *Water Cond. Purif.* 47, 13–17.

[6] WHO (2011). *Desalination for Safe Water Supply, Guidance for the Health and Environmental Aspects Applicable to Desalination*. Geneva: WHO.

[7] Vanhems, U.S. Clean Water Act, 1998.

[8] Abu Qdais, H. (2008). Environmental impacts of the mega desalination project: the red–dead sea conveyor. *Desalination* 220, 16–23.

[9] California State University, Sacramento Center for Collaborative Policy (2008). *California Desalination Planning Handbook*. Available at: www.owue.water.ca.gov/recycle/docs/Desal_Handbook.pdf

[10] Lattemann, S. (2010). *Development of an Environmental Impact Assessment and Decision Support System for Seawater Desalination Plants*. Ph.D. thesis, Delft University of Technology and of the Academic Board of the UNESCO-IHE Institute for Water Education, Delft.

[11] Heather, C., Gleick, P. H., and Wolff, G. (2006). *Desalination, with a Grain of Salt*. Oakland, CA: Pacific Institute for Studies in Development.

[12] Damitz, D., Furukawa, D., and Toal, J. (2006). *Desalination feasibility study for the Monterey Bay region, final report prepared for the Association of Monterey Bay Area Governments (AMBAG)*. Available at: http://www.ambag.org/ [accessed September 12, 2009].

[13] Gulamhusein, A., Khalil, V., Fatah, I., Boda, R., and Rybar, S. (2008). "IMS SWRO Kindasa, two years of operational experience," in *Procee- dings of the EUROMED 2008, Desalination Cooperation among Mediterranean Countries of Europe and the MENA Region*, Dead Sea.

[14] UN ESCWA (2001). *Water Desalination Technologies in the ESCWA Member Countries*. Beirut: UN Economic and Social Commission for Western Asia.

[15] Hopner, T., and Windelberg, J. (1996). Elements of environmental impact studies on coastal desalination plants. *Desalination* 108, 11–18.

[16] MEDRC (2002). *Assessment of the Composition of Desalination Plant Disposal Brines (Project NO. 98-AS-026)*. Seeb: Middle East Desalination Research Center (MEDRC).

[17] Höpner, T. and Lattemann S., (2008). *Seawater Desalination & The Marine Environment*. Aquila: European Desalination Society intensive course.

[18] World Bank Group (1998). *Thermal Power: Guidelines for New Plants*. Available at: www.ifc.org/ifcext/sustainability.nsf/Content/Environment alGuidelines [accessed September 12, 2009].

[19] Jenner, H. (pers. comm.). KEMA, Netherlands, 2009.

[20] Abdel-Jawad, M., and Al-Tabtabaei, M. (1999). "Impact" of current power generation and water desalination activities on Kuwaiti marine environment," in *Proceedings of the mIDA World Congress on Desalination and Water Reuse,* San Diego, CA, 231–240.

[21] Shams, A. E., and Mohammed, R. (1998). Kinetics of reaction between hydrogen peroxide and hypochlorite. *Desalination* 115, 145–153.

[22] Paulsen K., and Hensel, F., (2007). Design of an autarkic water and energy supply driven by renewable energy using commercially available components. *Desalination* 203, 455–462.

[23] WHO (2007). *Desalination for Safe Water Supply: Guidance for the Health and Environmental Aspects Applicable to Desalination*. Geneva: World Health Organization.

[24] ISO 14040 (2006). *Environmental Management, Life Cycle Assessment, Principles and Framework*. Geneva: ISO.

[25] Zhou, J., Victor, W., Chang, C., Fane, A. G., (2014). Life cycle assessment for desalination: a review on methodology feasibility and reliability. *Water Res*. 61, 210–223.

[26] Guinee, J. B., Gorree, M., Heijungs, R., Huppes, G., Kleijn, R., De Koning, A., et al. (2002). *Handbook on Life-cycle Assessment. Operational Guide to the ISO Standards*. Dordrecht: Kluwer Academic Publishers.

[27] Reap, J., Roman, F., Duncan, S., and Bras, B. (2008). Asurvey of unresolved problems in life cycle assessment. Part 1: goal and scope and inventory analysis. *Int. J. Life Cycle Assess*. 13, 290–300.

[28] Rebitzer, G., Ekvall, T., Frischknecht, R., Hunkeler, D., Norris, G., Rydberg, T., et al. (2004). Life cycle assessment part 1: framework, goal and scope definition, inventory analysis, and applications. *Environ. Int*. 30, 701–720.

[29] ISO 14044 (2006). *Environmental Management – Life Cycle Assessment – Requirements and Guidelines*. Geneva: ISO.

[30] Raluy R.G., Serra L., Uche J. (2005). Life cycle assessment of desalination technologies integrated with renewable energies. *Desalination* 183, 81–93.

[31] Wang, S., Leung, E., Cheng, R., Tseng T., Vuong D., Carlson D., et al. (2007). "Ocean floor seawater intake and discharge system," in *Proceedings of the IDA World Congress on Desalination and Water Reuse*, Maspalomas.

PART V

Social and Commercial Issues

CHAPTER 17

Rural Desalination

Leila Karimi[1] and Abbas Ghassemi[1]

[1]Institute for Energy and the Environment, New Mexico State University, Las Cruces, New Mexico
E-mail: lkarimi@nmsu.edu; aghassem@ad.nmsu.edu

17.1 Introduction

As discussed in previous chapters, clean freshwater resources are limited, but demand for water has continuously increased, driven by factors such as population growth and industrialization. Freshwater scarcity is a significant challenge. Around the world, 1.2 billion people are experiencing water scarcity, and the scenario is expected to worsen. According to United Nations estimates, 1.8 billion people will face absolute water scarcity by 2025, while 2/3 of the global population will face water stress [1]. The most severe water scarcity in 2025 will be in the rural areas of Asia, most of Africa, the "bread baskets" of northwest India, and northern China [2].

Remote rural areas often face dire conditions already. As of 2002, 1.1 billion people relied on unsafe water sources, resulting in 1.7 million deaths annually – mainly among children [3]. The lack of safe water also impairs living conditions by obstructing personal hygiene, reducing the quality of food production, and impeding education. The water-related barriers to education especially harm girls and women, who typically bear most of the burden for collecting water and caring for children who fall ill (often from water-borne diseases) [3]. It is almost universally agreed that there will be a worsening trend of water scarcity with respect to quantity and continued degradation in quality in the coming years. In rural areas, water will be a controlling factor for food production, quality of life, and even survival.

Desalination and water treatment systems will play a large role in alleviating water scarcity and promoting human well-being. In areas where contaminated surface water is available, that water can be treated to produce potable freshwater; in areas where surface freshwater is not available, there may be brackish groundwater or seawater, both of which can be treated by desalination systems to provide life-giving freshwater. In urban and industrial areas, desalination systems are already producing approximately 90 million cubic meters of freshwater per day, and similar technologies may be applied in rural areas, too.

Rural desalination systems bring many benefits including:

1. Desalination plants offer health improvements and eliminate disease vectors from polluted water sources; they also prevent injuries to women and children who previously had to carry heavy water containers over long distances.

2. Freshwater from desalination brings new job opportunities and agricultural improvements, and it allows women to earn money through cottage industry in the time they previously spent carrying water. Female-led cottage industries can significantly bolster rural economies: for example based on statistics from 1999, 80% of Iran's carpet producers are women and young girls in rural areas of the country, who provide a high percentage of the total household income [4]. In scenic areas such as islands, desalination plants can also increase revenue from tourism by supporting the temporarily higher populations that tourism brings.

3. Water from desalination eliminates the need for girls to fetch water, increasing their attendance at schools and improving educational outcomes [5].

4. Adequate freshwater reduces migration rates out of rural areas, stabilizing rural economies and social life [6, 7]. Emigration from rural areas typically deprives rural communities of working-age people, separating families and worsening poverty because the very young and the very old are left behind. The economic effects of desalination can reduce unemployment in rural areas, thereby encouraging working-age people to stay.

Despite the many similarities between urban and rural desalination systems (which use the same fundamental technologies, after all), there are significant differences between desalination systems in urban and rural areas. Before we discuss those differences, let's differentiate urban areas from the rural areas that are the focus of this chapter.

Although urban and rural areas are two known concepts and widely applied, they do not have universal definitions. By the standards of the US Census Bureau, any area with a population of more than 50,000 people is considered an urban area, and any area with a population of less than 50,000 but more than 2,500 is called an urban cluster. By these criteria, an urban area is any densely developed area, including residential, non-residential, and commercial areas. Any area outside these standards can be called a rural community [8]. However, based on the UN recommendations, a universal definition is not possible for urban and rural areas due to their different characteristics across the globe [9]. Therefore, the published data on urban and rural areas by the UN relies on their national definitions [9–11].

As countries develop their own classifications for differentiating urban communities form rural ones, they can use several criteria. In industrialized countries, population size and population density are the most important factors for distinguishing urban communities from rural ones [12]. In developing countries, however, the standard of living may be the most important factor, with the size of the population merely a supplementary parameter. By this perspective, both a Brazilian *favela* and an Amazonian subsistence farming community may be classified as rural. Other factors that may be considered include ease of access to electricity, drinkable water, medical care, and education, as well as the percentage of the population whose occupation is agriculture. So in general, there are no established international criteria for differentiating between rural and urban areas, and there are often no defined criteria even among countries in a specific region [13].

Although desalination systems can address water scarcity in both urban and rural areas, the disparate characteristics of urban and rural areas require that the desalination systems be different. From a technical perspective, rural systems are smaller than urban systems, may need to operate without a reliable source of conventional energy, may need to manage brine disposal without existing sewers or secondary water treatment systems, and may need to function with a reduced need for skilled engineers, technicians, and operators. These technical issues are also interconnected with sociocultural factors: in remote, rural areas, providing a clean, reliable water source involves far more than optimizing a technical desalination problem. Although large desalination plants in developed, urban societies can be engineered and optimized with precision and confidence, in rural areas, water management is "a social-technological process" in a "social-ecological system" [14]. This social-ecological system contains, at the very least, a water source, infrastructure and technology, governance systems, and local actors, all contained within a larger environment where each component of the social-ecological system may be repeated at a grander scale [14]. Therefore, in remote rural areas, it is essential to understand not only the technical aspects of desalination, but also the physical and social environment in which the technical system will operate.

In the body of this chapter, the technical and social challenges facing remote rural desalination systems will be discussed; guidelines that can increase the likelihood of developing a successful and sustainable socio-cultural-technological desalination system will be established; and real-world case studies of failure and success will be presented.

17.2 Factors that Affect the Success of Rural Desalination Systems

The successful design and operation of a rural desalination system depends on overlapping technical, social, environmental, religious, cultural, and economic aspects as shown in Figure 17.1. Although these factors are likely to be different at different sites and may have different levels of importance, all of them, should be considered to increase the chances of developing and sustaining a successful rural desalination system. The following material presents an overview of the factors that should be considered, in an order that could be applied in actual projects. This order is by no means absolute, and the characteristics of a specific site may affect the order and the relative importance of each factor.

Figure 17.1   Schematic of effective factors on success of rural desalination systems.

17.2.1 Water Resources

One of the first technical steps for developing a rural desalination system is performing a hydrogeological study to determine water resources [15, 16]. As part of the hydrogeological study, the quality of the available water resources should be determined, since the quality of the available water sources will determine what pretreatment processes should be considered to prevent scaling and fouling during the operation of the desalination system [15].

Additionally, the meteorological characteristics of the area should be determined since the amount of precipitation affects the availability of freshwater in rural areas. In some rural areas, the inhabitants have their own rainwater storage, or there is central rainwater storage that people can access [5, 17, 18]. Communities in south and southeast Asia, the highlands of East Africa, and semi-arid areas of the Jordan Valley and other similar places rely on rainwater harvesting methods for domestic water needs [18].

Even in areas with obvious water resources, a complete evaluation of available water is valuable because the most obvious resources are not always the best. For example, islands and coastal villages may have access to practically unlimited amounts of seawater, but adequate brackish water resources in these areas could be desalinated with a lower energy consumption [19], making the brackish water resources preferable [15]. From an optimization perspective, the obvious resource of seawater would be used only if brackish water was unavailable or present in quantities that could not meet the area's water needs [15].

17.2.2 Water Needs

A rural area's need of water—in terms of both water quantity and water quality – significantly affects the process of choosing the most applicable desalination system for that area [16]. The requisite quantity and quality of the treated water will affect the capital cost of the system, the optimal type of system, the amount of energy required, and the operating costs [20–23].

Different areas have different water demands, partly because of population, partly because of agriculture, and partly because of differences in the quality of the available water (which you will have determined in the hydrogeological study). In some rural areas, undrinkable brackish water resources can be used for washing and cleaning purposes, even if they contain certain toxic heavy metals: some heavy metals, such as arsenic, are not easily absorbed through the skin [24]. However, if water resources in an area contain toxic components that can be absorbed through the skin – such as inorganic mercurial salts – these water resources cannot be used for drinking, cooking, washing, or cleaning [25], and treated water must be provided to meet all of these needs. A rural area may also have seasonally variable water needs based on population changes from tourism or seasonal agriculture. In some areas with tourist attractions, such as certain Mediterranean villages, desalination systems are needed not to support local people, but to support tourism during peak seasons. Such seasonal changes and population changes should be considered in the selection and design of an appropriate desalination system [15].

17.2.3 Energy Sources

Communities need both water and energy, and communities with desalination systems need energy to get water. In general, desalination is energy intensive, so the availability of energy resources is one of the main factors that should be taken into account in designing and choosing a desalination system in remote rural areas. Design parameters should account for the availability of the electrical grid, the availability of fossil fuels, and the availability and type of renewable energy resources (Figure 17.2).

Figure 17.2   Energy source considerations for rural desalination plants.

In large desalination plants, the required energy can be provided from grid electricity or from a power plant built specifically for the needs of the large desalination plant. In some cases, the power plant for a desalination system may draw on renewable energy, as does the 80 MW Emu Downs Wind Farm, whose 48 turbines power the Perth Seawater Reverse Osmosis Plant in western Australia [26]. Renewably powered plants this large are relatively rare, but renewable energy sources are common for small desalination systems in remote rural communities that do not have reliable access to grid electricity.

In remote rural areas not connected to the electrical grid, diesel generators can power desalination systems if it is practically and economically viable to transport the fossil fuel, but otherwise, renewable energy sources are the only solution. The choice of renewable energy source will be based on the availability and reliability of renewable energy sources, some of which may have been identified during the meteorological study, which can provide information about average wind speeds and solar insolation.

In many areas, solar energy is one of the most abundant energy resources, and it can be harvested and transformed into usable electricity by photovoltaic panels [27]. In areas with adequate solar energy, photovoltaic panels can readily power electrodialysis and reverse osmosis desalination systems; the solar energy can also be used directly through solar distillation. In such cases, the preferences for the different technological options will be determined by factors such as the quality and quantity of the feedwater and treated water, the systems' ease of operation and maintenance, environmental conditions, and results from technical and economic analyses.

The relative reliabilities of the different renewable energy sources should also be considered, and this factor can strongly affect which renewable energy resource is used [15]. For example, although solar energy is often readily available, it is also variable, perhaps making the more reliable geothermal energy a more attractive option. In some cases, the best option may be a hybrid renewable energy system, such as a combination of wind and solar energy sources.

When different types of renewable energy resources are available, it is important to consider the adaptability of each desalination technology to each power source [15]. For example, thermal desalination technologies are well suited to thermal energy sources (such as geothermal energy), while reverse osmosis and electrodialysis systems are well matched with wind turbines and photovoltaic panels, which produce directly usable electrical energy.

In general, when an electrical form of energy is available (for example, from photovoltaic panels or wind turbines), the optimal desalination technology will be either reverse osmosis, electrodialysis reversal, or mechanical vapor compression; thermal desalination systems can be excluded. Reverse osmosis is usually preferred when seawater is the water resource, but otherwise electrodialysis/electrodialysis reversal is more desired in remote rural areas because of its robustness and lower maintenance requirements [15]. Electrodialysis/electrodialysis reversal is also preferred when the brackish water has a high silica content and a high risk of scaling from less soluble salts [19, 28]. Mechanical vapor compression is disadvantaged by its comparatively high energy consumption, but it does require less pretreatment than reverse osmosis [15], potentially making mechanical vapor compression a suitable choice when pretreatment would be problematic.

When low-temperature thermal resources are available, the optimal desalination system is usually multi-stage flash, multi-effect distillation, a solar still, membrane distillation, thermal vapor compression, or humidification-dehumidification. Solar stills, membrane distillation, and humidification-dehumidification are especially appropriate for small-scale desalination systems with a maximum capacity of 5 m^3/day [15].

Different desalination technologies and different saline water sources also require different levels of energy. Thermal-based desalination processes, such as multi-stage flash and multi-effect distillation (used mostly for seawater desalination) are much more energy intensive than membrane-based desalination processes, such as reverse osmosis [29, 30]. As seen in Chapter 14, regardless of the type of process, desalinating brackish water requires less energy than desalinating seawater, due to the lower salinity of brackish water [31].

Daily or seasonal variation in the availability of renewable energy resources should also be considered in the design of desalination plants, which should have backup systems in place to provide water even if the primary energy source becomes unavailable. For instance, although solar energy is easily available during the day, the lack of solar energy at night and on cloudy days can shut a desalination system down. To keep the system operating during these times, the system can be connected to the electrical grid (if the grid is available); use energy storage systems, such as a battery bank; or employ another energy source, such as diesel generators. Each of these options has its own set of challenges, including environmental impacts and a significant increase in capital costs.

In light of these challenges, it may be more efficient to let the system shut down, but use holding tanks to store water produced when energy is available. The stored water can then be distributed even when the desalination system is not operational. Holding tanks are often viable for remote rural desalination systems, which usually produce relatively small quantities of water [15]. However, the water storage tanks are susceptible to contamination problems, so they should be maintained and monitored carefully [32].

Ultimately, the most feasible desalination system and energy source will be determined not only by the technical match between desalination system and energy source, but also by economic factors, the skill needed to operate and maintain each alternative, and the quality and quantity of needed water. In every case, a detailed technical, social, and economic analysis should be carried out to determine the most feasible solution.

17.2.4 Technological Factors

Large-scale urban desalination plants and remote rural desalination systems share several technological considerations, including pre- and post-treatments, capital and operating costs, blending options for treated water with feedwater, energy supply, brine discharge solutions, water storage and distribution options, and attention to raw water quality, including its salinity and potential for scaling and fouling [33–36]. However, these factors carry different weights in urban desalination plants than they do in remote rural areas. For example, the operating costs for a rural desalination system may be very low if the desalination system uses a renewable energy source, making operating costs a small consideration; on the other hand, the up-front capital costs associated with the renewable energy system may carry significant weight. In contrast to this hypothetical rural desalination system, a desalination plant in an urban area may be designed to minimize operating costs, since the costs for grid electricity and fossil fuels can be significant.

Despite these differences, most of the technological factors for urban and rural desalination plants are similar, and the same principles and techniques apply for designing both urban and rural desalination systems. However, there are key differences in the technologies used for environmental protection and water distribution; in remote rural areas, system design may also be influenced by the availability or lack of trained personnel.

One major technological factor for all desalination systems is environmental protection. All desalination plants carry environmental risks, whether from brine discharge or the environmental footprint of the energy system.

When desalination systems are powered by fossil fuels, the systems' energy consumption can cause power plants to release huge volumes of greenhouse gases [37]. Reverse osmosis, which on average consumes 4 kWh of electricity to produce 1 cubic meter of treated water [31, 38], therefore, also produces 1.8 kg of CO_2 per cubic meter of treated water [39]. The main environmental concern, though, is usually the safe disposal of rejected brine, a saline byproduct of desalination. This issue affects all desalination plants, regardless of their capacity: large plants often have significant resources for dealing with rejected brine, but they also have a large volume of rejected brine with which to deal; small plants may have a smaller volume of rejected brine, but they also have fewer resources for managing the brine.

As discussed in Chapter 15, several brine management methods have been developed. However, these methods are not always available, and the brine management methods seldom eliminate all risk from the brine. In coastal desalination plants, the brine may be diluted by mixing it with seawater [40]. This approach is cheap, but it carries risks from the brine stream's heightened temperatures (if thermal desalination methods are employed) and elevated concentrations of salts and chemicals. Ocean discharge is not available for inland desalination plants, which must manage their rejected brine through other processes, such as deep well injection; salt production; evaporation ponds; discharge to either municipal sewers or surface water sources; or land applications, such as the irrigation of plants tolerant to high salinities. Another option is zero liquid discharge, a process that completely separates the liquid from the brine but is very expensive and often energy-intensive [41]. Generally, all of the inland brine management methods are more expensive than ocean discharge. They can add 40% to the capital costs of desalination plants and increase desalination costs by 5–30% [42].

For rural desalination plants, evaporation ponds are often an attractive option because of their low cost and ability to manage the relatively small quantities of brine produced by most rural desalination plants [15]. Evaporation ponds have simple structures, are built easily, and can be operated at low costs. The only mechanical instrument used in this method is the pump used to transport the brine from the desalination plant to the pond. This mechanical simplicity reduces the skill and effort needed to operate and maintain the brine management system as long as the pond is properly constructed [15]. In warm, dry climates with a high evaporation rate, this method is often the most feasible brine management approach for remote rural areas.

Despite the advantages of evaporation ponds, they do have several dis-advantages, including the use of land for the ponds, the risk of water contamination from pond leaks, and the costs for materials, such as poly(vinyl chloride) piping and high density polyethylene pond lining [15]. Therefore, other brine management techniques can also be considered.

In addition to evaporation ponds, another promising method for brine management in remote rural areas is using the brine stream for agriculture and aquafarming. Although this method requires more downstream infrastructure, it can be an optimal brine disposal method for inland desalination plants, since it not only manages a large volume of brine in a sustainable way, but also creates new job opportunities for local people [15]. As an example of how this approach can work, the brine can be used as a growth medium for algae, which decrease the salinity of the brine; then, the overflow stream from the algae can be sent to a fish farm where certain fish, such as Tilapia, can thrive; lastly, the water can be used to irrigate halophytic plants, which are able to tolerate elevated salinities [15]. The use of brine in agriculture and aquaculture should be performed carefully, though, since the irrigation water increases undesirable soil salinity due to salt accumulation and can also contaminate freshwater aquifers.

Just as the brine must be managed responsibly, so too must the desalinated water, which must safely reach the people who need it. In most urban areas, a water distribution network has already been developed, and desalination plants do not need to consider this factor; however, rural areas often lack a water distribution network, and developing one can increase the cost and complexity of the desalination system. Nevertheless, a complicated distribution system is not necessary in many cases, since desalinated water can often be stored and accessed in a common tank for the whole community.

A final technical difference between urban and rural desalination systems is the availability or lack of adequate operation and maintenance skills. In urban areas, the availability of skilled and professional people is not a serious issue; in rural areas, the lack of such personnel is often a central design consideration. In rural areas far from trained personnel, the desalination systems should be very reliable, and it may be preferable to trade efficiency for reliability and a simpler design that is easier to maintain. Additionally, in order for local people to understand the desalination system, they need to be involved throughout the entire process, from design and decision making to plant building and system installation. A limited number of local people should be trained to safely operate and maintain the system, and financial planning should include professional technical support to handle special and complicated issues [15].

Desalination plants affect societies in both urban and rural areas, primarily for the better, but also for the worse. Desalination plants have significantly improved economic and social prospects for people around the world, but the desalination of water at an affordable price can encourage unsustainable water management. With subsidized (below full cost) freshwater from desalination, people in arid regions may mistakenly consider water abundant and needlessly increase their water demand [37].

Social effects are even more pronounced in rural areas, where the desalination system interacts with small communities; traditional cultural and religious beliefs; a closely-knit society; existing and prospective businesses; and socially prescribed duties for specific demographic groups, such as women and girls or young people. Researchers have only recently begun to consider the demographical, cultural, and social aspects of desalination plants in rural areas, but these factors can significantly affect the success or failure of the desalination plant [15]. For desalination technologies to be sustainable, societal factors must be taken into account. These factors – community involvement, institutions and social power, cultural issues, gender issues, religious issues, and economic issues – are the focus of this section. Although these factors are discussed under separate headers, they are strongly interrelated and often overlapping.

Because societal factors determine the success of any rural desalination project as much as technological factors, it is best to involve the community from the planning phase onward; if this is impossible, the community members should at least be active participants during the construction, operation, and maintenance phases. The inhabitants should be clearly informed about the project and should become familiar with how the process works, covering topics including water sources, brine management, capital and operation and maintain costs, final water cost, and the similarities between desalination technologies and the natural water cycle. The community members must also know that the project will be unsuccessful if they do not take an active and significant role in the planning, building, and operating phases of the project [15].

A lack of local support is one of the key reasons why rural desalination systems in developing countries fail, which is explained by one of the unsuccessful case studies later in this chapter. After the external funding and activities stop, the community members must be willing and able to sustain the system; therefore, the community members should be involved throughout the entire process, from the early design steps to building, installing, operating, and maintaining the plant. Throughout the process, the community's ownership of the project should be emphasized [43], and the community's power over the process should reflect this ownership.

In the effort to involve communities, several special considerations should be kept in mind. First, women should be involved in the decision-making process for water policies [15]; although men are primarily responsible for decision-making in most rural communities, women have a significant and often primary role in providing water for living needs, and their input is indispensable. Second, in any project, some individuals or groups can be affected negatively. Even with a technology as beneficial as a desalination system, this holds true: water merchants may lose profits from transporting and selling water [44], and in coastal regions, fishers could rightly worry about brine discharge harming fish [15]. These stakeholders should be involved in the project as well, and their concerns should be understood and addressed. Third, although technically trained outside engineers may prefer a particular technical solution, they cannot impose their preferred solution on a community that has an opposing perspective. The community members must have the ultimate choice about decisions regarding their lives and needs [15]. Fourth, from the very early phases of the project, the community's people should be trained, and made aware of the benefits of desalination and some of its technical aspects. This not only helps community members make informed choices, but it also trains eventual operators of the desalination system, whose participation will play a key role in the success or failure of the project.

Throughout the process of working with the community, several steps can be taken to avoid or minimize adverse social effects. First, be sure to understand the numerous perspectives within the community by conducting on-site interviews and discussions. Synchronously, communicate with the community members about how desalination can improve their health and quality of life, and determine the community members' ability and inclination to respond to potential problems and difficulties. Additionally, reinforce community ownership of the desalination system by involving the local people in the decision-making process, with the goal of ensuring that the community members appreciate the plant's presence and do not abandon it during difficulties or even temporary failures [16]. The feedback and acceptance of the community can determine the social sustainability of the system [16].

17.2.5.2 Institutions and social power

The technology of desalination does not exist in isolation, but in relation to people and institutions. Water is often managed by overlapping institutions of various formalities, and existing institutional structures for water management may be hundreds of years old and integrated into the fabric of society. Longstanding traditions about water resources, management, and use often developed for good reason, and therefore these traditional thoughts should be combined with the new strategy of desalination. Wherever possible, desalination technologies should be integrated as complements to existing water management systems and institutions, whose representatives should be involved in the development of the desalination system. This is especially true when religious aspects are involved, as discussed later in this chapter.

Informal institutions and informal social power should also be taken into account. In small communities, it is common for singular individuals or groups of people to significantly affect how others think, and sometimes these socially powerful individuals or groups may directly or indirectly oppose any change in the existing social structure that has afforded them power and prestige. For example, it would be difficult to install a desalination system in a remote rural area where an influential community member transported and sold bottled water at a reasonable rate. Traditionally minded people may also resist desalination systems as imprudent outside meddling that is incompatible with their heritage and potentially damaging to sacred sites.

Such challenges can be resolved to some extent by respecting the traditions and the customs of the community and involving all community members in the planning and implementation activities. Conflicts should be detected and addressed [43], and insights from traditional cultures should be considered.

17.2.5.3 Cultural issues

In the implementation of any new technology, there are some objections due to cultural issues. Such objections are more prevalent in small rural communities than in large urban areas because of rural community members' traditional beliefs and thoughts. Traditional cultural beliefs are especially common among older generations, but traditional perspectives may also be transferred to members of younger generations, especially young people with low levels of education. Although traditional beliefs are sometimes deleterious, it should be emphasized that traditional beliefs and practices may beneficially guide water management, especially when traditional beliefs emphasize ecological stewardship and are tailored to particular environments.

Some traditional cultural beliefs can be harmful, though, and cultural differences between the system developer and the community members can cause non-technical problems, disagreement about the control and operation of the desalination system, and finally system failure. An approach that considers cultural issues and communicates with end users can improve the likelihood of developing a sustainable desalination system [43].

For instance, before installing any desalination system in a rural community, it should be determined whether the local people would welcome or even tolerate the desalinated water as a new water source. Attitudes toward feedwater sources should also be determined; some feedwater sources may be technically suitable but culturally taboo. For instance, the Nyungar Aboriginal tribe near Perth, Australia, have a special antipathy toward brackish groundwater desalination plants, which have dropped the water table and destroyed sacred sites [45]. Cultural factors like this can play a decisive role in the success or failure of desalination plants, and since cultural factors vary from culture to culture and community to community [43], investigative approaches from the social sciences can be beneficial [33]. Research should be conducted to identify the social benefits from installing and operating the plant, determine the community's attitudes, discern freshwater requirements, quantify the availability of private wells, identify the technical knowledge of the community members, and evaluate their willingness for training [16].

In some instances, education and training can alleviate cultural conflict, since community members' objections to the technology may be based on a misunderstanding of the technology or the desalination process in general. Sometimes, even individual taste preferences may play a role, such as when a difference in the taste between desalted and locally available water causes community members to regard the desalted water as unsuitable for human consumption and use [43].

17.2.5.4 Gender issues

Although water is not a resource that male and female organisms need differently for life and health, females in remote rural areas typically face water issues more directly than males. Women often have the responsibility for water use in food production, cooking, cleaning, sanitation, washing, and waste removal. Women are also responsible for procuring water; therefore, they usually collect information about water resources, treatment processes, and safe storage procedures. Some societies have also given women primary responsibility for managing water resources, but in other societies, women have no official voice on water related issues [46].

The crucial significance of both male and female participation in water supply management and sanitation was decreed in the 1977 United Nations Water Conference at Mar del Plata. This importance has been reinforced in later conferences and meetings, such as the 1992 International Conference on Water and the Environment in Dublin and the 1992 United Nations Conference on Environment and Development in Rio de Janeiro [46].

In urban areas with a centralized water network, women are not responsible for providing household water supplies, so the installation of a new desalination plant does not disproportionately affect any gender or demographic group. However, rural desalination plants often affect women more than men because women are primarily responsible for providing the family with water. For instance, 1/3 of women in Egypt walk almost 1 km a day to get water. In other African countries, women may walk for more than 6 km and spend almost 8 h daily to collect water for the family. In India too, women spend a significant part of the day for providing water for the family from remote sources as shown in Figure 17.3 [47]; in contrast, based on standards from the World Health Organization, everyone should have access to at least 20 L of water per capita per day within 1 km [48]. The task of transporting water is a heavy burden for women as shown in Figure 17.4: carrying the heavy water containers can cause serious health problems, especially during

Figure 17.3 Indian women in rural areas spend a significant part of the day for providing water for the family.

Source: Environmental Engineering and Earth Sciences (2009). Photo courtesy of Dr. Stephen Moysey, Clemson University.

Figure 17.4 Women carry heavy water containers.

Source: GE Reports [50].

growth and pregnancy; the need to walk long distances exposes women to severe natural risks, such as floods and animal attacks; and the responsibility of providing water can keep girls out of school [49].

Reliable water supplies can significantly alleviate all of these problems. In Morocco, a water supply project conducted in six rural provinces increased female school attendance by 20% over 4 years and reduced females' time fetching water by 50–90% [5]. Although these results are based on a water supply sanitation project rather than a desalination plant, desalination projects should produce similar results by providing similar water; since water availability has contributed to the statistics regardless of the approaches

17.2.5.5 Religious issues

As mentioned briefly in the preceding section on culture, religious issues and water management issues can overlap. In our previous example of the Nyungar Aboriginal tribe, we learned of religious objections to brackish groundwater desalination that depletes aquifers, disrupts natural water cycles, and destroys sacred sites. Such interrelations between water management and religion are frequent.

There may also be religious obligations about the purposes to which water may be applied. In Islam, water has a sort of blessed status, and is used for the purification of that which is *najis*, or unclean. There may also be religious teachings about whether treated wastewater can be considered pure for religious purposes. Islam, Christianity, Judaism, and the other religions use ritually pure water, making this issue a common one. In Islam, the Council of Leading Islamic Scholars issued a Fatwa in 1978 stating that "impure wastewater can be considered as pure water and similar to the original pure water, if its treatment using advanced technical procedures is capable of removing its impurities with regard to taste, color and smell, as witnessed by honest specialized and knowledgeable experts" [51]. Notably, ritually pure water may be considered as pure and safe, but not be safe according to physiochemical standards.

In rural areas, traditional religious and cultural beliefs may far outweigh scientific perspectives. For example, of the people surveyed in rural Pakistan, only about 31% of the population "linked diarrheal diseases with the ingestion of contaminated drinking water"; most (about 60%) believed diseases to be punishments from God. Of the people who recognized contaminated water as a disease vector, most (67%) considered turbidity the primary risk factor and only about 17% of the surveyed villagers recognized the risks from microbes[1]. The people surveyed still exhibited great enthusiasm for affordable, accessible water treatment methods, but the grounds for that enthusiasm (turbidity reduction) were narrower than what might be expected in a developed, industrial society. The possibility of such different perspectives should be taken into account when proposing water treatment systems [52] and developing educational and training programs.

17.2.6 Economic Aspects

Clean and safe water is a right of all humans, but to keep a desalination plant operational, the cost of the water must cover expenses through sales prices,

---

[1]*Editor's note:* philosophical debates are not within the scope of this book. However, it must be said that although the cited study seems to dissociate the effects of microbes and a God, logically speaking, the effects of microbes and a God needn't be mutually exclusive. For example, the microbes could be agents acting within the bounds imposed on them by an omnipotent, omnipresent God. In that way, both the microbes (secondary cause) and God (first cause) could be influential on the observed impacts of drinking contaminated water. For more information refer to H. Beebee, C. Hitchcock, and P. Menzie (2009). "The Oxford Handbook of Causation".

subsidies, or a mixture of both. At minimum, funding sources should cover operation and maintenance costs.

In general, the cost to produce water through desalination is not cheap therefore, desalination systems should only be installed in rural areas when research confirms that other freshwater resources are not available at lower costs [15]. Transported water can also be an option [16], so economic assessments should compare the price of desalinated water to the price of transported water, which may be less expensive. In some cases, wastewater treatment and water reuse are more economical than desalination from an energy standpoint [30], so these options should be considered as well.

If desalination proves to be the most economical option, research should be conducted to determine community members' ability and willingness to pay for the desalinated water. Available government subsidies should also be explored, since rural communities may face financial challenges, such as low average income, limited financial resources, and high capital costs. Governmental or intergovernmental funding may be necessary to guarantee the right to safe drinking water by subsidizing capital costs or operation and maintenance costs. Nonetheless, some communities may be able to pay for operation and maintenance costs [15], reducing the project's dependence on external funding while increasing the economic viability of the system and community ownership of the project.

In another approach for ensuring that water is available to meet basic needs at an affordable rate, water costs can be adjusted through a "life-line rate" that charges very little for basic water needs, but that charges progressively higher prices for progressively higher water use [15]. In such a system, more affluent community members who consume higher quantities of water effectively subsidize their poorer neighbors. Should the entire community be financially unable to cover the expenses associated with producing water, financial supports can be provided to system operators, thereby allowing the life-line rate to be extended to the entire community.

The preceding challenges apply to both grid-tied and renewable-energy desalination systems, but for renewable-energy desalination systems, the financial challenges are even greater. As discussed in the section on energy sources, renewable energy resources are often the most technically viable energy option for rural desalination systems, but their high capital costs can render renewable-energy systems economically prohibitive. As with grid-tied systems, these challenges can be alleviated through the use of life-line rates and direct subsidies [15, 34, 43]; the financial burden is typically greater for renewable-energy systems than grid-tied systems, though. Also, even if renewable-energy desalination systems are the most economical power source before subsidies are considered, grid-tied desalination plants may be economically preferable if subsidies are provided for on-grid desalination systems in rural areas [15]. Such regulatory and governmental factors must therefore be considered alongside conventional market forces.

17.3 Case Studies

The theoretical principles for developing remote rural desalination systems have practical and decisive impacts on the success or failure of rural projects. The social factors are especially important, as illustrated by an unsuccessful project in Kimolos, Greece, and a successful project in Ksar Ghilène, Tunisia.

17.3.1 Case Study of an Unsuccessful Project: Solar Stills on the Greek Island of Kimolos

Freshwater is scarce on many Greek islands, which typically lack ground-water [44] and rely heavily on rainwater collection. To meet water needs, residents typically collect rainwater in household cisterns, or the community as a whole builds and maintains a collective reservoir. Rain is more common in the winter than in the summer, so water scarcity is most pronounced during the dry summer months. The Greek islands do, however, have abundant seawater and solar resources, which Prof. Delyannis of Greece's National Center for Scientific Research decided to exploit through a solar still on the island of Kimolos.

In general, solar stills can be installed simply and operated easily and inexpensively, without any dependence on fossil fuels. Once installed, they provide the community not only with clean and drinkable water, but also hot water. The only critical issue is that the system must be operated and maintained properly [44].

To provide such an advantageous system to the people of Kimolos, Delyannis designed an asymmetrical solar still system with a surface area that could range between 2008 and 8600 m^2 and a production capacity that could be adjusted between 12 and 40 m^3/day. Based on Delyannis's calculations, the solar stills could provide water for a third to a quarter of the price charged by local merchants who transported water by tanker, and Delyannis arranged for the system to be given to the local community.

However, the operator assigned to prevent scaling and fouling by flushing the stills every two to three days performed the work only infrequently during the summer months and not all during the winter, when rainwater could meet local needs and the solar stills were not needed. The lack of maintenance quickly produced salt build-up and scaling problems, reducing the heat absorption rate and decreasing the efficiency of the system. The water in the solar stills, left for long periods without flushing, also grew large amounts of algae and other microorganisms, worsening scaling problems and introducing fouling problems. The scaling and fouling problems rapidly disabled the pumps that brought seawater into the stills, rendering the system inoperable. Furthermore, the plant was not protected by a fence or any regular personnel, so many of the glass panels covering the solar stills were destroyed by children throwing stones. In short, the solar stills were supposed to be operated under the supervision of the local community, but improper operation and lack of maintenance quickly destroyed the system.

As Delyannis notes, the project is an example of what not to do [44]. Although the project was compatible with the local environment and made good use of the local area's abundant seawater and solar resources, the project was designed and installed without significant community involvement; the solar stills were given to the community, which was not poor but was not asked to take ownership through investment; the community members were not adequately trained to operate and maintain the system; the benefits of the system were not emphasized to community members; the project did not consider or involve the adversely affected water merchants; and the system, which had to operate and be maintained year-round, was not well-tailored to the actual demand, which was nonexistent in the rainy winter months. Ultimately, the technically adept solar still project on Kimolos was doomed by these social factors.

17.3.1.1 Case study of a successful project: Photovoltaic reverse osmosis in Ksar Ghilène, Tunisia

Ksar Ghilène is a village of 300 people located in Tunisia's Saharan desert, 60 km from the closest freshwater well and 150 km from the closest electrical grid. Orange sand dunes stretch to the horizons, broken only by the village and an oasis with geothermal brackish groundwater.

To provide freshwater for the people of Ksar Ghilène without the need to transport it from 60 km away, the governments of Spain and Tunisia entered into a cooperative agreement to develop a remote rural desalination system for the village [53]. In cooperation with local associations, researchers proposed a photovoltaic reverse osmosis plant that would desalinate the locally available brackish groundwater. From 2004 to 2006, the researchers and

Figure 17.5   The photovoltaic reverse osmosis desalination system in Ksar Ghilène.
Source: Peñate et al. [53].

community members decided on a design tailored to the local environment, conducted training and education to let community members operate and maintain the plant, and conducted a local public awareness campaign about the benefits of desalination.

The project result (Figure 17.5) is a photovoltaic reverse osmosis plant completely powered by a 10.5 kW photovoltaic system, a 110-kW charger/inverter, and a 660-Ah backup battery bank. Feedwater is provided by the brackish groundwater beneath the nearby oasis, which is stored in a 30-m^3 tank in the plant. The capacity of the desalination system is about 2 m^3/h, with a recovery of 70%. The brine discharge rate is 0.9 m^3/h, and the brine is used to irrigate the forested side of the oasis.

During the ongoing successful operation of the plant, more than 15,000 m^3 of freshwater has been produced in over 8,000 h of operation. The demand for freshwater, 15 m^3/day as of summer 2006, is less than 40% of the design capacity for the plant. Therefore, the system needs to be operated for only 5 h/day to support the local population, and production times can be extended as necessary to support temporary population increases from visitors [53].

The village of Ksar Ghilène recieves several benefits from its desalination system [53]:

1. Drinking water does not need to be brought from 60 km away.
2. The standard of living improves, and the availability of clean freshwater reduces health risks.
3. Agriculture and cattle ranching expand sustainably.
4. The village can support more visitors and tourists, improving social life and economic activities in the village.
5. The photovoltaic panels reduce the need for fossil fuels, protecting the environment.

These benefits illustrate the success of the system, which was achieved by attending to technical factors as well as the social and environmental aspects of the community where the technology is used [54]. The project also follows the guideline of being "compatible with or rea- dily adaptable to the natural, economic, technical, and social environment" [55]. The lead researcher for the implementation of the desalination plant, Peñate, gives special credit to the social factors: the project was developed in close coordination with local associations; local technicians were trained to operate and maintain the system; and the local community members were familiarized with the benefits of desalination, making them familiar with and protective of the desalination system that still serves them.

Acknowledgement

The authors would like to thank Patrick DeSimio for helping edit this chapter.

References

[1] Water Scarcity Factsheet (2013). *Un Water.Org*. Available at: http://www.un.org/waterforlifedecade/scarcity.shtml [accessed December 11, 2015].
[2] Rijsberman, F. R. (2006). Water scarcity: fact or fiction? *Agric. Water Manag.* 80, 5–22. doi: 10.1016/j.agwat.2005.07.001
[3] Prüss-Üstün, A., and Corvalán, C. (2006). *Preventing Disease through Healthy Environments: Towards an Estimate of the Environmental Burden of Disease*. Geneva: World Health Organization.
[4] Kracht, U., and Schulz, M. (1999). *Food Security and Nutrition: The Global Challenge*. New York, NY: St Martin's Press, Inc.

[5] The World Bank (2003). *Implementation Completion Report on a Loan in the Amount of 10 US$ Million Equivalent to the Kingdom of Morocco for a Rural Water Supply and Sanitation Project.* Available at: http://www.wds.worldbank.org/external/default/WDSContentServer/ WDSP/IB/2003/06/17/000090341_20030617084733/Rendered/PDF/25 9171MA1Rural1ly010Sanitation01ICR.pdf

[6] Daniel, H., and Martin-Bordes, J. L. (2009). "Water related migration, changing land use and human settlements," in *Proceedings of the 5th World Water Forum Top,* Japan, 1–78.

[7] Kazem Beck, M. (2007). A comprehensive solar electric system for remote areas. *Desalination* 209, 312–318. doi: 10.1016/j.desal.2007. 04.045

[8] U.S. Census Bureau (2010). *2010 Urban and Rural Classification. Geography.* Suitland, MD: United States Census Bureau.

[9] Dijkstra, L., and Poelman, H. (2014). A Harmonised Definition of Cities and Rural Areas: the New Degree of Urbanisation, Regional and Urban Policy.

[10] The Wye Group Handbook (2011). *Statistics on Rural Development and Agricultural*, 2nd Edn. New York City, NY: United Nations.

[11] United Nations (2005). Definition of "Urban." *Demogr. Yearb.* 2005 1–3.

[12] Mondal, P. (2015). *Rural Community: Top 10 Characteristics of the Rural Community – Explained! [WWW Document].* Available at: http://www.yourarticlelibrary.com/sociology/rural-sociology/rural-com munity-top-10-characteristics-of-the-rural-community-explained/34968/

[13] Department of Economic and Social Affairs (2014). *Principles and Recommendations for a Vital Statistics System*, 3rd Edn. New York, NY: Department of Economic and Social Affairs.

[14] Ostrom, E., Lam, W. F., Pradhan, P., and Shivakoti, G. P. (2011). Improving irrigation in asia: sustainable performance of an innovative intervention in Nepal. Northampton: Edward Elgar Publishing.

[15] Bushnak, A. (2012). Assessment of Best Available Technologies for Desalination in Rural/Local Areas (Revised Final Report).

[16] Fath, H. E. S., El-Shall, F., and Subiela, V. J. (2008). "Installation guide for autonomous desalination system (Ads): task by task planning and implementing-part i: planning," in *Proceedings of the Twelfth International Water Technology Conference, IWTC12 2008,* Alexandria, 1–17.

[17] Boberg, J. (2005). *Liquid Assets: How Demographic Changes and Water Management Policies Affect Freshwater Resources.* Santa Monica, CA: RAND.

[18] Gender and Water Alliance (2006). *Resource Guide: Mainstreaming Gender in Water Management.* Dieren: Gender and Water Alliance.

[19] Karimi, L., Abkar, L., Aghajani, M., and Ghassemi, A. (2015). Technical feasibility comparison of off-grid PV-EDR and PV-RO desalination systems via their energy consumption. *Sep. Purif. Technol.* 151, 82–94. doi: 10.1016/j.seppur.2015.07.023

[20] California Department of Water Resources (2013). *California Water Plan, California Water Plan Update 2013—Public Review Draft.* Sacramento, CA: California Department of Water Resources.

[21] El-Dessouky, H. T., and Ettouney, H. M. (2002). *Fundamentals of Salt Water Desalination,* 1st Edn. Amsterdam: Elsevier Science.

[22] Greenlee, L. F., Lawler, D. F., Freeman, B. D., Marrot, B., and Moulin, P. (2009). Reverse osmosis desalination: water sources, technology, and today's challenges. *Water Res.* 43, 2317–48.

[23] Mezher, T., Fath, H., Abbas, Z., and Khaled, A. (2011). Techno-economic assessment and environmental impacts of desalination technologies. *Desalination* 266, 263–273. doi: 10.1016/j.desal.2010.08.035

[24] Monachese, M., Burton, J. P., and Reid, G. (2012). Bioremediation and tolerance of humans to heavy metals through microbial processes: a potential role for probiotics? *Appl. Environ. Microbiol.* 78, 6397–6404. doi: 10.1128/AEM.01665-12

[25] Gochfeld, M. (2003). Cases of mercury exposure, bioavailability, and absorption. *Ecotoxicol. Environ. Saf.* 56, 174–179. doi: 10.1016/S0147-6513(03)00060-5

[26] Water Corporation (2010). *Perth Seawater Desalination Plant [WWW Document].* Available at: http://www.watercorporation.com.au/D/desalination.cfm [accessed November 10, 2015].

[27] Tzen, E., Perrakis, K., and Baltas, P. (1998). Design of a stand alone PV – desalination system for rural areas. *Desalination* 119, 327–334. doi: 10.1016/S0011-9164(98)00177-5

[28] Hanrahan, C., Karimi, L., Ghassemi, A., and Sharbat, A. (2015). High-recovery electrodialysis reversal for the desalination of inland brackish waters. *Desalin. Water Treat.* 57, 11029–11039. doi: 10.1080/19443994.2015.1041162

[29] Elimelech, M., and Phillip, W. A. (2011). The future of seawater desalination: energy, technology, and he environment. *Science* 333, 712–717. doi. 10.1126/science.1200488

[30] Siddiqi, A., and Anadon, L. D. (2011). The water-energy nexus in middle East and North Africa. *Energy Policy* 39, 4529–4540. doi: 10.1016/j.enpol.2011.04.023

[31] Semiat, R. (2008). Critical review energy issues in desalination processes. *Environ. Sci. Technol.* 42, 8193–8201. doi: 10.1021/es801330u

[32] Missouri Department of Natural Resources (2006). *Microbial Contamination of Water Storage Tanks.* Jefferson City, MO: Missouri Department of Natural Resources.

[33] Assaf, S. A. (2001). Existing and the future planned desalination facilities in the gaza strip of palestine and their socio-economic and environmental impact. *Desalination* 138, 17–28. doi: 10.1016/S0011-9164(01)00240-5

[34] EU (2008). ADIRA Handbook, A Guide to Desalination System Concepts, Euro-Mediterranean Regional Programme for Water Management (MEDA).

[35] Isaka, M. (2012). "Water desalination using renewable energy, technology brief," in *Proceedings of the International Energy Agency (IEA)-Energy Technology Systems Analysis Program (ETSAP),* Paris.

[36] Seibert, U., Vogt, G., Brennig, C., Gebhard, R., and Holz, F. (2004). Autonomous desalination system concepts for seawater and brackish water in rural areas with renewable energies—potentials, technologies, field experience, socio-technical and socio-economic impacts—ADIRA. *Desalination* 168, 29–37. doi: 10.1016/j.desal.2004.06.166

[37] Meerganz von Medeazza, G. L. (2005). "Direct" and socially-induced environmental impacts of desalination. *Desalination* 185, 57–70. doi: 10.1016/j.desal.2005.03.071

[38] Liang, S., Liu, C., and Song, L. (2009). Two-step optimization of pressure and recovery of reverse osmosis desalination process. *Environ. Sci. Technol.* 43, 3272–3277. doi: 10.1021/es803692h

[39] Raluy, G., Serra, L., and Uche, J. (2006). Life cycle assessment of MSF, MED and RO desalination technologies. *Energy* 31, 2025–2036. doi: 10.1016/j.energy.2006.02.005

[40] Jenkins, S., Paduan, J., Roberts, P., Schlenk, D., and Weis, J. (2012). *Management of Brine Discharges to Coastal Waters Recommendations of a Science Advisory Panel Management of Brine Discharges to*

Coastal Waters Recommendations of a Science Advisory Panel. Technical Report, No. 694. Sacramento, CA: State Water Resources Control Board.

[41] Bleninger, T., and Jirka, G. H. (2010). *Environmental Planning, Prediction and Management of Brine Discharges from Desalination Plants.* Muscat: Middle East Desalination Research Center Muscat.

[42] De Vito, C., Mignardi, S., Ferrini, V., and Martin, R. F. (2011). "Reject brines from desalination as possible sources for environmental technologies," in *Expand: Issues Desalination,* ed. Y. R. Ning (Rijeka: In Tech), 953–978.

[43] Michael, E., Marcel, P., and Charlotte, W. (2010). *Roadmap for the Development of Desalination Powered by Renewable Energy.* Available at: http://www.prodesproject.org/fileadmin/Files/ProDes_Road_map_on_line_version.pdf

[44] Delyannis, E.-E., and Belessiotis, V. (1995). Solar application in desalination: the Greek Islands experiment. *Desalination* 100, 27–34.

[45] McDonald, E., Coldrick, B., and Christensen, W. (2008). The green frog and desalination: a nyungar metaphor for the (mis-)management of water resources, swan coastal plain, Western Australia. *Oceania* 78, 62–75.

[46] Brewster, M. M., Herrmann, T. M., Bleisch, B., and Pearl, R. (2006). A gender perspective on water resources and sanitation. *Wagadu* 3, 1–23.

[47] Environmental Engineering and Earth Sciences (2009). *Highlighted Projects: Using Geophysics to Improve Water Resource Management in Rural India [WWW Document]. Clemson.* Available at: http://www.clemson.edu/ces/departments/eees/research/highlight.html [accessed December 20, 2015].

[48] United Nations Population Fund (2002). *Water: A Critical Resource.* New York, NY: United Nations Population Fund.

[49] Water Supply and Sanitation Sector Board (2004). *The World Bank's Program for Water Supply and Sanitation.* Mandaluyong: Water Supply and Sanitation Sector Board.

[50] GE Reports (2015). *Sun-Powered Desalination for Villages in India [WWW Document]. Gen. Electr. Co.* Available at: http://www.gereports.com/post/99044029438/sun-powered-desalination-for-villages-in-india/ [accessed December 25, 2015].

[51] Morgan, R. A., and Smith, J. L. (2013). Rethinking Clean: Historicising Religion, Science and the Purity of Water in the Twenty-First Century.

[52] Mahmood, Q., Baig, S. A., Nawab, B., Shafqat, M. N., Pervez, A., and Zeb, B. S. (2011). Development of low cost household drinking water

treatment system for the earthquake affected communities in Northern Pakistan. *Desalination* 273, 316–320. doi: 10.1016/j.desal.2011.01.052

[53] Peñate, B., Subiela, V. J., Vega, F., Castellano, F., Domínguez, F. J., and Millán, V. (2014). Uninterrupted eight-year operation of the autonomous solar photovoltaic reverse osmosis system in Ksar Ghilène (Tunisia). *Desalin. Water Treat.* 3994, 1–8. doi: 10.1080/19443994.2014.940643

[54] Schumacher, E. (1989). *Small is Beautiful Economics as if People Mattered, Space Policy*. New York, NY: Harper & Row Publishers, Inc. doi: 10.1016/0265-9646(91)90030-L

[55] Balkema, A. J., Preisig, H., Otterpohl, R., and Lambert, F. J. D. (2002). Indicators for the sustainability assessment of wastewater treatment systems. *Urban Water* 4, 153–161. doi: 10.1016/S1462-0758(02)00014-6

CHAPTER 18

Society, Politics, and Desalination

David Zetland[1]

[1]Leiden University College, The Hague, The Netherlands
E-mail: d.j.zetland@luc.leidenuniv.nl

18.1 Introduction

This chapter will discuss the factors affecting the decision to increase desalinated water supplies and the resulting social and political impacts of that decision. On the one hand, desalination can strengthen the bonds within or between communities. On the other, it can damage these bonds, weakening relations with local and global neighbors.

Success or failure arrives on two margins. On the intensive margin of relations within a community, desalination will be helpful when costs are allocated in proportion to benefits but harmful when they are not. On the extensive margin of relations between neighboring communities, desalination will be helpful when shared/collective goods are augmented or strengthened but harmful if they are weakened.

These themes will be explored by sketching a basic theory of how to manage goods efficiently, outlining the role of political mechanisms in classifying and managing waters as different types of goods, and evaluating the social impacts of allocating the costs and benefits of water. First a framework for discussing "good" (fair, efficient, and sustainable) water management will be formed by clarifying where political decisions create the rights and responsibilities that social groups must claim or bear. Once this framework is in place, we will consider whether technology can be "innocent" of its impacts before exploring a few case studies in which desalination brings a mix of helpful and harmful social, political, and economic impacts. The chapter concludes with some principles for "no regrets" desalination.

18.2 Separating and Mixing Water

The properties of water as a good depend on local conditions. Scarce water may be rival in consumption, as when one person's use reduces the supply for others to use. Water can also be open access if it is difficult to restrain others from using it. These two characteristics (rivalry and excludability) can be combined in four ways when it comes to describing what "type" of good water resembles (Table 18.1).

The key insight is that the type of good can change with local conditions. It is possible to turn a public good—a lake, for example, whose waters are non-rival and non-excludable—into a common-pool good after farmers start to divert its waters to their fields, thereby introducing rivalry with those who want to swim. That same lake can then be converted into a club good (or toll good) by the creation of a "lake users' association" whose limited membership protects the lake from crowding or depletion.

What about private lakes? Yes, they can be drained (rival), but the costs and benefits of such actions fall squarely on the owner, not others (excludable). If such an action harms others by reducing lake discharges, disrupting local ecosystems, or ruining the view, then the lake was not really a private good but a common-pool good. A good's type does not depend on ownership or what people call it, but how it is managed and how those management decisions affect others. This difference explains how a common-pool *situation* in which a good is available to many may or may not turn into a common-pool *dilemma* in which many users deplete the good [1].

Using these definitions, we can also label water as a "collective" or "separate" good, according to excludability. Collective waters are common-pool or public waters shared among whichever groups use them. Separate waters are private or club waters whose use is reserved for a particular group.

Turning from goods to groups, we will define a "social group" as a self-identified set of people, voluntarily assembled, and a "political group" as a set of people inside exogenous or administrative boundaries. Individuals belong to groups in different ways. Someone can be administratively assigned to several nested political groups (e.g., city, region, and nation) at the same time

Table 18.1   A good should be managed according to its nature (rival goods are consumed in use) and accessibility (people can be prevented from using excludable goods)

	Excludable	Non-Excludable
Rival	Private Good	Common-pool Good
Non-Rival	Club Good	Public Good

Source: Based on Cornes and Sandler (1986) [2].

as they voluntarily affiliate with embedded or overlapping social groups (e.g., a local parent but a global environmentalist).

The peer-to-peer versus top–down difference between social and political groups is magnified by water's physical nature. Social groups might form around shared interests in drinking water, environmental flows, aquifers, and so on. The definition of political groups, in contrast, often ignores standing or flowing waters, creating conflict among such groups when they manage water crossing political boundaries in conflicting ways.

These differences say less about the quality of water management than what techniques might be used or abused. Among neighbors in communities, there are informal and social institutions for directing water [3]. Between neighboring communities, there are formal legal and political institutions for dividing water quantities and maintaining water qualities [4]. These existing institutions will play a big role in outcomes. Social groups that constrain rivalry will preserve their common-pool goods. Those that fail will experience a common-pool dilemma that destroys the good, i.e., a "tragedy of the commons" [1, 5–7]. Political groups, likewise, can create useful exclusionary rules that protect resources, but they can also be manipulated into delivering goods to special interests at the expense of the majority [8, 9]. Desalination can change the status quo, but existing political and social institutions will decide if the change is helpful or harmful. The next section will focus on political groups and how they separate waters for their exclusive management. The following section will look within social groups to examine how they manage collective waters.

18.2.1 Political Groups and Separate Waters

Start with a river that crosses from one political jurisdiction into another. At some time this river might have been a public good for both places, in the sense that neither side could stop the other from diverting, using or polluting the river but also that such use would not harm the other. The river (as a public good) was "not worth" managing because everyone could use as much of it as they wanted.

Such a situation usually ends when one or both neighbors affect the quantity or quality of water in a way that is detrimental to the other, thereby introducing rivalry that turns the river into a common-pool good. This moment is usually noticed by some groups before others in each jurisdiction. Fisherfolk and drinking water operators are closer to the river than politicians or citizens who indirectly interact with it.

The next step depends upon (and eventually affects) the relations between neighbors. If they are in constant contact and cooperating across multiple issues ("close"), then we can assume that the emergent water issue will get speedy attention. If they are hostile to each other or preoccupied with other concerns ("distant"), then the issue may be ignored as conditions deteriorate.

The difference between close and distant neighbors explains success or failure in managing water. Close neighbors with multiple ongoing discussions will find it easy to put water on the agenda and agree on how to reduce harm. Distant neighbors with weak links and perhaps hostile relations will find it difficult to discuss water when other issues are more important to political leaders and the public prefers to blame neighbors for water woes [10].

In economic terms, we can say that the transaction costs for addressing problems in managing water as a common-pool (or collective) good are low for close neighbors who can find easy ways to "exclude" their impacts from each other (e.g., limiting diversions or pollution) such that they share common waters without problems. Distant neighbors, to the contrary, will face higher transactions costs and thus difficulty in cooperating to protect their collective waters.

The addition of desalination in these conditions can strengthen or weaken relations among neighbors. On the one hand, desalination can make it easier to share with a neighbor, thereby building trustful relations. On the other, it might replace interdependence with independence, thereby making it easier to neglect cooperation. The direction of impact (helpful or harmful) depends more on political intent than baseline cooperation, but momentum makes it more likely that desalination will make a good situation better but a bad situation worse.

Desalination might, therefore, strengthen cooperation, efficiency, resiliency, and equity between neighbors that share its costs and benefits within a joint portfolio of waters. By the same logic, desalination might also weaken cooperation, efficiency, resiliency, and equity as neighbors fight over the impacts of its costs and benefits on portfolios they believe to be separate. Desalination might, for example, make it easier for close neighbors to cooperate in recycling joint wastewater or anger a distant neighbor opposed to the plant's pollutants.

Put differently, desalination technology can expand the space for action, but the impact of that expansion depends on how the space is being used by the desalinating party and neighboring parties. An example from the workhorse of cooperation games—the Prisoner's Dilemma (PD)—will illustrate this point.

In the PD's most common form, two players do not cooperate because each—fearing the other's betrayal—betrays "first"—leaving both worse off than if they had been cooperating. This (lose, lose) result depends on caveats about information, communication, and the game's "one shot" nature, but it results because cooperation is hard without trust. Few human interactions fall into the PD paradigm, but the idea is useful for explaining how unrelated groups do not form beneficial relationships. The result of a repeated-PD, on the other hand, is often cooperation for mutual benefit [11]. Why do we see more cooperation in repeated PD games? The returns to continuing cooperation are much greater than the returns to isolation and strife [12].

These two structures—one-shot and repeated—give us a hint of how desalination will impact collective waters and political relations. If neighbors are "playing one-shot," then desalination can prolong that defect strategy by increasing independence. Desalination can also initiate or perpetuate a "repeated game" if it reduces risk and thereby makes it easier to cooperate. Morrissette and Borer [13] miss these points when they claim "many future conflicts throughout the [Middle East and North Africa] will be directly linked to" water scarcity because they forget to consider desalination as an alternative source that can reduce pressure on shared waters.

This section explains how desalination's impact of relations between political neighbors will depend on whether they are already cooperating or not. In most cases, desalination will strengthen a positive or negative relationship, but it can also initiate a shift from lower to higher cooperation (or vice versa). The exact move will depend on how each side perceives the other's regard for their separate and collective waters, i.e., whether the costs and benefits of desalination fall within the same political jurisdiction. This same logic applies to the impact of desalination on social groups, to which we next turn.

18.3 Social Groups and Collective Costs and Benefits

Each political group contains different social groups that exist within or across political entities. These social groups have informal links and powers that may exceed or fall short of formal political powers. Political decisions over the allocation of benefits and costs from infrastructure projects often affect different social groups differently [14]. Uneven allocations tend to encourage wasteful projects because they allow political agents in a principal–agent relationship to allocate benefits to special interests (including themselves) at the expense of the principals they supposedly serve [8, 15, 16].

Thus, it is possible to see how a politician might ignore a social group of citizens opposed to desalination (for environmental reasons) in favor of a pro-desalination, pro growth and development lobby. Even more typical is the case where desalination costs are blended with other costs to determine an average cost water price ("postage stamp pricing," or PSP) that burdens existing customers with the cost of a desalination project that will serve new customers [10, 17]. These structural features, along with desalination's high cost relative to naturally occurring freshwater sources [18], explain why desalination projects are often controversial, but they are insufficient to rule out desalination as a tool for addressing water scarcity. On the one hand, it may make sense to charge its putative beneficiaries the full cost of their new water. On the other, there may be a social consensus that everyone should share those charges. These opposing reasons do not recommend sticking with the status quo of PSP, as few water customers realize they are enrolled in a PSP system. Stepping back, one might ask whether desalination is cost-effective for addressing various water problems. Is there, perhaps, not a community solution involving trade-offs that might work better than a desalination solution [19, 20]? In some cases, the answer is no, but it is also common for water managers to prefer supply-side, engineering solutions to an extended dialogue with citizens who need effort to reach effectively [21].

Like political groups, social groups can also experience mismatched costs and benefits. The difference is that the political calculus within a jurisdiction has to take far more care in balancing among social groups than when acting in relation (or against) another jurisdiction. As a simple example, consider the debate over a division of water rights within a country versus the debate over how much water to "give" to a downstream country. Cooperation is easier when interdependent sides play a repeated game.

18.4 Society, Politics, and Technology

Most of the last section focused on the magnitude and distribution of desalination's costs and benefits among different social (or interest) groups. The claim of this chapter is that the greater the mismatch between costs and benefits on various groups, the larger the likelihood of inefficient (and unjust) desalination.

Technology improvements can reduce (non)monetary costs or increase benefits. More efficient reverse-osmosis membranes, for example, lower energy use (and greenhouse gas emissions if power is supplied via fossil-fuel driven sources). Better technology can also increase benefits. Memsys Inc.

makes membranes that desalinate using waste heat from electrical generators, which increases motor efficiency by reducing operating temperatures. The net result—given high fuel costs—is that the membranes desalinate water at a "negative cost" because savings from lower fuel consumption are larger than the desalination unit's capital costs [22].

These examples do not mean that technology always offers the best answer to a problem, nor that technology can scale to address larger problems such as poor water management [23]. Desalination is an expensive source of supply and not necessarily the only option for addressing water scarcity. On the demand side, water use can be reduced by increasing water prices to the full cost of service, i.e., operating plus capital and risk-related costs. An increase in prices can reduce consumption by enough to eliminate scarcity— especially if it is accompanied by conservation publicity [24]. In some cases, awareness alone can reduce demand by enough to eliminate the need for desalinated supplies—as happened recently in San Diego [25]. Note that higher prices and conservation messages need to be salient and actionable if they are to impact customer behavior [26].

Desalination can also be useful as a cash solution to a bigger, non-cash problem. Local desalination can eliminate the need to defend a distant natural water source. It can replace environmentally-sensitive water sources that can be left to resume their natural flows. In some cases, desalination can even secure food production from market disruptions or unreliable irrigation, but those cases are not as common as examples where desalinated boondoggles merely subsidize farmers. Desalination must be deployed with care if it is to help social and political relations.

18.5 Examples of Harmful and Helpful Desalination

The proceeding discussion might leave readers without a clear definition of the social and political impacts of desalination. This generality is intentional, as the existing institutional environment determines whether desalination is useful. This section's examples will explore how desalination's impacts vary with deeper social and political forces as the technology itself is neither necessary nor sufficient for good water management.

10.5.1 Israel and Singapore

Israel and Singapore have managed water from a national security perspective since Israel's formation in 1948 and Singapore's independence in 1965.

Both countries had limited domestic supplies and less-than-brotherly relations with neighbors—facts that deeply affected their views on desalination.

In Israel's case, insecurity was lessened after the 1967 Six Day War in which Israel captured the West Bank (and thus the Jordan River) from Jordan and the Golan Heights (and thus the Sea of Galilee, or Kinneret) from Syria. These territorial gains, combined with Israel's power to veto any water-related projects by Palestinians in the Occupied West Bank, might have placed Israel in a position of water security except for the power of Israel's agricultural lobby [27, 28].

It is not hard to see expanding irrigation demand as one of the causal factors behind the Six Day War, as anxious neighbors watched Israel's irrigation network expand towards their territories, and agricultural consumption grew from around 1,000 MCM (million cubic meters) before 1967 to remain above 1,200 MCM since the war [28, 29]. Many casual observers of Israel's "water productivity miracle" might be confused by this statistic on total use due to their familiarity with falling irrigation per capita, but—as Kislev [27] explains—that drop is misleading. First, it reflects population increases. Second, agricultural policy does not target food security: Israel imports over half its food and exports a sizable share of its domestic production [27].

Others disagree on a connection between water and conflict, arguing that irrigation water is worth too little [30, 31] or that farmers can cut back on their demand if charged higher prices (i.e., elasticities of –0.30 to 0.46), but these arguments are trumped by political calculation [32]. As Stukki [33] points out, Israel's mid-1990s agreements with Jordan and the Palestinian government did not reduce Israeli supplies as much as promise new supplies to the others. Those agreements explain why increases in water recycling (since the 1950s) and 100% urban desalination (since the 2010s) have not reduced pollution, aquifer overdrafting, or the 1.2 m/year drop in Dead Sea levels [29]. Increasing water scarcity is not driven by inelastic urban demand, but agricultural and nationalist lobbies that see existing and additional water supplies as theirs [27].

Tal [29] calls for new supplies to support peace and environmental restoration, but neither goal may be met. Cities on 100% desalinated water have probably abandoned natural sources to farmers. The Red-Dead desalination project promises to add water to the Dead Sea and increase supplies to Jordan without reducing Israeli supplies, but the World Bank-subsidized project will not improve water conditions for Palestinians [34, 35]. Israeli desalination has done nothing to reduce the agricultural demand that imperils neighbors and ecosystems. According to OECD [36], irrigation consumes

57% of Israel's water, and agricultural consumption takes priority over environmental flows.

Israel has used desalination to divide its domestic, common-pool water into separate private goods allocated to cities or farmers. The end of domestic debate over agricultural water policies has made it harder for neighbors, e.g., worsening relations with Palestinians over common-pool groundwater in a manner similar to that seen over surface wastewater flows [37]. Relations have perhaps improved with Jordan over the common-pool Jordan River, but this has only occurred with the aid of subsidies from the World Bank that made it possible to add private water that would not threaten Israeli irrigation. Desalination is not going to improve water management in the region until deeper social and political changes occur.[1]

The contrast between Israel and Singapore is stark for political and social reasons, not for technological or cost-benefit reasons. In the case of Israel, the division and management of common-pool water is poorly defined with respect to Palestinian and Jordanian neighbors. Singapore, in contrast, has access to a legally defined quantity of water from the Malaysian "mainland" until 2061 [39]. This sole source of regional water leaves the city state of Singapore in charge of finding the rest of its supply domestically, turning the focus from transnational relations to the domestic distribution of costs and benefits. That distribution is fair and efficient for domestic users (there are no farmers) who all pay the same, full cost price for water because desalination is used to improve portfolio reliability [39]. Although exact capacities and shares of Singapore's supply sources are not public, PUB [40] reports that desalination and recycling ("NEWater") will provide 80% of the city state's water—the rest will come from rainfall capture—when Malay deliveries end in 2061. This example shows how desalination can be efficient, fair and timely. It also shows how "good fences make good neighbors" by removing a possible obstacle for cooperation with a neighbor with whom relations are sometimes strained.

---

[1]Editor's note: Israel's blatant violations and oppressive behavior is not within the scope of this book. For more information, refer to *'Amnesty International [38], Troubled Waters—Palestinians Denied Fair Access to Water'* which states: "The inequality in access to water between Israelis and Palestinians is striking . . . the amount of water available to Palestinians is restricted to a level which does not meet their needs . . . Israeli settlers face no such challenges—as indicated by their intensive-irrigation farms, lush gardens and swimming pools . . . Israel has over-exploited Palestinian water resources, neglected the water and sanitation infrastructure in the OPT [Occupied Palestinian Territories], and used the OPT as a dumping ground for its waste." (pages 3–4).

18.5.2 San Diego and Monterey

For examples related to domestic use (rather than international sharing), let us look at desalination projects in California where, despite political unity, we must still consider political issues related to formal claims on the commons and social issues of how to divide costs and benefits. In San Diego, for example, problems arise over access to imported water from the regional water supplier (The Metropolitan Water District of Southern California, or Met) and testy relations with local environmentalists over the costs and benefits of desalination.

San Diego is willing to pay $900 million for a desalination plant because it does not consider Met's water to be fairly priced or reliably supplied, and managers see desalination as a private source of water that will reduce their reliance on Met's common-pool supplies [10, 41]. The decision to desalinate is not without controversy because social interest groups have different perspectives on the relative costs and benefits of the project. On the one hand, you have a pro-growth group that sees desalination as necessary for planned urban expansion [42]. On the other, you have environmental groups worried about sustainability, and fiscal conservatives upset that politicians are paying for growth that benefits land developers by charging existing residents [43–46].

Free desalination would allow San Diego to walk away from Met, but costly desalination means the two sides must continue to wrestle over the division of Met's common-pooled costs and waters [10]. Turning to local relations, the focus of private benefits and diffusion of social costs means that desalination neither drives efficiency in water use nor contributes to equity in water management [47]. Indeed, the harshest irony is that conservation efforts during California's recent drought (residents still use a whopping 120 gallons, or 455 L, per capita per day) have reduced demand by enough to make the plant's supply superfluous to the region's current population [25, 48]. Further growth is likely to spark conflict over the allocation of the desalination plant's costs between existing and arriving residents.

The case in the Monterey region of California is different in all respects. CalAm, the water provider in the region, depends on the Caramel River and local precipitation for most of its supply, but the State Water Resources Control Board ruled in 2009 that most of these diversions were unlawful [49]. This political action re-defined the river's common-pool flows into a private share for CalAm and a remaining common-pool share to smaller users and the environment. That action combined with low consumption (by California standards) of 440 liters per capita per day for municipal and industrial uses—triggered CalAm's application to build a desalination plant [49]. In this case, the social discussion of costs and benefits is simplified by low consumption and the lack of growth in the area. The battle has thus shifted to one between residents willing to pay for desalination to meet 80% of the demand and environmentalists (many of them local) who want to see the area do more with less. A decision to build a desalination plant would not solve local problems, but it would make it easier to restore river flows without depopulating the region.

18.5.3 Saudi Arabia and United Arab Emirates

Many people associate desalination with oil-rich Persian Gulf states, but that stereotype ignores some important facts and policy differences. In both the Kingdom of Saudi Arabia (KSA) and the United Arab Emirates (UAE), for example, most water is supplied from fossil (non-renewable) ground-water and used for agriculture and landscaping [50, 51]. In both countries, water prices were subsidized for years, far below the cost of desalinating (20–30% of supply) or pumping water from underground [52, 53]. The resulting high water consumption (as well as population growth) was costly to the environment and political relations with neighbors [54].

Although it is easy to claim that desalination is necessary for life in these countries, the technology's impacts vary with social-political circumstances. In the KSA, cheap water was often seen as an entitlement worthy of subsidizing with oil revenues. In the UAE, citizens and residents in a diversified economy put higher value on fiscal prudence and reliable water (the KSA has frequent service interruptions). Abu Dhabi, the wealthiest of the UAE's sheikdoms (and the only one with significant remaining oil reserves) plans to stabilize groundwater levels, reduce water consumption (via higher prices), and reduce the drain on government finances [55, 56]. The situation in the KSA was more worrying on every margin—vanishing groundwater, conservation ignorance and costly subsidies—but a December 2015 announcement of higher domestic prices for water, energy and gasoline indicates that massive fiscal deficits have forced the KSA's rulers to reconsider subsidies [57–59].

Perhaps more important, however, is that the UAE's policies on water, finance and sustainability are part of a larger social discussion that includes related matters, such as the irrigation that consumes most of the water [55]. The lack of policy debates in the KSA—price increases were implemented overnight, without warning—does not mean other water policies make sense. With irrigation, for example, Saudi farmers—like Israeli farmers—are likely to continue with practices that threaten the environment, water reliability, and diplomatic relations.

18.6 Desalination without Regret

These examples have shown how the social, political, and economic impacts of desalination depend a lot less on the technology than the existing institutions for managing political relations with neighbors and distributing costs and benefits within social groups. We have seen how Israeli desalination can worsen complex, vague relations with Palestinian neighbors at the same time as Singapore's desalination further clarifies mutual rights and obligations with Malaysia. In the case of San Diego and Monterey, desalination allows business-as-usual on two different trajectories: San Diego seems destined to trade its "water independence" for reliability-sapping (and community splitting) growth. Monterey's desalination, on the other hand, promises to help ecosystems without increasing growth. In our final pair of examples, Saudi Arabia treats desalinated water and groundwater as separate gifts to interest groups while the United Arab Emirates manages water within the same portfolio.

Several clear principles emerge from this discussion. The first is that desalination cannot fix deeper social, political or economic problems. The second is that desalination is beneficial if it increases collective security within, between and among nations. The third is that desalination is likely to improve water management efficiency if its costs and benefits are transparently allocated to users. Fourth, the costs of desalination must be considered not only within a diversified water portfolio but also in comparison to the benefits of using desalinated water for basic human activities, ecosystem protections, and so on. Finally, we have to remember that desalination—as perhaps the most expensive source of additional supply—should be considered within a range of options for addressing water scarcity and security, many of which are far cheaper and easier to implement.

Acknowledgements and Disclosure

I am an adviser to The Aquiva Foundation, which uses Memsys membranes at their facilities in the Maldives. I thank Richard Scott and the referees for their helpful comments.

References

[1] Ostrom, E., Gardner, R., and Walker, J. (1994). *Rules, Games, and Common-Pool Resources*. Ann Arbor, MI: Ann Arbor Books.

[2] Cornes, R., and Sandler, T. (1986). *The Theory of Externalities, Public Goods, and Club Goods*. Cambridge: Cambridge University Press.

[3] Williamson, O. E. (2000). The new institutional economics: taking stock, looking ahead. *J. Econ. Lit.* 38, 595–613.

[4] Wolf, A. T. (1997). "'Water Wars' and Water Reality: Conflict and Cooperation along International Waterways," in *Environmental Change, Adaption, and Security*, ed. S. Lonergan (Boston, MA: Kluwer), 251–265.

[5] Ostrom, E. (1965). *Public Entrepreneurship: A Case Study in Ground Water Basin Management*. Ph.D. thesis, UCLA, Los Angeles, CA.

[6] Hardin, G. (1968). The tragedy of the commons. *Science* 162, 1243–1248.

[7] Lansing, J. S. (1991). *Priests and Programmers: Technologies of Power in the Engineered Landscape of Bali*. Princeton, NJ: Princeton University Press.

[8] Olson, M. (1965). *The Logic of Collective Action*. Cambridge, MA: Harvard University Press.

[9] Shepsle, K. A., and Weingast, B. R. (1984). Political Solutions to Market Problems. *Am. Polit. Sci. Rev.* 78, 417–434.

[10] Zetland, D. (2008). *Conflict and Cooperation Within an Organization: A Case Study of the Metropolitan Water District of Southern California*. Ph.D. thesis, UC Davis, Davis, CA.

[11] Axelrod, R., and Hamilton, W. D. (1981). The evolution of cooperation. *Science* 211, 1390–1396.

[12] Smith, A. (1776). *An Inquiry into the Nature and Causes of the Wealth of Nations*. New York, NY: P. F. Collier & Son.

[13] Morrissette, J., and Borer, D. A. (2004). Where oil and water do mix: environmental scarcity and future conflict in the Middle East and North Africa. *Parameters* 34, 86–101.

[14] Flyvbjerg, B., Bruzelius, N., and Rothengatter, W. (2003). Megaprojects and risk: an anatomy of ambition. Cambridge: Cambridge University Press.

[15] Gottlieb, R., and FitzSimmons, M. (1991). *Thirst for Growth: Water Agencies as Hidden Government in California*. Tucson: University of Arizona Press.

[16] Hall, D. C. (2000). "Chapter-9 Public choice and water rate design," in *The Political Economy of Water Pricing Reforms*, ed. A. Dinar (New York, NY: Oxford University Press), 189–212.

[17] Zetland, D. (2014). *Living with Water Scarcity*. Mission Viejo, CA: Aguanomics Press.

[18] Gleick, P. H. (2003). "A soft path: conservation, efficiency, and easing conflicts over water," in *Whose Water Is It? The Unquenchable Thirst of a Water Hungry World*, eds B. McDonald and D. Jehl (Washington, DC: National Geographic).

[19] Ciriacy-Wantrup, S. V. (1944). Taxation and the conservation of resources. *Q. J. Econ.* 58, 157–195.

[20] Wahl, R. W., and Davis, R. K. (1986). *Satisfying Southern California's Thirst for Water: Efficient Alternatives, Scarce Water and Institutional Change*. Washington, DC: Resources for the Future.

[21] Zetland, D. (2013). "Chapter-14: Water managers are selfish like us," in *Handbook on Experimental Economics and the Environment*, eds J. List and M. Price (Northampton, MA: Edward Elgar), 407–433.

[22] Aquiva Foundation (2016). *Membrane Distillation: A Possible Solution to the World's Water Crisis*. Available at: www.youtube.com/watch?v= X9XWH6KPUVs [accessed Aug 10, 2017].

[23] Schenkeveld, M., Morris, M., Budding, B., Helmer, J., and Innanen, S. (2004). *Seawater and Brackish Water Desalination in the Middle East, North Africa and Central Asia: A Review of Key issues and Experience in Six Countries*. Final Report. Washington, DC: World Bank.

[24] Loaiciga, H. A., and Renehan, S. (1997). Municipal water use and water rates driven by severe drought: a case study. *J. Am. Water Resour. Assoc.* 33, 1313–1326.

[25] Rivard, R. (2015). Desal deal leaves San Diego with extra water in drought. *Voice of San Diego*, 27th May.

[26] Kahneman, D. (2011). *Thinking, Fast and Slow*. New York: Farrar, Straus and Giroux.

[27] Kislev, Y. (2001). *The Water Economy of Israel*. Jerusalem: Hebrew University of Jerusalem.

[28] Schwarz, R. (2004). "The Israeli-Jordanian water regime: A model for resolving water conflicts in the Jordan River Basin?," in *Proceedings of the PSIS Occasional Paper 1/2004, Programme for Strategic and International Security Studies*, Geneva.

[29] Tal, A. (2006). Seeking Sustainability: Israel's Evolving Water Management Strategy. *Science* 313, 1081–1084.

[30] Beaumont, P. (1994). The myth of water wars and the future of irrigated agriculture in the Middle East. *In. J. Water Resour. Dev.* 10, 9–21.

[31] Fisher, F. M., Huber-Lee, A., Amir, I., Arlosoroff, S., Eckstein, Z., Jarrar, A., et al. (2005). *Liquid Assets: An Economic Approach for Water Management and Conflict Resolution in the Middle East and Beyond.* Washington, DC: RFF Press.

[32] Bar-Shira, Z., Finkelshtain, I., and Simhon, A. (2006). Block-rate versus uniform water pricing in agriculture: an empirical analysis. *Am. J. Agric. Econ.* 88, 986–999.

[33] Stukki, P. (2005). "Water wars or water peace? Rethinking the nexus between water scarcity and armed conflict," in *Proceedings of the PSIS Occasional Paper 3/2005, Programme for Strategic and International Security Studies,* Geneva.

[34] Al-Khalidi, S. (2015). Jordan, Israel agree $900 million Red Sea-Dead Sea project. *Reuters,* 26th February.

[35] Melhem, A. (2015). Canal project from Dead Sea to Red Sea makes waves. *Palestine Pulse,* 16th March.

[36] OECD (2015). *Water Resources Allocation: Sharing Risks and Opportunities, OECD Studies on Water.* Paris: OECD Publishing.

[37] Fischhendler, I., Dinar, S., and Katz, D. (2011). The politics of unilateral environmentalism: cooperation and conflict over water management along the Israeli-Palestinian border. *Glob. Environ. Polit.* 11, 36–61.

[38] Amnesty International (2009). *Troubled Waters – Palestinians Denied Fair Access To Water. Report MDE 15/027/2009.* London: Amnesty International Publications.

[39] Tortajada, C. (2006). Water Management in Singapore. *Int. J. Water Resour. Dev.* 22, 227–240.

[40] PUB (2013). *Four National Taps.* Available at: https://www.pub.gov.sg/watersupply/fournationaltaps [accessed Aug 10, 2017].

[41] Zetland, D. (2017). Desalination and the commons: Tragedy or triumph? *Int. J. Water Resour. Dev.* 33, 890–906.

[42] RWMG (2013). *San Diego Integrated Regional Water Management Plan. Technical report, Regional Water Management Group.* Available at: http://sdirwmp.org/2013-irwm-plan-update [accessed Aug 10, 2017].

[43] Larson, T. (2013). What Became of San Diego's Newspaper. *The Awl,* 18th January.

[44] Keatts, A. (2013). The other developer donation. *Voice of San Diego,* 1st July.

[45] Yerardi, J. (2014). Developer money follows circuitous route to San Diego county supervisor's race. *Inewsource* 16th May.

[46] Jennewein, C. (2015). Could we fight inequality by building more homes? *Times of San Diego*, 31st March.

[47] Zetland, D. (2012). *The SDCWA-Poseidon water purchase agreement does not serve the people of San Diego*. Technical Report. San Diego, CA: Surfrider Foundation.

[48] SDCWA (2017). *Water Use*. Available at: http://www.sdcwa.org/water-use [accessed Aug 10, 2017].

[49] Mueller, C. (2015). *Monterey Peninsula Water Supply Project, Draft Environmental Impact Report Water Demand, Supplies, and Water Rights*. San Francisco, CA: Environmental Science Associates.

[50] Szabo, S. (2011). *The Water Challenge in the UAE*. Dubai, AE: Dubai School of Government.

[51] GWI (2014b). Thirsty Riyadh places its hope in groundwater. *Glob. Water Intell.* 15:23.

[52] Ouda, O. K. (2013). Towards Assessment of Saudi Arabia Public Awareness of Water Shortage Problem. *Resour. Environ.* 3, 10–13.

[53] GWI (2014a). Kingdom concessions to offer generous return. *Glob. Water Intell.* 15:22.

[54] Economist (2014). Inside the sausage factory. *The Economist*, 10th May.

[55] EAD (2014). *The Water Resources Management Strategy for the Emirate of Abu Dhabi 2014–2018*. Strategy Report, Environment Agency Abu Dhabi. Abu Dhabi: UAE.

[56] Abdul, B. (2014). Abu Dhabi revises water and electricity tariff. *Gulf News*, 5th November.

[57] Al-Saud, M. B. I. (2013). *National Water Strategy: The Road Map for Sustainability, Efficiency and Security of Water Future in the KSA*. Riyadh: Ministry of Water and Electricity.

[58] Samad, N. A., and Bruno, V. L. (2013). The urgency of preserving water resources. *EnviroNews*, 21, 3–6.

[59] Al-Otaibi, M. (2015). New water and electricity tariff to rationalize consumption: Minister. *Saudi Gazette*, 29th December.

CHAPTER 19

Desalination Costs and Economic Feasibility

Jadwiga R. Ziolkowska[1]

[1]Department of Geography and Environmental Sustainability,
The University of Oklahoma, 100 E. Boyd St., Norman,
OK 73019-1018, USA
Phone: (405) 325-9862
E-mail: jziolkowska@ou.edu

19.1 Introduction

As discussed in the introductory chapters of the book, factors such as increasing population, urbanization, and extreme and exceptional droughts in many regions of the world have sparked intense discussions about current water shortages and the threat of potential widespread water scarcity in the future. Water shortages can be managed with mechanisms and policies on the demand (consumer) side and the supply (provider) side. On the demand side, pricing mechanisms, conservation incentives, payments and subsidies, as well as improvements in water usage efficiency and management have proven to be effective. On the other hand, on the supply side, the following measures can be implemented:

1. creation of new water supplies, through methods such as desalination, wetland restoration or rainwater harvesting,
2. economic incentives, e.g., water markets, cap and trade systems, or financial support for new water supply and distribution, and
3. efficiency improvements in infrastructure, e.g., dams, levees, and canals.

While the above-mentioned incentive-driven or policy-enforced mechanisms to conserve water resources have been applied in different countries, managing water through demand-side measures only is rarcly sufficient in the face of drought and expanding water scarcity. Furthermore, steadily growing population generates an additional water demand. As discussed in the previous chapters, desalination has proven to be a viable solution to this problem, providing an additional water supply not only in emergency situations, but also on a regular basis for growing populations.

Global desalination capacity (production of desalinated water) has been increasing over years from roughly 10 million m^3/day in 2000 up to almost 80 million m^3/day in 2013 [1]. This has been a result of a strong period of investment in desalination, which is expected to record a compound annual growth rate of 8.4% for the sector in the span of 2000–2018 [2].

One of the major challenges for desalination is economic feasibility and price affordability for the final product—desalinated water. Several authors have analyzed the economics of desalination in different countries and for different desalination facilities [3–6] as well as the effect of socio-economic impacts on the desalination sector [7]. Uchenna et al. [8] evaluated the economic feasibility of different desalination technologies (such as thermal processes versus membrane technology) and potential changes in the process efficiency subject to CO_2 taxation. Shahabi et al. [9] focused their analysis on different supply-management systems of desalination by using life-cycle methodology. Determining the costs and long-term feasibility of desalination processes is crucial for an efficient implementation and operation of desalination facilities that are—in many cases—financed with public funds. The financing model of desalination plants determines the cost-effectiveness of a plant throughout its entire lifetime.

Desalination facilities can be financed by both public and private funds, and different models are applied depending on the plant size (and the corresponding production capacity), location, and the client (public water utility or a private company). The most common models fall into the following four categories:

Design–Bid–Build (DBB)

A design engineer employed by the client (such as a municipal water utility) prepares drawings and specifies necessary equipment for a plant. Drawings are put out to bid to a separate contractor. The client bears the construction costs of the plant and is responsible for the operation of the plant.

Design–Build (DB)

The client tenders for a contractor to design and build a desalination plant. This model is similar to the model known as an 'engineering, procurement, and construction' (EPC) contract. The plant is usually financed through direct funding from the client, which is responsible for operating the plant.

Design–Build–Operate (DBO)

The client tenders both the complete EPC/DB contract and the operating and maintenance in a single procurement process. In some cases, the EPC part and the O&M (operation and maintenance) part might be contracted to different companies.

Build–Operate–Transfer (BOT)

The client tenders for a development company based on a price for the contracted amount of water. The development company is responsible for appointing an EPC contractor to build the plant, and an O&M company to operate the plant. The EPC contract is financed though funding raised by the developer via debt or equity. At the end of the contract duration, the plant is transferred to client ownership. There are a number of variations of the BOT-type model, for example the Build–Own–Operate (BOO) model, in which the plant is not transferred back, and is owned by the development company at the end of the contract.

The selection of a financing model influences desalination costs in both the short and long run. Contracting cost, overheads, equipment, and material costs will differ among different building and operating companies and their subcontractors. Also, the risk premium and interest rates will vary depending on the length of the contract. A detailed analysis of desalination costs and the respective cost components is presented in the next section.

19.2 Definition and Breakdown of Desalination Costs

Total desalination costs can be calculated as a sum of capital expenditures (CAPEX) and operational and maintenance costs (OPEX). While capital costs are incurred mainly in the construction and commissioning stages of establishing a plant, the operational and maintenance costs occur steadily over the entire period of operating a plant.

19.2.1 Capital Costs

Capital costs encompass the following components:

1. Land acquisition
    - Production and injection wells
    - Architectural and landscape related costs

2. Construction costs

   - Preliminary costs (design & construction)
   - External works
   - Structural costs
   - Civil costs (design of production and injection wells)
   - Mechanical and process work costs
   - Electrical costs
   - Solar panels (if any, for additional green energy supply)

3. Development costs

   - Labor work and management costs
   - Permitting
   - Overheads
   - LEED design (Leadership in Energy and Environmental Design – establishing energy efficiency of the plant)
   - Costs of controlling and instrumentation
   - Pre- and post-construction testing & studies (including commissioning)

Construction costs account for approximately 50–85% of the total capital cost. A large percentage of that constitutes indirect capital costs, including interest during construction, working capital, freight and insurance, contingencies, import duties, project management, and architectural and engineering fees. These costs range on average between 15% and 50% [3, 10, 11].

A capital cost breakdown for a typical RO (reverse osmosis) desalination plant on the global scale reveals that equipment and materials account for the highest percentage (25.4%) of all capital expenditures, followed by civil engineering costs (15.8%), and piping costs (12.3%). Other capital cost components (such as design, installation, feedwater intake, and brine outfall) each make a similarly equal share of 5–7% of the total capital costs [1]. Thus, equipment and material availability on national markets (or import prices) will be a determining factor for the feasibility of a plant in the first place. Further, as capital costs are amortized over the entire lifetime of a desalination plant, they will directly impact the final cost of desalinated water delivered to the consumers.

Capital expenditures on desalination have generally been increasing over time with a projected peak of $18 billion in 2016. This constitutes nearly a doubling from $9.5 billion within the last 10 years (Table 19.1). The overall increase in expenditures between 2007 and 2016 was reflected in the

Table 19.1 Global capital expenditures breakdown (all desalination) in 2007–2016

Expenditures ($ million)	2007	2008	2009	2010	2011	2012	2013	2014	2015	2016
Design/ prof cost	616	402	399	477	779	932	720	760	1100	1352
Equipment and materials	2752	1725	1409	2212	3034	3619	3088	3336	4444	5402
Thermal plant fabrication costs	449	141	120	286	310	225	274	421	374	425
Piping/high grade alloy	1039	718	450	893	982	1227	1155	1191	1510	1961
Membranes	317	256	161	289	327	447	399	378	522	711
Energy recovery	99	74	68	95	108	154	130	120	168	234
Pressure vessels	87	70	45	79	90	124	110	104	143	197
Pumps	760	482	325	607	715	789	759	865	1012	1328
Civil engineering	1542	953	702	1287	1494	1745	1578	1722	2104	2739
Pretreatment	467	345	235	419	475	645	564	545	740	1024
Intakes/Outfalls	689	428	574	509	905	1055	717	764	1093	1444
Legal/prof costs	99	61	69	76	140	165	125	134	199	233
Installation and services	660	415	334	534	662	752	668	733	906	1207
Total	9581	6074	4898	7770	10024	11885	10292	11081	14322	18264

Source: Author's presentation based on [1].

respective cost components as well, and shows the amount of investment in desalination. A decrease in capital expenditures in 2008 and 2009 can be attributed to the global financial crisis.

A comparative analysis of capital expenses for all desalination around the globe shows that the capital expenditures on most components nearly doubled between 2007 and 2016 (Figure 19.1) with the exception of thermal plant fabrication costs. This change indicates a rapid increase in desalination investments, and the increasing prominence of membrane processes compared to thermal desalination.

In 2007, roughly 55% of the total capital expenditures at the global scale were attributed to SWRO (seawater reverse osmosis), only about 4% to BWRO (brackish water reverse osmosis), and 16% to distillation processes – MSF (multi-stage flash distillation) and MED (multi-effect distillation) combined. The remaining 9% can be attributed to other RO plants, small thermal

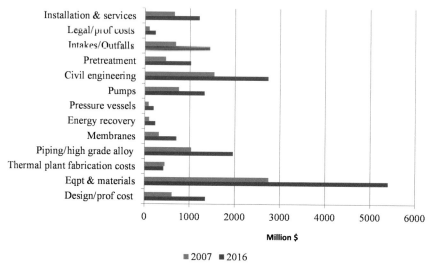

Figure 19.1   Capital cost breakdown for all desalination plants around the world in 2007 and 2016.

Source: Author's presentation based on [1].

plants, and ED/EDR (Electrodialysis/Electrodialysis Reversal) desalination. The visible shift in capital expenditures over time shows a technological change from distillation to membrane technologies. In 2016, almost 77% of all capital expenditures on the global scale were spent on SWRO, while the share of distillation processes in total CAPEX decreased by more than 50% compared to a decade earlier [1] (Figure 19.2).

19.2.2 Operational and Maintenance Costs

Operational and maintenance (O&M) costs include, among others, the following components:

- Energy costs
- Chemical costs (reagents, lubricants)
- Labor and management costs
- Upkeep & membrane exchange costs
- Disposal costs
- Institutional charges (compliance and regulatory costs)
- Amortization of fixed charges
- Distribution costs to water utilities (in some cases)
- Financing costs (repayment of project financing, bonds or other amortized capital funding costs).

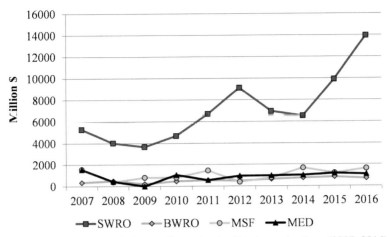

Figure 19.2 Global capital expenditures by desalination technology (2007–2016).

Source: Author's presentation based on [1].

In 2010, global OPEX expenditures for all operating desalination plants amounted to $6.6 billion, with electricity, labor, and chemicals accounting for 34%, 28%, and 16% of the total costs, respectively (Figure 19.3).

Total global operational cost and each of the O&M cost components increased over time due to the growing number of operating plants worldwide (Table 19.2). This statement can be justified especially with rising

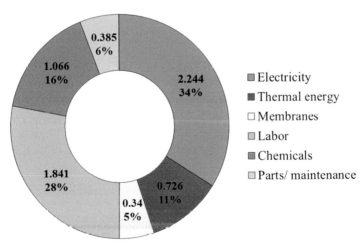

Figure 19.3 Breakdown of global desalination operating expenditures in 2010 (in billion US$).

Source: [1].

Table 19.2   Global operating cost breakdown (all desalination) in 2007–2016

Expenditures ($ million)	2007	2008	2009	2010	2011	2012	2013	2014	2015	2016
Electricity	1596	1837	2052	2244	2471	2778	3152	3529	3910	4376
Thermal energy	636	671	707	726	753	796	833	866	914	979
Membranes	242	280	313	340	368	409	463	519	576	646
Labor	1422	1580	1725	1841	1971	2154	2382	2612	2857	3156
Chemicals	793	898	993	1066	1149	1270	1421	1576	1741	1939
Parts and maintenance	281	321	357	385	416	461	519	578	639	714
Total	4972	5589	6150	6604	7131	7871	8771	9683	10639	11813

Source: Author's presentation based on [1].

employment in the sector as well as dropping marginal energy costs induced by technological improvements and the technological market shift from distillation processes to membrane desalination. The operational and maintenance costs did not experience any decrease over time due to their representation of constantly occurring expenses, regardless of market fluctuations. On the contrary, the economic crisis might have contributed to an increase in O&M costs. Total estimated O&M expenditures will reach $11.8 million in 2016, which is more than double the expenditures in 2007 [1].

Among the operating and maintenance expenditures, brine disposal, and energy costs have the highest impact on the final OPEX. For this reason, those two cost components will be discussed in-depth in the following sections.

19.2.2.1 Disposal costs

One of the most relevant OPEX components in the desalination process, for plants located inland, that is not listed in the above-mentioned global calculations (Table 19.2) is the disposal of brine (a highly saline solution and a by-product of desalination processes, also described in detail in Chapters 15 and 16). Different methods of brine disposal are used, with each of them generating different costs [12]:

- Surface water discharge (direct ocean outfall, shore outfall, co-located outfall, discharge to rivers, canals, lakes);
- Disposal to sewer (sewer line, direct line to waste water treatment plant [WWTP], tracking concentrate to WWTP);
- Subsurface injection (deep or shallow well injection);

- Evaporation ponds;
- Land application (percolation pond, rapid infiltration basin or irrigation of salt-resistant plants).

Other unconventional methods of brine disposal include:

- Landfill (usually at a dedicated monofill or industrial landfills);
- Recycling to the front end of a WWTP

Seawater desalination plants generally dispose of brine directly to the ocean after dilution. Most brackish water desalination plants dispose of brine into surface waters and/or the sewer. While this approach can be harmful to the environment (environmental flows and ecosystems), it is the most cost-effective way of dealing with brine from the perspective of a brackish water desalination facility. Due to the lack of consistent environmental standards in most countries around the world or no stringent enforcement of the existing rules, brine from brackish water desalination facilities is often disposed of at no significant cost. This might be a reason for brine disposal not being listed consistently as a component of the global O&M costs. Deep injection wells may cost up to $3/m^3 of brine, but have been criticized by environmentalists for a potential risk of causing earthquakes. Evaporation ponds are another common solution; however, under certain circumstances and tight regulations, the overall cost may become unbearable. The disposal costs to surface waters and sewer can be significantly lower ($0.03–0.30/m^3 of brine and $0.30–0.66/m^3 of brine, respectively) [13–17].

It can be anticipated that possible future legislation changes, strengthening and enforcing environmental standards and regulations could result in a potential spike in the overall operational and maintenance costs of desalination.

19.2.2.2 Energy costs

Energy costs account for the highest percentage of O&M costs in the desalination process. Thus, variations in oil and gas prices will be directly correlated with and have a direct impact on the costs of desalination and the final price of desalinated water.

The energy requirements for desalination differ, largely depending on the applied technology. Distillation processes (MSF and MED) use for the most part thermal energy (and electrical energy to some extent), while membrane filtration (RO) uses only electrical energy. Several studies analyzed the impact of energy requirements on the cost of desalinated water [18–20] and revealed significant variations in energy requirements subject to the desalination technology. Accordingly, distillation processes using mainly thermal energy can require up to 10 times more energy than SWRO and up to 20 times more than BWRO. Consequently, the energy use will significantly impact the final price of desalinated water. It needs to be emphasized that each desalination plant will generate different energy costs depending on several factors, as described above. In addition, the age of the plant is also likely to be a factor, with efficiency improving for newer plants. Recently opened plants with new equipment and most efficient membranes have reported lower energy requirements approaching 2 kWh/m^3 for SWRO processes [21].

To improve energy efficiency and thus reduce total desalination costs, desalination processes can be combined (creating the so-called hybrid system) and/or coupled with a power plant (referred to as a co-generation facility). The extent of benefits of a hybrid system (compared to a standalone RO, MED, or MSF plant) will depend on energy and feedwater availability, and other factors discussed above for a standard desalination plant. However, the system combination will allow for optimizing feedwater intake and outfall and thus reduce energy requirements and civil work. Moreover, in a hybrid system, the feedwater temperature for RO processes can be maintained at a constant level due to the waste heat available from thermal brine discharge. Similarly, when used within the RO process, water from thermal plants can be cooled down to acceptable temperatures before the brine discharge to the sea. The system combination can contribute to energy, chemicals and other costs reduction [5]. According to Asiedu-Boateng et al. [22] water desalination cost from a reverse osmosis plant coupled with a nuclear power plant operating on the steam cycle (NSC) is about 35% lower than the same desalination plant coupled with oil/gas power plant operating on the steam cycle (OSC). Also, the co-generation water costs of RO + NSC have been estimated to be 6% lower than the RO plant coupled with coal fired power plant operating on steam cycle (CSC). Water production cost of RO coupled with nuclear power plant operating on the gas cycle (NGC) is about 32% lower than RO coupled with oil powered plant operating on the gas cycle (OGC).

Another benefit of combined thermal distillation and membrane filtration is that the product water of both processes can be blended to improve the quality of drinking water. In some cases, two stages of filtration might be required, depending on the targeted water quality (and the final use). Based on a study by Awerbuch [23], Ghaffour et al. [5] calculated that at a plant producing 150,000 m^3/d with an optimum capacity ratio of SWRO to MSF/MED of 2 (i.e., 100,000 m^3/day RO and 50,000 m^3/day MSF/MED), the total water cost would decrease by $0.064/m^3. This would generate a total economic benefit of roughly $3.4 million/year.

19.3 Determining the Final Cost and Price of Desalinated Water

The price of desalinated water is normally calculated as a ratio of the sum of the amortized capital costs and annual O&M costs to the average annual water production volume:

$$\text{Water price} = \frac{\text{Amortized capital costs} + \text{Annual O\&M costs}}{\text{Annual water production volume}}.$$

Wittholz et al. [24] suggested using the concept of 'supply costs' (also called 'total water cost' or 'life-cycle cost') in determining the final cost of desalinated water which is calculated according to the Rogers' formula [25]:

$$\text{Supply costs} = \frac{(\text{Capital cost}/\text{Plant lifetime}) + \text{Annual O\&M cost}}{\text{Plant capacity} \times \text{Plant availability}}.$$

As emphasized by Ghaffour et al. [5], it is important to deliberately distinguish between the real cost of treating feedwater by desalination technology and the actual cost of providing water to the consumer that includes administrative and conveyance costs, and in many cases also a profit margin for the provider. It is not unusual that both cost descriptions are unintentionally confounded, which leads to a biased description of desalination cost. This might also be the case with distinguishing O&M costs and the total cost of desalinated water, with the latter including amortized capital cost over the lifetime of a desalination plant.

One of the main challenges in determining the final price of desalinated water is that standards are missing for reporting the costs of desalinated water both on the global and national level. Thus, a direct cost and price comparison among countries, states or even single desalination plants is impossible. To mitigate this shortcoming, and further assure viability of cost comparisons, a multitude of desalination variables need to be considered, such as: the project scope, technical and commercial aspects, geographical location, feed water source, infrastructure, financial background etc. Figure 19.4 shows the complexity of factors determining the final water cost that has been outlined specifically for desalination plants in Australia. The presented framework

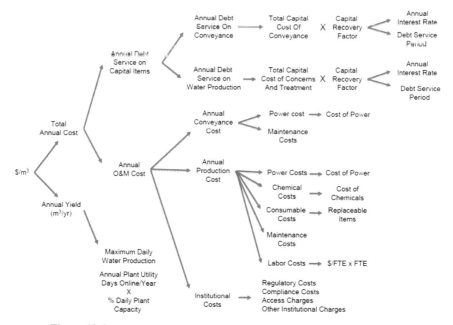

Figure 19.4 Logic tree for determining the unit price of desalinated water.

Legend: FTE, Full Time Equivalent
Source: [26].

can be generalized and transferred as a theoretical concept to any desalination plant in the world. However, as mentioned above, it is vital to adjust the price and cost analysis to geographical locations and other socio-economic conditions specific for the analyzed plant.

Capital costs and O&M costs are the main variables determining the final cost of desalination and the price of desalinated water. Some cost components are fixed costs that do not change with a changing production capacity (e.g., wages), while most of the cost components are variable costs that are varying in proportion to a changing production capacity (e.g., energy, management, and chemicals). The most important determinants of the final cost and price of desalinated water can be summarized as follows:

1. Size of the plant (production capacity)

It is straightforward that with an increasing production capacity, the final cost of water per unit (m^3) will decrease due to economies of scale. However, the operation of a desalination plant should be based on the existing water demand rather than solely on the economic necessity of cost reduction.

Large desalination plants with high capital costs and not operating at their design capacities might be cost-inefficient in the long-term and void the cost reduction that normally results from the economies of scale [5].

2. Salinity of the feed water and its biological conditions

Brackish groundwater has a salinity of several thousand mg/L TDS (total dissolved solids), while the salinity of seawater is tens of thousands mg/L TDS. In order to make this water drinkable and compliant with national drinking water quality standards, it needs to be desalinated down to the TDS level of at most 500 mg/L. The higher the salinity of feedwater, the higher the energy requirements and consequently the higher the final price of desalinated water. In addition, water quality needs to be examined to determine if additional pre-treatment is required. Poor water quality and the associated biofouling might reduce membrane life-expectancy and eventually increase the O&M costs [5].

3. Energy type (conventional versus renewable) and energy availability

As a stand-alone facility using electrical energy generated at a considerable distance, the RO process may have greater economic advantage over a thermal or hybrid process. Renewable energy is in many cases not competitive with conventional energy prices and thus cost prohibitive. However, consistently decreasing renewable energy prices might allow for a significant reduction of desalination costs in the future [5].

4. Energy prices

In most cases, energy prices are agreed upon and warranted in the contract agreement for the entire service period as part of the total water costs. Regardless of energy prices, however, all BOT contracts offer an energy adjustment cost provision to cover variations (mostly increases) in electric power costs. Thus, decreased energy consumption will have a considerable impact on the reduction of the final unit water cost [5].

5. Applied technology (distillation or membrane filtration)

Generally, thermal desalination requires higher energy inputs and uses more costly materials than RO facilities. Also, more chemicals are required in a thermal energy plant to control scaling, corrosion, and foam. On the other hand, the quality of purified water with distillation processes is better than the RO-treated water. Also, thermal processes can theoretically process water at any given salinity level, which is not possible with RO.

6. Brine disposal

Costs of brine disposal depend on the outfall, the disposal method (refer to Section 19.2.2.1), and environmental standards in place. In case of seawater desalination, brine may need to be diluted, its temperature may need to be reduced (when thermal technologies are used) by mixing it with cooling water, and chemicals (e.g., chlorine) may need to be neutralized before the discharge to the sea [27, 28]. The brine dilution ratio differs from country to country and ranges from 1 ppt (parts per thousand) salinity in Japan and Australia to 40 ppt in California, which represents the discharge dilution ratio of 7.5:1 [29].

7. Geographical location

To minimize costs and avoid unnecessary water transfer, desalination plants are generally located in a proximity to water resources (seawater or brackish groundwater). If water resources are not readily available (or intermittent), there is a risk of operating the plant far below its production capacity which would result in a significant cost increase. Geographical location is also crucial in terms of land acquisition and its prices. A combination of both factors will impact the final cost of desalination [5].

8. Post-treatment of desalinated water

The major costs at this stage refer to improving the pH, alkalinity, and hardness of desalinated water, as well as enrichment with minerals according to water quality standards specific to each country.

9. Purchase of materials, equipment, and chemicals, as well as the replacement of equipment such as memrbanes.

10. Financing, discount rate, amortization period, and inflation

Depending on the contract agreement, financing details and economic growth of a country, the price of desalinated water may vary. Potential market risks and instability of an economy will induce higher and thus unfavorable interest rates that will be rolled over toward the final cost of desalinated water, and consequently the final consumer [5].

11. Contract type and performance standard

Many BOT tenders include a minimum performance reliability standard (typically 95%) in the contract agreements. If a 98% standard is chosen and applied, a substantial increase in capital expenses can be anticipated due to the increased requirement for stability of production and additional equipment to secure the stability in an emergency situation (redundancy measures) [5].

12. Costs of environmental compliance (if any).

Due to the multitude of factors impacting desalination costs, there is no single and straightforward answer in terms of price of desalinated water. Each desalination plant produces water at a different price, and even if similarities can be found among plants, they are based on a combination of varying factors. The literature and studies conducted in this field show the diversity of estimates. Miller [15] and Sethi [16] found the unit water price of brackish water RO to range between $0.10–1.00/m^3, while the price of seawater RO varied between $0.53–1.50/m^3. The review studies by Younos [30] and Miller [15] showed water prices varying between $0.80–5.36/m^3 for MSF seawater facilities, between $0.45/m^3 and $6.56/m^3 for RO seawater plants, and between $0.18–0.70/m^3 for RO brackish groundwater plants.

A regional analysis and examples from the US and Australia confirm large discrepancies in the final price of desalinated water, varying from $0.90/m^3 at the desalination plant in Tampa Bay, FL and $1/m^3 at the Kwinana Desalination Plant in Perth up to the estimated price of $5.3/m^3 at the Wonthaggi Desalination plant in Melbourne [31–34]. Other completed desalination plants report final water prices in the range of $1.6/m^3 at the Gold Coast plant and up to $2.3/m^3 at the Southern Seawater Desalination Plant. Proposed plants (Santa Cruz, California American Water, Deep Water Desal projects) estimate the final water price to average around $1.9–2.6/m^3 [35].

The variability in the final price of desalinated water subject to changing technologies and different desalination capacities (10,000–500,000 m^3/day) has been studied extensively by Wittholz et al. [24]. According to the study, with increasing capacity, the capital costs rise regardless of the technology. However, for all the scenarios of varying production capacities and all technologies, the final cost of water is decreasing with growing capacity, which indicates the economies of scale (Table 19.3). According to Wittholz et al. [24], the price of water treated with SWRO can range between $0.45–0.95/m^3. The price of water treated with BRWO is less than half of SWRO, which confirms the findings by GWI [1].

According to Gude et al. [36] and Karagiannis and Soldatos [37], desalinating water using conventional energy sources offers the lowest cost range of $0.2–1.3/m^3 for desalinated brackish water and of $0.2–3.2/m^3 for desalinated seawater. Brackish groundwater desalination in remote areas is often more cost-effective than long-distance water transfers to those areas.

Table 19.3 Costs of desalinated water by different desalination technologies subject to changes in production capacities (these values are not universal, and should be considered as generalized)

	Capacity (m³/day)	Capital Cost ($ million)	Unit Price Cost ($)
SWRO	10,000	20.1	0.95
	50,000	74.0	0.70
	275,000	293.0	0.50
	500,000	476.7	0.45
BWRO	10,000	8.1	0.38
	50,000	26.5	0.25
	275,000	93.5	0.16
	500,000	145.4	0.14
MSF	10,000	48.0	1.97
	50,000	149.5	1.23
	275,000	498.1	0.74
	500,000	759.6	0.62
MED	10,000	28.5	1.17
	50,000	108.4	0.89
	275,000	446.7	0.67
	500,000	734.0	0.60

Source: Modified from [24].

Concentrated Solar Power (CSP) technology can be used for water desalination as well as for electricity, heat and cooling, and shows a promise for reducing water desalination costs in the future [38]. In general, at the current stage of technology development, employing renewable energies for desalination would increase the final cost of water [36, 37, 39], and thus is not plausible for desalination at this point in time. However, as renewable energy prices have indicated a sharp decline over the last decade and are further anticipated to drop, renewable energies for desalination might become competitive with conventional energy sources in the future.

Currently, in many regions of the world, prices of desalinated water are still higher than prices of water from conventional sources. However, desalination should be understood as a technology that can mitigate water scarcity and provide an additional water supply to lessen pressure on scarce water resources. At the current prices of desalinated water, it is not economically feasible to consider desalination as a complete substitute of groundwater and surface water resources, but rather as a mix with other conventional water supplies in a sustainable water portfolio.

19.4 Predictions about Future Desalination Costs

Different notions have been expressed by experts about the future of desalination and a potential cost decrease.

Ghaffour et al. [5] do not anticipate the prices of desalinated water to decrease further at the same rate as they have been decreasing recently, due to unstable prices of crude oil, currency fluctuations, and anticipated increases in membrane prices [34]. The authors further claim that costs of energy, shipping, raw materials, equipment and chemicals combined with more stringent environmental regulations will lead to an inevitable cost increase in desalination. Moreover, intake and brine disposal costs and a potential short supply of skilled manpower for plant construction and O&M could result in a cost spike in the years to come [40–42]. Potential currency fluctuations and inflation might cause constructing companies to impose higher risk protection premiums, which again would lead to a cost increase.

Other authors, e.g. Voutchkov [43], are more optimistic and anticipate a fast growth of the desalination market in the future. Improving membrane quality and productivity as well as a longer membrane life span (10–15 years in 20 years compared to 5–7 years as of 2010) could allow for a significant decrease of energy use by as much as 40–50% at SWRO plants. Also, low crude oil prices in 2015 resulted in lower than usual energy prices, and thus lower desalination costs. Consequently, the final cost of desalinated water could be reduced by 50% in the next 20 years [6].

Despite inconsistent and incongruent opinions among researchers, it is undisputable that new desalination technologies and membranes (nanofiltration, forward osmosis) and/or coupling desalination technologies in hybrid systems has a potential to increase membrane efficiency [44, 45], which normally translates to a decrease in the production costs [17, 46]. Also, improvements in thermal desalination processes are predicted through the implementation of nanofiltration as a pre-treatment stage and the application of renewable energy [41, 42, 47–49]. It also needs to be considered that the anticipated positive developments will take several years until a full commercial implementation can occur. In the long term, distillation processes are expected to be further displaced by membrane technologies, which will consequently lead to a decrease in the total energy requirements and O&M costs of desalination at the global scale.

Hence it is inconclusive and difficult to predict future price and cost developments, as past trends have proven highly variable. According to GWI [1], the cost variation depends on several economic and financial factors that can be hardly predicted or leveraged. They include: the increasing/falling value of the dollar, rising environmental costs of building a desalination plant, decrease/increase in energy costs, and desalination growth outside the traditional Persian Gulf market that involves greater risk premiums.

References

[1] Global Water Intelligence [GWI] 2013. *Market Profile and Desalination Markets*. Oxford: GWI.

[2] Gasson, C. (2013). *Desalination Market Update. Fourth quarter Assessment*. Oxford: Global Water Intelligence (GWI).

[3] Ettouney, H. M., El-Dessouky, H. T., Faibish, R. S., and Gowin, P. (2002). Evaluating the economics of desalination. *Chem. Eng. Prog.* 98, 32–40.

[4] Lapuente, E. (2012). Full cost in desalination. a case study of the Segura River basin. *Desalination* 300, 40–45.

[5] Ghaffour, N., Missimer, T. M., and Amy, G. L. (2013). Technical review and evaluation of the economics of water desalination: current and future challenges for better water supply sustainability. *Desalination* 309, 197–207.

[6] Ziolkowska, J. R. (2015). Is desalination affordable? – Regional cost and price analysis. *Water Resour. Management* 29, 1385–1397.

[7] Ziolkowska, J. R., and Reyes, R. (2016). Impact of socio-economic growth on desalination in the US. *J. Environ. Manag.* 167, 15–22.

[8] Uchenna K., Milne, N., Aral, H., Cheng, C. Y., and Duke, M. (2013). Economic analysis of desalination technologies in the context of carbon pricing, and opportunities for membrane distillation. *Desalination* 323, 66–74.

[9] Shahabi, M. P., McHugh, A., Anda, M., and Ho, G. (2015). Comparative economic and environmental assessments of centralized and decentralised seawater desalination options. *Desalination* 376, 25–34.

[10] Moch, I., and Moch, J. (2003). A 21st century study of global seawater reverse osmosis operating and capital costs, in *Proceedings of the IDA World Congress on Desalination and Water Reuse*. Atlanta GA.

[11] Wilf, M., Awerbuch, L., Bartels, C., Mickley, M., Pearce, G., and Voutchkov, N. (2007). *The Guidebook to Membrane Desalination Technology*. L'Aquila: Balaban Desalination Publications.

[12] Mickley, M. (2012). US municipal desalination plants: number, types, locations, sizes, and concentrate management practices. *IDA J.* 4, 44–51.

[13] Koyuncu, I., Topacik, D., Turan, M., Celik, M. S., and Sarikaya, H. Z. (2001). Application of the membrane technology to control ammonia in surface water. *Water Sci. Technol.* 1, 117–124.

[14] Mickley, M. C. (2001). Membrane Concentrate Disposal: Practices and Regulation, U.S. Department of the Interior, Bureau of Reclamation, Mickley & Associates.

[15] Miller, J. E. (2003). *Review of Water Resources and Desalination Technologies. Sandia National Laboratories.* Available at: http://www.prod.sandia.gov/cgi-bin/techlib/access-control.pl/2003/0308 00.pdf (05/09/2014).

[16] Sethi, S. (2007). *Desalination Product Water Recovery and Concentrate Volume Minimization (DRAFT). American Water Works Association Research Foundation Project #3030.* Washington, D.C: U.S. Department of the Interior.

[17] Greenlee, L. F., Lawler, D. F., Freeman, B. D., Marrot, B., and Moulin, P. (2009). Reverse osmosis desalination: water source, technology, and today's challenges. *Water Res.* 43, 2317–2348.

[18] Darwish, M. A. (2007). Desalting fuel energy cost in Kuwait in view of $75/barrel oil price. *Desalination* 208, 306–320.

[19] Banat, F., and Jwaied, N. (2008). Economic evaluation of desalination by small-scale autonomous solar-powered membrane distillation units. *Desalination* 220 566–573.

[20] Qtaishat, M. R., and Banat, F. (2013). Desalination by solar powered membrane distillation systems. *Desalination* 308, 186–197.

[21] Elimelech, M. (2012). "Seawater desalination," in *Proceedings of the 2012 NWRI Clarke Prize Conference*, Newport Beach, CA.

[22] Asiedu-Boateng, P., Nyarko, B. J. B., Yamoah, S., Ameyaw, F., and Tuffour-Acheampong, K. (2013). Comparison of the Cost of Co-Production of Power and Desalinated Water from Different Power Cycles. *Energy Power Eng.* 5, 26–35.

[23] Awerbuch, L. (2004). "Hybridization and dual purpose plant cost considerations," in *Proceeding of the MEDRC International Conference on Desalination Costing*, Lemesos.

[24] Wittholz, M. K., O'Neill, B. K., Colby, C. B., and Lewis, D. (2008). Estimating the cost of desalination plants using a cost database. *Desalination* 229, 10–20.

[25] Rogers, P., Bhatia, R., and Huber, A. (1998). *Water as a Social and Economic Good: How to Put the Principle into Practice.* Stockholm: Global Water Partnership.

[26] National Water Commission [NWC] (2008). *Emerging Trends in Desalination: A Review*. Waterlines Report No. 9. Canberra, ACT: National Water Commission.

[27] Höpner, T., and Lattemann, S. (2002). Chemical impacts from desalination plants – a case study of the northern Red Sea. *Desalination* 152, 133–140.

[28] Lattemann, S., and Höpner, T. (2008). Environmental impact and impact assessment of seawater desalination. *Desalination* 220, 1–15.

[29] Jenkins, S., Paduan, J., Roberts, P., Schlenk, D., and Weis, J. (2012). *Management of Brine Discharges to Coastal Waters – Recommendations of a Science Advisory Panel. Southern California Coastal Water Research Project*. Technical Report No. 694. Sacramento, CA: State Water Resources Control Board.

[30] Younos, T. (2005). The economics of desalination. *J. Contemp. Water Res. Educ.* 132, 39–45.

[31] Wilf, M., and Bartels, C. (2005). Optimization of seawater RO systems design. *Desalination* 173, 1–12.

[32] Cooley, H., Gleick, P. H., and Wolff, G. (2006). *Desalination with a Grain of Salt*. Oakland, CA: A California Perspective.

[33] Dreizin, Y. (2006). Ashkelon seawater desalination project – off-taker's self costs, supplied water costs, total costs and benefits. *Desalination* 190, 104–116.

[34] Pankratz, T. (2008). *Total Water Cost Discussion Rages On*. Water Desalination Report No. 44. Oxford: GWI.

[35] Cooley, H., and Ajami, N. (2012). *Key Issues for Desalination in California: Cost and Financing*. Oakland CA: Pacific Institute.

[36] Gude, V. G., Nirmalakhandan, N., and Deng, S. (2010). Renewable and sustainable approaches for desalination. *Renew. Sustain. Energy Rev.* 14, 2641–2654.

[37] Karagiannis, I. C., and Soldatos, P. G. (2008). Water desalination cost literature: review and assessment. *Desalination* 223, 448–456.

[38] International Energy Agency [IEA] (2012). *Water Desalination. Using Renewable Energy*. Paris: IEA.

[39] Baten, R., and Stummeyer, K. (2013). How sustainable can desalination be? *Desalination Water Treat.* 51, 44–52.

[40] Ghaffour, N. (2009). The challenge of capacity-building strategies and perspectives for desalination for sustainable water use in MENA. *Desalination Water Treat.* 5, 48–53.

[41] Mahmoudi, H., Abdellah, O., and Ghaffour, N. (2009a). Capacity building strategies and policy for desalination using renewable energies in Algeria. *Renew. Sustain. Energy Rev*. 13, 921–926.

[42] Mahmoudi, H., Saphis, N., Goosen, M., Sablani, S., Sabah, A., Ghaffour, N., et al. (2009b). Assessment of wind energy to power solar brackish water greenhouse desalination units—a case study. *Renew. Sustain. Energy Rev*. 13, 2149–2155.

[43] Voutchkov, N. (2010). *How Much Does Seawater Desalination Cost? Water Globe Consulting*. Austin: Texas Innovative Water.

[44] Elimelech, M., and Phillip, W. A. (2011). The future of seawater desalination: energy, technology and the environment. *Science* 333, 712–717.

[45] Li, Z., Yangali-Quintanilla, V., Valladares-Linares, R., Li, Q., Zhan, T., and Amy, G. (2011). Flux patterns and membrane fouling propensity during desalination of seawater by forward osmosis. *Water Res*. 46, 195–204.

[46] Park, H. B., Freeman, B. D., Zhang, Z. B., and McGrath, J. E. (2008). Highly chlorine-tolerant polymers for desalination. *Angew. Chem. Int. Ed. Engl*. 47, 6019–6024.

[47] Gastli, A., Charabi, Y., and Zekri, S. (2010). GIS-based assessment of combined CSP electric power and seawater desalination plant for Duqum-Oman, Renew. *Sustain. Energy Rev*. 14, 821–827.

[48] Ghaffour, N., Reddy, V. K., and Abu-Arabi, M. (2011). Technology development and application of solar energy in desalination: MEDRC contribution, Renew. *Sustain. Energy Rev*. 15, 4410–4415.

[49] Zejli, D., Ouammi, A., Sacile, R., Dagdougui, H., and Elmidaoui, A. (2011). An optimization model for a mechanical vapor compression desalination plant driven by a wind/PV hybrid system. *Appl. Energy* 88, 4042–4054.

Tom Scotney[1] and Simeon Pinder[2]

[1]Deputy editor, Global Water Intelligence, Suite C, Kingsmead House,
Oxpens Road, Oxford OX1 1XX, United Kingdom
E-mail: ts@globalwaterintel.com
[2]Co-founder, GMR Data Ltd, 2 Ardbeg Park, Artane, Dublin 5, Ireland
E-mail: simeon.pinder@gmrdata.com

20.1 Introduction

International spending on desalination infrastructure exploded in the 21st century, driven by growing populations and their thirst for water, climate change and improving desalination technology. In the year 2000, around 30.4 million m^3/day of desalination capacity had been commissioned Figure 20.1 shows contracted capacity, rather than commissioned globally since the first municipal-scale uses of the technology in the 1960s. By 2014, this figure had nearly tripled to just under 90 million m^3/day installed capacity. By 2030, this figure is expected to rise to more than 200 million m^3/day (Figure 20.1).

However, the constant rise in commissioning of desalination facilities hides a large amount of volatility in the market. The years 2007–2010 marked a high point in the commissioning of new desalination projects, as large seawater desalination facilities in the Middle East came online at a regular pace. However, the effects of the global financial crisis, and the knock-on effect on water demand and government spending meant that fewer plants were commissioned in the years that followed. 2014 was the lowest point in terms of newly commissioned desalination capacity since nearly the turn of the century, although there have been some signs of improvement in the market since then, with 2015 marking an improvement on the previous year. Going forward, the effects of the crash in oil prices on government spending in water-scarce parts of the Middle East may be the key factor in the speed at which the market recovers.

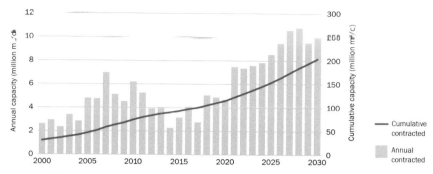

Figure 20.1   Desalination capacity history and forecast, 2000–2030.

Source: [1].

Trying to predict the future of the desalination market is a difficult prospect, as the industry is highly responsive to a wide array of factors. Both long-term and short-term factors play a role in driving the desalination market, with a number of complex factors coming into play.

Short-term factors driving the desalination market include:

- **Real estate development:** While growing population and urbanization is a long-term ongoing trend, the expansion of cities and other urban developments is not always a steady process. A drop in real estate development can lead to a short-term downturn in the amount of desalination capacity being planned and built.
- **The state of public finances:** The effects of the global financial crisis had a noticeable knock-on effect on the desalination market. On a more focused level, the state of public finances has a direct impact on the ability of countries to roll out desalination facilities, even with the growing role of private finance through public–private partnership (PPP).
- **Commodity prices:** With many of the world's richest energy reserves and natural resource deposits lying in arid areas, the energy and minerals markets have been key contributors to spending on desalination. Hence high prices and high income in commodities markets can translate into higher spending in the desalination sector.
- **Environmental crises:** Many countries are investing in desalination to deal with cyclical drought conditions, rather than permanent water shortages. However, the political and financial will to spend on desalination is rarely present until water shortages become an immediately

pressing issue. So incoming drought conditions and water crises can be an immediate driver for the desalination market, while improving conditions can mean a drying up of spending.

- **Political crises:** Planning and commissioning desalination facilities requires stable financial and business systems. Country-wide instability can disrupt this and put a freeze on desalination markets. Revolutions in the Middle East such as in Libya and Egypt have drastically impacted spending on desalination in these countries.

While these factors are very changeable, the longer-term issues that have driven the market are expected to continue, meaning that more and more desalination is likely to be needed, even if the year-by-year market fluctuates. Longer-term factors that need to be taken into account when predicting the desalination market include:

- **Urbanization:** Increasing concentration of populations puts more strain on naturally occurring sources of freshwater. In addition, the majority of fast-growing urban areas are located near coastal regions. With desalination the only way of creating 'new' drinking water sources, increasing urban populations are driving the expansion of the market.
- **Competition for resources:** With agricultural usage rising as a result of population and economic growth, urban and industrial water users are forced to look beyond existing natural water sources for a secure and high-quality source of supply water.
- **Climate change:** Global warming is likely to increase the frequency and impact of droughts around the world, meaning more parts of the world have to look beyond their traditional sources of water.
- **Depleting groundwater:** Over-taxed groundwater sources are becoming less reliable for urban, industrial or agricultural usage. In addition, as groundwater is used more heavily, it often tends to become more saline, creating a growing need for the deployment of brackish water desalination as part of the groundwater abstraction and treatment system.

While the Middle East and North Africa (MENA) region has been the largest market for desalination, it is not the only region where desalination is a serious market. This is particularly true when looking beyond seawater desalination (for example with the desalination of brackish surface or groundwater), and when looking at the development of desalination for non-municipal purposes, such as industrial brine concentration, feedwater treatment and industrial reuse.

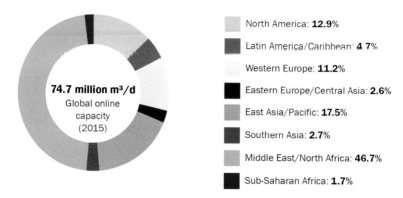

North America: **12.9%**

Latin America/Caribbean: **4 7%**

Western Europe: **11.2%**

Eastern Europe/Central Asia: **2.6%**

East Asia/Pacific: **17.5%**

Southern Asia: **2.7%**

Middle East/North Africa: **46.7%**

Sub-Saharan Africa: **1.7%**

74.7 million m³/d
Global online
capacity
(2015)

Figure 20.2   Online desalination capacity by region, 2015.

Source: [1].

Markets in Asia have seen strong growth, largely driven by industrial growth in China, with desalination deployed both for feedwater treatment and produced water handling. In the Americas, mining projects in Latin America have proved to be a growth market. However, a long-expected boom in municipal spending on desalination in the USA has not yet materialized, with many municipal authorities opting for other water solutions, or putting off spending (Figure 20.2).

The most significant ongoing technological trend in the procuring and commissioning of new desalination facilities over the last decade has been the shift from the use of thermal-based technologies toward membrane desalination. This has largely been a result of constant improvement in membrane performance reducing the energy requirements of membrane-based desalination processes.

With energy costs accounting for the most significant variable portion of the operating cost of a desalination plant, the energy usage of a facility is a crucial figure in determining how cost-effective it is. In general, thermal processes of desalination such as multi-stage flash (MSF) and multi-effect distillation (MED) require significantly more energy to produce water than membrane processes. However, until recently they had kept market share, particularly for large facilities in the Middle East (see technology market section). Recently, this picture has changed, with even the largest projects in the Middle East opting for membrane-based desalination, except in a limited and specialized number of cases (Figure 20.3).

The huge amount of desalination capacity contracted since the turn of the century has created a massive market for private plant suppliers. The world's

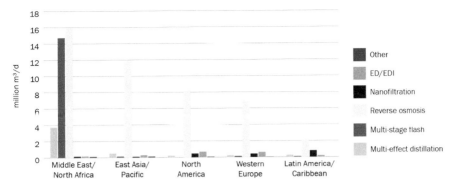

Figure 20.3 Regional online capacity by main technology type, 2015.

Source: [1].

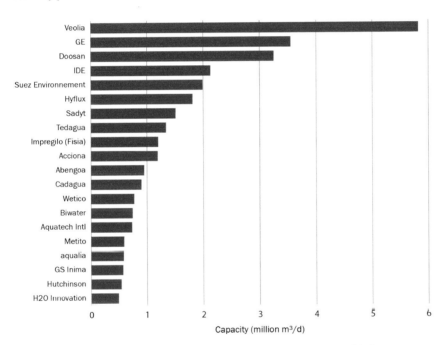

Figure 20.4 Top 20 desalination plant suppliers, 2005–2015.

Source: [1].

top desalination plant suppliers alone have between them installed more than 30 million m³/d desalination capacity in the last 10 years (Figure 20.4).

The huge capital requirement of installing desalination infrastructure has also meant a growing role for the private sector in designing, operating and

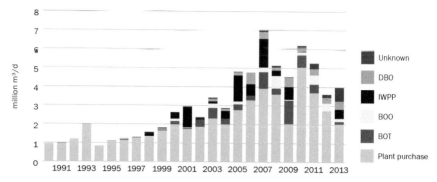

Figure 20.5   Global annual desalination capacity* by delivery model, 1990–2013.
*plants > 5,000 m³/day only
Source: [1].

funding desal facilities through various types of PPPs. This includes both the involvement of private operators through operations and maintenance (O&M) or design–build–operate (DBO) contracts, and the wielding of private finance through build–own–transfer (BOT) or other concession-type contracts.

The use of private finance to deliver desalination projects has been a growing trend in the market, and has also been a major driving force behind improvements in technology and operating efficiency (Figure 20.5).

20.2 Key Markets

The high cost of desalination – both in terms of capital expense and operating/ energy costs – means that it is generally only seen as a feasible option on a mass scale under certain situations, i.e., when arid conditions and a lack of natural water resources are combined with access to brackish or sea-water sources and the availability of sufficient finance to fund the necessary infrastructure. This is most noticeable in the use of brackish groundwater desalination in the USA, and the wide adoption of seawater desalination for municipal supply purposes in the MENA region.

In other areas, desalination has been used as a method to combat cyclical drought conditions, for example with facilities in California, the UK and Australia. This can be a politically sensitive move, as the ending of drought conditions can mean authorities are left with facilities that are either standing idle, or producing water at above market cost.

Desalination is also seen as a solution to water security, most notably in the cases of Israel and Singapore. In both these cases, tensions with neighbors over water sources led the countries to pursue seawater desalination as a method of ensuring the countries have the ability to be self-sufficient in terms of their water supply.

As a result, a large proportion of the desalination market is strongly concentrated in a number of key markets. Nearly half of all global online desalination capacity is in the MENA region, and of the top 20 countries by installed desalination capacity, 11 are in the MENA region (Figure 20.6).

Globally, plants producing drinking water for domestic customers have accounted for the majority of desalination projects, both in terms of numbers of plants built, and the size of the plants. That said there has been a growing demand for desalination plants serving industrial clients in areas where natural water resources are too scarce, too unreliable or too poor in quality. In these cases, desalination is often deployed as an alternative. Major industrial clients for desalination plants include the oil and gas and petrochemicals industries in the Middle East and the mining industry in Latin America.

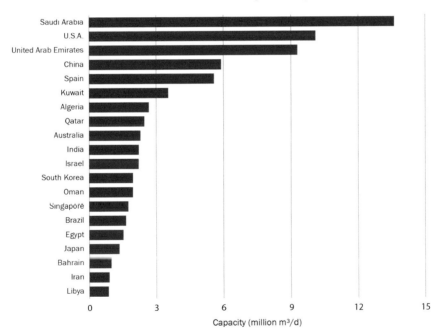

Figure 20.6 Top 20 countries by desalination capacity, 2016*.

*includes both operating and under-construction facilities.
Source: [2].

20.2.1 The Middle East and North Africa (MENA)

Historically, the Middle East has been by far the most active region in terms of procurement of desalination infrastructure as a source of water. This is most notable in the six countries that comprise the Gulf Cooperation Council (GCC) – Saudi Arabia, Kuwait, Oman, Bahrain, Qatar, and the United Arab Emirates. In the smaller and more urbanized of these countries – Kuwait, Bahrain, Qatar, and the UAE – the municipal and industrial water supply is almost exclusively supplied by seawater desalination. The heavy reliance on seawater desalination in these countries comes as a result of a number of factors, including:

- A lack of alternative potable water sources;
- Growing populations and industrial/commercial development putting more pressure on established water sources;
- Low water tariffs for consumers making demand management difficult and leading to high per-capita usage levels;
- National budgets boosted by income from petrochemicals making heavy capital investment programs more feasible;
- Subsidised fuel costs making the energy cost burden of desalination less intensive.

As a result of these factors, the GCC countries between them have both some of the largest installed desalination capacities, and also the largest individual plants. Many of these large plants were commissioned alongside power facilities, often through the privately financed project model known as the Independent Water and Power Project (IWPP).

Fourteen of the world's 20 largest operating desalination facilities are accounted for by the GCC countries (Table 20.1).

Elsewhere in the MENA region, desalination has also become a serious business, and there are a large number of major installations in places like Morocco, Algeria and Tunisia, plus growing interest in countries such as Egypt, Jordan and Iran, as populations and economies grow beyond the scope of the water resources that have traditionally sustained the countries. However, the development of desalination elsewhere in the region has not been nearly as rapid or complete as it is in the Arabian Peninsula, where an almost total lack of surface water resources is combined with oil-boosted national economies that make the price tag for desalination less of an issue.

Table 20.1 The world's 20 largest operating desalination plants

Project Name	Country	Capacity (m³/day)
Ras Al-Khair[a]	Saudi Arabia	1,034,700
Shoaiba 3	Saudi Arabia	880,000
Jubail	Saudi Arabia	800,000
Jebel Ali M Station	United Arab Emirates	636,440
Fujairah 2	United Arab Emirates	591,000
Soreq	Israel	540,000
Magtaa	Algeria	500,000
Az Zour North 1 IWPP	Kuwait	486,400
Jubail (Phase 2)	Saudi Arabia	472,794
Shuweihat 2	United Arab Emirates	459,146
Shuweihat 1	United Arab Emirates	454,200
Shoaiba 2	Saudi Arabia	454,000
Fujairah 1	United Arab Emirates	454,000
Wonthaggi (Melbourne, Victoria)	Australia	444,000
Doha – West	Kuwait	392,400
Sulaibiya	Kuwait	375,000
Hadera	Israel	368,000
Al Taweelah B1	United Arab Emirates	340,950
Ashkelon	Israel	330,000
Ashdod	Israel	320,000
Tuaspring	Singapore	318,500

[a]Partially commissioned.
Source: [2].

The major exception to this situation is in Israel, which has been a major proponent of desalination throughout the country's history, and where desalination provides roughly half the water used nationally. Five major facilities were procured through build-own-operate contracts with the private sector at Palmachim, Ashkelon, Hadera, Sorek, and Ashdod. Israel was an early proponent of mass-scale use of reverse osmosis for seawater desalination, and the five seawater reverse osmosis (SWRO) facilities remain among the largest of their type in the world. Domestic firm IDE Technologies is the key mover behind the Sorek, Ashkelon, and Hadera plants.

20.2.2 Europe

Historically, Spain has been the leading market for desalination in Europe, with desalination plants being built to match the country's construction boom. However, the collapse in construction contracts caused by the global financial crisis had a knock-on effect on the market for desalination in Spain, with a number of high-profile projects abandoned or left incomplete. The country still has nearly 6 million m³/day of installed desalination capacity in operation in total.

Another notable European deployment of desalination was the 150,000 m³/day Beckton facility in London, treating brackish water from the Thames estuary. However, this was not designed to run on a permanent basis, instead supplying an extra option for drinking water supplies in times of drought.

20.2.3 Americas

The US is the second-largest market for desalination in the world. The bulk of capacity installed to date has been through a decentralized network of plants treating brackish groundwater for potable water purposes.

Seawater desalination has often been mooted as a solution to water shortages in the USA, particularly in California following the drought in 2014–2015. The industry had a major success with the completion of the 189,250 m³/d Carlsbad SWRO plant in 2015. However, this success only came after more than a decade of negotiations, and the range of environmental and permitting legislation in the state has so far blocked any further moves to pursue seawater desalination on a large scale.

The situation in California for desalination planners is so difficult that one option being explored is to build a plant across the border in Mexico, with a pipeline to supply homes in the US. This project – a 378,500 m³/day facility that would also serve Rosarito and Tijuana in Mexico – would become the largest plant of its type in the Americas.

In Latin America, the largest driver for desalination has been the demand from the mining industry in Chile. Plants serving tourist developments and small municipal areas in the Caribbean are also popular as a way of guaranteeing what can often be an otherwise intermittent supply. However, the majority of plants in the Caribbean are relatively small.

20.2.4 Asia Pacific

China and India have both explored desalination as a targeted solution to water shortages, both in the industrial and municipal water sectors. This has led to a number of high-profile plants being built in the countries.

In India, the cost of desalination – both in terms of capital requirements and ongoing operating costs has proved a burden for planners in a country where income from water users is often very low.

In China, the completion of the massive south-north water transfer scheme has taken off the edge from the desalination market, combined with an imminent expected slump in growth. However, clients in the industrial sector have proved to be a rich source of desalination contract opportunities, as environmental regulations clamp down on their use of natural resources.

Elsewhere in Asia, desalination has not yet established itself on the same scale as in China or India. One notable exception is Singapore, where seawater desalination is known as the "fourth national tap" (alongside natural water from local catchments, water imported from Malaysia, and recycled wastewater). In Singapore, a self-sufficient water supply is seen as a major security issue after disputes in the past with Malaysia, which had historically supplied the majority of water to the city-state. The country has a number of major seawater desalination plants, as well as using reverse osmosis facilities as part of its wastewater treatment and recycling process.

In Australia, local water utilities made the decision to move into desalination as a response to the decade-long drought that begun at the end of the previous century. This resulted in six major projects being completed around the country. However, the ending of drought conditions in 2010 proved to be a major blow to backers of desalination, with some plants ending up standing idle, even as the authorities were forced to make payments to the developers that had built them under long-term PPP contracts.

20.2.5 Sub-Saharan Africa

Outside North Africa, there have not yet been many desalination projects on a large scale. However, in recent years there have been some projects completed in countries including Ghana, Namibia and South Africa. In many cases the low levels of income and lack of management ability at local water utilities make dealing with the private sector for building desalination plants a difficult prospect. This means that industrial clients can be the ones to make the push.

This was the case with the Areva desalination plant in Namibia. The plant was originally built to supply uranium mines, but following the arrival of drought conditions in the country, became the subject of a takeover bid by local water utility NamWater, which was looking to shore up its water supply options.

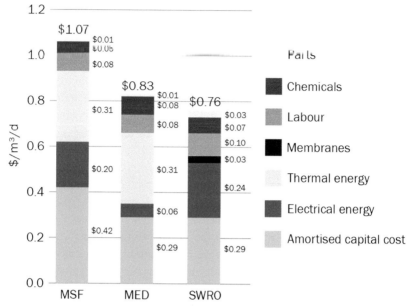

Figure 20.7   Desalination lifetime cost breakdown.

Source: [3].

20.3 Technology and Energy

As mentioned earlier, the most prominent trend in desalination technology over the past decade has been the widespread switch to membrane-based technologies. This is largely due to the prominence of the cost of energy when calculating lifetime costs of desalination plants.

The following figure is a generalized explanation of the lifetime costing breakdown of a desalination plant (Figure 20.7).

Membrane desalination is generally cheaper, both overall and in terms of energy requirements, than the two main thermal competitors – a situation that has been the case for some time. However, thermal technologies kept a hold of the majority of market share, particularly when looking at the largest plants around the world. This was due to a number of factors:

- The perception that reverse osmosis could not be scaled up to the same level as thermal desalination, due in part to the reduced efficiency and hence the need for a proportionally larger intake and outfall system.
- The fact that energy costs in many countries – particularly in the MENA region – were heavily subsidised, meaning that the true cost of energy was not passed on to the cost of desalination.

- The potential for the thermal portion of the energy requirement to be provided by steam from co-located industrial facilities, such as oil- or gas-fired thermal power plants.

Continuing improvements in the efficiency of membrane desalination systems, as well as the global shift toward renewable energy sources rather than fossil fuels, have diminished the dominance of thermal desalination systems. In the last decade, ever-growing membrane desalination installations have proved that reverse osmosis can compete on price with even the largest thermal desalination facilities (Table 20.2).

The use of membrane desalination technology, which does not need access to a source of thermal power, also ties in better with plans by governments and other procuring bodies to look for alternative sources of power. Sources of energy such as wind and photovoltaic (PV) solar power are better suited to supply membrane desalination facilities, which require only a source of electrical energy.

Table 20.2 The world's largest individual desalination projects, by technology and year commissioned

Project Name	Capacity (m^3/day)	Technology	Online Date
Shoaiba 3	880,000	MSF	2009
Al Jubail	800,000	MED	2010
Ras Al-Khair (MSF)[a]	728,000	MSF	2016
Jebel Ali M Station	636,440	MSF	2013
Yanbu 3	550,000	MSF	2016
Soreq	540,000	SWRO	2013
Magtaa	500,000	SWRO	2014
Az Zour North 1 IWPP	486,400	MED	2016
Al Jubail (Phase 2)	472,794	MSF	1983
Shuweihat 2	459,146	MSF	2011
Shuweihat 1	454,200	MSF	2004
Al Fujairah 2 MED	454,200	MED	2010
Shoaiba 2	454,000	MSF	2002
Wonthaggi (Melbourne, Victoria)	444,000	SWRO	2012
Doha – West	392,400	MSF	1985
Sulaibiya	375,000	SWRO	2004
Hadera	368,000	SWRO	2009
Al Taweelah B1	340,950	MSF	1996
Ashkelon	330,000	SWRO	2005
Ashdod	320,000	SWRO	2015

[a]Note that the Ras Al-Khair plant will include both MSF and RO units when it is fully commissioned.
Source: [2].

The shift from thermal-based to membrane-based desalination technology has been one of the key factors behind the downturn in the use of combined power and water projects – previously a common option in countries looking to build out their power and water generating capacities at the same time. Without the need for a ready source of waste heat, power and water generation facilities can be planned and procured apart from each other, with no need for co-location.

20.3.1 Equipment and Innovation

Following the trend for the desalination market as a whole, the market for technologies which enable the removal of dissolved solids also entered 2016 on a low ebb. Dissolved solids removal encompasses a broad range of desalination-related technologies, which can usefully be split into two major areas: those used for the production of fresh water – including seawater desalination – and those used for volume reduction, such as in waste-water brine streams, where the volumes of water being disposed of poses an environmental or financial risk. Dissolved solids removal plays a role in almost every sector of the water market, due to growing environmental restrictions on wastewater disposal, and increasing urbanization and competition for increasingly scarce natural water resources.

Until recently, seawater and brackish water desalination has accounted for the largest portion of spending on dissolved solids removal, as shown in Figure 20.8 below.

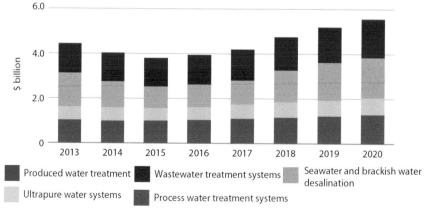

Figure 20.8   Capital expenditure on dissolved solids removal technology.

Source: [4].

However, this market dropped by more than 35% between 2013 and 2015, as a result of a number of factors, including low commodity prices and their knock-on effect on industrial water demand, the emergence of alternative sources of water such as wastewater reuse, and financial weaknesses in Middle Eastern countries that had accounted for the bulk of the world's major seawater desalination projects. While predictions about the desalination market are difficult to make, the market for desalination technologies is expected to grow by 87% from its 2015 trough to reach nearly $1.8 billion by 2020.

Away from seawater and brackish water desalination, the market for dissolved solids removal will continue to grow – and to fragment. The market for wastewater treatment technologies is expected to grow steadily at an average annual rate of 7.6%, driven by demand for municipal reuse schemes as a consequence of water scarcity.

Innovation in the sector has led to a plethora of emerging technologies aimed at lowering the energy costs of treating brine and other wastewater streams. While dropping commodity prices have hit the industrial markets that provided the incentive for much of the innovation in this sector, the market for technologies is likely to become ever more complex as time goes on. Meanwhile, reducing the volumes of wastewater streams will continue to become more pressing, given increasing levels of environmental regulation.

The Figure 20.9 below outlines key players and the way the dissolved solids removal market breaks down.

20.4 Finance and Companies

The complexity and cost of desalination means a wide range of expertise is needed to plan, design, build, operate and fund facilities. This has created a global industry worth billions of dollars. In the period of 2015–2020, an estimated $82.2 billion is set to be spent on desalination. This figure covers both the capital cost of building new desalination plants, and the cost of operating them.

The majority of the spending in the desalination market – particularly in the MENA region – is accounted for by municipal contracts generating potable water for household use. This is a reflection of the scale of demand. Household usage outstrips the demand from industry, and is also more politically sensitive. While agricultural water usage is generally an order of magnitude larger than domestic or industrial usage in the majority of countries, the fact that agricultural users rarely pay a price for their water

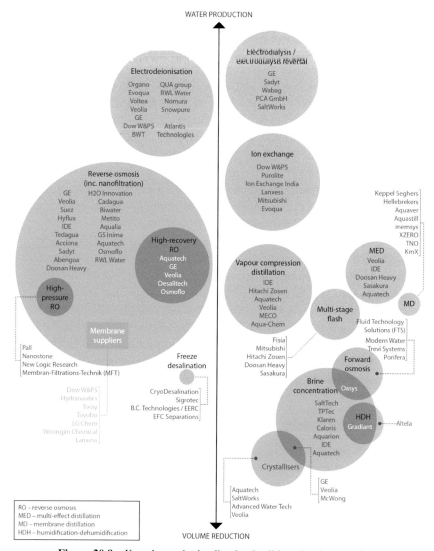

Figure 20.9 Key players in the dissolved solids technology market.

Source: [4].

that reflects the cost of production means that desalination is generally too expensive an option to supply farmers.

In the case of the municipal supply contracts that make up the bulk of desalination spending, the client that procures a desalination facility is likely

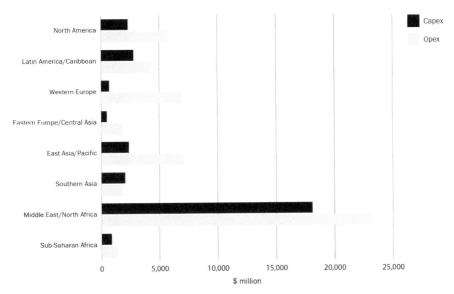

Figure 20.10 Desalination capex and opex by region, 2015–2020.

Source: [1].

to be a government ministry, state-owned water body or municipal water utility. In the majority of the world this is likely to be a group owned by the public sector. However, the private sector has played an increasingly important role in the business of desalination over the years.

Procuring bodies deal with the private desalination sector through a number of different contract models. At its most basic level, this involves the private sector as a contractor in the construction of desalination facilities. However, there has been a growing trend toward the use of contract models that represent different types of PPPs (Figure 20.11).

Essentially, PPP involves the participation of the private sector in the funding and/or operation of desalination facilities, as well as just their construction. There is a wide array of different contract types that fall under the PPP banner, and each market defines their contracts in a different way. The following table broadly outlines the different contract types and how the responsibilities break down.

Public sector clients see a number of advantages to the use of PPP contracts to deliver desalination facilities. The following list covers a number of these perceived advantages.

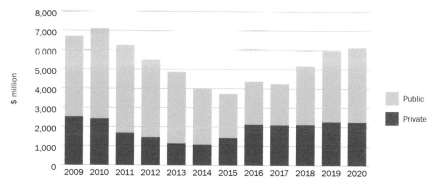

Figure 20.11   Desalination capex by source (public/private).

Source: [1].

Table 20.3   PPP responsibility breakdown

Contract Type	Responsibility for Construction	Responsibility for Operation of Assets	Responsibility for Capital Investment
DB, DBB, EPC	Private sector	Public sector client	Public sector client
DBO, EPC + O&M	Private sector	Private sector	Public sector client
BOT, BOO, Full concession	Private sector	Private sector	Private sector

- Combining construction and operations in a single contract provides an extra incentive for the constructor to supply a high-quality asset, rather than deliver a low-quality facility in order to bid the lowest price for a construction-only contract.
- Allowing the private sector to own or operate assets takes operating risk away from the public sector.
- Putting the burden of capital investment on a private sector project developer means that the public sector client is not faced with a large one-off capital bill, and can instead spread the cost of the facility over the lifetime of the agreement with the developer.
- Allowing the private sector to own and operate assets in exchange for a long-term service contract pushes them to build a high-quality asset that operates as efficiently as possible, because finding operating efficiencies improves their profit margin.

20.4.1 Desalination Companies

A wide variety of companies have been involved in the building of desalination facilities. In some cases, these are global water companies with a construction wing focusing on water treatment facilities. In other cases, they are general-purpose construction or engineering companies that have established a water-specific business.

The growth of reverse osmosis as the technology option of choice in the majority of cases has opened up the desalination market to new players. This is due to the modular 'plug and play' nature of reverse osmosis that means contractors can more easily buy in technology from OEMs without the need for so much in-house water expertise and access to technology.

The infrastructure boom in Spain and the corresponding growth in demand for water led to the country's major construction groups establishing water specialist wings, many of which have become established names outside their domestic markets. Companies like Acciona, Aqualia, Abengoa Water, Tedagua and Valoriza have all established a long list of desalination references both inside and outside the country.

20.4.2 Finance

As mentioned above, the heavy capital burden of desalination has meant that many water utilities struggle to cover the initial cost of installation, even if over the lifetime of the plant it makes financial sense as the best option.

This has left a gap for other sources of funding outside water utilities to support the development of new facilities.

- Direct government subsidy: where utilities have capital budgets directly supported by central government funding.
- Private debt investors: funding through long-term debt agreements, either commercial debts from banks or other private lenders, or through the issuance of bonds, with returns linked to the income from sales of water at the plant.
- International development funding: Institutions such as the World Bank, Islamic Development Bank, Asian Development Bank and African Development Bank have all supported desalination projects in the past through debt funding. In addition, international export credit agencies – particularly those in Japan and Korea – have been keen to support the actions of their own domestic water engineering companies by supporting projects they are involved in overseas.
- Private sector water developers: where independent companies fund the capital cost of projects through private finance contracts. This usually involves a mix of equity from the developer's balance sheets, alongside debt support from private lenders, development agencies or export credit banks.

The availability of finance to fund desalination projects has been a major decider in the successful implementation of desalination facilities. It can also have a major influence on the type of contract deployed to build desalination plants, and the type of company that becomes involved in developing the project.

For example, the presence of project developers and desalination construction companies from the Far East has been boosted by the availability of capital funding in those markets, both through commercial bank lenders, and through government-supported export credit funds. Lenders in Japan, Korea and more recently China, Malaysia and elsewhere, have all been active investors in water projects in the Middle East. This has been a major factor in the success of Far Eastern developers and contractors in securing contracts in the world's largest market for desalination.

20.5 Looking Forward

Spending on desalination is highly responsive to fluctuations in the global economy – a fall in national growth can mean a downward revision of water demand predictions, as well as a reduction in the ability to fund often capital-intensive facilities.

Given that it is the largest market for desalination by some distance, the fortunes of the Middle East will be the largest single factor affecting the overall global market for desalination. The price of oil, and its effect on the spending abilities of petro-economies in the Persian Gulf will to a major extent determine the amount of money spent on desalination. That said, the ongoing shortage of alternative water resources in the region, combined with rising demand for water, is likely to mean there is no alternative to increased investment in desalination. This should mean an increasing reliance on privately financed models of desalination, with international development agencies, export credit banks and private-sector developers taking an ever more active role in the market.

In terms of technology, the most important factor is likely to be development in alternative sources of energy. While the efficiency of reverse osmosis desalination is constantly improving, the cost of energy remains the most important factor in determining the lifetime cost of a desalination plant. This means that integration with alternative sources of energy, such as solar, wind and nuclear, will become crucial for desalination developers in an age when the world is becoming less reliant on fossil fuels for energy generation.

Table 20.4 Global desalination products and services revenue forecast 2010–2020

Region	2010	2015	2020
Asia Pacific	$2,092 m	$2,646 m	$3,271 m
Europe	$267 m	$315 m	$389 m
Latin America	$637 m	$1,008 m	$1,246 m
Middle East	$3,138 m	$5,040 m	$6,232 m
North Africa	$889 m	$1,137 m	$1,402 m
North America	$661 m	$2,079 m	$2,570 m
ROW	$181 m	$378 m	$467 m
Total	$7,865 m	$12,603 m	$15,577 m
Compound Annual Growth Rate (CAGR) 2010–2020		7.1%	

Source: GMR Data 2016 [5].

Proposed desalination projects and plants under construction are heavily focused in the Middle East, with 2.45 million m³/day, the equivalent of 64% of global planned large scale capacity. This follows a trend, across the last decade, of large scale desalination plants being installed across the Middle East (Table 20.4).

Going forward, GMR Data forecasts additional capacity of varying sizes being installed in the energy rich/financially robust Middle Eastern countries, a market that has grown from $3.1 billion in 2010, to reach $5 billion in 2015. Further growth in this key desalination sector is forecast, to reach $6.2 billion in 2020, displaying a 10 year (2010–2020) Compound Annual Growth Rate (CAGR) of 7.1%. Other additional Middle Eastern desalination projects could include Turkey, Iraq, and Iran where the lifting of harsh sanctions, allied with a government that appears to be very much interested in attracting foreign investment, is worth further regard.

Other key markets in the global desalination sector include Asia Pacific; a sector that includes China and India, countries with huge potential growth opportunities for this sector. The Asia Pacific desalination sector reached $2.1 billion in 2010, growing to reach $2.6 billion in 2015. Further growth is forecast in this key desalination sector, to reach $3.3 billion in 2020, displaying a moderate 10 year (2010–2020) CAGR of 4.6%.

In the US, significant notices of drought conditions have been seen in many of the western states; most notably California. Drought often has significant impacts on agriculture and municipal water supplies and many authorities see desalination as a way to alleviate water shortages. GMR Data expects that mothballed plants will come back online. Existing desalination plants are likely to have their capacities extended as well as totally new projects launched.

Investment in desalination is shared between municipalities, which often include both governments and private companies, and the industrial sector. Investment decisions from these sectors are very much dependent on the wider economic performance of the country in question. So for many Latin American countries, for example, significant municipal investment in desalination is not expected; while mining operations are likely to drive the sector, as seen at the two projects of Antofagasta and Arica in Chile. Many island nations in the Latin American desalination sector require plants with only moderate desalination capacities. Since 2007, 68 new desalination plants have been built across the Caribbean, which now boasts a combined installed capacity of 782,000 m^3/d.

North African countries on the Mediterranean coast are some of the most water stressed regions after the Middle East and due to even more arid landscape on the southern parts of these nations, they will have few alternatives but to increase their water desalination capacity for increased water supply. Political stability defines many North African states' desalination sectors, i.e countries that can attract investment have driven the market in this area; Morocco and Tunisia, leading the desalination market worth $1.1 billion in 2015.

The Global Water Awards[1]

Each year, the Global Water Awards are presented at the Global Water Summit, one of the major annual business conferences of the water industry. Since 2006, the Awards have been recognizing those initiatives in the water, wastewater and desalination sectors that are moving the industry forward. Nominations for the Global Water Awards are shortlisted by a panel of industry experts based on their operating performance, innovative technology adoption and sustainable financial models. The winners are voted for by *Global Water Intelligence* and *Water Desalination Report* subscribers, reflecting the views of the international water community. In this section, from among the various awards, a recap of the "Desalination Company of the Year" and "Desalination Plant of the Year" awards will be provided for the span of 2012–2017.

---

[1]Note from the editor: with permission, the following are excerpts from the Global Water Awards website (http://www.globalwaterawards.com/) with minimal editing [6].

Year	Desalination Company of the Year (Winner)	Desalination Company of the Year (Distinction)
2012	GE Water & Process Technologies	Doosan Heavy Industries and Construction
2013	Abengoa	Degremont
2014	Qatar Electricity & Water Company	IDE Technologies
2015	Saline Water Conversion Corporation	Degremont
2016	Acciona Agua	Black and Veatch
2017	Doosan Heavy Industries and Construction	Utico

Year	Desalination Plant of the Year (Winner)	Desalination Plant of the Year (Distinction)
2012	Southern Seawater Desalination Plant, Australia	Fujairah 2, UAE
2013	Victorian Desalination Plant, Australia	Adelaide Desalination Plant, Australia
2014	Soreq, Israel	Tuaspring, Singapore
2015	Ras Al-Khair SWRO, Saudi Arabia	Cambria BWRO, California
2016	The Claude "Bud" Lewis Carlsbad Desalination Plant, USA	Ghalilah SWRO, Ras Al Khaimah, UAE
2017	Divided into two categories; Municipal: Yanbu 3, Saudi Arabia Industrial: Escondida SWRO, Chile	Divided into two categories; Municipal: Az-Zour North 1, Kuwait Industrial: Sadara SWRO, Saudi Arabia

Recap of the 2012 Global Water Awards

Desalination Company of the Year (Winner): GE Water & Process Technologies

What is it?

The water technology subsidiary of General Electric, housed within the GE Power & Water division.

What has it done?

After buying Ionics in 2005, the industrial giant seemed to spend years struggling to find a way of engaging with the desalination industry. [In the year leading up to the award], however, everything finally seemed to come together. Putting its imagination to work, the company established itself as the dominant technology supplier in two of the fastest growing market niches: industrial wastewater evaporation for the oil and gas market, and containerized modular desalination plants available for rapid deployment.

What makes it special?

- Over the space of a year, GE Water's evaporation business secured five projects recycling water from the Canadian oil sands. This is the kind of tough environmental challenge that the company's Ecomagination initiative was designed to address, and its success is undoubtedly a triple win: for GE's customers, for GE itself and for the delicate environment of Northern Canada.
- Customers for small to medium-sized desalination plants were once tied to poorly performing assets which never quite met their needs. Then GE developed its modular containerized desalination plants, revolutionizing that sector of the market. It brought together cutting-edge technology, lean manufacturing, innovative ownership models and short lead times to deliver a knock-out proposition which was all-conquering in 2011.
- Whether it is lowering the operating pH for its industrial wastewater evaporators, using its UF technology for seawater pretreatment, or investing in proprietary energy recovery systems, GE Water has been quietly pushing the frontiers of technology to meet the needs of its core industrial customers.

Desalination Company of the Year (Distinction): Doosan Heavy Industries and Construction

What is it?

Doosan Heavy Industries and Construction is a division of South Korean conglomerate Doosan Corporation. It is involved in a range of industry sectors from power generation to construction, as well as desalination and water treatment.

What has it done?

2011 was another barnstorming year for Doosan in the Persian Gulf. The company dramatically scaled up its MED backlog, bagging projects to supply Saudi Arabia's Saline Water Conversion Corporation with a 68,190 m^3/day unit for the Yanbu Power plant, and for a 54,552 m^3/day two unit plant for Marafiq – and all the while commissioning one of the largest ever MSF plants, the 459,146 m^3/day Shuweihat 2 project in the UAE. Meanwhile, its membrane arm Doosan Hydro continued its foray into the North American RO market with the commissioning of a 4,353 m^3/day BWRO plant for Bayonne Energy Centre in New Jersey.

What makes it special?

- Shuweihat 2 was the largest desalination plant to be commissioned anywhere in the world in 2011, while its auxiliary load consumption of less than 3.9 kWh/m^3 is the lowest of any MSF installation to date [at the time. It underlines Doosan's commanding position in the MSF market, and proves that it has effectively wrestled the initiative away from former top dog Fisia.
- Doosan pulled off a stunning coup with a direct award from SWCC for the Yanbu plant – the company had never built an MED unit even one tenth that size before. The fact that this was quickly followed up with a second large MED win at Marafiq was an impressive display of chutzpah, underscoring confidence in the outlook for large-scale thermal projects.
- Not content with its achievements in the Persian Gulf thermal market, Doosan continued to make progress with large-scale RO projects, commissioning the 136,260 m^3/day Shuwaikh SWRO facility in Kuwait in 2011.

Desalination Plant of the Year (Winner): Southern Seawater Desalination Plant, Australia

What is it?

The second large-scale SWRO desalination plant to be built in Perth, Western Australia, with a total contracted capacity of 280,000 m^3/day. The first 140,000 m^3/day phase was commissioned in August 2011.

Who is responsible?

Phase one of the plant was built for client Water Corporation by the Southern Seawater Alliance, a consortium of Técnicas Reunidas, Valoriza Agua, AJ Lucas, Worley Parsons, and Water Corporation. The same consortium is currently [at the time of the award] building the second phase, which will bring the capacity up to the full 280,000 m^3/day. Dow provided the RO membranes, with UF membranes supplied by Siemens Memcor. The energy recovery devices were furnished by Energy Recovery Inc.

What makes it special?

- Going from initial concept to producing first water in 4 years, the plant was ordered to be doubled in size before it was even finished, a testament to the inspiration and dedication of the project team. The SSDP represents a milestone in terms of drought-proofing Western Australia, satisfying 17% of the water demand of 1.6 million people in one of most isolated and climate-vulnerable cities in the world.

- Construction was carried out with an absolute minimum of environmental impact – the bulk of the facility is housed in a disused limestone quarry, with tunneling methods rather than blasting being used to construct the marine pipelines, thus minimizing the impact on the coastal dune system, and keeping the neighboring beach open for recreational activities during construction. An 8-m high vegetated berm will reduce noise pollution and provide a visual screen to the east and south of the plant site.
- The plant is 100% powered by renewable energy, showing that the client and its team on the ground were willing to work together to enhance the reputation of the desalination industry as a whole. It is a glimpse of what all plants will one day look like.

Desalination Plant of the Year (Distinction): Fujairah 2, UAE

What is it?

A hybrid desalination complex, commissioned in January 2011, which combines a 2,000 MW power generating station with a 454,200 m^3/day MED plant and an SWRO plant with a capacity of 136,000 m^3/day.

Who is responsible?

International Power and Marubeni developed the project in consortium with the client, Abu Dhabi Water and Electricity Authority (ADWEA). The EPC contract was awarded to an Alstom-led consortium, with Sidem (Veolia) supplying the MED plant and OTV (Veolia), providing the SWRO. Toray supplied the RO membranes, with Flowserve Calder provided the energy recovery devices.

What makes it special?

- Power demand fluctuates wildly from season to season in the UAE, while water demand is more consistent. Fujairah 2 uses an elegant balance of MED and SWRO technologies to consistently provide the largest output, at the lowest energy cost, of any hybrid desal plant anywhere in the world [at the time of the award].
- The plant was one of the first working facilities of its scale in the region to use a Dissolved Air Flotation (DAF) pretreatment system for the SWRO phase, vital in dealing with harmful algal bloom (HAB) events.

During February 2011, the newly commissioned plant maintained its production capacity during a severe algal bloom which forced neighboring desal plants to shut down or reduce their throughput – a credit to the resilience and ingenuity of its design.

- At 38,640 m³/day (8.5 MIGD) each, Fujairah 2 has the largest MED units of any desal plant commissioned thus far [at the time of the award], showing that in the right hands, even the most mature desalination technology can be pushed to new limits.

Recap of the 2013 Global Water Awards

Desalination Company of the Year (Winner): Abengoa

What is it?

A publicly traded Spanish construction group involved in the development, financing and operation of water and wastewater infrastructure through the Abengoa Water group, supported by the Abeinsa engineering division.

What has it done?

The business reshuffle that distributed the water empire Befesa/Abengoa between Abeinsa and Abengoa Water in 2011 has paid off in spades, freeing up the construction department to develop its strengths in an increasingly competitive global desal market. In 2012, Abengoa Water continued Befesa's record of excellence in the BOT market, securing a flagship African project that broke new ground for the company in the form of a 60,000 m³/day BOT at Nungua in Ghana.

What makes it special?

- Completion of the Nungua SWRO contract – a project that had defeated earlier attempts by other developers – was a huge feather in the cap for Abengoa and its partners. Securing financing from South African lenders was the key enabler for a low-cost, long-term desal option that will assuage local water worries, and set the stage for further expansion by Abengoa in rapidly developing sub-Saharan markets.
- Nungua was the fifth BOT project to join the Abengoa Water stable. The start of operations at the fourth, the 100,000 m³/day Qingdao SWRO plant in China, cemented Abengoa Water's position as a truly global developer.
- The EPC business of Abeinsa continued to push the boundaries of design and cost-effectiveness last year, securing the 10 MIGD (45,460 m³/day)

expansion contract at the Barka 1 IWPP in Oman – the first of its kind in the region to reuse brine reject from the existing desalination process for potable water production.

Desalination Company of the Year (Distinction): Degremont

What is it?

The 2-billion-dollar water treatment plant arm of Paris-listed Suez Environnement.

What has it done?

Degrémont's desalination business spread to encompass and define every aspect of the global desal market in 2012, winning contracts and completing projects in a huge array of diverse markets, with differing technical demands. It made significant progress in municipal desal, completing one of its largest references to date – the 444,000 m^3/day Wonthaggi plant in Melbourne – as well as winning a major emergency supply program in Saudi Arabia. It also extended the reach of its successful and growing industrial business.

What makes it special?

- The 444,000 m^3/day Wonthaggi desalination project in Melbourne – commissioned in 2012 – will act as a showcase for the company's top-of-the-range technical abilities for years to come. The design of the plant – built with an almost invisible footprint and offsetting emissions through the use of renewable energy – shows that even mega-projects can deliver an environmentally sustainable solution to water scarcity.
- Degrémont also showed last year [2012] that size is not always everything, mobilizing dozens of small modular RO units as part of the Riyadh Water Supply Enhancement Program. This 220,000 m^3/day emergency groundwater desalting project, aimed at solving a looming water shortage crisis in the Saudi capital, was procured at an unprecedented speed, amply demonstrating the French company's ability to be nimble when it matters most.
- The company's technical acumen saw it spread it's desal expertise into the most challenging parts of the industrial water treatment market in 2012. The groundbreaking Chengdu industrial reuse plant in China deploys Degrémont desalination technology as part of its array of solutions, and is the first industrial water reuse plant in China to supply municipal customers.

Desalination Plant of the Year (Winner): Victorian Desalination Plant, Australia

What is it?

A 444,000 m^3/day SWRO project at Wonthaggi, supplying the city of Melbourne in Australia.

Who is responsible?

The Aquasure consortium – consisting of Suez Environnement, Thiess and Macquarie Capital – developed the project, with a Thiess-Degrémont joint venture designing, constructing and operating the plant. The energy recovery devices were supplied by ERI, and the RO membranes by Hydranautics.

What makes it special?

- Starting up the largest desal plant in the southern hemisphere presented an immensely complicated challenge. Degrémont found an ingenious way of circumventing the risk with its "virtual plant." A computer program simulated the action of every single valve and pump in the facility, enabling over 20,000 tests to be conducted under safe conditions, enabling enviable cost and time savings to be realized during the commissioning phase.
- It is a boldly ambitious project in every sense: as Australia's first privately financed desal plant, it was conceived in 2007 while the city labored under harsh stage-4 water restrictions. The project involved tunneling under the sea for over a kilometer to minimize the environmental effects of both the intake and outfall.
- Greenhouse gas emissions from the plant's 90 MW power requirement are entirely offset by the purchase of renewable energy, while earthworks make the finished plant virtually invisible within the landscape. The plant's "living roof" contains more than 98,000 plants and shrubs, and at 6.4 acres (2.6 ha) is one of the largest of its kind in the world. In terms of environmental sustainability, the Wonthaggi plant is a genuine pioneer.

Desalination Plant of the Year (Distinction): Adelaide Desalination Plant, Australia

What is it?

A 300,000 m^3/day SWRO plant, built to supply the city of Adelaide in South Australia.

Who is responsible?

The plant was designed and built by the Adelaide Aqua consortium of Acciona Agua, Trility, McConnell Dowell, and Abigroup. The plant is operated by Acciona Agua and Trility. The energy recovery devices were supplied by ERI, and the RO membranes by Hydranautics. The UF membranes were supplied by Siemens Memcor. The client is SA Water.

What makes it special?

- The ADP is one of the most technically accomplished desal plants ever built, taking the search for energy efficiency to an entirely new level. Turning the plant's 52 m elevation to its advantage, an innovative energy recovery system at the outfall recovers approximately 2.5% of the plant's total energy consumption.
- Operating costs are reduced by using the largest submerged UF pre-treatment system in the world, enabling an 80% reduction in sludge generation. A unique two-pass hybrid blind split array in the RO configuration gives the plant operational flexibility, increased membrane life and a high recovery rate.
- The plant reduces Adelaide's reliance on the Murray River, while generating a minimal impact on the environment through the use of renewable energy and careful outfall management – tests showed no excess salinity just 100 meters from the brine discharge point.

Recap of the 2014 Global Water Awards

<u>Desalination Company of the Year (Winner): Qatar Electricity & Water Company</u>

What is it?

The largest power and desalination developer in Qatar, holding a desalination portfolio with a total capacity of 1.65 million m^3/day. QEWC is majority government-owned, with a portion of its shares traded on the Qatar Exchange.

What has it done?

Having negotiated investments in power generation facilities in Oman and Jordan, QEWC formalized its international water ambitions in 2013, launching the $1 billion Nebras Power Company – in which it will hold a 60% stake – to invest in power generation and desalination opportunities overseas. Closer to home, financial close was achieved on a complex deal for the 163,656 m^3/day Ras Abu Fontas A2 independent water project.

What makes it special?

- The financing of the RAF A2 complex was a groundbreaking deal which marked the first time a privately developed project of this scale had been fully financed by domestic Qatari banks. As well as being a significant coup for the Qatari financial system, it also marked a bold step forward for Islamic project financing in the Persian Gulf.
- QEWC's reputation as the desal developer of choice in its home market was underscored by the decision – by national water utility Kahramaa – for QEWC to take an equity lead in the next major greenfield power and water project to be procured in the state. It will hold a 60% stake in the 2,000 MW and 120 MIGD (545,520 m³/day) Facility D plant, which is [was] under tender.
- Focusing on its home market has given QEWC an enviable position in the Qatari power and desal sector, and allowed it to accumulate the cash and the expertise to seek out opportunities elsewhere. The creation of a fund through which to consolidate its international power and water interests demonstrates a towering ambition for expansion, and cements the group's commitment to developing its reputation as a regional powerhouse.

<u>Desalination Company of the Year (Distinction): IDE Technologies</u>

What is it?

An Israel-based EPC contractor and developer of desalination plants.

What has it done?

Bringing the largest SWRO plant in the world online is no small achievement, and yet the commissioning of the Soreq facility, with a maximum capacity of 624,000 m³/day, was only one of a number of milestones passed by IDE last year. It successfully started up China's largest desalination plant at the Tianjin SDIC power station (200,000 m³/day), as well as a 60,000 m³/day facility at the Vasilikos Power Plant in Cyprus. Its business development efforts did not skip a beat, meanwhile, as it secured the contract to build Reliance Industries' first SWRO plant at Jamnagar in India, whilst starting work on the 189,250 m³/day Carlsbad project in California.

What makes it special?

- Not content with 400 references in 40 countries over four decades, IDE continues to innovate in order to meet the demands of its clients.

Last August, its eco-friendly PROGREEN reverse osmosis technology was chosen for a new desalination plant serving the five-star Hayman resort on Australia's Great Barrier Reef. The sensitivity of the surrounding natural environment meant that a chemical-free, low-energy system was the ideal choice to replace the island's outdated and energy intensive thermal unit.

- The raging drought in California has brought the issue of rainfall-independent water resourcing infrastructure to the top of the state's political agenda as never before. The completion of more than 25% of the ground-breaking Carlsbad desalination project within budget last year [2013] is proof that IDE and its construction partners are taking the challenge seriously. When completed in 2016, the plant will offer residents of the Golden State a truly drought-proof water source.

- IDE's research and development program is the envy of the desalination industry. The company signed a strategic technology partnership with Beijing Enterprises Water Group last year, while embarking on a research partnership with Clean Harbors to increase the reliability of mechanical vapor compression (MVC) evaporators to treat oil sands produced water in Canada. It is initiatives like these which will ensure that IDE remains at the cutting edge of the desalination industry for at least another four decades.

Desalination Plant of the Year (Winner): Soreq, Israel

What is it?

Israel's largest seawater reverse osmosis plant, with a total contracted capacity of 150 million m^3 per year, and a maximum daily capacity of 624,000 m^3.

Who is responsible?

The project was developed on a 25-year build-operate-transfer basis by Sorek Desalination Ltd., a partnership between main EPC contractor IDE Technologies (51%) and Hutchison Water (49%). Dow Filmtec and Hydranautics supplied the membranes, and the plant employs Calder's DWEER energy recovery devices, supplied by Flowserve.

What makes it special?

- Following its successful commissioning in 2013, Soreq is the largest SWRO plant operating anywhere in the world today. The plant provides clean potable water to over 1.5 million people, satisfying 20% of Israel's municipal water demand.

- In a revolutionary departure from other large-scale SWRO facilities, the Soreq plant employs 16-inch membranes arranged in a vertical formation. This innovation reduces the number of elements, pressure vessels and piping headers required by a factor of four. It also allows for the safer operation of the membranes, as well as fast installation and increased accessibility.
- Smart design and construction within the boundaries of a uniquely challenging site footprint has enabled the plant to be environmentally sensitive, despite its gargantuan size, with no shoreline impacts due to underground pipe-jacking techniques. A sophisticated 'pressure center' minimizes energy consumption by coordinating the high-pressure pumps and energy recovery devices, and enables the feed pressure to the RO trains to be finely controlled, which means fluctuating production demand is handled as efficiently as possible.

Desalination Plant of the Year (Distinction): Tuaspring, Singapore

What is it?

A 318,500 m³/day seawater reverse osmosis plant in Singapore. The site will also accommodate a 411 MW power plant.

Who is responsible?

Hyflux developed the plant under a design-build-own-operate model, contracting to deliver treated water to Singapore's Public Utilities Board for 25 years. Black & Veatch served as PUB's engineer on the project. The RO membranes were supplied by Toray, and Flowserve provided 68 Calder DWEER energy recovery devices.

What makes it special?

- Hyflux's genius in creating Asia's first independent water and power project at Tuaspring single-handedly addresses Singapore's dual challenges of having no natural groundwater or energy sources. The commissioning of the RO plant last year means that desalination now provides 25% of Singapore's water needs, greatly strengthening one of the nation's 'four taps' in anticipation of a much greater level of water independence in the future.
- Far from being a carbon copy of the SingSpring desalination plant built by Hyflux in Singapore in 2005, Tuaspring is the first of a new breed

of Asian megaplants, featuring a compact design which reduces the physical footprint per cubic meter by more than 30%. This is a vital pre-requisite in a country where urban infrastructure growth is reaching its natural limits.

- As well as featuring one of the world's largest ultrafiltration pre-treatment systems for seawater desalination, a partially split SWRO membrane train configuration and a self-sufficient on-site power plant all combine to ensure that Tuaspring is one of the most energy-efficient large-scale desalination plants on the planet.

Recap of the 2015 Global Water Awards

Desalination Company of the Year (Winner): Saline Water Conversion Corporation

What is it?

The bulk water supply agency for the Kingdom of Saudi Arabia, and the largest producer of desalinated water in the world, generating 4.6 million m^3/day of water and 7,400 MW of power across 27 production centers. It celebrates its 50th anniversary in 2015.

What has it done?

Under the leadership of AbdulRahman Al-Ibrahim, SWCC has not only accelerated its investment in new plants and extended the life of its existing assets – it has also transformed itself as a learning organization, empowering its staff, simplifying its processes, and bringing a new focus to operational efficiency, reliability and economic returns.

What makes it special?

- SWCC has the toughest job in desalination. It has the largest fleet of desalination plants in the world, with the least margin for error: some cities have just 48 hours of water reserves. The problem gets tougher as plants reach the end of their productive lives and demand for water increases inexorably. In 2014, SWCC not only stepped up its production to a new high, but did so while increasing its commitment to environmental stewardship, fuel efficiency, and the employment and development of Saudi staff.
- In 2014, SWCC's Saline Water Desalination Research Institute cemented its reputation as a global center of desalination expertise,

working in partnership with organizations including Saudi Aramco, King Abdullah University of Science and Technology, Doosan, Dow, and Singapore PUB, and bringing its total haul of complete applied research projects to 29, with ten patent approvals over the past decade. Developing a new approach to multiple-effect distillation technology that reduces energy consumption by 40% and extending the life of desalination plants from 25 years to 40 years have been key foci of the institute's research.

- 2014 saw the commissioning of the 309,128 m^3/day Ras Al-Khair reverse osmosis plant, which is specially designed to treat the difficult waters of the Persian Gulf. With nearly 50 years of experience of every kind of desalination challenge behind it, SWCC effortlessly rose to the challenge.

Desalination Company of the Year (Distinction): Degrémont

What is it?

The [...] water treatment arm of Paris-listed Suez Environnement.

What has it done?

The company demonstrated a mastery of every aspect of the process of membrane desalination in 2014, from the ability to design and commission some of the largest membrane facilities in the world, to pushing the boundaries of the technology through exciting new tech link-ups in the lab.

What makes it special?

- Degrémont reigned supreme in 2014 as the master of large-scale membrane desal, securing the contract to build the desalination element of Abu Dhabi's Mirfa independent water and power project – the emirate's first Persian Gulf Coast membrane facility. As the changing energy mix forces the Middle East away from thermal desal technology, Degrémont's forward-thinking approach means it is strongly positioned to take advantage of the shift to reverse osmosis.
- At the opposite end of the scale, Degrémont is looking to create the future of desalination in an energy-conscious world through its link-up with the Masdar renewable energy desalination program in Abu Dhabi. The pilot plant it is [was] building will provide a vital glimpse into the future of a constantly changing industry that is becoming more important than ever.

- Over the last two years [2012–2014], Degrémont has developed an uncanny knack for supplying filtration systems to floating production, storage and offloading (FPSO) installations for the oil and gas industry This trend offers compelling evidence of the company's increasingly sophisticated and dominant position in the world of industrial desalination – a position cemented by the acquisition of the Australia-based Process Group in June 2014.

Desalination Plant of the Year (Winner): Ras Al-Khair SWRO, Saudi Arabia

What is it?

A 68 MIGD (309,128 m³/day) membrane desalination installation on the Persian Gulf coast of Saudi Arabia. It forms part of the world's largest desalination facility, and along with a new water transportation pipeline transforms the picture for potable water in Riyadh, one of the world's fastest-growing and most water-stressed cities.

Who is responsible?

The plant was procured and is owned and operated by Saudi Arabia's Saline Water Conversion Corporation, the world's largest desalination infrastructure operator. It was built under an EPC contract signed with Korean contractor Doosan, alongside civil works contractor Saudi Archirodon and design consultant Pöyry. Toyobo supplied the RO membranes, while FEDCO took responsibility for the supply of energy recovery devices.

What makes it special?

- The sheer scale of the project – it is the largest membrane facility ever to be built in the Persian Gulf, and the largest in the world to feature DAF pre-treatment – amply demonstrates that reverse osmosis can easily cope with the difficult-to-treat feedwaters of the Persian Gulf, where high salinity and red tides are the norm. The use of heavy-duty DAF/DMF pretreatment to combat the oppressive environmental conditions proved once and for all that membrane desal is a serious contender in the GCC.
- The growing confidence in, and appetite for membrane desalination in Saudi Arabia is paving the way for an energy-efficient regional desal portfolio ready to withstand the changing approach to energy generation in the Middle East. As countries diversify away from oil as a feedstock, the establishment of excellence in membrane desalination opens up further potential for exploring solar and other renewable sources of energy for desalination.

- The speedy completion of the membrane element of the project, along with dedicated features like the installation of a dedicated wastewater treatment plant to deal with DAF sludge, proved that even the most complex and extensive of projects can be delivered effectively in the Kingdom. The fact that the project notched up 24 million hours of accident-free construction activity in 2014 is testament to the fact that a project's size can easily be matched by its commitment to health and safety, and earned it a special commendation from SWCC.

Desalination Plant of the Year (Distinction): Cambria BWRO, California

What is it?

A 400 GPM (2,180 m^3/day) brackish water treatment plant featuring a three-step treatment process: ultrafiltration, reverse osmosis, and UV/advanced oxidation.

Who is responsible?

The prime design-build contractor was CDM Smith. H$_2$O Innovation supplied the RO and ultrafiltration units, which use Toray UF membranes and Hydranautics RO membranes. Trojan provided the UV disinfection unit. The client is the Cambria Community Service District.

What makes it special?

- A 50-gallon-per-day consumption limit for local residents, a long-standing moratorium on new water connections, and a ban on the outdoor use of potable water meant that the coastal community of Cambria (pop. 6,000) ranked among the worst casualties of California's raging drought. Although the city had looked at seawater desalination before, Governor Brown's declaration of a drought emergency freed up the possibility of developing an alternative brackish water option, which was exempt from a burdensome environmental review process, enabling it to move ahead in record time.
- Fast-tracking the construction of a desalination project such as this is unprecedented in California, and sets a new benchmark for what is achievable in the face of severe water stress. Following the decision to move ahead in January 2014, an emergency coastal development permit was granted in May, and construction began in August. The use of pre-fabricated processing units and above-ground plumbing reduced the capital cost, and ensured that the plant was granted an operational permit in November 2014 – less than a year after the process began.

- The feedwater is a unique mix of groundwater, brackish water and secondary treated effluent, and the two-stage RO system results in a 92% permeate recovery rate – close to double that of a standard seawater desalination plant. The high level of acceptance from local residents for what is ultimately an indirect potable reuse project conclusively demonstrates that Californians are willing to retain their pioneering spirit when faced with long odds.

Recap of the 2016 Global Water Awards

Desalination Company of the Year (Winner): Acciona Agua

What is it?

A Spanish EPC contractor and project developer active in the desalination, water and wastewater markets globally.

What has it done?

Acciona Agua had a stellar year in the international desalination market in 2015, commissioning its second-largest project ever – at Torrevieja in Spain – while winning a contract to supply a pair of reverse osmosis plants in Cape Verde. It also delivered on its commitment to revolutionize desalination in the Persian Gulf, bringing the 136,383 m^3/day Fujairah F1 expansion in Abu Dhabi online in November, whilst securing a pair of flagship contracts in Qatar.

What makes it special?

- Acciona Agua's rise in the Persian Gulf desalination market has been nothing short of meteoric. From a standing start in 2012, it succeeded in bringing the 136,383 m^3/day Fujairah F1 expansion online last year, and secured two important EPC contracts in Qatar in May. It is now the Spanish company with the single largest presence in the Persian Gulf desalination market.
- The GCC's historical resistance to membrane desalination means that most EPC contractors would have been happy to be awarded Qatar's first large-scale reverse osmosis facility (the 164,000 m^3/day Ras Abu Fontas A3 plant). Acciona made it a double whammy by simultaneously winning the contract to build the 284,000 m^3/day RO component of the Facility D plant, also in Qatar.

- The success of Acciona Agua's desalination business doesn't just come down to skillful negotiation and forward-thinking process engineering. It is backed by a robust R&D team, which is developing the UltraDAF-Evo pre-treatment system to deal with algal blooms, the Hiflus membrane-based pre-treatment application, an energy-efficient back-wash treatment system (Vetra), and HydroBionets, a wireless sensor to detect membrane soiling.

Desalination Company of the Year (Distinction): Black and Veatch

What is it?

The water division of a global employee-owned engineering, consulting and construction company, with expertise covering membrane, thermal and hybrid systems for both brackish and seawater desalination systems.

What has it done?

Last year marked a new dawn in the company's efforts to target dynamic high-value international markets, securing a dazzling array of marquee desalination contracts in crucial markets such as Saudi Arabia and Singapore. At the same time, it continued to prove its advanced water credentials at home – with the successful commissioning of the Orange County groundwater recharge project showing that desal expertise can help close the water cycle – all the while pushing the boundaries of holistic water treatment and management through work with the WateReuse Research Foundation.

What makes it special?

- The company has long been keen to rebalance its water business with a larger emphasis on international contracts. The securing of the contract to advise the Saline Water Conversion Corporation on Jeddah 4 – the first mass-scale membrane plant in Saudi Arabia – plus major consultancy wins with PUB in Singapore and the government of the Hong Kong SAR proves that B&V's desalination credentials are making their mark on the largest clients in the most significant markets around the world.
- As independent engineer on the Carlsbad desalination project, B&V was faced with a mind-boggling array of technical and permitting issues on one of the most complex and long-running projects in desal history. The successful commissioning of the plant in 2015 marks a staggering achievement for B&V, the Carlsbad team, and the desal industry as a whole.

- The sharing of technological solutions between desalination and wastewater reuse is becoming more and more commonplace, partly thanks to the pioneering work undertaken by Black & Veatch's pragmatic engineers. The company's design blueprint for the expansion of the Orange County Groundwater Replenishment System in California – which came online last year – demanded a similar suite of technologies to brackish water desalination to achieve the same end result.

Desalination Plant of the Year (Winner): The Claude "Bud" Lewis Carlsbad Desalination Plant, USA

What is it?

A 50 MIGD (189,250 m³/day) SWRO desalination plant serving nearly 400,000 people in San Diego County, California.

Who is responsible?

The project was developed by a joint venture of Poseidon Resources and Stonepeak Infrastructure Partners under a 33-year build-own-operate contract with the San Diego County Water Authority (SDCWA). EPC work was carried out by a Kiewit/JF Shea team, while IDE Technologies was responsible for the design and supply of the desalination equipment. The pressure vessels were provided by Protec Arisawa, while Dow Water & Process Solutions supplied the reverse osmosis membranes. ERI furnished the energy recovery devices.

What makes it special?

- The successful completion of the largest desalination plant in North America followed years of seemingly insurmountable technical, financial and legal hurdles. The tenacity shown by the developer team is matched only by the importance of seawater desalination as a key part of the solution to California's water crisis.
- A canny combination of state-of-the art energy recovery technology with an external energy offsetting program makes Carlsbad the first major infrastructure project in the state of California to completely neutralize its carbon footprint. The carbon offsetting program helped fund the regeneration of forest areas decimated by wildfires in 2007.
- The repurposing of an existing seawater intake pipe formed a crucial plank of the developer's plans to minimize the impact on the surrounding environment. At the same time, Poseidon retained its green credentials by partnering with the US Fish and Wildlife Service to create, restore and enhance 66 acres of vulnerable local wetland.

Desalination Plant of the Year (Distinction): Ghalilah SWRO,
Ras Al Khaimah, UAE

What is it?

A low-energy seawater desalination plant in the United Arab Emirates, with
a capacity of 15 MIGD (68,190 m^3/day). The $82 million contract to build
the plant was awarded in 2011 and completed in 2015.

Who is responsible?

The plant was designed and built by Aquatech International under an engineering, procurement and construction (EPC) contract with the client, the
Federal Electricity and Water Authority (FEWA). SWRO membranes were
provided by Toray, with UF membranes from X-Flow. ERI supplied the
energy recovery devices.

What makes it special?

- The Ghalilah plant rewrites the rules for energy consumption at large-
  scale desalination installations. Aquatech secured the contract to build
  the plant with an audacious energy performance bid of 3.14 kWh/m^3.
  In reality, the plant now operates at under 3 kWh/m^3, an unprecedented
  figure for full-scale membrane desalination. It is FEWA's largest desal
  plant to date, and sets a new global benchmark for performance in
  membrane desalination.
- By coupling a game-changing technical design with a highly competitive construction cost of just over $82 million, Aquatech has proved
  that innovation in desalination does not need to come with a sky-high
  price tag.
- The operation of the Ghalilah plant takes full account of one of the most
  hostile feedwater sources around. The design features advanced pretreatment and monitoring systems to protect against the risk of seasonal
  red tides, whilst coping with salinity levels as high as 42,000 ppm.
  A veritable all rounder.

Recap of the 2017 Global Water Awards

Desalination Company of the Year (Winner): Doosan Heavy
Industries and Construction

What is it?

A Korean contractor with an established strength in thermal desalination technology, both multi-stage flash and multiple effect distillation. Its US-based
division Doosan Hydro Technology specialises in membrane desalination.

What has it done?

Last year saw the culmination of Doosan's blossoming as a full-service desalination firm, as the company added an enviable array of skills in operations and new technologies to its robust roster of major plant engineering references.

What makes it special?

- 2016 marked a new peak of success for Doosan's membrane desalination business. The completion of the 216,000 m³/d Escondida SWRO plant in Chile and the securing of the contract for the 227,300 m³/d Doha RO plant in Kuwait bear testament to a company with a membrane capability to match its decades of excellence at the forefront of the thermal market.
- Doosan's willingness to think big, push the boundaries of contracting quality, and take on projects of any scale put paid to worries over the demise of thermal desalination. The completion of the massive 550,000 m³/d Yanbu 3 plant in Saudi Arabia in 2016 was an impressive achievement of construction excellence, showing that thermal desal can still punch its weight.
- The value proposition of Doosan's core engineering business has been complemented by new capabilities in zero liquid discharge and dissolved air flotation, while the securing of major operating contracts in Kuwait and Ras Al-Khair in Saudi Arabia completed the company's successful transition from engineering leader to full-service desalination giant.

Desalination Company of the Year (Distinction): Utico

What is it?

An Emirates-based power and water systems integrator and investor. It has developed a unique independent water and power project (IWPP) model which involves the concessionaire acting as a private water utility dealing directly with end-users.

What has it done?

In 2016, Utico took its private utility model to the next level, lining up a string of power and water projects around the Persian Gulf, while spreading its wings abroad as it negotiated the acquisition of the Nemmeli desalination plant in Chennai, India. The year was capped by the company securing a $147 million equity and project finance commitment from an Islamic Development Bank-backed infrastructure fund.

What makes it special?

- Utico's unique private IWP model has gone from success to success in a notoriously slow market. Having secured a contract to build a major facility in Ras Al Khaimah in partnership with Spain's Grupo Cobra, Utico spent 2016 negotiating a string of further privately financed projects around the Persian Gulf. Further afield, the deal to fully integrate the Chennai plant will mark the arrival of Utico as a global force in desalination.
- The IDB fund's investment ranks as an astonishing mark of confidence in Utico's business model by one of the world's largest development agencies – and more importantly by the fund's impressive array of sovereign backers.
- Utico's strategy of integrating privately financed desalination projects with solar energy goes to the heart of the water-energy nexus without the need for state intervention. Commercial renewable desalination has arrived.

Municipal Desalination Plant of the Year (Winner): Yanbu 3, Saudi Arabia

What is it?

A 550,000 m^3/d multi-stage flash (MSF) plant serving the Medina region of Saudi Arabia.

Who is involved?

The plant was designed, built and commissioned by Korean contractor Doosan, on behalf of its client, the Saline Water Conversion Corporation (SWCC). Following commissioning, the plant will be operated by SWCC.

What makes it special?

- Doosan's mastery of thermal desalination technology resulted in a plant which amply demonstrates that MSF can still compete on a global scale, despite being one of the oldest desalination technologies still in mass operation. By relentlessly pushing up the size of MSF units, the company has gone a long way to keeping the physical footprint of some of the world's largest desalination plants under control.
- Deft work during the delivery process avoided a potential pitfall when problems arose with the completion of a neighbouring power plant that was meant to drive the desalination facility. Doosan ensured timely

delivery of the desalination plant by adapting the auxiliary boiler to produce the necessary steam required to run the MSF units.

- Tight collaboration on design and commissioning resulted in the flawless delivery of a valuable asset that will ensure the smooth transition of the plant into the portfolio of the world's largest desalination company. It will go a long way to meeting rapidly rising local water demand from the growing numbers of pilgrims to Mecca.

<u>Municipal Desalination Plant of the Year (Distinction):</u>
<u>Az-Zour North 1, Kuwait</u>

What is it?

A 486,400 m³/d multiple effect distillation plant in Kuwait. The first privately owned desalination plant in the country, it forms the water desalination element of the first stage of the Az-Zour North independent power and water project.

Who is involved?

The desalination plant was delivered by EPC contractor Sidem (Veolia) on behalf of the plant's owner, a consortium comprising Engie (17.5%), Sumitomo (17.5%), A. H. Al-Sagar & Brothers (5%) and the Kuwaiti government (60%). Water is supplied to Kuwait's Ministry of Energy and Water.

What makes it special?

- By showcasing the very pinnacle of what MED has to offer, the contractor delivered a massive asset that requires a minimum of handling. Combining low O&M costs with a limited requirement for scaling treatment, the plant allows its owners to push the margins of performance and profitability – a key condition for the country's pathfinder water PPP.
- The plant redefines efficiency in thermal desalination. An ultra-low electrical consumption of around 1 kWh/m³ keeps its reliance on external power sources to a minimum. Sidem's in-house MED expertise allows the plant to use relatively low-pressure steam, freeing up energy and resources at the attached power facility.
- The location of the plant meant it had to be configured to handle extreme levels of seawater salinity and a wide range of feedwater temperatures ranging from 13°C to 38°C. The delivery of a truly flexible plant of this size is a stunning paean to engineering excellence.

Industrial Desalination Plant of the Year (Winner): Escondida SWRO, Chile

What is it?

A 216,000 m³/d seawater reverse osmosis plant serving the Escondida copper mine in Chile's Atacama Desert. The plant is the largest desalination facility in Latin America.

Who is involved?

The plant was built by Bechtel for client Minera Escondida, whose major shareholders are BHP Billiton and Rio Tinto. Doosan Heavy supplied nine SWRO units, while the pre-treatment system, consisting of 60 dual media filters, was furnished by Doosan Enpure. Flowserve supplied the pumps as well as 27 DWEER energy recovery devices. BEL supplied the pressure vessels. Black & Veatch acted as the engineer of record.

What makes it special?

- The expansion of the world's largest copper mine, at Escondida in Chile's parched Atacama Desert, was entirely dependent on securing a reliable water supply. By drought-proofing its operations via a seawater desalination plant, Minera Escondida can now go a long way to meeting the rising global demand for copper, driven by anticipated economic expansion in China and the US.
- State-of-the-art slurry tunnelling machines were used at the intake and outfall locations so as to minimize the impact on the coastal zone, which supports abundant marine life. The largest diameter offshore drill in the world was used to bore shafts in the sea floor, eliminating the need for more disruptive offshore construction techniques.
- Making the physical connection between the ocean and the mine, some 180 km distant and at an altitude of 3,100 metres, was a true engineering feat, carried out in the world's most seismically active region. The successful commissioning of the Escondida desal plant proves that large-scale desal for mining operations is viable in Chile, and paves the way for a raft of similar facilities to serve other mine sites.

Industrial Desalination Plant of the Year (Distinction): Sadara SWRO, Saudi Arabia

What is it?

A 178,000 m³/d seawater reverse osmosis plant serving the Sadara Chemical Company's massive manufacturing facility in the industrial city of Jubail, on the Persian Gulf coast of Saudi Arabia. The facility reached full commercial operation in 2016.

Who is involved?

The plant is owned and operated under a 20-year build-own-operate contract by Marafiq, the power and water utility company for the cities of Jubail and Yanbu. The offtaker for the water is Sadara, the Dow/Aramco joint venture based in Jubail. The plant was built by Veolia, which also supplied the DAF pre-treatment system. UF membranes were supplied by Pentair and RO membranes by Dow. ERI supplied energy recovery systems.

What makes it special?

- The delivery of the project through a dedicated single-user build-own-operate contract leverages the financial strength of a secure utility to guarantee a supply of high-quality water on a performance-linked basis. The choice of contract structure demonstrates a high degree of confidence in Marafiq as a reliable provider of utility services.
- The site deploys an array of high-end pre-treatment technologies – dissolved air flotation followed by self-cleaning microfiltration and ultrafiltration stages – to allow the plant to handle large volumes of water from a feedsource at the extreme reaches of salinity and temperature for a desalination plant.
- The installing of a unique dual-train SWRO-then-BWRO process allows for a water recovery level approaching 50%, while a rotary isobaric pressure exchanger gives a specific power consumption of just 5.1 kWh/m^3, an impressive achievement for a plant of this scale dealing with hostile feedwater conditions.

References

[1] Global Water Intelligence. (GWI) (2015). *Desalination markets 2016: Global perspectives and opportunities for growth.*
[2] DesalData. published by Global Water Intelligence. http://www.desal data.com
[3] Global Water Intelligence (GWI). (2010). *MSF rules the roost at Ras Azzour.* Global Water Intelligence magazine, Vol. 11, issue 9.

[4] Global Water Intelligence (GWI). (2016). *Global Water Market 2017: Meeting the world's water and wastewater needs until 2020* (2016).

[5] Recap of the 2017 Global Water Awards should be added here as per the attached file. Keep formatting/indentations/fonts for 2017 similar to the previous recaps (2016, 2015 etc.)

[6] Global Water Awards by Global Water Intelligence http://www.global waterawards.com/

Dr. Alireza Bazargan, Ph.D., began his academic career with a BSc in Chemical Engineering at Sharif University of Technology, Iran. He was subsequently awarded the prestigious TOTAL Scholarship for MSc studies at Ecole des Mines de Nantes, France, in Project Management for Environmental and Energy Engineering. Dr. Bazargan continued his PhD studies at the Hong Kong University of Science and Technology with a research attachment at the University of Cambridge, UK. Upon completing his PhD, Dr. Bazargan became a faculty member at the Environmental Engineering Group at the Department of Civil Engineering at K.N. Toosi University of Technology. He is the author of numerous scientific papers and book chapters, and has received multiple honors and awards, including the "Kazemi Ashtiani" prize from the National Elites Foundation in 2017. Since 2015, Dr. Bazargan has been the R&D Manager and Business Development Advisor at Iran's premier desalination firm, Noor Vijeh Company. He is well known within Iran's desalination ecosystem, and as a member of various think tanks, he has advised policy-makers regarding the strengths, weaknesses, opportunities, and threats facing the industry.